A to Z
of
Physicists

Darryl J. Leiter, Ph.D.

Former Senior Research Associate,
NASA, Goddard Space Flight Center, Greenbelt, Maryland

with

Sharon L. Leiter, Ph.D.

Research Consultant, RAND Corporation,
Washington, D.C., and Santa Monica, California

☑®

Facts On File, Inc.

For Jacob,
who will see further

A TO Z OF PHYSICISTS

Notable Scientists

Copyright © 2003 by Darryl J. Leiter, Ph.D.

Facts On File, Inc.
132 West 31st Street
New York NY 10001

Library of Congress Cataloging-in-Publication Data

Leiter, Darryl J.
  A to Z of physicists / Darryl J. Leiter with Sharon L. Leiter.
    p. cm. — (Notable scientists)
  Includes bibliographical references and index.
  ISBN 0-8160-4798-7 (hardcover : acid-free paper)
  1. Physicists—Biography—Dictionaries. I. Leiter, Sharon. II. Title.
III. Series.
QC15.L45 2003
530′.092′2—dc21   2002014709

Facts On File books are available at special discounts when purchased in bulk quantities for businesses, associations, institutions, or sales promotions. Please call our Special Sales Department in New York at (212) 967-8800 or (800) 322-8755.

You can find Facts On File on the World Wide Web at http://www.factsonfile.com

Text design by Joan M. Toro
Cover design by Cathy Rincon
Illustrations by Jeremy Eagle

Printed in the United States of America

VB Hermitage 10 9 8 7 6 5 4 3 2 1

This book is printed on acid-free paper.

If I have seen further than other men, it is because I have stood on the shoulders of giants.

—Isaac Newton

The most beautiful thing we can experience is the mysterious. It is the source of all true art and science.

—Albert Einstein

A great truth is a truth whose opposite is also a great truth.

—Niels Bohr

If you want to learn about nature, to appreciate nature, it is necessary to understand the language that she speaks in.

—Richard P. Feynman

CONTENTS

LIST OF ENTRIES

ACKNOWLEDGMENTS

My primary debt of gratitude is to my wife and colleague, Dr. Sharon L. Leiter, for her invaluable contributions to the creation of this work. The essays in this book took shape as a result of the ongoing dialogue between us, between scientist and layman, which challenged me to find clear and precise expression for the complex ideas of modern physics.

Thanks go to my editor, Frank K. Darmstadt, for his patience and guidance in preparing the manuscript, and to my agent, Jodie Rhodes, for making the opportunity to write this book available to me.

I would also like to thank Heather Lindsay at the Emilio Segrè Photo Archives of the American Institute of Physics for her kindness and competence in helping us obtain the photographs in this volume.

Finally, I am grateful to my colleagues in the physics community for many insightful discussions involving conceptual and historical developments in physics.

INTRODUCTION

A to Z of Physicists is an up-to-date biographical dictionary containing profiles of 150 of the most illustrious figures in the history of physics. It provides a comprehensive and accessible guide to the men and women behind the ideas that have shaped the modern vision of physical reality.

These biographical essays, focusing on individual struggles and achievements, never lose sight of the communal nature of physics: the ways in which the lives and works of physicists are interconnected. They give brief yet detailed explanations of each physicist's key discoveries and of how they grew out of previous findings and established the ground for future breakthroughs. Anyone wanting to know at a glance "who did what" in physics will find the answer as well as in-depth explanations of the work itself in this volume. Each entry concludes with a brief list of suggested further reading, enabling the student to delve more deeply into the life and contributions of a given subject.

The stories of the men and women in this volume tell of astounding insights, happy accidents, and perseverance rewarded, as well as fruitless quests, misinterpretations, and frustrations. Whatever their individual differences, however, physicists are people with "wind in their sails." The common desire of physicists from all lands and ages to play a part in the evolution of an ultimate explanation of physical reality—as well as the belief that such an explanation is possible, that is, that nature is fundamentally orderly and comprehensible—imbues their lives with a rare spiritual and intellectual excitement.

The history of physics is one of repeated "paradigm shifts," in which an established picture of the universe is replaced by another, more far-reaching vision. Thus, Galileo's understanding of physical laws was modified and partially replaced by the discoveries of Sir Isaac Newton, whose "classical" physics was overturned by the "quantum" physics of Niels Bohr and the "relativistic" universe of Albert Einstein. In each instance, the prior physical model was not so much invalidated as shown to apply to a circumscribed level of reality, beneath which a deeper one was revealed. The direction of these revolutionary shifts has been toward the world of the infinitesimally small. When at last their existence was accepted by the majority of physicists, atoms were found to consist of elementary particles, which in turn comprise even smaller "quarks."

Few if any physicists living today believe that the "final answer" to the question of nature's fundamental structure has been found. Indeed, the quest for a "unified field theory" or "theory of everything" is the hallmark of contemporary physics. For this reason, the next edition of this volume will doubtless contain the story of individuals and discoveries that fundamentally modify what we know now.

Limits of time and resources made it impossible for me to include a great many figures who belong in this book; the fact that this is an age when physicists work as large teams, writing 35-author papers, only heightened the problem of inclusiveness. Regretfully, I have omitted physicists whose work is well established, advocates of promising theories that have yet to be accepted by the physics community, as well as physicists whose careers were not marked by a lasting discovery, but whose dedication to research and teaching created the vital medium out of which such discoveries emerge.

THE ENTRIES

Entries are arranged alphabetically by surname, with each entry listed under the name by which the entrant is most commonly known. The typical entry provides the following information:

Entry Head: Name, birth/death dates, field(s) of specialization, and nationality.

Essay: Essays range in length from 500 to 2000 words, with most averaging around 1200 words. Each contains basic biographical information—date and place of birth, family information, educational background, positions held, prizes awarded, and so on—but the greatest attention is given to the entrant's work. Names in small capital letters within the essays provide easy reference to other people represented in the book. All direct quotations, unless otherwise noted, are the physicist's own words. In addition, entries conclude with Further Reading, sources the reader who is interested in following up on a person or his/her work can check.

In addition to the alphabetical list of physicists, readers searching for names of individuals from specific countries can consult the nationality index, which organizes entrants by country of birth and/or citizenship. The subject index lists scientists by field. Finally, the Chronology lists entrants by their birth and death dates.

A

Alferov, Zhores Ivanovich
(1930–)
Russian
Experimentalist, Solid State Physicist

Zhores Alferov is a major figure in solid state physics, whose groundbreaking work on semiconductors led to the creation of the physics of modern miniaturized electronic devices such as cell phones, pagers, and compact disc (CD) players. For this work he shared the 2000 Nobel Prize in physics with two other pioneers in information and communications technology, HERBERT KROEMER and JACK ST. CLAIR KILBY.

He was born on March 15, 1930, in Vitebsk, Belorussia, in what was then the Soviet Union, to Anna Vladimirovna, a librarian and head of a public organization of homemakers, and Ivan Karpovich, a factory director and a Communist Party member, who told Zhores and his brother tales of his exploits in the civil war. He attended a boys' school in Minsk, which had been devastated in World War II, where an inspiring physics teacher helped him find his calling and advised him to apply for admission to the celebrated V. I. Ulyanov [Lenin] Electrotechnical Institute in Leningrad (now Saint Petersburg). Although he found theoretical physics "easy enough," he was attracted to laboratory work and began the research on semiconductors, that

is, materials such as silicon or germanium that have a resistivity midway between that of conductors and that of insulators, that would become his life's work. In 1952 he graduated from the Department of Electronics, after completing a thesis on the problem of obtaining thin films.

He was offered the opportunity to stay on but instead accepted a position at the Physico-Technical Institute in Leningrad, under the leadership of Abram F. Ioffe in 1953 as a junior researcher in the recently organized semiconductor laboratory. He considered this "lucky chance" to work with the elite of his chosen field the cause of his "happy scientific career." By May 1953, the first Soviet transistor receivers had been developed and Alferov began to comprehend the significance of the technology for electronic devices as well as basic research. Within the next few years his group developed the first Soviet high-power germanium rectifiers and germanium and silicon photodiodes, used in modern electronic devices. He was part of the team that developed a special semiconductor device for the first Soviet atomic submarine in 1958.

He received his candidate's degree (somewhere between American master's and doctoral degrees) from the Ioffe Institute (formerly the Physico-Technical Institute) in 1961. At this point, he became involved with heterostruc-

1

tures and semiconductor lasers. Because early transistors were relatively low-powered and slow, semiconductor transistors based on heterostructures were proposed as a way of increasing amplification and achieving higher frequencies and power. Such a heterostructure consists of two semiconductors whose atomic structures fit one another well, but that have different electronic properties. Kroemer had formulated these ideas theoretically in a 1957 paper. Alferov understood that semiconductor physics would be developing on the basis of hetero- rather than homostructures and led his group at the Ioffe Institute in the race to develop this technology before Bell Telephone, IBM, and RCA could do so.

During this period of intense research, he rose steadily in rank at the Ioffe Institute, becoming a senior researcher in 1964 and head of his laboratory in 1967. That year he married Tamara Darskaya, a researcher at a large space enterprise, with whom he would have two children. By 1969 his group had mastered all the ideas on control of the electron and light fluxes in classical heterostructures based on the arsenid gallium–arsenid aluminum heterostructure.

During his first trip to the United States in 1969, Alferov spoke about his group's recent development of low-threshold room temperature lasers. He electrified his audience when he explained how they had obtained the continuous wave regime by developing an optical fiber with low losses—a breakthrough that resulted in the discovery and rapid development of optical fiber communication. He was able to visit Bell Labs and IBM and later wrote of his relationship with the giant American firms as "a rare example of open and friendly competition between laboratories belonging to the antagonistic great powers."

Semiconductor heterostructures have been important to the development of lasers, light-emitting diodes, modulators, and solar panels. The semiconductor laser is based on the recom-

bination of electrons and holes, emitting photons (particles of light). If the density of these photons becomes sufficiently high, they may begin to move in rhythm with each other and form a phase-coherent state, that is, laser light. The first semiconductor lasers had low efficiency and could only shine in short pulses. Both Kroemer and Alferov had suggested in 1963 that the concentration of electrons, holes, and photons would become much higher if they were confined to a thin semiconductor layer between two others—a double heterojunction. Despite a lack of the most advanced equipment, in May 1970, a few weeks earlier than their American competitors, Alferov's group succeeded in producing a laser that operated continuously and that did not require troublesome cooling.

After receiving his doctorate from the Ioffe Institute in 1970, Alferov spent six months in 1971 in the United States, working in the semiconductor devices lab at the University of Illinois. In 1973, he became professor of optoelectronics at the Saint Petersburg State Electrotechnical Institute (the new name of the Ulyanov Institute). In 1987 he was elected director of the Ioffe Institute and in 1988 was appointed dean of the faculty of physics and technology at Saint Petersburg Technical University. He was elected a corresponding member of the Union of Soviet Socialist Republics (USSR) Academy of Sciences in 1972 and a full member in 1979. Since 1989 he has been vice president of the USSR (now Russian) Academy of Sciences and president of its Saint Petersburg Scientific Center.

In the tumultuous period since the Soviet Union collapsed in 1991, Alferov has been deeply concerned with the plight of Russia's severely underfunded scientific community. In 1995, to protect the Academy of Sciences, he became a deputy of the State Duma (Russian parliament) and won fame as a leading advocate for educational funding programs. When he won the 2000 Nobel Prize he donated a third of his winnings for the support of Russian education and science. He believes that

> If Russia is to be a great power, it will be, not because of its nuclear potential, faith in God or the president, or Western investment, but thanks to the labor of the nation, faith in knowledge and science and the maintenance and development of scientific potential and education.

Alferov's work has led to spectacular scientific breakthroughs in which the advanced materials and tools of microelectronics are being used for studies in nanoscience and investigations of quantum effects. The impact of Alferov's research on the modern world of electronics and communication has been enormous. Lasers and light-emitting diodes (LEDs) have been further developed in many stages. Without the heterostructure laser, today we would not have had optical broadband links, CD players, laser printers, bar code readers, laser pointers, and numerous scientific instruments. LEDs are used in displays of all kinds, including traffic signals, and may eventually replace lightbulbs altogether. In recent years, it has been possible to make LEDs and lasers that cover the full visible wavelength range, including blue light. Today, high-speed transistors are found in cellular phones and in their base stations, in satellite dishes and links. There they are part of devices that amplify weak signals from outer space or from a faraway cellular phone without drowning in the noise of the receiver itself.

Alferov is currently editor in chief of the Russian journal *Technical Physics Letters* and a member of the editorial board of the Russian journal *Science and Life*. He is the author of four books, 400 articles, and 50 inventions involving semiconductor technology.

Further Reading

Noll, A. Michael. *Principles of Modern Communications Technology*. Boston and London: Artech House, 2001.

⊠ **Alfvén, Hannes Olof Gösta**
(1908–1995)
Swedish
Plasma Physicist, Astrophysicist

Hannes Alfvén was the founder of the modern field of plasma physics, the study of electrically conducting gases, and the father of the branch of plasma physics known as magnetohydrodynamics (MHD), the study of plasmas in magnetic fields. He was honored for this work with the 1970 Nobel Prize in physics.

Alfvén was born on May 30, 1908, in Norrkoping, Sweden, to Anna-Clara Romanus and Johannes Alfvén, both physicians. He attended the University of Uppsala and earned a Ph.D. in 1934 for a dissertation on ultrashort electromagnetic waves; in that year, he was appointed lecturer in physics at Uppsala. In 1935, he married Kerstin Maria Erikson, with whom he would share a 67-year marriage that would produce five children.

He became a research physicist, in 1937, at the Nobel Institute in Stockholm, where he began his groundbreaking work in plasma physics. Plasmas are highly ionized gases containing both free positive ions and free electrons. The dominant state of matter in the universe, they are rare on Earth, but abundant in stars, galaxies, and intergalactic space. Alfvén studied plasma physics primarily within the context of astrophysics, beginning with an attempt to explain the phenomenon of sunspots by investigating the interaction of electrical and magnetic fields with plasmas. He formulated the frozen-in-flux theorem, which postulates that under certain conditions, a plasma is bound to the magnetic lines of flux passing through it. On the basis of this theorem he then postulated the existence of the galactic magnetic field, which now forms the basis for cosmic magnetism, using the idea to explain the origin of cosmic rays.

In the early 1930s, most physicists believed that cosmic rays were gamma rays that permeated the whole universe. But when cosmic rays were discovered to be charged particles, Alfvén made a unique proposal: if the galaxy contained a large-scale magnetic field, then cosmic rays could move in spiral orbits within the galaxy, because of the forces exerted by the magnetic field. He argued that there could be a magnetic field pervading the entire galaxy if plasma were spread throughout the galaxy. This plasma could carry the electrical currents that would then create the galactic magnetic field. Whereas this intuitive hypothesis was first dismissed on the grounds that interstellar space was known to be a vacuum incapable of supporting electrical currents and particle beams, it was later accepted by physicists and became very fashionable in the 1980s and 1990s.

Alfvén was the first to devise the guiding center approximation, a widely used technique that enables the complex spiral movement of a charged particle in a magnetic field to be calculated with relative ease. In 1939, using this technique, Alfvén proposed a theory to explain auroras and magnetic storms that would exert a profound influence on future attempts by physicists to understand the Earth's magnetosphere. At the time, the renowned space scientist Sydney Chapman argued that the currents involved were restricted to flow only in the ionosphere with no downflowing currents. Alfvén challenged this widely accepted theory by championing the ideas of the Norwegian scientist Kristian Birkeland, who believed that electric currents flowing down along the Earth's magnetic fields into the atmosphere were the cause of the aurora and polar magnetic disturbances. Alfvén won the debate decades later, in 1974, when Earth satellites were able to measure and observe the downflowing currents for the first time. In a similar manner many of Alfvén's theories about the solar system were disputed for many years and only vindicated as late as the 1980s, through measurements of cometary and planetary magnetospheres by artificial satellites and space probes.

In 1940, Alfvén joined the Royal Institute of Technology, Stockholm, and became professor of electronics at the Royal Institute in 1945. In 1942, he hypothesized that a form of electromagnetic plasma wave (now called the Alfvén wave) could propagate through plasma, a phenomenon that was, in fact, later observed by other physicists in plasmas and in liquid metals. On purely physical grounds he concluded that an electromagnetic wave could propagate through a highly conducting medium such as the ionized gas of the Sun, or in plasmas anywhere. Since his hypothesis contradicted JAMES CLERK MAXWELL's theory of electromagnetism, initially no one took it seriously. It was, after all, "well known" that electromagnetic waves could penetrate only a very short distance into a conductor and that as the resistance of a conductor became smaller and smaller, the depth of penetration of an electromagnetic wave would approach zero. Thus, with an ideal electrical conductor, there could be no penetration of electromagnetic radiation. Alfvén's proposal of a form of electromagnetic wave that could propagate in a perfect conductor with no attenuation or reflection was ignored. However, in 1948, when Alfvén lectured on it at the University of Chicago, ENRICO FERMI agreed with the conclusions of Alfvén's work and it became widely accepted.

Also in 1942 Alfvén put forth a theory of the origin of the planets in the solar system, sometimes called the Alfvén theory, which hypothesizes that planets were formed from the material captured by the Sun from an interstellar cloud of gas and dust. The theory envisages the following series of events: As atoms were drawn toward the Sun, they became ionized and influenced by the Sun's magnetic field. Then, in the plane of the solar equator, the ions condensed into small particles, which, in turn, coalesced to form the planets. Although Alfvén's theory did not adequately explain the formation of the inner planets, it was important in suggesting the role of MHD in the origin of the solar system.

In 1950, together with his colleague N. Herlofson, Alfvén was the first to identify non-thermal radiation from astronomical sources as synchrotron radiation, which is produced by fast-moving electrons in the presence of magnetic fields. The recognition that the synchrotron mechanism of radiation is important in celestial objects proved extremely productive in astrophysics, since nearly all the radiation recorded by radio telescopes derives from this mechanism.

It was in the 1960s that Alfvén formulated his opposition to the big bang theory of cosmology. He wrote, "I have never thought that you could obtain the extremely clumpy, heterogeneous universe we have today, strongly affected by plasma processes, from the smooth, homogeneous one of the Big Bang, dominated by gravitation."

Instead of the big bang theory, in which the universe is created out of nothing, in a fiery explosion, at a fixed moment in time, he postulated the plasma universe, an evolving universe without beginning or end. He believed that the appeal of the big bang was rooted in its mythological approach, in which a perfect principle is sought, on the basis of which "the gods" created the universe. He juxtaposed to this, what he called a scientific, empirical approach, reasoning that, since we never observe something emerging from nothing, there is no reason to assume that this occurred in the distant past. On the other hand, since we now see an evolving universe, plasma cosmology assumes that the universe has always existed and evolved and will continue to do so for an infinite time to come. Today big bang theory continues to be more persuasive to the great majority of physicists, in large part as a result of the observation of the cosmic background radiation that permeates the universe, believed to be a remnant of the initial explosion. However, Alfvén's theory continues to attract a small dissident minority.

Around this time, in addition to his scientific debates, Alfvén became embroiled in political controversies. He was a writer of popular science books, sometimes with the collaboration of his wife, Kerstin Maria Erikson Alfvén, and in 1966 he published, under the pseudonym Olaf Johannesson, *The Great Computer*, a pointed political–scientific satire in which the planet is taken over by computers. In Alfvén's hands, this popular science fiction theme became a vehicle for ridiculing the growing infatuation of government and business with computers, as well as for attacking a large part of the Swedish scientific establishment. The book succeeded in greatly antagonizing the objects of its pointed critique. By the following year, Alfvén's quarrel with the Swedish government, particularly his condemnation of Sweden's nuclear research program for allocating insufficient funds for projects on peaceful uses of thermonuclear energy, became bitter enough for him to decide to leave Sweden. He was immediately offered two positions—one in the United States and the other in the Soviet Union. After two months in the Soviet Union, he became a professor at the University of California, San Diego. Eventually, he reconciled his differences with the Swedish scientific bureaucracy and divided his time between the Royal Institute and the University of California.

In 1970, he shared the Nobel Prize with Louis-Eugène-Félix Néel. He was president of the Pugwash Conference on Science and World Affairs and became a leading advocate of arms control.

In spite of such recognition, for much of his career his ideas were dismissed or treated with condescension, forcing him to publish in obscure journals.

Alfvén's contentious career was balanced by his private life—his happy marriage and large, accomplished family; his zest for travel; and his physical vitality. He died on April 2, 1995, in Stockholm, at the age of 86.

Alfvén was a controversial figure, regarded by a few as "the Galileo of the late 20th century" and by many as a heretic for his rejection of the big bang theory. Yet his contributions to physics are many and undeniable and are today being applied in the development of particle beam accelerators, controlled thermonuclear fusion, rocket propulsion, and the braking of reentering space vehicles. Applications of his research in space science include explanations of the Van Allen radiation belt, the reduction of the Earth's magnetic field during magnetic storms, the magnetosphere (a protective plasma envelope surrounding the Earth), the formation of comet tails, the formation of the solar system, the dynamics of plasmas in our galaxy, and the fundamental nature of the universe itself. His numerous contributions to this field are reflected in the concepts that bear his name, including the Alfvén wave, Alfvén speed, and Alfvén limit.

See also GAMOW, GEORGE; PENZIAS, ARNO; VAN ALLEN, JAMES ALFRED.

Further Reading

Alfvén, Hannes. *On the Origin of the Solar System.* Westport, Conn.: Greenwood Press, 1973.

———. *Worlds-Antiworlds: Antimatter in Cosmology.* New York: W. H. Freeman, 1966.

⊠   Alvarez, Luis W.

(1911–1988)
American
Experimental Physicist, Particle Physicist, Electronic Engineer

Luis W. Alvarez was an experimentalist of extraordinary scope and ability, who won the 1968 Nobel Prize in physics for his development of the hydrogen bubble chamber, which led to great advances in the discovery of high-energy unstable states of nuclear particles.

He was born in San Francisco on July 13, 1911, the son of Walter C. and Harriet Smyth Alvarez. His father was a physician and researcher in physiology; by the time he was 10, Luis could use all the tools of his father's shop and wire up

electrical circuits. Alvarez would later credit his father's role in nurturing his scientific creativity:

> He advised me to sit every few months in my reading chair for an entire evening, close my eyes and try to think of new problems to solve. I took his advice very seriously and have been glad ever since that I did.

In 1925, the family moved to Minnesota, when Walter Alvarez joined the staff of the Mayo Clinic in Rochester. Luis spent two high school summers as an apprentice in the clinic's instrument shop.

When he enrolled at the University of Chicago, he intended to study chemistry but found himself fascinated by physics instead. He was particularly drawn to optics; his native talent was nourished by ALBERT ABRAHAM MICHELSON's optical technicians. He earned his B.S., M.S., and Ph.D. in physics from the University of Chicago in 1932, 1934, and 1936, respectively. He would later describe his training at Chicago as "atrocious," adding, however, "I could build anything out of metal or glass, and I had the enormous confidence to be expected of a Robinson Crusoe who had spent three years on a desert island."

In 1934, he began flying, soloing with just three hours of dual instruction. Over the next 50 years, he would log more than a thousand hours as a pilot. In 1936, he married Geraldine Smithwyck, with whom he had two children, Walter and Jean. Beginning in 1936, he spent his entire career at the University of California at Berkeley as professor of physics; there he would later become Professor Emeritus in 1978. In 1936, Ernest Lawrence, who would become a close friend, invited him to join the Berkeley Radiation Laboratory. There he spent a year immersing himself in the literature on nuclear physics and attended Lawrence's weekly journal club—a tradition he would continue for decades in his own home.

In 1937 Alvarez gave the first experimental demonstration of the existence of the K-shell electron capture process by nuclei. Another early development was a method for producing beams of very slow neutrons, which led to investigation of neutron scattering and to the first measurement of the magnetic moment of the neutron. Just before the beginning of World War II, he discovered 3H (tritium), best known as an ingredient of thermonuclear weapons, and He3 (helium three), which became important in low-temperature physics.

During the World War II years, Alvarez played a key role in radar research and development at the Massachusetts Institute of Technology. He invented the system known as Vixen, which permitted radar-equipped aircraft to destroy surfaced German submarines. He and his group developed ground-controlled approach (GCA) radar, which allowed ordinary aircraft and pilots to land at night and in poor visibility. He also made important contributions to the microwave early warning system and the Eagle blind bombing system.

Then, after working with ENRICO FERMI at the Metallurgical Laboratory at the University of Chicago on the first nuclear reactor, Alvarez joined the Manhattan Project team at Los Alamos under J. ROBERT OPPENHEIMER. He made one of his most important contributions in the critical 1944–1945 period before the end of the war, when he invented capacitor-discharge bridgewire detonators. These allow the simultaneous initiation of the multiple high-explosive "lenses" required to generate the implosion system needed for the development of the plutonium version of the atomic bomb. After the war, he returned to Berkeley and designed and constructed the first operational linear accelerator. He was also deeply involved in the effort to build a large deuteron accelerator for the production of plutonium for nuclear weapons at the Lawrence Laboratory, where he was associate director of the Lawrence Radiation Laboratory

in 1949–1959 and 1975–1978. At this point in his life he married his second wife, the former research assistant Janet L. Landis; they had two children, Donald and Helen.

During this period Alvarez began to concentrate on the development and use of large, liquid hydrogen bubble chambers, in order to track unstable nuclear particles. Such particles disintegrate rapidly into other particles and are so minute that they can be identified only by the tracks they leave behind them as they move. In high-energy accelerators, these particles move at near light speeds. Although the particle life span might be only 10,000th of 1 millionth of a second, the track acquires a length of several centimeters. The pattern of tracks thus becomes very complicated. Interpreting them correctly requires advanced experimental techniques. Alvarez extended the idea of the liquid helium bubble chamber invented by DONALD ARTHUR GLASER and developed the hydrogen bubble chamber, an invaluable instrument for this kind of investigation.

The hydrogen bubble chamber contains many hundreds of liters of liquid hydrogen reduced to a temperature of $-2500°C$. When the particle passes through the chamber, the liquid hydrogen is warmed to its boiling point along the particle's track. In the particle's wake are a trail of bubbles that can be photographed while still very small. The photos reproduce the path of the particle. Because the chamber contains only hydrogen, all reactions must occur with hydrogen nuclei, which for high-energy processes are essentially protons. This fact considerably simplifies the interpretation of the particle production and decay phenomenon. Alvarez and his group constructed a series of increasingly delicate automatic scanning and measuring instruments for transferring the information from the photographic film into a state that can be analyzed by computer. Using these newly developed hydrogen bubble chamber techniques, they were able to discover a large

number of "fundamental nuclear particle resonances." This work on particle physics with
hydrogen bubble chambers garnered Alvarez the
1968 Nobel Prize in physics.

As an inventor, Alvarez was often the one
who, many years after their inception, took his
own ideas into production. This was true for his
stabilized optical system for binoculars or cameras, invented in 1963 and produced 20 years
later as a stabilizing zoom lens for shoulder-held
video cameras, and for his variable-power lens,
invented in 1971 and first marketed by
Polaroid in 1986. He realized profits from his
more than 40 patented inventions only a few
years before his death in his Berkeley home on
August 31, 1988.

Alvarez was a path breaker in high-energy
particle physics, a consummate engineer and
technologist who made vital contributions to
civil and military aviation, as well as a radical
thinker whose intellectual curiosity and talent
for experiment continually led him in new
directions. Among his more exotic ventures
was the x-raying of the great pyramid of Cheops
by the use of cosmic X-ray muons, which
revealed that the pyramid had no hidden
chamber. He was also the first person to suggest
that the extinction of the dinosaurs 65 million
years ago was due to a collision of a giant comet
with the Earth. Alvarez's controversial
dinosaur extinction theory was later confirmed
by geological observations, which discovered
high levels of minerals characteristic of the
comet over the surface of the Earth. This, in
turn, led to the discovery of the location of the
giant comet's crater under the waters of the
Yucatan peninsula.

See also GELL-MANN, MURRAY.

Further Reading

Alvarez, Luis W. *Adventures of a Physicist*. New York:
Basic Books, 1978.
Trower, W. Peter, ed. *Discovering Alvarez*. Chicago
and London: University of Chicago Press, 1987.

⊠ **Ampère, André-Marie**
(1775–1836)
French
*Theoretical and Experimental Physicist
(Electrodynamics), Mathematical
Physicist*

André-Marie Ampère founded a new branch of
physics that he named *electrodynamics*: the study
of the relationship between mechanical forces
and electric and magnetic forces. In his honor
the unit of electric current is called the ampere
or amp.

He was born in Polémieux, near Lyons,
France, on January 22, 1775. The man who
would later teach physics, mathematics, and
chemistry had no formal education. Ampère's
father was a wealthy merchant, who, in addition
to having his son privately tutored, was personally involved in his education and inspired in
him a passionate desire to learn. Ampère's caring
and protective family imbued in him both an
optimistic belief in the fruits of scientific investigation, which was characteristic of the Enlightenment, and a devotion to Catholicism. He
was apparently something of a prodigy, mastering mathematical texts independently, at an
early age.

When he was 18, however, his peaceful
world of study and family affection was shattered
by the advent of the French Revolution. In
1793, Lyons was captured by the Republican
Army, and his beloved father, who was a wealthy
city official, was guillotined. A devastated
André-Marie put down his books on mathematics and would not return to them for 18 months.
When he met his future wife, Julie, he began to
come to life again.

In 1802 Ampère published his first treatise,
The Mathematical Theory of Games, an early
contribution to probability theory, and was
appointed professor of physics and chemistry at
the École Centrale in Bourg. Later that year, he
became professor of mathematics at the Lycée in

Lyons. Ampère's career was thriving, but another of the personal tragedies that would haunt his life was about to occur: his young wife, whose health had been steadily declining, died in 1804. The grieving Ampère left Lyon with its sad memories for the intellectual excitement of Paris.

He quickly found a position as an assistant lecturer in mathematical analysis at the École Polytechnique in Paris, and, four years later, he was promoted to professor of mathematics. Once more his personal life formed a dark counterpoint to his professional activities. He remarried in August 1806 but had already separated from his second wife by the time their daughter was born 11 months later. Meanwhile, his talent had been recognized by Napoleon, who in 1808 appointed him inspector-general of the newly formed university system, a post he held until his death. He also taught philosophy at the University of Paris in 1918, became assistant professor of astronomy in 1920, and was appointed to the chair of experimental physics at the Collège de France in 1824.

Ampère's intellectual voracity and versatility continued unabated; between 1805 and his famous work on electrodynamics in the 1820s, he studied psychology, philosophy, physics, and chemistry. In 1820, the Danish physicist HANS CHRISTIAN ØRSTED demonstrated his discovery that a magnetic needle is deflected when the current in a nearby wire is flowing, thereby showing evidence of a relationship between electricity and magnetism. Ampère witnessed a restaging of Ørsted's demonstration in Paris that same year. Within a week he had prepared the first of several papers that would describe a new branch of physics revealed by Ørsted's work. Ampère called it electrodynamics, to differentiate it from the electrostatics of CHARLES AUGUSTIN COULOMB. Central to his thinking was a relationship that came to be known as Ampère's law, a mathematical description of the magnetic force between two electric currents. It shows that two parallel wires carrying electric currents in the same direction attract each other, whereas two parallel wires carrying electric currents in opposite directions repel one another. Specifically, Ampère's law relates the magnetic force produced by two parallel current-carrying conductors to the product of their currents divided by the square of the distance between the conductors. Today, Ampère's law is usually stated in the form of calculus: the line integral of the magnetic field around an arbitrarily chosen path is proportional to the net electric current enclosed by the path.

Ampère also predicted and demonstrated that a helical "coil" of wire (which he named a solenoid) behaves as a bar magnet when carrying an electric current. The numerous experiments he performed enabled him to explain known electromagnetic phenomena and predict new ones. In addition, Ampère pioneered the development of measuring techniques for electricity, inventing an instrument using a free-moving needle to measure the strength of the current. This was the prototype of what we know today as the galvanometer.

He also tried to develop a theory to explain electromagnetism, proposing that magnetism is merely electricity in motion. Prompted by his close friend AUGUSTUS FRESNEL, one of the originators of the wave theory of light, he suggested that molecules are surrounded by a perpetual electric current.

Ampère culminated his groundbreaking studies with the publication in 1827 of his *Memoir on the Mathematical Theory of Electrodynamic Phenomena, Uniquely Deduced from Experience*, in which he enunciated precise mathematical formulations of electrodynamics, notably Ampère's law.

He died of pneumonia at the age of 61, on June 10, 1836, while on an inspection tour of Marseille. The epitaph on his gravestone reads, *Tandem felix* (Happy at last).

Ampère was honored by election as a fellow of the Royal Society in 1827. His name is inscribed, along with those of 71 other prominent French scientists, on the Eiffel Tower. The Rue Ampère, a street in the 17th arrondisement in Paris, and the Mons Ampère, a feature of the lunar landscape, have been named for him.

More than any other scientist, Ampère was responsible for creating the discipline of electrodynamics. Decades later Ampère's law became an integral part of JAMES CLERK MAXWELL's unified theory of electrodynamics.

Further Reading

Darrigol, Olivier. *Electrodynamics from Ampère to Einstein*. Oxford: Oxford University Press, 2000.

Hofmann, James R. *André-Marie Ampère: Enlightenment and Electrodynamics*. London: Cambridge University Press, 1966.

Anderson, Carl David
(1905–1991)
American
Experimentalist, Particle Physicist

Carl David Anderson built an enhanced cloud chamber with which he discovered the positive electron or positron, an achievement that garnered him the 1936 Nobel Prize in physics, as well as the positive and negative muon (or mu meson).

He was born in New York City on September 3, 1905, the son of Swedish immigrants, and attended the California Institute of Technology, where he received a B.Sc. in physics and engineering in 1927. He was awarded a Ph.D. in 1930 for a dissertation on the space distribution of photoelectrons emitted from various gases as a result of irradiation with X rays. Although his nominal dissertation director was ROBERT MILLIKAN, who had accurately determined the charge of the electron, Anderson complained, "Not once in the three years of my graduate thesis work did he visit my laboratory or discuss the work with me."

Anderson would remain at Caltech for the rest of his career. In 1930 he became a research fellow working with Millikan's research team and began to study gamma rays and cosmic rays. Anderson built and ran the Caltech Magnet Cloud Chamber: a special type of cloud chamber that was divided by a lead plate in order to slow the particles sufficiently for their paths to be accurately determined. He used it to measure the energies of cosmic and gamma rays, by measuring the curvature of their paths, in strong magnetic fields (up to about 24,000 gauss or 2.4 tesla). In 1932, he announced that he had obtained "dramatic and completely unexpected" results: approximately equal numbers of positively and negatively charged particles, where only electrons had been expected. He also noted that in many cases several negative and positive particles were simultaneously projected from the same center. At first he assumed that the positive particles were protons; however, it turned out that their mass was identical to that of electrons (as opposed to the much heavier protons), and so he called them positive electrons. He went on to suggest the name *positrons* for these antimatter particles. The previous year PAUL ADRIEN MAURICE DIRAC, on the basis of his relativistic theory of the atom, had predicted the existence of a positive electron when he discovered that the mathematical description of the electron contained twice as many states as were expected. Working with his first graduate student, Seth Neddermeyer, Anderson also showed that positrons can be produced by irradiation of various materials with gamma rays. In 1932 and 1933, LORD STUART BLACKETT, SIR JAMES CHADWICK, and the Joliot-Curies independently confirmed the existence of the positron and later elucidated some of its properties.

In 1933, Anderson was promoted to assistant professor of physics at Caltech. Three years later, when he shared the 1936 Nobel Prize with Victor Hess, the discoverer of cosmic rays, he was the youngest physicist ever to be so honored.

That same year he contributed to the discovery of another elementary particle, the muon. He and Neddermeyer transported their magnet cloud chamber to the summit of Pikes Peak, Colorado, in order to obtain more intense, higher-energy cosmic rays. After a summer at Pikes Peak, they analyzed their results and found that the positive and negative tracks were different from those made by electrons and protons and appeared to have been made by a particle with an intermediate mass. They proposed that the high-altitude tracks were new, unknown particles and called them mesotrons, because of their "middle" mass; the name was later shortened to *meson*. At first Anderson thought that this was the particle previously predicted by HIDEKI YUKAWA, which was supposed to hold the nucleus together and carry the strong nuclear force. Further studies, however, showed that it did not readily interact with the nucleus and therefore could not be Yukawa's particle. Anderson's particle is now called the mu meson to distinguish it from the pi meson.

When World War II broke out, Anderson turned down an offer to direct development of the atomic bomb, "on purely economic grounds," and the job went to J. ROBERT OPPENHEIMER. In the autobiography he would write in his 70s, he observes:

I believe my greatest contribution to the World War II effort was my inability to take part in the development of the atomic bomb. Thinking so brings me peace of mind.

Instead, he worked on the development of rocket launchers at Caltech and, in 1944, supervised the installation of the first rocket launchers on Allied planes.

After the war, in 1946, he married Lorraine Bergman, with whom he had two sons, Marshall and David. He served as full professor at Caltech until his retirement in 1976, when he became

Carl David Anderson discovered the positive electron, or positron, as well as the positive and negative muon. *(NARA, courtesy AIP Emilio Segrè Visual Archives)*

professor emeritus. He died on January 11, 1991, after a brief illness, at the age of 85.

Carl Anderson played a major role in the discovery of new elementary particles and pointed the way to the existence of antimatter. His discovery of the positron led to the prediction of other antiparticles, such as the antiproton discovered by EMILIO SEGRÈ in 1955. Today the existence of antimatter—an antiparticle for all particles—is universally accepted.

See also WILSON, CHARLES THOMSON REES.

Further Reading
Weiss, Richard J., ed. *The Discovery of Anti-Matter: The Autobiography of Carl David Anderson, the Youngest Man to Win the Nobel Prize.* Series in Popular Science, Vol. 2. Singapore: World Scientific, 1999.

⚛ Anderson, Philip Warren

(1923–   )
American
Theoretical Physicist, Solid State Physicist

Philip Warren Anderson shared the 1977 Nobel Prize in physics with JOHN HOUSBROOK VAN VLECK and NEVILL FRANCIS MOTT for his theoretical work on the behavior of electrons in magnetic, noncrystalline solids.

He was born in Indianapolis on December 13, 1923; shortly afterward, Harry Warren Anderson, his father, became a professor of plant pathology at the University of Illinois in Urbana, where Philip spent his childhood and adolescence. He spent his happiest hours with the families of his parents' friends, enjoying hiking and camping and developing the political consciousness that would make him an opponent of McCarthyism, a supporter of liberal causes, an opponent of the Vietnam War, and a scientist who refused to engage in classified research. Although the physicists among his father's friends encouraged him, he initially intended to major in mathematics when he entered Harvard University, in 1940, on a full-support National Scholarship. The world was at war and physics students were urged to concentrate on a field with immediate applications, such as "electronic physics." After earning his B.A., summa cum laude, in 1943, he spent the next two years doing antenna engineering work at the Harvard Naval Research Laboratories in Washington, D.C.

Returning to Harvard in 1945, he plunged into a series of stimulating courses, including those of JULIAN SEYMOUR SCHWINGER. He earned his M.S. in 1947 and that same year married Joyce Gothwaite, who soon presented him with a daughter, Susan. He received his Ph.D. in 1949 from Harvard, after completing his dissertation under Van Vleck, on the pressure broadening of spectral lines in microwave, infrared, and optical spectroscopy. (Pressure broadening refers to the increase in the width of a spectrum line resulting from collisions between atoms and molecules in a gas that occur as the gas pressure increases.)

In 1949, he went to work at Bell Laboratories in Murray Hill, New Jersey, where he joined a stellar theoretical group, which included JOHN BARDEEN, the coinventor of the transistor. From his Bell colleagues, Anderson learned about ferromagnetism (the magnetism of substances caused by a domain structure, that is, a material region in which all the atomic magnetic fields point the same way), crystallography, and solid state physics. At the Kyoto International Physics Conference in 1953 he gained a lasting admiration for Japanese culture and met Nevill Mott, whose work he admired.

In the late 1950s, Anderson developed a theory that explained superexchange: the coupling of spins of two magnetic atoms in a crystal through their interaction with a nonmagnetic atom located between them. He was then able to apply the Bardeen–Cooper–Schrieffer (BCS) theory of superconductivity to explain the effects of impurities on the properties of superconductors. Working with a French graduate student, Pierre Morel, Anderson studied the Josephson effect, an electrical effect associated with pairs of superconductors. He went on to develop the theoretical treatments of antiferromagnetics (paramagnetic substances with a small susceptibility to the external magnetic field, which behave as ferromagnetic substances when their temperature is changed), ferroelectrics (crystalline compounds having natural spontaneous electric polarization that can be reversed by the application of an electric field), and superconductors.

During this same period, Anderson did his important work on disordered systems. In crystalline materials, the atoms form regular lattices, which greatly facilitate the theoretical treatment. In disordered materials, the regularity is lacking so that there is no lattice whatsoever, as, for instance, in glass; this makes it very hard to treat such materials theoretically. In 1958, Anderson published a paper in which he showed under what conditions an electron in a disordered system can either move through the system as a whole or be more or less tied to a specific position as a localized electron. Mott drew the attention of solid state physicists to this paper, which became one of the cornerstones in the understanding of the electric conductivity in disordered systems. Anderson would return to the subject of disordered media in the early 1970s, working on low-temperature properties of glass and later studying spin glasses.

In the early 1960s, Anderson developed a model of the interatomic effects that influence the magnetic properties of metals and alloys (now called the Anderson model) to describe the effect of the presence of an impurity atom in a metal. He also devised a method of describing the movements of impurities within crystalline substances, a method now known as Anderson localization. He also studied the relationships among the phenomena of superconductivity, superfluidity, and laser action, all of which involve coherent waves of matter or energy, and predicted the possibility of superfluid states of helium 3, an isotope of helium. On the more practical side he performed research on the semiconducting properties of inexpensive, disordered glassy solids. His studies of these materials indicated the possibility that they could be used in place of the expensive crystalline semiconductors now used in many electronic devices, such as computer memories, electronic switches, and solar energy converters.

In 1967, through the efforts of Mott, Anderson obtained a "permanent visiting professorship" at the Cavendish Laboratories at Cambridge University. For the next eight years he and Joyce divided their time between Cambridge and New Jersey. Anderson headed the Theory of Condensed Matter Group at Cambridge, which he

recalls as "eight productive and exciting years, spiced with warm encounters with students, visitors, and associates from literally the four corners of the Earth."

In 1975, the Cambridge appointment was replaced by a part-time appointment as Joseph Henry Professor of Physics at Princeton University. The following year he became Consulting Director of Research at Bell, and he would later assist ARNO ALLAN PENZIAS in the difficult years of restructuring that followed the breakup of the Bell Laboratory system. In 1977, Anderson received the Nobel Prize in physics for developing Van Vleck's ideas about how local magnetic moments can occur in metals, such as silver or copper, that in pure form are not magnetic at all.

The years following the Nobel Prize were productive ones for Anderson. He retired from Bell in 1984 and took up full-time duties at Princeton. During this fertile period he and his colleagues at Princeton revitalized localization theory in solid state physics by developing a scaling theory that made it into a quantitative experimental science.

A longtime proponent of "small science," in the late 1980s Anderson became a controversial figure in the physics community when he argued before Congress against funding for the proposed superconducting super collider to be built in Texas at a cost of $8 billion. He believed the project would yield neither practical benefits nor any fundamental truths that could not be gained elsewhere and more cheaply. When Congress killed the plan in 1993, Anderson said he was only sorry that Congress had allowed the project to go on for so long. He was also an outspoken critic of "Star Wars," the Reagan administration plan to build a satellite-based missile defense system.

In 1986, he became deeply involved with the Sante Fe Institute, a new interdisciplinary institution dedicated to emerging scientific syntheses, especially those involving the sciences of complexity. The following year, news of a new class of "high-temperature" superconductors galvanized the world of many-body quantum physics, leading Anderson to reexamine older ideas and search for new ones. He found that he was able to account for most of the wide variety of unexpected anomalies observed in these materials by invoking a new two-dimensional state of matter and a new mechanism for electron pairing called deconfinement.

The amazing range of Anderson's research in solid state physics has spanned the topics of spectral line broadening, exchange interactions in insulators, the Josephson effect, quantum coherence, superconductors, and nuclear theory. Experimental confirmation continues to support the predictions of Anderson's theory of high-temperature superconductors, which is expected to find many new scientific applications in the 21st century.

See also JOSEPHSON, BRIAN DAVID; SCHRIEFFER, JOHN ROBERT.

Further Reading

Vidali, Gianfranco. *Superconductivity: The Next Revolution?* Cambridge: Cambridge University Press, 1993.

⊠ **Ångstrom, Anders Jonas**
(1814–1874)
Swedish
Spectroscopist, Astronomer

Anders Ångstrom is known as the father of spectroscopy, the branch of physics that studies light by using a prism or diffraction grating to spread out the light into its range of individual colors or spectrum.

Ångstrom was born in Logdo, Sweden, on August 13, 1814, the son of a chaplain. He attended the University of Uppsala, where he obtained his doctorate in physics in 1839. That same year, he became a lecturer at the university and, four years later, an observer at the Uppsala Observatory. His first significant research project involved the development of a method of measuring heat conductivity, which enabled him to demonstrate that it was proportional to electrical conductivity.

He published *Optical Investigations*, his most important work, in 1853; in it he presented his principle of spectral analysis. He had studied electric arcs and discovered that they yield two spectra, one superimposed on the other. The first was emitted from the metal of the electrode itself, the second from the gas through which the spark passed. He was also able to demonstrate that a hot gas emits light at the same frequency as it absorbs it when it is cooled. Ångstrom had established his reputation and was elected to the chair of physics at the University of Uppsala in 1958.

Ångstrom's early work provided the basis for the spectrum analysis that would occupy him for the rest of his life. In 1862 he announced his hypothesis—which would later be confirmed—that the Sun's atmosphere contains hydrogen. The solar spectrum was his primary interest, but in 1867 he also became the first person to study the spectrum of the aurora borealis, the spectacular display of blue, pink, red, orange, and yellow light seen in the night sky in the Northern Hemisphere during the fall and winter. He used a spectroscope consisting of a simple triangular prism, which broke up the white light passing through it into a rainbow-like band of colors known as a spectrum. Looking through this prism with the aid of a telescope, he showed that the light of the aurora borealis differed from that of the Sun. He thus reached the momentous conclusion that no two substances have the same spectrum and, therefore, that any substance can be identified by its spectrum.

In 1868 he published his famous *Researches on the Solar Spectrum*, which contained measurements of the wavelengths of more than 1,000 lines in the Sun's spectrum, known as Fraunhofer lines, measured to six significant figures in units of 100 millionths of a centimeter (10^{-8} cm).

Another of his important achievements was his atlas of the normal solar spectrum, published in 1869, which became a standard reference tool. He remained at the University of Uppsala until his death on June 21, 1874.

Ångstrom's legacy is visible in the work of contemporary astronomers and astrophysicists, who still identify elements found in stars through the use of spectroscopy.

The unit of measure for the wavelength of light, officially adopted in 1907, is called the angstrom in his honor. Signified by Å and equal to one hundred millionth of a centimeter (10^{-8} cm), the angstrom serves as a convenient unit that can be used to specify radiation wavelengths, which enables physicists to avoid writing large numbers of zeroes, when discussing the wavelength of light.

See also FRAUNHOFER, JOSEPH VON.

Further Reading

Beckman, Olof. Ångstrom: Father and Son. Acta Universitatis Upsaliensis, C. Organisation Och Historia, No. 60. Uppsala, Sweden: Coronet Books, 1997.

⊠   **Archimedes**
(c. 287–212 B.C.)
Greek
*Theoretician and Experimentalist
(Mechanics, Hydrostatics),
Mathematician, Astronomer*

Archimedes is widely viewed as the greatest scientist of the ancient world. As a physicist, he is credited with establishing the fields of statics, a branch of mechanics dealing with the forces on an object or in a system in equilibrium, and hydrostatics, the study of fluids (liquids and gases) in equilibrium. He is most famous for Archimedes' principle, which offered the first scientific explanation of what makes solid objects float.

Only a handful of facts about Archimedes' life have been established with certainty. The biography of him written by his friend Heracleides has been lost. What remains for historians to draw on are Archimedes' nine surviving mathematical treatises, which he published in the form of correspondence with the leading mathematicians of his time (including the Alexandrian scholars Conan of Samos and Eratosthenes of Cyrene); the accounts of his life left by his Greek contemporaries; and stories from Plutarch, Livy, and others. He was widely known during his lifetime, mainly because of his inventions that were used in war. The impression his mechanical genius made on the popular imagination gave rise to numerous legends, which today are viewed by most historians as apocryphal.

Archimedes is believed to have been born in 287 B.C., in Syracuse, Sicily, then a Greek colony. The date has been based on the claim of a 12th-century historian that he died at age 75, and then working backward from the date of his death in 212 B.C., which is reliably established. His father was Phidias, an astronomer; the family was a noble one, possibly related to that of King Hieron II of Syracuse. As a young man, he traveled to Alexandria, then a great capital of learning for mathematics, and studied under Conan and other mathematicians, who had been students of Euclid. Most disciples would remain in Alexandria, but Archimedes returned to Syracuse, where he spent the rest of his life studying mathematics and physics, diverting himself by designing the numerous mechanical devices that earned him widespread renown.

The boldness and originality of Archimedes' mathematical work are tempered by its extreme rigor and adherence to the highest standards of the geometry of his time. Among his most important results was the determination of the value of π. He made the most accurate prediction of his time about the value of π, bracketing its value within the upper and lower limits given by 223/71 = 3.14085 > π > 3.1429 = 220/70. What is more amazing is the fact that the average of Archimedes' upper and lower limits on the value of π is 3.1419, less than three parts in ten thousand different from the modern approximation given: π = 3.1416. In an attempt to improve Greek numerical notation, he devised an ingenious system for the expression of very large numbers. In his treatise *The Sandreckoner,* he proposed a number system capable of expressing very large numbers. He then used this number system to estimate the number of grains of sand in the universe as given by the number 10^{63}. In this manner he showed that large numbers could be considered and handled effectively. He also invented methods to solve cubic equations and to determine square roots by approximation. Most impressively, he devised formulas to determine the surface areas and volume of curved surfaces and solids, a topic that anticipated the development of the integral calculus 2,000 years later by Newton and Leibniz.

As a physicist Archimedes is credited with establishing the fields of statics of objects and hydrostatics of fluids. In statics he worked out the rigorous mathematical proofs behind the principle of the lever and the compound pulley—mechanical devices that can multiply the effects of forces. Although the scientists of his time were familiar with the use of the lever, Archimedes was the first to show that the ratio of the effort applied to the load raised by a lever is equal to the inverse ratio of the distances of the effort and load from the pivot or fulcrum about which the lever rotates. He is said to have claimed that if he could stand at a great enough distance, he could use a lever to move the world. In response to this, King Hieron purportedly issued him the lesser challenge of showing that he could move a very heavy object with ease. Archimedes allegedly responded by easily moving a ship, laden with passengers, crew, and

cargo, which a number of men had struggled mightily to lift out of the harbor onto dry land. Sitting at a distance from the ship, he is said to have used a compound pulley to pull it over the land as if it were gliding through water.

His two-volume treatise on hydrostatics, *On Floating Bodies,* is the first known work on the topic and survives only partly in Greek, the rest in medieval Latin translation from the Greek. The first book contains his most famous result, the Archimedes' principle, which states that the upward force on an object totally or partly submerged in a fluid is equal to the weight of fluid displaced by the object. He is said to have become engrossed in the problem of floating bodies when King Hieron ordered that his new crown be evaluated to see whether it was pure gold, without damaging the object. In what is probably the most famous Archimedes story, the great scientist is said to have been watching water overflow from the bath he was immersed in when the idea now known as Archimedes' principle dawned on him. So jubilant was he, legend relates, that he ran through the town naked, crying, "Eureka!" ("I've got it!"). What he had grasped was that if the gold of the king's crown had been mixed with silver, which is less dense, then in order to have an equal weight to that of a purely golden crown, the king's crown would have to have a greater volume and therefore would displace more water than that of a purely golden crown. Unfortunately, as the story goes, this method proved that the crown contained silver and the unlucky goldsmith was put to death by the king.

In antiquity Archimedes was also known as an astounding astronomer, although little is known of this side of his activities. According to the Greek biographer Plutarch, Archimedes chose to publish only the results of his theoretical researches because he deemed only these worthy of serious consideration. But his interest in mechanics deeply influenced his mathematical thinking. In this context he wrote works on theoretical mechanics and hydrostatics, and his treatise *Method Concerning Mathematical Theorems* shows that his intuitive mechanical reasoning was an essential tool leading to his discovery of new mathematical theorems. He wrote:

> Certain things first became clear to me by a mechanical method, although they had to be proved by geometry afterwards because their investigation by the said method did not furnish actual proof. But it is of course easier, when we have previously acquired, by the [mechanical] method, some knowledge of the questions, to supply the proof than it is to find without any previous knowledge.

Inventions attributed to him include a design for a model planetarium able to show the movement of the Sun, Moon, planets, and possibly constellations across the sky. Cicero reports that when Marcellus sacked Syracuse, he took this planetarium as booty. He is also credited with the Archimedes screw, an augur pump used to raise water for irrigation, which is still used in many parts of the world. He reportedly invented it during his days in Egypt, although it is also possible that he borrowed the idea from others in Egypt.

The most dramatic of his inventions, however, were instruments of war. At the urging of King Hieron, he transformed his playful mechanical diagrams into viable machines. Some of these proved invaluable during the Roman siege of Syracuse from 212 to 215 B.C., when Archimedes' weapons allegedly set fire to the ships of the Roman fleet under Marcellus and made them capsize. This held the Romans at bay for a long time, although they eventually succeeded in sacking the city, an operation in which Archimedes met his end. Plutarch gives three different versions of his death, all of which picture him killed while absorbed in scientific pursuits. In one version, despite orders to spare him, Archimedes was killed on the spot by a Roman soldier when he was ordered to leave his study, where he was contemplating a mathematics problem. He left instructions for his tomb to be marked with a sphere inscribed in a cylinder, together with the formula for the ratio of their volumes—since he considered this discovery his greatest achievement. Cicero found the tomb, overgrown with vegetation, a century and a half after Archimedes' death.

Given the magnitude and originality of Archimedes' achievements, it is ironic that his influence remained so small and undeveloped in ancient times. His work was not widely known in antiquity and did not lead directly to other advances at that time. However, his legacy was preserved by Byzantium and Islam, inspiring important work by medieval Islamic mathematicians, and from there it spread to Europe from the 12th century onward. The greatest impact of his work on later mathematicians occurred in the late 16th and early 17th centuries, after which it had a profound impact on the history of science. In particular, Archimedes' method of finding mathematical proof to substantiate experiment and observation became the method of modern science introduced by GALILEO GALILEI. Galileo published a study of the behavior of bodies in water, *Bodies That Stay Atop Water or Move Within It,* in which he championed Archimedes' law of buoyancy, which states that the buoyant force on a body in water is equal to the weight of the water displaced.

Further Reading

Archimedes. *The Works of Archimedes.* New York: Dover Publications, 2002.

Ipsen, D. C. *Archimedes: Greatest Scientist of the Ancient World.* Hillside, N.J.: Enslow, 1988.

Lafferty, Peter. *Archimedes.* New York: Bookwright Press, 1991.

Stein, Sherman. *Archimedes: What Did He Do Besides Cry Eureka?* Washington, D.C.: The Mathematical Association of America, 1999.

B

Bardeen, John
(1908–1991)
American
*Theoretical Physicist, Solid State
Physicist, Electrical Engineer*

John Bardeen was a giant of solid state physics, whose greatest achievements were the invention of the transistor, with WILLIAM BRADFORD SHOCKLEY and Walter Battrain, and the development of a comprehensive theory of superconductivity, with JOHN ROBERT SCHRIEFFER and Leon Cooper. These contributions made him the first person to win two Nobel Prizes in physics, in 1956 and 1972.

He was born on May 23, 1908, in Madison, Wisconsin, the second of five children born to Charles Russell Bardeen and Althea Harmer. His father was the first graduate of the Johns Hopkins Medical School and founder of the Medical School of the University of Wisconsin; his mother had studied art at the Pratt Institute in Brooklyn, New York, and practiced interior design in Chicago. John attended elementary and secondary schools in Madison. His brilliance was immediately apparent, and he was skipped from third grade to junior high. The death of his mother of cancer when he was 12 was a devastating blow. He managed to continue his studies, however, and entered the University of Wisconsin at age 15. He chose to major in engineering, both because of his love of mathematics and because he had no desire to be an academic as his father was.

He received a B.S. in electrical engineering in 1928 and an M.S. the following year, at the University of Wisconsin. Between 1928 and 1930, he was a graduate assistant, examining mathematical problems of antennas and working on applied geophysics. The Great Depression had begun and jobs were scarce. He managed to get hired by Gulf Research in Pittsburgh, where he worked on mathematical modeling of magnetic and gravitational oil prospecting surveys. This was an exciting period when geophysical methods were first being applied to oil prospecting. But Bardeen, who kept abreast of advances in physics, was increasingly drawn to pure science. In 1933, he gave up his industrial career and enrolled for graduate work at Princeton University with EUGENE PAUL WIGNER. At Princeton he was introduced to the rapidly developing field of solid state physics. Bardeen was fascinated by the work of such physicists as Wigner and Frederick Seitz, who were using the new quantum mechanics to help understand how semiconductors worked. He finished his dissertation on the theory of the work function of metals in 1935.

From 1935 to 1938 he was a junior fellow at Harvard University, where he worked with JOHN HOUSBROOK VAN VLECK and PERCY WILLIAMS BRIDGMAN. In 1938, he married Jane Maxwell, a biologist who taught at a girls' high school near Boston. It was to be an enduring union that would produce three children and six grandchildren. From 1938 to 1941, he was an assistant professor at the University of Minnesota. During World War II, between 1941 and 1945, he returned to applied physics at the Naval Ordnance Laboratory in Washington, D.C., where he investigated ways to protect U.S. ships and submarines from magnetic mines and torpedoes. In 1945 he joined the newly formed research group in solid state physics, which included Walter Brittain and was directed by William Shockley, at the Bell Telephone Laboratories in

John Bardeen invented the transistor and developed a comprehensive theory of superconductivity. *(AIP Emilio Segrè Visual Archives)*

Murray Hill, New Jersey, where his research on semiconductors led in 1947 to the development of the transistor. Physicists first understood the electrical properties of semiconductors in the late 1930s, when they became aware of the role of low concentrations of impurities in controlling the number of mobile charge carriers in materials. Current rectification (i.e., the conversion of oscillating current into direct current) at metal–semiconductor junctions had long been known, but the next step required was to produce amplification analogous to that achieved by vacuum tube technology. Shockley's group began a program to control the number of charge carriers at semiconductor surfaces by varying the electric field.

Bardeen and Brittain worked together harmoniously, as Brittain designed the experiments and Bardeen worked out theoretical explanations for the results. In the spring of 1947, Shockley asked them to investigate the reason for the failure of an amplifier he had designed, which was based on a crystal of silicon, later replaced by germanium. By observing Brittain's experiments, Bardeen realized that the assumption they had been making—that electrical current traveled through all parts of the germanium in the same way—was incorrect. On the contrary, electrons behave differently at the surface of the metal. If they could control what was happening at the surface, the amplifier should work. They demonstrated the effects of amplification of two metal contacts 0.05 mm apart on a germanium surface. Large variations of the power output through one contact were observed in response to tiny changes in the current through the other. On December 23, 1947, they succeeded in building the first point-contact transistor, the forerunner of the many complex devices now available through silicon chip technology. Bardeen, Brittain, and Shockley were awarded the Nobel Prize for this work in 1956.

In 1951, Bardeen left Bell and moved to the University of Illinois, where, with the graduate student Bob Schrieffer and postdoctoral student Leon Cooper, he developed the microscopic theory of superconductivity, known as the Bardeen–Cooper–Schrieffer (BCS) theory. In 1911, HEIKE KAMMERLINGH ONNES had first observed zero electrical resistance in some metals below a critical temperature. Since then physicists had looked for a microscopic interpretation of this phenomenon of superconductivity. The methods that were successful in explaining the electric properties of normal metals were unable to predict the effect. At very low temperatures, metals were still expected to have a finite resistance due to scattering of mobile electrons by the ions in the crystal lattice. Bardeen's solution to this problem was to show that electrons pair up through an attractive interaction, and that zero resistivity occurs when the thermal energy available is insufficient to break the pair apart.

Thus, for electrons embedded in a crystal, the normal Coulomb repulsion can be compensated for by this pairing effect when the temperature is below the critical value. The ion cores in the crystal lattice respond to the presence of a nearby electron, and the motion may result in the attraction of another electron to the ion. The net effect is an attraction between two electrons through the response of the ions in the solid. The BCS theory was based on the idea that the interaction between the electrons and the lattice leads to the formation of bound pairs of electrons, called Cooper pairs. The different pairs are strongly coupled to each other; this leads to a complex collective pattern in which a considerable fraction of the total number of conduction electrons are coupled to form a superconducting state. Because of the characteristic coupling of all the electrons, one cannot break up a single pair of electrons without also perturbing all the others, and this process requires an amount of energy that must exceed a critical value. Many of the remarkable qualities of superconductors can be understood qual-

itatively from the structure of this correlated many-electron state.

The comprehensive BCS theory has the ability to explain all known properties associated with superconductivity. Although applications of superconductivity to magnets and motors were possible without the BCS theory, the theory is important for strategies to increase the critical temperature as much as possible, since, if it could be raised above liquid nitrogen temperature, the economics of superconductivity would be transformed. In addition, the theory was an essential prerequisite for the prediction of Josephson junction tunneling, which has important applications in magnetometers and computers and in determination of the fundamental constants of physics. The BCS theory has had profound effects on nearly every field of physics from elementary particle to nuclear physics and from helium liquids to neutron stars. Bardeen and his two colleagues shared the Nobel Prize in physics for their theory in 1972.

During this period of intense theoretical work, he continued to be actively interested in engineering and technology. In 1951 he became a consultant for the Xerox Corporation (called Haloid at the time), and he continued work with them throughout their development as a technological giant. From 1961 to 1974, he was a member of the Board of Directors of the Xerox Corporation. He also was a consultant for General Electric Corporation for many years and for other technology firms.

In 1975 he became professor emeritus at Illinois, where he began working on theories for liquid helium 3 that have analogies with the BCS theory. He lived out the rest of his life in Urbana, Illinois, teaching, researching, and playing his favorite game, golf. He died in 1991 at the age of 82.

During his 60-year scientific career, Bardeen made important contributions to virtually every aspect of condensed matter physics, from his early work on the electronic behavior of metals, the surface properties of semiconductors, and the theory of diffusion of atoms in crystals to his later work on quasi-one-dimensional metals. In his 83d year he continued to publish original scientific papers. Both of Bardeen's Nobel Prize–winning achievements have had a revolutionary impact on computer technology. The invention of the transistor led directly to the development of the integrated circuit and then the microchip.

See also JOSEPHSON, BRIAN DAVID; KILBY, JACK ST. CLAIR.

Further Reading

Bernstein, Jeremy. *Three Degrees Above Zero: Bell Labs in the Information Age.* New York: Scribner & Sons, 1984.

Riordan, Michael, and Lillian Hoddeson. *Crystal Fire: The Invention of the Transistor and the Birth of the Information Age.* New York: W. W. Norton, 1998.

⊠　**Barkla, Charles Glover**
(1877–1944)
British
Experimentalist (X-ray Scattering), Solid State Physicist, Quantum Physicist

Charles Glover Barkla made important experiments involving the scattering of X rays, proving that they are a form of transverse electromagnetic radiation like that of visible light. He received many honors for this work, including the 1917 Nobel Prize in physics "for his discovery of the characteristic X-radiation of the elements."

He was born on June 7, 1877, in Widnes, Lancashire, where his father, J. M. Barkla, was secretary to the Atlas Chemical Company. He first studied at the Liverpool Institute and then entered University College in Liverpool in 1895; there he studied physics under Oliver Lodge, a pioneer in radio. He earned a bachelor's degree in 1898 with highest honors and, a year later, a master's degree. A prestigious research scholarship enabled him to attend Trinity College, Cambridge, in the autumn of 1899 and to work at the Cavendish Laboratory under JOSEPH JOHN (J. J.) THOMSON, who had recently discovered that cathode rays consisted of electrons. Barkla's first original experiment measured the velocity of electromagnetic waves traveling along wires of different thickness and composition.

He transferred to King's College, Cambridge, in 1900, to pursue another passion—choral singing—but also continued his physics research. It had been known since 1897 that when X rays fall on any substance, whether solid, liquid, or gas, they cause a secondary radiation to be emitted. As were many physicists of the day, Barkla was drawn to investigating this phenomenon, which would become the primary focus of his career. In 1902, he returned to University College in Liverpool, as Oliver Lodge Fellow. The following year he published his first paper on secondary X radiation, in which he announced his discovery that the secondary radiation emitted from gases of elements with a low atomic mass has the same average wavelength as that of the primary X-ray beam impinging on the gas. More importantly, he showed that the amount of secondary radiation produced is proportional to the atomic mass of the gas. This work represented a significant early advance in the development of the concept of atomic number.

In research completed in 1904, for which he received a doctorate, he found that the heavy elements produced secondary radiation of a greater wavelength than that of the primary X-ray beam, a process that would later be explained by ARTHUR HOLLY COMPTON. Further, by demonstrating that X rays can be partially polarized, as can visible light, Barkla proved that they are a form of transverse electromagnetic radiation that obeys JAMES CLERK MAXWELL's equations.

In 1907, after being appointed as a lecturer at Liverpool, he married Mary Esther Cowell, with whom he had two sons and one daughter. That year he did his most important research, with his colleague C. A. Sadler. They discovered, in the process of X-ray scattering, that the secondary radiation is homogeneous and that the radiation from the heavier elements is of two kinds: one consisting of X rays scattered unchanged and the other a fluorescent radiation whose properties were related to the particular substance involved. They were also able to show that these characteristic X-ray radiations (i.e., of a specific wavelength associated with a particular substance) were monochromatic, that is, contained one frequency. Barkla discovered two types of characteristic X-ray emissions: the K series (for the more penetrating emissions) and the L series (for the less penetrating). A later prediction that other series of secondary emissions might exist was justified when an M series with even lower penetrating power than the K series was discovered.

In 1909 Barkla became professor of physics at Kings College, London, and in 1913 he became professor of natural philosophy at Edinburgh University. In the years that followed he received many honors for this work, all culminating in the 1917 Nobel Prize in physics.

The last years of his life were saddened by the death of his youngest son, Flight Lieutenant Michael Barkla, a brilliant scholar, who was killed in action in 1943. Barkla remained at Edinburgh University until his death at his home, Braidwood, in Edinburgh, on October 23, 1944, at the age of 67.

Barkla made valuable contributions to our understanding of the absorption and photographic action of X rays. Building on Barkla's later work, Compton demonstrated the relationship between the characteristic X radiation, a wave, and the corpuscular radiation, a particle, accompanying it. His work also showed both the applicability and the limitations of quantum theory in relation to X rays.

See also RÖNTGEN, WILHELM CONRAD.

Further Reading

Knoll, Glenn Frederick. *Radiation Detection and Measurement.* 3d ed. New York: John Wiley & Sons, 1999.

⊠   **Becquerel, Antoine-Henri**
(1852–1908)
French
Experimentalist (Radioactivity)

Antoine-Henri Becquerel changed the course of modern physics by his accidental discovery in 1896 of natural radioactivity, the spontaneous emission of radiation by a material. For this achievement he shared the 1903 Nobel Prize with Pierre Curie and MARIE CURIE, who further investigated Becquerel's discovery.

He was born in Paris on December 15, 1852, into a family of physicists: Antoine César, his grandfather, had made important contributions in electrochemistry; Alexander Edmond, his father, was a professor of applied physics who had devoted his career to the study of luminescence. The young Becquerel was trained as an engineer at the École Polytechnique and École des Ponts et Chaussées and, in 1875, began doing research into the behavior of polarized light in crystals. He married the daughter of a civil engineer, with whom he had a son, who would also become a physicist. From 1878 on, he held the post of assistant at the National Museum of Natural History in Paris and took over his father's job as chair of applied physics at the Conservatoire des Arts et Métiers. In 1888, he received the doctor of sciences degree for a dissertation on the absorption of light by crystals. He became a member of the French Academy of Sciences in 1889 and a professor of applied physics at the museum in 1892 and at the École Polytechnique in 1895.

In his research, Becquerel had taken up his father's studies of luminescence, which encompasses fluorescence, the emission of light only during stimulation by external radiation, and phosphorescence, the light that persists after the external radiation ceases. An exciting new path of inquiry was suggested to him in 1896, the year that WILHELM CONRAD RÖNTGEN discovered X rays. At a meeting of the academy, Becquerel heard the mathematician Henri Poincaré describe how X rays were emitted from a fluorescent spot on the glass cathode ray tube used by Röntgen. Becquerel had the idea that X rays might be produced naturally by fluorescent crystals. To test his hypothesis he placed some crystals of a double sulfate of uranium and potassium on a photographic plate wrapped in paper and put it in sunlight to make the crystals fluoresce. Observing that the plate he developed was fogged, Becquerel concluded, "The phosphorescent substance in question emits radiation which penetrates paper opaque to light." He assumed that the Sun's energy was being absorbed by the uranium, which then emitted X rays.

Becquerel conscientiously set about repeating his experiment, only to find himself hindered by the Paris weather: the sky was cloudy and the crystals could not fluoresce without sunlight. There was nothing to do but put the wrapped plate and the crystals into a drawer and wait. When for several days the clouds refused to part, Becquerel, whether from impatience or curiosity, decided to develop the plates anyway. Instead of the faint images he expected to appear, he was amazed to discover images that were quite distinct; the plates had been strongly exposed to radiation. He concluded that this phenomenon was not related to fluorescence, but was a continuous and natural emission by the crystals.

Studying the radiation, he found that it behaved in the same way as X rays: that is, it penetrated matter and ionized air. He showed that the radiation was due to the presence of uranium in the crystals and subsequently found that a disk of pure uranium is also highly radioactive. Searching for other radioactive materials, the Curies discovered polonium and radium, which are even more radioactive, in 1898. In 1900, Becquerel subjected the radiation from radium to a magnetic field and demonstrated that radiation emitted by uranium shared certain characteristics with X rays; however, unlike X rays, it could be deflected by a magnetic field and therefore must consist of charged particles. He also showed that radioactivity transforms one element into another. He discovered that radioactivity could be removed from a radioactive material by chemical action, but was subsequently regained by the material.

In 1900, Becquerel was made an officer of the Legion of Honor, and he won the Nobel Prize in physics in 1903. Subsequently, he became vice president (1906) and president (1908) of the Academy of Sciences. He died on August 25, 1908, in Le Croisic, France.

The implications of Becquerel's revolutionary discovery of radioactivity would be realized in 1902–1903, when ERNEST RUTHERFORD explained it as the spontaneous transmutation of elements. The study of radioactivity itself was thereby transformed into the modern science of nuclear physics, which would give birth to nuclear medicine, nuclear reactors, and nuclear weapons.

See also THOMSON, JOSEPH JOHN (J. J.).

Further Reading
Badash, Lawrence. "The Discovery of Radioactivity." *Physics Today* February 1996, pp. 21–26.

⊠   **Bernoulli, Daniel**
(1700–1782)
Dutch/Swiss
Theoretical and Experimental Physicist (Hydrodynamics), Mathematical Physicist

Daniel Bernoulli was an early pioneer of hydrodynamics, who also laid the groundwork for the later development of the kinetic theory of gases. He is famous for discovering what has come to be called the Bernoulli effect, the relation between the pressure in a steadily flowing fluid and its velocity, which is crucial to an understanding of how airplanes fly.

He was born in Groningen, the Netherlands, on February 8, 1800, into a family of brilliant, fiercely competitive mathematicians, which included his father, Johann; his uncle; and two brothers. When Daniel was five, the family returned to Basel, Switzerland, their native city, where Johann took over the chair in mathematics after the death of its previous occupant, his brother Jakob. Johann was determined to steer Daniel away from a career in mathematics, allegedly because it paid poorly, and planned to prepare him to be a merchant. First, however, he sent him at age 13 to the University of Basel to study logic and philosophy. When his son's passion for mathematics persisted, Johann agreed to tutor him. It was Johann who introduced him to the theory of conservation of energy, which would later lead Daniel to his important work on the mathematical theory of fluid flow. When he had given up on making his son a merchant, Johann ordered him to study medicine. Daniel dutifully pursued a medical degree in Heidelberg, Strasbourg, and Basel, where he obtained a doctorate in 1721 with a thesis on respiration, in which he applied mathematical physics to medicine. Attracted to the work of the British physician William Harvey on blood as a fluid, he did research on blood flow and pressure, which combined his interests in mathematics and fluids.

Upon graduation, Bernoulli hoped to be an academician as his father was but was unable to obtain a position. Instead, at age 23, he set out for Padua, Italy, to study practical medicine. When illness led to enforced solitude, he spent his time studying mathematics and in 1724 published his first mathematical work, on probability. The following year he designed an hourglass for a ship that would flow even in stormy weather and won the first of a series of 10 prizes awarded by the Paris Academy for papers on such diverse topics as marine technology, navigation, oceanography, astrology, and medicine. He returned home to Basel that year and found a letter from Empress Catherine the Great, inviting him to become professor of mathematics in Saint Petersburg, Russia. To make the proposal more attractive, she offered a second position for his brother Nicolaus. The brothers accepted, but within eight months, Nicolaus was dead of tuberculosis. Grieving and oppressed by the harsh climate, Bernoulli wrote to his father, expressing his intention to return home. At this point Johann sent him his star student Leonard Euler to assist him in his work. Bernoulli remained in Russia, embarking on what would be a lifelong collaboration with the brilliant young mathematician.

From 1727, when Euler arrived, until 1733, when Bernoulli left Saint Petersburg, he enjoyed his most fruitful period. Probing the relationship between the speed at which blood flows and its pressure, he performed experiments that involved puncturing the wall of a pipe with a small open-ended straw and noting that the height to which the fluid rose in the straw was related to the fluid's pressure in the pipe. Soon European physicians were measuring blood pressure by sticking point-ended glass directly into their patients' arteries. In 1896 an Italian doctor would discover the less painful blood pressure cuff used today. However, Bernoulli's original method of measuring the pressure in a flowing fluid is still used in modern aircraft to measure the speed of the air passing the plane, that is, its air speed.

In Saint Petersburg, Bernoulli completed his most famous work, *Hydrodynamica*, a theoretical and practical study of equilibrium, pressure, and velocity in fluids, which relied on his earlier work on conservation of energy; it would be published, after years of polishing, in 1738. When

Bernoulli wrote his magnum opus, scientists knew that a moving body exchanges its kinetic energy for potential energy when it gains altitude. Bernoulli had the insight that in a similar way, a moving fluid exchanges its kinetic energy for pressure. From these principles he developed the principle of what is now known as the Bernoulli effect, which states that the pressure of a fluid depends inversely on its velocity: the pressure decreases as the velocity increases. Thus, the Bernoulli effect, which governs fluid flow and has many applications, is a consequence of conservation of energy. Applied to aerodynamics, it explains how a moving wing whose cross section has the shape of an airfoil (curved on the top, flat on the bottom) experiences the lifting force that allows an airplane to fly. The curve of the wing is designed to create a faster flow of air over the top of the wing than over the bottom. As a result of the Bernoulli effect, the air pressure over the top of the wing is lower than the air pressure beneath the wing. In this way, the wing is pushed upward, enabling an airplane to fly.

In his *Hydrodynamica* Bernoulli also attempted to construct a mathematical description of the behavior of gases, based on the assumption that they are composed of tiny particles. By producing an equation of state, that is, an equation that relates the pressure, temperature, and volume of a gas, he was able to relate atmospheric pressure to altitude. This was the first step toward the kinetic theory of gases that would be developed a century later.

Despite his productive collaboration with Euler, Bernoulli was unhappy in Saint Petersburg. In 1734, he returned to Basel to lecture, first on botany and later on physiology—the only post he could get at the time. He would continue to correspond with Euler, who put many of Bernoulli's physical insights into rigorous mathematical form. When Bernoulli and his father were declared joint winners of the Paris Academy's Grand Prize that year, the father,

enraged that his son had been judged his equal, broke off relations with him. Bernoulli stayed in Basel, barred from his father's house. A year after publication of *Hydrodynamica*, his father published *Hydraulica*, based on his son's work but written as if his son's work had been based on his. Europe, however, gave the younger Bernoulli the credit he was due, electing him to most of the leading scientific societies of his day. In 1750, Bernoulli at last became professor of physics at Basel, where he would remain for the next 26 years, giving a series of memorable physics lectures during which he performed actual experiments. In research that advanced the field of mathematical physics, he developed SIR ISAAC NEWTON's theories and used them together with the more powerful calculus of Gottfried Leibnitz.

Bernoulli died on March 17, 1782, in Basel, where he was buried. An imaginative scientist with broad interests, Bernoulli helped launch the field of hydrodynamics, anticipated the kinetic theory of gases, and discovered a fundamental principle of aerodynamics.

Further Reading

Vischer, D. "Daniel Bernoulli and Leonard Euler: The Advent of Hydromechanics," in Gunther Garbrecht, ed., *Hydraulics and Hydraulic Research: A Historical Review*. Rotterdam and Boston: Balkema, 1987, pp. 145–156.

⊠ Bethe, Hans Albrecht
(1906–)
German/American
Theoretician, Nuclear Physicist, Astrophysicist, Solid State Physicist

During his long and brilliant career, Hans Bethe has made numerous contributions to both fundamental and applied science. His early work laid the foundation for nuclear physics and explained the dynamics of the energy of stars. During World War II, he headed the theoretical group at Los Alamos, New Mexico, that developed the atomic bomb. He later came to epitomize the socially responsible scientist, working tirelessly in the interests of nuclear disarmament.

Hans Albrecht Bethe was born on July 2, 1906, in Strasbourg, Germany (now France), the son of a university professor. He studied at the University of Frankfurt for two years and then transferred to the University of Munich, where he worked under the eminent ARTHUR SOMMERFELD. He earned a Ph.D. in 1928 for a dissertation on electron diffraction, which continues to be used by physicists for interpreting observational data.

During the next five years, Bethe's work flourished within the exciting European physics world, which was discovering the secrets of the atom through the insights of the new quantum mechanics. He lectured at the universities of Frankfurt, Munich, and Tubingen and worked with ENRICO FERMI in Rome. His research on atomic physics and scattering theory resulted in a highly successful theory of inelastic collisions between fast particles and atoms. Bethe's theory determined the stopping power of matter for fast charged particles, thereby enabling nuclear physicists to study radiation effects in matter. He then looked at more energetic collisions, calculating the *Bremsstrahlung* (braking radiation), that is, the radiation emitted by relativistic electrons (electrons moving nearly as fast as the speed of light) as they are slowed during the process of colliding with charged particles in matter. He also studied the production of electron–positron pairs by high-energy gamma rays.

As were those of so many outstanding physicists of his generation, Bethe's career in Germany was derailed by the rise of Hitler. In 1933 he moved to England, where he would remain for two years, first at Manchester University and then at Bristol University. In 1935 he emigrated to the United States, where he would become a naturalized citizen. Cornell University in Ithaca, New York, was to become his academic home

from 1937 until his retirement in 1975, when he became professor emeritus.

During the mid- and late 1930s, Bethe continued his research on atomic nuclei. In 1934, he first formulated his theory of the deuteron (the nucleus of an atom of deuterium: a combination of a proton and a neutron. (Deuterium is a naturally occurring stable isotope of hydrogen; it is known as heavy hydrogen.) The following year he went on to resolve contradictions in the nuclear mass scale. His theoretical investigations of nuclear reactions enabled him to predict many reaction cross sections. He also gave a more precise quantitative form to NIELS HENRIK DAVID BOHR's theory of the compound nucleus. Bethe summarized his own work and other theoretical and experimental studies in nuclear physics in three articles in the *Reviews of Modern Physics*, which became the basic textbook for nuclear physicists.

Bethe's fundamental work on nuclear reactions led him to the discovery of the nuclear processes that supply stellar energy, a problem that had remained unsolved since LORD KELVIN (WILLIAM THOMSON) and HERMANN VON HELMHOLTZ had drawn attention to it 75 years earlier. The riddle was an old one: how has it been possible for the stars, including our Sun, to emit light and heat without exhausting their energy sources? Bethe determined that the most important nuclear reaction in the brilliant stars is the carbon–nitrogen cycle, whereas the Sun and fainter stars mainly use the proton–proton reaction. Equally important, he was able to exclude other possible nuclear reactions. He would receive the 1967 Nobel Prize in physics for this groundbreaking work.

The years of Bethe's great discoveries also saw the onset of World War II. From 1943 to 1946, the emphasis of his work would be on the military applications of modern physics. After working on the development of radar at the Massachusetts Institute of Technology in Cambridge, he joined the Manhattan Project, which Pres.

Hans Bethe laid the foundation for nuclear physics, explained the dynamics of stars, and headed the theoretical group that developed the atomic bomb. *(AIP Emilio Segrè Visual Archives, Marshak Collection)*

Franklin Delano Roosevelt had charged with the mission of developing on atomic bomb. As director of the Theoretical Physics Division at Los Alamos, Bethe played an important role in the work that led to the first successful atomic bomb test in the New Mexico desert in July 1945.

Throughout his long career, Bethe has been an amazingly versatile theoretical physicist. In the late 1940s, he was the first to explain the Lamb shift, a frequency shift in atoms that revealed the fundamental quantum nature of electrodynamic processes in the hydrogen spectrum. This work laid the foundation for the modern development of quantum electrodynamics. Later on, he collaborated on the scattering of pi mesons (the quanta of the nuclear force) and on their production by photon processes. He also worked on solid state theory, investigating the splitting of atomic energy levels when an atom is inserted into a crystal. He made studies of the theory of metals and developed a theory of order and disorder in alloys. In astrophysics, he made important contributions to the understanding of the explosions of giant stars called supernovae and helped to formulate the theory of neutron stars, the dense remnants of supernova explosions in which the gravitational force is so strong that protons fuse with electrons to form neutrons.

In addition to continuing his scientific research, Bethe has devoted himself to fighting for a sane international policy for controlling the nuclear weapons he had helped to create. He was instrumental in founding the *Bulletin of Atomic Scientists* and frequently contributed to it. Bethe served as delegate to the first International Test Ban Conference in Geneva in 1958. His efforts were recognized in 1961 when he was chosen for the Atomic Energy Commission's prestigious Enrico Fermi Award. He then became the principal science adviser to the U.S. government during the negotiations on the 1963 Nuclear Test Ban Treaty. Through his writings and lectures on the nuclear threat, he has raised public awareness and been a leader in the efforts of scientists to take social responsibility for the fruits of their inventions.

In the latter decades of his career, Bethe has written and spoken extensively on aspects of the energy crisis: the finite amount of energy on Earth that will be available to a growing population and the more immediate problem of maintaining a supply of gasoline to power automobiles. He is a supporter of a nonnuclear future, when people will rely on solar energy in all its forms.

See also LAMB, WILLIS E.

Further Reading

Bernstein, Jeremy. *Hans Bethe, Prophet of Energy.* New York: Basic Books, 1980.

Bethe, Hans Albrecht. *The Road from Los Alamos, Masters of Modern Physics.* New York: Springer Verlag, 1991.

Schweber, Silvan S. *In the Shadow of the Bomb.* Princeton, N.J.: Princeton University Press, 2000.

⊠ Bhabha, Homi Jehangir
(1909–1966)
Indian
Theoretician, Particle Physicist

Homi Jehangir Bhabha was a brilliant theoretician whose best-known contribution to particle physics is an accurate expression of a cross section (i.e., the probability) of the quantum electrodynamic scattering of positrons by electrons, what is now called Bhabha scattering. He was also the chief architect of India's atomic energy program, a seminal figure in the development of a world-class cadre of Indian scientists after World War II, and a prominent advocate for the cause of peaceful uses of atomic energy.

He was born on October 30, 1909, in Bombay, into a wealthy Parsi family with a long tradition of learning and service to the country. The Bhabha family cultivated interests in the fine arts, particularly Western classical music and painting, all of which became an essential part of Homi Bhabha's quest for a vivid, rich existence. As a young man, he wrote:

> I know quite clearly what I want out of my life. Life and my emotions are the only things I am conscious of. I love the consciousness of life and I want as much of it as I can get. But the span of one's life is limited. What comes after death no one knows. Nor do I care. Since, therefore, I cannot increase the content of life by increasing its duration, I will increase it by increasing its intensity.

He attended the Cathedral and John Connom Schools in Bombay. Then, after passing the entrance exam at age 16 in 1927, he traveled to England, where he planned to study mechanical engineering at Gonville and Caius College at Cambridge University. Both his father and his uncle, Sir Dorab J. Tata, wanted him to become an engineer so that he could eventually join the

Tata Iron and Steel Company in Jamshedpur. But events turned out quite differently. At Cambridge, PAUL ADRIEN MAURICE DIRAC, who discovered the quantum mechanical equation that described relativistic electrons and that led to the prediction of antimatter electrons or positrons, was his tutor in mathematics. In 1928, Bhabha wrote to his father:

> I seriously say to you that . . . a job as an engineer is . . . totally foreign to my nature and radically opposed to my temperament and opinions. Physics is my line. I know I shall do great things here . . . I am burning with a desire to do physics.

After the young man earned his first-class honors degree in engineering in 1930, his father relented and Bhabha joined the Cavendish Laboratories in Cambridge, where he received a Ph.D. in theoretical physics in 1935. During this time the Cavendish was at the height of its golden age under the directorship of ERNEST RUTHERFORD. In the single year of 1932, SIR JAMES CHADWICK demonstrated the existence of the neutron; JOHN DOUGLAS COCKCROFT produced the artificial disintegration of light elements by bombarding them with high-speed protons; and LORD PATRICK MAYNARD STUART BLACKETT invented a counter-cloud chamber that would demonstrate the production of positron and electron pairs. Working in this revolutionary atmosphere, Bhabha made major contributions to the early development of quantum electrodynamics (QED), the study of the interaction of electrons and the electromagnetic field. His first paper was on the absorption of high-energy gamma rays in matter. A primary gamma ray dissipates its energy in the formation of electron showers. In 1935, he became the first physicist to use the QED formalism to determine the cross section (the probability) of electrons scattering on positrons—a phenomenon that is now known as Bhabha scattering.

After this important work, Bhabha focused on the study of cosmic rays and, in a classic paper written with W. Heitler in 1937, suggested that the highly penetrating particles detected at ground level could not be electrons (nine years later these highly penetrating particles were in fact found to be mu mesons). In his paper with Heitler he described how primary cosmic rays from space interact with the upper atmosphere to produce particles observed at the ground level and was able to explain how the cosmic ray showers were produced by the cascade production of gamma rays and electron–positron pairs. In 1938, he proposed a classic method of confirming the time dilation effect of the special theory of relativity by measuring the lifetimes of cosmic ray particles striking the atmosphere at very high speeds. Their lifetimes were found to be prolonged by exactly the amount predicted by relativity. This impressive body of research led to his election to the prestigious Royal Society in 1940, when he was only 30.

Bhabha remained at Cambridge until 1939, when he returned to India for what was to be a brief visit. He was in India when World War II broke out, however, and found it impossible to return to Cambridge. A readership was created for him as head of the cosmic rays department at the Bangalore Institute of Science, under the directorship of the great Indian physicist SIR CHANDRASEKHARA VENKATA RAMAN. Raman was influential in Bhabha's decision to remain in India to advance the development of science and technology there.

In 1943, he became president of the physics section of the Indian Science Congress. The following year, he proposed the establishment of an institute for training scientists in advanced physics and stimulating physics research for application to industrial development. Realizing the need for an institute devoted entirely to fundamental research, he wrote for funding to a close family friend, J. D. R. Tata, whose family had pioneered projects in metallurgy, power generation, and science and engineering education in the early 20th century. In 1945, he became the first director of the Tata Institute of Fundamental Research (TIFR) in Bombay, a post he held for the rest of his life. TIFR became the training ground for a team of experts for India's atomic energy program. Since Bhabha was also an accomplished artist and architect, under his supervision not only did science prosper, but so did the edifice itself: aesthetically designed, it was surrounded by beautiful lawns and gardens, at Land's End in Bombay, facing the Arabian Sea.

Homi Jehangir Bhabha formulated the quantum electrodynamic theory of electron-positron scattering and was the chief architect of India's atomic energy program. *(AIP Emilio Segrè Visual Archives, Physics Today Collection)*

The other great institute he established was the Atomic Energy Establishment in Trombay, renamed after his death the Bhabha Atomic Research Center (BARC).

In 1948 he wrote to the Indian president, Jawaharlal Nehru, proposing the establishment of an atomic energy commission (AEC). In August 1948 the Indian Atomic Energy Act was promulgated and the AEC was established. For more than 20 years (1944 to 1966), he led India's atomic energy program, advancing his priorities: surveying of natural resources; development of basic sciences—physics, chemistry, and biology; and creation of a program for instrumentation, particularly in electronics. He was determined that India would be self-sufficient in supplying the experts necessary to nuclear energy production.

Bhabha also became an important figure in the international science community. He served as president of the first United Nations Conference on the Peaceful Uses of Atomic Energy, held in Geneva in 1955. He was president of the International Union of Pure and Applied Physics from 1960 to 1963.

He died tragically in a plane crash on Mount Blanc in the French Alps, on January 24, 1966, while on his way to Vienna to attend a meeting of the Scientific Advisory Committee of the International Atomic Energy Agency. He was mourned by his peers, not only as a distinguished theorist, but also as one of those rare beings in whom scientific and artistic excellence, as well as love of country and a strong sense of internationalism, harmoniously coexist.

See also EINSTEIN, ALBERT; MILLIKAN, ROBERT ANDREWS; WILSON, CHARLES THOMAS REES.

Further Reading

Abraham, Itty. *The Making of the Indian Atomic Bomb: Science, Secrecy, and the Postcolonial State.* New York: St. Martin's Press, 1998.
Venkataraman, G. *Bhabha and His Magnificent Obsessions.* Hyderabad, India: University Press, 1994.

Blackett, Patrick Maynard Stuart, Lord

(1897–1974)
British
Experimentalist, Atomic Physicist, Particle Physicist

Patrick Blackett made the first photograph of an atomic transmutation and in the early 1920s developed the cloud chamber, invented in 1911 by CHARLES THOMSON REES WILSON, into an invaluable instrument for studying nuclear reactions. These achievements garnered him the 1948 Nobel Prize in physics. His discovery of the phenomenon of pair production of positrons and electrons in cosmic rays provided the first evidence of ALBERT EINSTEIN's theory that mass and energy are equivalent.

He was born on November 18, 1897, in Croydon, Surrey, the son of Arthur Stuart Blackett. He joined the Royal Navy in 1912 as a naval cadet. During World War I, he took part in the battles of the Falkland Islands and Jutland and designed a revolutionary new gunsight. Attracted to science, he resigned from the navy when the war ended with the rank of lieutenant and began to study science at Cambridge University, where he received a B.A. in 1921. At Cambridge's Cavendish Laboratory, he began working under ERNEST RUTHERFORD on cloud chambers.

In 1924, he married Constanza Bayon, with whom he had a son and a daughter, and in the same year he obtained the first photos of an atomic transmutation, which was of nitrogen into an oxygen isotope. In 1919, he explained the transmutation of elements from experiments in which nitrogen was bombarded with alpha particles. The oxygen atoms and protons produced were detected by scintillations in a screen of zinc sulfide. He used a cloud chamber to photograph the tracks formed by these particles, taking more than 20,000 pictures and recording about 400,000 tracks. Of these, eight showed that a nuclear reaction had taken place, confirming Rutherford's explanation of the transmutation process. His accurate measurements showed that the course of a collision between atomic nuclei always follows the laws of conservation of momentum and energy, provided that the mass–energy relationship given by the theory of relativity is also taken into account.

Blackett continued to develop the cloud chamber and in 1932, with a young Italian physicist, Guiseppe Occhialini, designed the counter-controlled cloud chamber, a brilliant invention in which cosmic rays could photograph themselves. The counter-controlled cloud chamber was activated when simultaneous discharges occurred as the result of an electrically charged particle's passing through two geiger counters, placed above and below the cloud chamber. This made the cloud chamber vastly more efficient and reduced the huge number of required observations. Early in 1933, this device confirmed the existence of the positron (or positive electron) proposed by CARL DAVID ANDERSON and also demonstrated the existence of pair production, the generation of positrons and electrons directly from the interaction of gamma rays with heavy nuclei. The observation of pair production was the first evidence that matter may be created from energy, as Einstein had predicted in his theory of special relativity. Blackett and Occhialini also performed experiments demonstrating the reverse process of pair annihilation, in which a positron–electron pair are transformed into gamma radiation, thus converting mass back into energy. In the interpretation of these experiments, they were guided by PAUL ADRIEN MAURICE DIRAC's quantum theory of the electron.

In 1933, Blackett became a professor of physics at Birkbeck College, London, and in 1935, he moved to the University of Manchester. He continued his cosmic ray studies, attracting a distinguished group of researchers around him. This work was interrupted by World War II, when he joined the Instrument

Section of the Royal Aircraft Establishment. Early in 1940 he became scientific adviser to the Coastal Command and started the analytic study of the anti-U-boat war. He later continued this work as director of Naval Operational Research at the Admiralty. Blackett served as adviser to the Anti-Aircraft Command and, during the Blitz of London, was involved in antiaircraft operations.

After the war Blackett returned to university life. At Manchester, his research team made many important discoveries about cosmic rays, including the discovery, in 1947, of the first two of what became known as strange particles: a large family of intrinsically unstable particles with a life span of 10^{-10} second and a mass greater than that of a proton, which undergo a series of elementary particle decays.

In 1948 Blackett's speculations about the isotropy of cosmic rays led him to studies of the origin of the interstellar magnetic field and of the magnetic field of the Earth and Sun. These, in turn, led him to study the Earth's magnetic field and particularly the phenomenon of rock magnetism. In 1953, he was appointed head of the physics department at the Imperial College of Science and Technology, London, where he built up a research team to investigate the phenomenon. After his retirement in 1963, he was instrumental in the establishment of the Ministry of Technology. He became president of the Royal Society in 1965 and a life peer in 1969. He died on July 13, 1974.

Blackett's pioneering discoveries provided the first evidence of Einstein's theory that mass and energy are equivalent and laid the foundation for modern experimental high-energy particle physics.

Further Reading

Blackett, Patrick Maynard Stuart. *Fear, War, and the Bomb: Military and Political Consequences of Atomic Energy.* New York: McGraw-Hill, 1949.

⊠   **Bloch, Felix**
(1905–1983)
Swiss/American
Experimentalist, Quantum Theorist (Statistical Mechanics), Solid State Physicist

Felix Bloch was a major figure in 20th-century physics, who was among the first to demonstrate the power of quantum mechanics to illuminate many hitherto mysterious physical phenomena. He did groundbreaking work in the quantum theory of metals and solids, discovering what came to be called Bloch walls, which separate magnetic domains in ferromagnetic materials such as iron. He shared the 1952 Nobel Prize in physics with EDWARD MILLS PURCELL for their simultaneous independent discovery of the nuclear magnetic resonance.

He was born in Zurich, Switzerland, on October 23, 1905, to Gustav Bloch, a wholesale grain dealer, and Agnes Mayer Bloch, Gustav's cousin from Vienna. Living in Zurich, where he developed his lifetime love of mountains, from 1912 to 1918, he attended primary and secondary school. His childhood was marred by the death of his 12-year-old sister, who had been his main support in dealing with painful feelings of exclusion at school. He loved the clarity and beauty of mathematics and was also drawn to music. In 1924, he passed the Matura, the final examination, which was the passport to a higher education.

Since his ambition was to become an engineer, he enrolled at the Federal Institute of Technology in Zurich. After a year, however, he knew he wanted to study physics and transferred to the division of mathematics and physics, in which his professors, including ERWIN SCHRÖDINGER and Peter Debye, introduced him to the new wave theory of quantum mechanics. By now Bloch's interests were enthusiastically directed toward theoretical physics. When Schrödinger left Zurich in 1927, Bloch transferred to the University of Leipzig, where he studied with Schrödinger's rival in the quest for a theoretical formulation of quantum mechanics, WERNER HEISENBERG. He received a Ph.D. in 1928 for a dissertation in which he investigated the quantum mechanics of electrons in crystals and developed the theory of metallic conduction.

After spending the 1930–1931 academic year with Heisenberg at Leipzig, he wrote a major paper on ferromagnetism, in which he developed the concept that came to be called Bloch walls. Earlier researchers had done experiments on the process of magnetization in ferromagnets, including the phenomenon of magnetic domain structure, that is, localized magnetic regions in a metallic material. Understanding this process required understanding the boundary wall between domains and the manner in which it moved. Bloch was able to determine theoretically the thickness and structure of the magnetic domain boundary wall and predicted that in a distance as small as a few hundred angstroms the magnetization could reverse direction. Many years later it became possible to observe this progress experimentally.

In 1931, Bloch also worked with NIELS HENRIK DAVID BOHR at his institute in Copenhagen. Bohr had long been interested in stopping power: that is, the loss of energy of a charged particle as it passes through matter. Since writing an important paper on the topic in 1913, in which he presented a classical calculation, Bohr hoped to improve his theory to agree more closely with observed losses of alpha and beta particles. In a 1930 paper based on quantum mechanics, HANS ALBRECHT BETHE came up with a better fit between theory and observation. Bloch was able to explain the discrepancy between their results in a 1933 paper, showing that their calculations were opposite limiting approaches corresponding to the different ways in which the phase of the quantum wave function varied as the particle passed near an atom. The equation describing the stopping of charged particles in matter became known as the Bethe–Bloch formula.

As did many physicists of Jewish descent, Bloch chose to leave Germany in 1933, when Hitler's Nazi regime came to power. Learning that his name was on a list of "displaced scholars," that year he went to Rome on a Rockefeller Fellowship, and there he worked at ENRICO FERMI's institute at the University of Rome. He then accepted an offer from Stanford University, where he was able to realize what had been a growing inclination to do experimental physics.

The neutron had been discovered in 1932 and the physics of neutron interactions was just starting to be explored. Suspecting that the neutron had a magnetic moment, that is, a measure of the strength of a magnet, he drew on his expertise in ferromagnetism and devised a method for polarizing neutrons in a ferromagnetic material. In his first studies, using an extremely simple neutron source, he discovered that a direct proof of the magnetic moment of free neutrons could be obtained through the strength of its magnetic field observing magnetic scattering of neutrons in a ferromagnetic substance such as iron. In a 1936 paper, in which he first described the theory of magnetic scattering of neutrons, Bloch showed how the scattering could lead to a beam of polarized neutrons. Further, he demonstrated how it was possible to distinguish the atomic scattering from the nuclear scattering by temperature variations of the ferromagnetic material in which the neutrons were scattered. From such experiments on neutron scattering at small angles, he was able to determine the magnetic moment of the neutron experimentally.

Developing these ideas, in 1939 he collaborated with LUIS W. ALVAREZ in conducting a famous experiment at the Berkeley cyclotron, in which they were able to determine the magnetic moment of the neutron with an accuracy of about 1 percent. That year, on the way to a meeting of the American Physical Society in

Washington, D.C., Bloch met Dr. Lore Misch, a fellow physicist and refugee from Germany. They married in Las Vegas on March 14, 1940. Their lifelong union produced four children, who would present them with 11 grandchildren.

When World War II began, Bloch became involved in the initial work on atomic energy at Stanford. Invited to Los Alamos by J. ROBERT OPPENHEIMER, he worked on the implosion method suggested by Seth Neddermeyer. Later on, he worked on radar countermeasures at Harvard University, which exposed him to the latest developments in electronics. When he returned to Stanford after the war, in 1945, he applied this new knowledge to his earlier work on the magnetic moment of the neutron and developed innovative approaches to the study of nuclear moments.

By then, ISIDOR ISAAC RABI had theoretically determined that magnetic resonance, associated with groups of atoms shot through a region of strong magnet fields as a beam, was a measurable phenomenon. But instruments capable of measuring magnetic resonance in liquids or solids had not yet been designed. Bloch's solid grasp of quantum mechanics, particularly the quantum mechanics of solids, uniquely qualified him to solve this problem. With W. W. Hansen and M. E. Packard, he devised a new method that used an electromagnetic procedure (initially called nuclear magnetic induction) for the study of nuclear moments in solids, liquids, or gases. He developed a phenomenological description for the frequency of precession (i.e., rotation), of the nuclear magnetic moments of neutrons and the electromagnetic signals that would be emitted from them in the nuclear magnetic induction process, using formulas that became known as the Bloch equations.

Only a few weeks after successfully performing this work, he learned that Purcell and his Harvard collaborators had independently made the same discovery, using a resonance method involving energy absorption of radiation in a cavity. The name *nuclear magnetic resonance* (NMR) was given to both methods for measuring nuclear magnetic moments, and Bloch and Purcell shared the 1952 Nobel Prize in physics.

After his groundbreaking discovery, Bloch devoted himself to investigations using his new method. He succeeded in combining it with his earlier work on the magnetic moment of the neutron to allow him to measure that quantity with a high degree of accuracy. In 1954, Bloch spent a year as the first director general of the European Center for Nuclear Research (CERN) in Geneva. The work, which was largely administrative, was not to his liking, and he returned to Stanford eager to continue his studies of nuclear magnetism. He joined his colleagues in designing and building the Stanford Linear Accelerator Center.

In 1961, he was appointed Max Stein Professor of Physics at Stanford. Throughout the early 1960s, after the announcement of the Bardeen–Cooper–Schrieffer (BCS) theory of superconductivity, he returned to his old interest in the theory of superconductivity, which he wanted to simplify in order to clarify the physical processes involved. In his farewell speech as president of the American Physical Society in 1966, he spoke of his unfulfilled hopes to find that simplicity. His last research papers dealt with superconductivity and presented a simplified discussion of the Josephson effect.

On retiring, Bloch began writing a book on statistical mechanics. Although it was unfinished at the time of his death, it was completed by a colleague on the basis of his notes and published as *Fundamentals of Statistical Mechanics*, which is regarded as an insightful, elegantly written account of the subject. He died suddenly of a heart attack on September 10, 1983, and was buried on a mountainside overlooking Zurich.

Bloch's breakthrough discovery of the process of nuclear magnetic resonance led to many unanticipated scientific and practical applications: in addition to its intended role in evaluating nuclear magnetic moments, NMR became an essential spectroscopic tool used in structural and dynamic studies in chemistry. More importantly, in medicine, NMR was developed by P. C. Lauterburg and others into magnetic resonance imaging (MRI), which dramatically improved upon the traditional X ray and rivaled the successful computer-assisted tomographic effect known as computed tomography (CT) or CT scanning.

See also BARDEEN, JOHN; CHADWICK, SIR JAMES; SCHRIEFFER, JOHN ROBERT; JOSEPHSON, BRIAN DAVID.

Further Reading
Hofstadter, Robert. "Felix Bloch," in *Biographical Memoirs*. Vol. 64. Washington, D.C.: National Academy Press, 1994, pp. 34–71.

⊠   **Bloembergen, Nicolaas**
(1920–   )
Dutch/American
*Theoretical and Experimental Physicist,
Laser Spectroscopist*

Nicolaas Bloembergen made his mark on 20th-century physics through his contributions to the technology of the laser, the revolutionary device that creates and amplifies a narrow, intense beam of coherent light, and its forerunner, the maser. His three-level crystal maser proved strikingly more powerful than earlier gaseous masers and became the most widely used microwave amplifier. Bloembergen went on to develop laser spectroscopy, which allows high-precision observations of atomic structure. On the basis of his laser spectroscopic studies, he established the theoretical foundation for the science of nonlinear optics, a new theoretical approach to the analysis of how high-intensity coherent electromagnetic radiation interacts with matter. Bloembergen also did seminal work in nuclear magnetic resonance.

Bloembergen was born on March 11, 1920, in Dordrecht, the Netherlands, the second of six children born to Auke Bloembergen, who was a chemical engineer, and Sophia Maria Quint, who had a degree that qualified her to teach French but instead devoted herself to family duties. Before Nicolaas reached school age, the family moved to Bilthoven, a suburb of Utrecht, where, he recalls, "We were brought up in the Protestant work ethic characteristic of the Dutch provinces." Encouragement of intellectual attainments and a level of frugality beyond that required by family income were twin pillars of his childhood. However, his parents also urged him to participate in field hockey, water sports, and skating on the Dutch waterways, rather than sit constantly over his books. At the age of 12 he entered the municipal gymnasium in Utrecht, whose rigid curriculum emphasized the humanities; he did not discover his love of science until the last years of secondary school. He describes his choice of physics as "probably based on the fact that I found it the most difficult and most challenging subject."

In 1938, he entered the University of Utrecht, where he was allowed to assist in the research of a graduate student, G. A. W. Rutgers, with whom he published his first paper, "On the Straggling of Alpha Particles in Solid Matter," in 1940. That same year, however, the Nazis invaded the Netherlands, and Bloembergen's mentor, Professor L. S. Ornstein, was forced to leave the university in 1941. He managed to obtain the Dutch equivalent of a master's degree in science degree just before the university was closed completely in 1943. For the next two years, until the war's conclusion in 1945, he went into hiding, "eating tulip bulbs" and reading about quantum mechanics "by the light of a storm lamp."

At war's end, he was accepted to do graduate work by Harvard University, and, with the help of his father and the Dutch government, he left behind the devastation of Europe. His arrival at

Harvard coincided with the detection, by E. M. PURCELL, R. V. Pound, and Henry Torrey, of nuclear magnetic resonance (NMR) in condensed matter (an effect that results when waves of radio frequency are absorbed by the nuclei of matter in a strong external magnetic field). Bloembergen was accepted as a graduate assistant to develop the early NMR apparatus. At the time the field of NMR in solids, liquids, and gases was unexplored, and Bloembergen and his Harvard colleagues gathered a rich harvest of findings, which they published in 1948 in a landmark paper, "Relaxation Effects in Nuclear Magnetic Resonance," in *The Physical Review*. Bloembergen did part of this work back in the Netherlands, at the Kammerlingh Onnes Laboratory in Leiden, in 1947 and 1948; there he discovered the nuclear spin relaxation mechanism by conduction electrons in metals and by paramagnetic impurities in ionic crystals, the phenomenon of spin diffusion, and the large shifts induced by internal magnetic fields in paramagnetic crystals. He was the first physicist to measure NMR relaxation times, that is, the decay time of the NMR process, accurately and discovered that NMR relaxation times could be measured in seconds or fractions of seconds. This result made NMR a practical research tool, with applications in chemistry, medicine, and physics, including a method for analyzing the structure of molecules and for producing the contrast needed to create images of tissues in the human body. Bloembergen wrote his doctoral thesis, "Nuclear Magnetic Relaxation," using these same materials, and submitted it, in 1948, to the University of Leiden, where he had already filled all his other requirements for the Ph.D.

During a vacation trip with the Physics Club in Leiden, in the summer of 1948, he met Huberta Deliana Brink (known as Deli), a premed student, whom he married in Amsterdam in 1950. The young couple immigrated to the United States, where they would raise three

children and become citizens in 1958. With Deli as "a source of light in my life," he picked up his career at Harvard. While a junior fellow there he broadened his experimental background to include microwave spectroscopy and some nuclear physics at the Harvard cyclotron; however, he preferred "the smaller-scale experiments of spectroscopy, where an individual or a few researchers at most, can master all aspects of the problem." Thus, in 1951, he returned to NMR research; his group made a number of significant discoveries that led Bloembergen to propose a three-level solid state maser in 1956.

He did not try to build a working laser, after CHARLES HARD TOWNES and ARTHUR LEONARD SCHAWLOW published their proposal for an optical maser, doubting that a small academic lab with no previous experience in optics could succeed. He did, however, recognize in 1961 that his laboratory could take advantage of some of the new research opportunities made possible by laser instrumentation. His group started a program in a field that became known as nonlinear optics, the study of the behavior of high-intensity coherent radiation in matter, and published their early results in a 1965 monograph. Nonlinear optical methods of spectroscopy are based on the mixing of two or more high-intensity coherent light waves in an optically active medium for which the principle of superposition, valid for ordinary electromagnetic radiation, breaks down. Bloembergen and his group demonstrated this type of nonlinear optical phenomenon shortly after the laser was introduced and comprehensively explored the theory describing it around the same time. Bloembergen's method of nonlinear four-wave mixing, in which three coherent light waves act together in generating a fourth coherent light wave, made it possible to generate laser light far outside the visible range, in both the infrared and the ultraviolet spectra. This greatly extended the range of wavelengths accessible to laser spectroscopy studies. One example of this is a special form of four-wave mixing called coherent anti–Stokes Raman scattering (CARS), which has been applied in studies of widely differing kinds—from optimization of combustion processes in automobile engines to the study of the transport of chemical elements in biological tissues. For this seminal work Bloembergen shared the 1981 Nobel Prize in physics with Schawlow, the coinventor of the laser.

While performing this work Bloembergen enjoyed a rich academic life at Harvard, including innumerable travels abroad.

He served on a 1986 committee to study President Ronald Reagan's "Star Wars" program and concluded that, in order for it to work, a decade of laser weapon research would be needed. The committee's findings confirmed the scientific community's "gut feeling" that it was not practical.

In addition to writing monographs on nuclear magnetic relation and nonlinear optics, he has published over 300 papers in scientific journals. In June 1990 he retired from the faculty at Harvard and became professor emeritus. He and his wife had a special feeling for Tucson, so after his retirement from Harvard, in 1991 he became an unpaid professor of physics at the University of Arizona, Tucson, where he continues to pursue research at the university's optical institute. Also in 1991, he served as president of the American Physical Society.

Bloembergen's theoretical and experimental work with masers and lasers led to a vast spectrum of practical applications, from surgical techniques to boring and cutting of metal to the development of fiber optics. In a 1998 talk, he foresaw an increasing laser use in scientific applications, noting that there is currently an "enormous push" for high-powered semiconductor lasers.

See also RABI, ISIDOR ISAAC; RAMAN, SIR CHANDRASEKHARA VENKATA; STOKES, GEORGE GABRIEL.

Further Reading

Bloembergen, Nicolaas. *Encounters in Magnetic Resonances: Selected Papers of Nicolaas Bloembergen*. River Edge, N.J.: World Scientific, 1996.

———. *Encounters in Nonlinear Optics: Selected Papers of Nicolaas Bloembergen (with Commentary)*. River Edge, N.J.: World Scientific, 1996.

———. *Nonlinear Optics*. 4th ed. River Edge, N.J.: World Scientific, 1996.

⊠ Bohr, Niels Henrik David
(1885–1962)
Danish
Theoretical Physicist, Quantum Theorist, Philosopher of Science

Among the revolutionary geniuses of 20th-century physics, the name of Niels Bohr, whose model of the atom laid the basis for quantum mechanics, is second only to that of ALBERT EINSTEIN. By wedding MAX PLANCK's notion of discrete quanta of energy to ERNEST RUTHERFORD's nuclear model of the atom, Bohr opened the door to a description of material processes at odds not only with the determinism of classical physics, but with what many, Einstein included, considered to be a coherent, existentially palatable vision of nature.

Niels Henrik David Bohr was born on October 7, 1885, in Copenhagen, Denmark, into a family remarkable for its intellectual attainments: his mother, Ellen Adler Bohr, was a member of a wealthy Jewish family, prominent in Danish banking and parliamentary circles, and his father, Christian Bohr, was a professor of physiology at the University of Copenhagen, known for his work on the physical and chemical aspects of respiration. Niels was considered less brilliant than his younger brother, Harold, who became a renowned mathematician. But he wasted no time in distinguishing himself: three years after entering the University of Copenhagen in 1903, he won a gold medal from the

Royal Danish Academy of Sciences for his theoretical analysis of vibrations of water jets as a means of determining surface tension. He remained in Copenhagen until 1911, when he received a doctorate for a theory explaining electron behavior in metals.

In 1912, he traveled to England, to continue his research in Cambridge with JOSEPH JOHN (J. J.) THOMSON, the discoverer of the electron. When Thomson proved indifferent to his ideas, Bohr moved to Manchester to work with Ernest Rutherford, who was making important contributions to the theory of the atom. Rutherford had proposed that atoms consist of electrons orbiting a positively charged nucleus. But it was not understood how electrons could continually

Niels Henrik David Bohr laid the foundation for the theory of quantum mechanics, an enormously successful description of physical processes that is at odds with classical determinism. *(Princeton University, AIP Emilio Segrè Visual Archives)*

orbit the nucleus without radiating energy, as classical physics demanded. According to JAMES CLERK MAXWELL's equations, orbiting electrons would be accelerating and continuously emitting electromagnetic radiation; this process would cause them to spiral into the nucleus in about a trillionth of a second. In contradistinction to the classical prediction, however, the hydrogen atom was extremely stable.

With Rutherford as his inspiration and mentor, Bohr set about explaining this discrepancy. He began with MAX ERNEST LUDWIG PLANCK's 1900 theory that energy is emitted in discrete packets or quanta and applied it to Rutherford's nuclear atom. Bohr postulated that electrons are confined to a certain number of stable orbits, in which they neither emit nor absorb energy. Only when it jumps from one discrete orbit to a lower one does the electron lose energy: it sends off an individual photon (particle of light). Since an electron in the innermost orbit has no orbit with less energy to jump to, the atom remains stable. Bohr's theory explained many of the spectral lines for hydrogen and helium, but he hesitated to publish his results, fearing that no one would take him seriously unless he explained the spectra of all the elements. It was Rutherford who persuaded him that the ability to explain hydrogen and helium would be quite enough to make his model credible. Indeed, when Bohr's three papers on the structure of the hydrogen atom and on heavier atoms appeared in 1913, they had a profound, unsettling effect. Many of Bohr's contemporaries balked at accepting so bizarre a picture of the atomic world. But new spectroscopic measurements and other experiments confirmed Bohr's theory, and in 1914 direct evidence for the existence of such discrete states was found.

In 1916, the University of Copenhagen appointed him to the chair of theoretical physics. When he made known his plans to return to "more ideal" research conditions in England, the Danish university created the Institute of Theoretical Physics for him (now the Niels Bohr Institute for Astronomy, Physics and Geophysics). In 1921, a year before he received the Nobel Prize in physics, he was appointed its director, a post he would retain for the rest of his life. The institute became a mecca for theoretical physicists, who traveled from all over the world to debate the meaning of the new physics, and the birthplace of what came to be called the Copenhagen school. While still in his 20s, Bohr found himself at the very center of the quantum mechanical revolution. Bohr and his colleagues, including WOLFGANG PAULI and WERNER HEISENBERG, brainstormed tirelessly in search of a physical interpretation of the new mathematical description of nature. The result was the Copenhagen interpretation of quantum mechanics, which introduced a radical assumption into physical thinking: because the quantum interaction between the "observer" and the "objects to be observed" can never be ignored at the microscopic level, microphysical processes are fundamentally random and probabilistic. Bohr enunciated one of the startling implications of this hypothesis in his complementarity principle, which states that an electron can be regarded as a particle or wave phenomenon, and both characterizations are equally valid, depending on the experimental circumstances:

> Evidence obtained under different experimental conditions cannot be comprehended within a single picture but must be regarded as complementary in the sense that only the totality of the phenomena exhausts the possible information about the phenomenon.

In the early 1920s, Bohr sought to develop a consistent quantum theory that would supersede classical mechanics and electrodynamics at the atomic level. During this period of intense and wide-ranging exploration, he formulated his principle of correspondence, a philosophical guideline for selection of new physical theories, requiring that they explain all the phenomena for which a preceding theory is valid. Since classical mechanics had met all challenges until physicists began to examine the atom itself, Bohr insisted that quantum mechanics, to be valid, must do what the old physics did—and more: it must describe atomic phenomena correctly and be applicable to conventional phenomena, as well. In 1923, he announced that the new physics could do just that:

> Notwithstanding the fundamental departure from the ideas of the classical theory of mechanics and electrodynamics involved in these postulates, it has been possible to trace a connection between the radiation emitted by the atom and the motion of the particles which exhibits a far-reaching analogy to that claimed by the classical ideas of the origin of radiation. Indeed, in a suitable limit the frequencies calculated by the two very different methods would agree exactly.

If Bohr's view of quantum theory gradually won almost universal acceptance, one convert he never succeeded in winning, though not for lack of trying, was Einstein. The Bohr–Einstein debates of the 1920s and 1930s are legendary. Einstein, who could never accept the probabilistic nature of quantum mechanics, produced a series of *gedanken* (thought) experiments designed to disprove the theory. Bohr would then attempt to expose the flaws in Einstein's reasoning. To Einstein's insistence that "God does not play dice with the universe," Bohr would counter, "Einstein, stop telling God what to do!" But if Bohr accepted the strangeness of the theory he had helped bring into being, his attitude toward it was anything but complacent. "Anyone who is not dizzy after his first acquaintance with the quantum of action," he said, "has not understood a word."

In the 1930s, Bohr's interests turned to nuclear physics and in 1939 he proposed the liquid-droplet model for the nucleus, which proved a key to understanding many nuclear processes: Nucleons (neutrons and protons) behave as molecules do in a drop of liquid. If given enough extra energy (by absorbing a neutron), the spherical nucleus may be distorted into a dumbbell shape and then split at the neck into two nearly equal fragments, releasing energy. In this way Bohr, working with JOHN A. WHEELER, was able to explain why a heavy nucleus could undergo fission after the capture of a neutron. Bohr validated his theory when he correctly predicted that the nuclei of uranium-235 and uranium-238 would behave differently, since the number of neutrons in each nucleus is odd and even, respectively.

During World War II, Bohr was active in the Danish resistance movement; in 1943, he escaped with his family to Sweden and then to England, where he and his son, Aage, took part in the project for making a nuclear fission bomb. He accompanied the British research team to Los Alamos and made significant contributions to physical research on the U.S. atomic bomb. After the war, Bohr became a prominent advocate for control of nuclear weapons, pleading, in a famous 1950 open letter to the United Nations, for an "open world" and exhorting Roosevelt and Churchill to strive for international cooperation. His passionate advocacy won him the first U.S. Atoms for Peace Award in 1957. He was instrumental in creating the European Center for Nuclear Research (CERN), Geneva, in 1952.

In his last years Bohr remained a staunch defender of the Copenhagen interpretation of quantum mechanics and published articles in which he related the quantum mechanical idea of complementarity to aspects of human life and thought. Bohr's unique approach to science and philosophy, his openness to new ideas, and willingness to learn from even the most junior of his colleagues left a lasting imprint on the generation of physicists who followed him. Until his death in Copenhagen on November 18, 1962, he remained a spirited participant in the great physics debates he had played so central a role in initiating.

Further Reading

Pais, Abraham. *Niels Bohr's Times: In Physics, Philosophy, and Polity.* Oxford: Oxford University Press, 1991.

Ruhla, Charles. *The Physics of Chance: From Blaise Pascal to Niels Bohr.* Oxford: Oxford University Press, 1992.

Spangenburg, Ray, and Diane K. Moser. *Niels Bohr: Gentle Genius of Denmark.* New York: Facts On File, 1995.

Whitaker, Andrew. *Einstein, Bohr and the Quantum Dilemma.* Cambridge: Cambridge University Press, 1996.

⊠ **Boltzmann, Ludwig**
(1844–1906)
Austrian
Theoretical Physicist (Statistical Mechanics, Electromagnetism, Thermodynamics)

Ludwig Boltzmann was an intuitive genius whose belief in the existence of an underlying atomic structure in nature, governed by the laws of probability, helped set the course for 20th-century physics. An embattled figure, whose ideas contradicted the scientific dogmas of his times and who died before his theories were validated, Boltzmann redefined the very nature of theoretical physics and paved the way for the statistical theory of quantum mechanics.

Boltzmann was born in Vienna on February 20, 1844, and received his early education in Linz and Vienna. He studied at the University of Vienna at a time when the fundamentals of thermodynamics and electromagnetism were being established. Over the next 40 years, he would make vital contributions to these fields. In 1859, while still a student in Vienna, he published his first paper, on the kinetic theory of gases. He went on to write a thesis on that subject, under the supervision of JOSEF STEFAN, the discoverer of a fundamental law of radiation, and received his Ph.D. in 1866.

The following year Boltzmann became Stefan's assistant at the Physikalisches Institut in Vienna. Building on the work of JAMES CLERK MAXWELL, in 1868 he published a groundbreaking paper on thermal equilibrium in gases. By examining the distribution of energy among colliding gas molecules, Boltzmann derived an exponential formula, later known as the Maxwell–Boltzmann distribution, to describe the distribution of molecules. This formula relates the mean total energy of a molecule to its temperature, in terms of a fundamental constant k, which came to be known as the Boltzmann constant. This seminal work formed the basis of the new field of statistical mechanics, which played a major role in the later discoveries of MAX ERNEST LUDWIG PLANCK and others.

Over the next several years Boltzmann would move restlessly from one academic post to another. From 1869 to 1873, he was professor of theoretical physics at the University of Graz; from 1873 to 1876, professor of mathematics at Vienna; in 1876, he returned to Graz, this time as professor of experimental physics, and remained there until 1879.

During this period, Boltzmann published a mathematical description of the tendency of a gas to reach a point of equilibrium as the most probable state. This famous equation, $S = k \log W$ (which was later engraved on his tombstone), describes the relationship between entropy and probability.

During the 1880s, he was director of the Physikalisches Institut in Vienna. During his tenure there, Boltzmann developed the work of his former mentor, Josef Stefan. Stefan had experimentally derived the law of blackbody radiation, showing that the energy radiated by a black body is proportional to the fourth power of its absolute temperature. In 1884, Boltzmann proposed a theoretical formulation for it, subsequently known as the Stefan–Boltzmann law, based on the second law of thermodynamics and Maxwell's electromagnetic theory.

Between 1889 and 1902, Boltzmann would occupy academic posts in Munich, Vienna, and Leipzig. He is said to have joked that he moved around so much because he had been born during the final hours of a Mardi Gras. A more likely explanation of his wandering is what we would today term a manic–depressive nature, aggravated by years of scientific jousting.

It was during these years that Boltzmann became embroiled in what is now known as the atomist–energeticist debate. With no technology available to verify their existence at the turn of the century, atoms had been relegated by most physicists to the realm of speculation. ERNST MACH, Boltzmann's leading opponent, was among these. For Mach the purpose of science was to measure and demonstrate only that which it could observe. Mach and his colleagues were content to measure the expansion of gases and empirically deduce a simple law relating temperature, pressure, and volume. They were not fazed by their inability to explain *why* these properties were related in this particular way. Boltzmann, on the other hand, believed that by hypothesizing a dynamic submolecular world of colliding atoms he could explain why gas expands and by how much. It was the atoms' incessant movement that produced the properties of gas, Boltzmann said: the greater the speed of the atoms, the higher the gas temperature; the greater the number of collisions of atoms against the walls of a container, the greater its pressure. Boltzmann's atomic model enabled him to understand the ability of hot gas to push on a piston in a steam engine, thereby converting energy into mechanical work. His statistical methods of evaluating the variations in the movements of molecules in a gas demonstrated that reliable physical laws could be built on probability. Boltzmann considerably broadened the definition of what it meant to be a theoretical physicist and paved the way for the nondeterministic microscopic description of nature soon to be proposed by the architects of quantum mechanics.

Despite continued opposition to his ideas, Boltzmann's career flourished. In Vienna, his lectures on the philosophy of science became so popular it was necessary to move them to the biggest lecture hall available; his fame even reached the palace of Franz Josef and earned him an invitation to visit. In 1904, he traveled to the United States. He lectured at the World's Fair in Saint Louis and visited Stanford and Berkeley, continuing to defend his belief in atomic structure. Sadly, he failed to realize that the new discoveries concerning radiation that he learned about on this visit were about to prove his theories correct.

Ludwig Boltzmann committed suicide, hanging himself while his wife and daughter were swimming, at Duino, near Trieste, on September 5, 1906. A short time later, the discovery of Brownian motion led to a near-universal acceptance of his kinetic and statistical theories.

See also EINSTEIN, ALBERT.

Further Reading

Cercignani, Carlo. *Ludwig Boltzmann: The Man Who Trusted Atoms.* Oxford: Oxford University Press, 1998.

Lindley, David. *Boltzmann's Atom: The Great Debate That Launched a Revolution in Physics.* New York: Free Press, 2000.

⊠ **Born, Max**
(1882–1970)
German/British
Theoretical Physicist, Quantum Theorist, Solid State Physicist

Max Born is best known for his distinctive contributions to the revolution in subatomic physics of the 1920s. It was Born who, in 1924, anointed the new physics *quantum mechanics.* The following year, along with his students, WERNER HEISENBERG and Pascual Jordan, he formulated matrix mechanics, the first mathematical system capable of explaining the behavior of electrons in an atom. Most importantly, his statistical interpretation of subatomic events, which suggests that a fundamental randomness is inherent in the laws of nature, became the defining feature of the Copenhagen interpretation of what the equations of quantum mechanics actually "mean."

He was born on December 11, 1882, in Breslau, Germany, now Wroclaw, Poland, into a wealthy Jewish family. His mother, Margarete, née Kauffmann, was part of a family of Silesian industrialists. Gustav, his father, was a professor of anatomy at the University of Breslau, where Max would enroll after completing his studies at Konig Wilhelm's Gymnasium. Born's studies would take him from Breslau to Heidelberg, Zurich, and, finally, the University of Göttingen, where, in 1907, he was awarded a doctorate in physics and astronomy. He married Hedwig Ehrenberg in 1913 and had three children with her.

When World War I broke out, Born joined the German army and was assigned to a research unit in which he worked on the problem of sound ranging. He also found time to study the theory of crystals and publish his first book, *Dynamics of Crystal Lattices.* After briefly holding academic positions in Göttingen and Berlin, he was appointed professor of physics at Frankfurt-am-Main in 1919. While there, he applied his work on the lattice energies of crystals (the energy given out when gaseous ions join together to form a solid crystal lattice) to the formation of alkali metal chlorides. By calculating the energies involved in lattice formation, from which the properties of crystals may be derived, he made a seminal contribution to solid state physics.

In 1921, Born moved from Frankfurt to a more prestigious post in Göttingen and, during his 12 years there, made it an international center for theoretical physics rivaled only by NIELS HENRIK DAVID BOHR's Institute for Theoretical Physics in Copenhagen. It was in Göttingen, with Heisenberg as his assistant, that Born did his most important work, on the electronic structure of atoms. In 1913, Bohr had published his theory of the atom, based on MAX ERNEST LUDWIG PLANCK's 1900 theory that energy is emitted in discrete packets or quanta. In Bohr's model, electrons rotating around the nucleus of an atom are confined to a certain number of stable orbits, emitting quanta of energy only when jumping to a lower orbit. The model proved highly productive, explaining many of the spectral lines for hydrogen and helium, but it had an ad hoc quality, which left physicists wondering about the guiding principle behind it. In 1925, Heisenberg returned to Göttingen after a year working with Bohr in Copenhagen and offered an approach to the problem. Together with Born and Jordan, he developed a mathematical system that explained the features of the atom. Their famous "three-man paper," published that year, introduced matrix mechanics, the first precise mathematical description of the workings of the atom. By mathematical treatment of values within matrices or arrays, the frequencies of the lines in the hydrogen spectrum were obtained.

The next year, however, matrix mechanics was challenged by ERWIN SCHRÖDINGER's wave function, a mathematical expression for LOUIS-VICTOR-PIERRE, PRINCE DE BROGLIE's, 1923 discovery that electrons do not occupy orbits but exist in standing waves around the nucleus. Schrödinger's approach, which he called wave mechanics, with its more familiar concepts and equations and its ability to allow visualization of the atom, rapidly became the theory of choice. Born made an important contribution to wave mechanics with his Born approximation method, a mathematical technique for solving the Schrödinger equation (i.e., computing the behavior of

subatomic particles) that continues to be used in high-energy physics.

However, his most fundamental work grew from the attempt to answer the "big question" posed by quantum mechanics: what is the physical reality described by the wave function? Born came up with the disturbing, revolutionary notion that the Schrödinger wave equation described not a matter wave existing in space and time, but a wave of probabilities. According to this statistical interpretation, when physicists measure the location of an electron, the probability of finding it in each region depends on the magnitude of its wave function there. The traditional, deterministic view of nature posited that someone shooting at a target can, in principle, aim the shot so that it will be certain to hit the target in the middle. In Born's view, that kind of certainty is no longer possible; we can only predict the probabilities that certain events will occur. On the basis of a large number of shots, we can determine that the average point of impact will be in the middle of the target. Whereas the implications of these quantum statistic mechanical effects may be trivial on the macroscopic level, they are fundamental for subatomic phenomena, suggesting a randomness inherent in the very laws of nature. Born's statistical interpretation of the wave function quickly became a cornerstone of the Copenhagen interpretation: along with the work of Jordan in Göttingen and PAUL ADRIEN MAURICE DIRAC in Cambridge, England, on the unified equations known as transformation theory, it appeared to complete the mathematical foundations of the new quantum mechanics.

Born's fruitful years in Göttingen came to an unnatural end, however, when he was forced to flee Nazi Germany in 1933. He continued his career in England, first as a lecturer at Cambridge and then at Edinburgh University, where, in 1936, he was appointed professor of natural philosophy. He became a British citizen in 1939

Max Born is best known for his statistical interpretation of subatomic events, which suggests that a fundamental randomness is built into the laws of nature. *(AIP Emilio Segrè Visual Archives, Gift of Jost Lemmerich)*

and returned to Germany when he retired in 1953. The following year he was awarded the Nobel Prize in physics for his contributions to quantum mechanics.

When he died in Göttingen on January 5, 1970, he was the recipient of numerous prestigious awards and the author of more than 360 publications, including textbooks and monographs for physics students and experts, as well as some popular science books.

Max Born's legacy to modern physics is a dynamic one, which continues to be debated. Schrödinger himself was never comfortable with Born's interpretation of his equation; he devised his famous Cat Paradox, a thought experiment whose outcome leaves the cat in question both dead and alive to show the absurdity of applying probability to the macroscopic world. And no student of physics is unfamiliar with Einstein's objection: "I cannot believe that God plays dice." Even Born, who declared in his autobiography that "theoretical physics is actual philosophy," speculated that, despite the predictive success of quantum mechanics, "something," although inaccessible to the observer, may yet exist beneath the laws of probability:

> If God made the world a perfect mechanism, He has at least conceded so much to our imperfect intellect that in order to predict little parts of it, we need not solve innumerable differential equations, but can use dice with fair success.

The inability of the Copenhagen interpretation to answer the question of "what lies beneath" left the field open to successive generations of physicists to probe this mystery, which lies at the heart of quantum mechanics.

Further Reading

Born, Max. *My Life*. New York: Scribner's, 1978.

Born, Max, R. J. Blin-Stoyle, and J. M. Radcliffe. *Atomic Physics*. New York: Dover, 1989.

Born, Max, Gunther Leibfried, and Walter Biem. *Einstein's Theory of Relativity*. New York: Dover, 1989.

Pagels, Heinz R. *Cosmic Code: Quantum Physics as the Law of Nature*. New York: Simon & Schuster, 1982.

⊠ Bridgman, Percy Williams
(1882–1961)
American
Experimentalist (High-Temperature and High-Pressure Physics), Philosopher of Science

Percy Williams Bridgman did pioneering work on the behavior of materials at high temperature and high pressure, for which he won the 1946 Nobel Prize in physics. He is, however, most famous for his work in the philosophy of modern physics as founder of the school of operationalism, an approach that stresses the primacy of the physical operations of measurement in physical research.

He was born on April 21, 1882, in Cambridge, Massachusetts, the son of a journalist who wrote on social and political issues. He received his early education at public schools in the nearby town of Newton. In 1900, he entered Harvard University, and four years later he received his bachelor's degree. In 1908, he received a Ph.D. from Harvard and joined its faculty. Harvard would remain his academic home for the rest of his life, making him an instructor in 1910, assistant professor in 1913, and full professor in 1919. In 1912, he married Olive Ware, with whom he had a daughter, Jane, and a son, Robert.

Working at Harvard's Jefferson Research Laboratory, Bridgman quickly demonstrated his unusual skill in operating machine tools and manipulating glass. He began doing experiments involving static high-pressure effects on materials at pressures of 6500 atmospheres (one atmosphere = 14.7 lb/in). In these experiments Bridgman had to invent his own equipment. This work led to his invention of a type of gasket seal in which the pressure in the gasket always exceeds that in the pressurized fluid within the gasket. This resulted in a gasket with a self-sealing closure property that made higher pressures possible. Using this new device in an experimental apparatus made of new types of steels and metals with heat-resistant compounds, he was able to increase the pressure in his experiments to 400,000 atmospheres. At these extraordinarily high pressures he measured the compressibility of liquids and the physical properties of solids, such as electrical resistance. He discovered new high-pressure forms of ice and new phenomena such as the

rearrangement of electrons in cesium at a certain transition pressure.

Bridgman also did pioneering work in devising a technique to synthesize diamonds, which he later used to synthesize many more minerals. Because the pressures and temperatures he achieved in his experiments simulated those deep beneath the ground, his discoveries yielded insights into the physical processes that take place within the Earth. This led to the development of a new school of geology based on his experimental work at high temperature and high pressure.

Bridgman was a loner, wholly dedicated to research, who disliked lecturing. His teaching experience, however, made him aware of the ambiguities of scientific language and caused him to shift his attention to the philosophy of science. In 1927, he became Hollis Professor of Mathematics and Philosophy at Harvard and published his most influential work, *The Logic of Modern Physics*, in which he developed his philosophy of operationalism. His fundamental argument was that in order to be meaningful, physical concepts must be defined in terms of the physical operations involved in their measurement:

> The concept of length involves as much as and nothing more than the set of operations by which length is determined. . . . [T]he concept is synonymous with a corresponding set of operations.

Bridgman was an influential and prolific writer, publishing more than 260 papers. In 1931, he published *Physics of High Pressure*, which became a classic in the field. His other books include *The Nature of Physical Theory*, 1936; *The Nature of Thermodynamics*, 1941, and *Reflections of a Physicist*, 1955.

From 1950 to 1954, Bridgman was Higgins Professor at Harvard. He then retired and became professor emeritus. On August 10, 1961, dying of cancer, he killed himself at his home in Randolph, New Hampshire.

A bold intelligence, Bridgman is famous for his aphorism "There is no adequate defense, except stupidity, against the impact of a new idea." For all his attempts to define the nature of physical discovery systematically, he never lost sight of its essentially unpredictable nature:

> In his attack on his specific problem [the working scientist] suffers no inhibitions of precedent or authority, but is completely free to adopt any course that his ingenuity is capable of suggesting to him. No one standing on the outside can predict what the individual scientist will do or what method he will follow. In short, science is what scientists do, and there are as many scientific methods as there are individual scientists.

Further Reading

Bridgman, Percy Williams. *The Logic of Modern Physics*. New York: Macmillan, 1961.

⊠ Broglie, Louis-Victor-Pierre-Raymond, prince de
(1892–1987)
French
Theoretical Physicist, Quantum Theorist

Louis-Victor-Pierre, prince de Broglie, made his groundbreaking contribution to modern physics while still a graduate student at the Sorbonne (University of Paris) in 1924. In 1905, ALBERT EINSTEIN had shown that light, long understood by physicists to be a wave, can behave as a particle under certain conditions. Building on this work, the young Broglie demonstrated that in addition to light, atomic particles have a dual nature. This wave–particle duality was accepted as a fundamental principle governing the structure of the atom and was subsequently used by ERWIN SCHRÖDINGER to develop the wave equation for the quantum mechanics of atomic structure. De Broglie's discovery earned him the 1929 Nobel Prize in physics.

The man whose full name and title was Louis-Victor-Pierre Raymond, duc de Broglie was born on August 15, 1892, in Dieppe, France, the second son of an ancient aristocratic family. At the age of 18 he entered the Sorbonne, intending to earn a degree that would help prepare him for a diplomatic career. But the lure of physics proved decisive; as did his older brother, Maurice, who was a pioneer in the study of X-ray spectra, de Broglie broke family tradition and earned a degree in science, in 1913. His studies were interrupted by World War I. He was fortunate, however, in his assignment to a radio unit stationed at the Eiffel Tower, where he had spare time to devote to the study of technical problems. At the end of the war he resumed his studies at the Sorbonne, specializing in theoretical physics. He took a special interest in the study of problems involving *quanta*, the term introduced by the German physicist MAX ERNEST LUDWIG PLANCK in 1900, when he discovered that certain experimental results could be understood only if energy were emitted and absorbed in discrete packets or quanta. This was a revolutionary vision, challenging the classical Newtonian view that energy could have any value and was transmitted in a continuous stream. ALBERT EINSTEIN was among the first physicists to accept this paradigm-shattering hypothesis. In 1905, he used the idea of quanta to explain the photoelectric effect. Light, he said, was composed not only of waves, but also of particles, which he named *photons*.

Broglie's breakthrough into the nature of matter built on Planck's and Einstein's work. In 1923, he published a brief note in the journal *Comptes Rendus*, containing an idea that was to revolutionize our understanding of the physical world at the most fundamental level. He began with the assumption that a particle of mass m is always associated with an internal clock ticking with frequency v. He noted that relativity theory predicts that when such a particle is set in motion, its total energy increases, tending to infinity as the speed of light is approached. Similarly, the frequency of the clock's ticking slows down. However, in quantum mechanics, a decrease in frequency is related to a decrease in the energy of the particle. This apparent contradiction between the tendency of the relativistic energy to increase and that of the quantum energy to decrease troubled de Broglie.

A year later, in 1924, Broglie published the resolution of this apparent contradiction in the first chapter of his doctoral thesis, by postulating that a wave always accompanies the motion of a particle. He argued that the wave aspect associated with quantum mechanical particles is an inherently relativistic phenomenon. Using the orbits of the electron in the quantum model of the hydrogen atom developed by NIELS HENRIK DAVID BOHR as his example, he suggested that relativistic effects are not inconsequential in determining the electronic properties of the hydrogen atom, even though the speed of the orbiting electron is significantly less than the speed of light.

In 1927, experimental evidence for de Broglie's theory was discovered independently by researchers in the United States and Great Britain. They performed experiments that produced electron wave diffraction patterns similar to the diffraction patterns associated with light waves, thus showing that particles can produce an effect that had until then been exclusive to electromagnetic waves such as light and X rays. Such waves are known as matter waves. For particles to be described as waves they must satisfy a partial differential equation known as a wave equation. De Broglie's attempt to develop such a wave equation in 1926 proved unsatisfactory. It fell to Schrödinger to succeed where de Broglie had failed; that same year, he unveiled his famous Schrödinger equation, a wave equation that predicted known experimental results on atoms.

After receiving his doctorate, de Broglie stayed on at the Sorbonne until 1928. In 1929, the Swedish Academy of Sciences conferred on him the Nobel Prize in physics for his discovery of the wave nature of electrons. He moved to the Henri Poincaré Institute in Paris in 1932 as professor of theoretical physics and remained there until 1962. In 1956, he received the gold medal of the French National Scientific Research Center. During this period he was also a senior adviser on the development of atomic energy in France.

From 1930 to 1950, de Broglie turned his attention to the extensions of wave mechanics: PAUL ADRIEN MAURICE DIRAC's electron theory, the new quantum theory of light, the general theory of spin particles, and applications of wave mechanics to nuclear physics. Not until 1951 did he return to the study of the wave equation he had developed in 1926, which attempted to give a causal interpretation to wave mechanics in terms of classical space and time. Throughout his life, de Broglie was concerned with the fundamental question of whether the statistical results of physics are all that there is to be known or whether there is an underlying, completely determined reality that experimental techniques are as yet inadequate to discern. Early in his career, he was inclined to accept the almost universal adherence of physicists to the purely probabilistic interpretation of MAX BORN, Bohr, and WERNER HEISENBERG. Like Einstein, however, he was constitutionally unable to live with it. In his later years he expressed the belief that "the statistical theories hide a completely determined and ascertainable reality behind variables, which elude our experimental techniques."

De Broglie's discovery of the wave–particle duality enabled physicists to view Einstein's conviction that matter and energy can be converted into one another as fundamental to the structure of nature. The study of matter waves led not only to a much deeper understanding of the nature of the atom, but also to explanations of chemical bonds and to the practical application of electron waves in electron microscopes.

See also DAVISSON, CLINTON; JOSEPH; THOMSON, JOSEPH JOHN (J. J.).

Further Reading

Achinstein, Peter. *Particles and Waves: Historical Essays in the Philosophy of Science*. Oxford: Oxford University Press, 1994.

Van der Merwe, Alwyn. *Waves and Particles in Light and Matter*. Boston: Kluwer Academic, 1994.

Wheaton, Bruce R. *The Tiger and the Shark: Empirical Roots of Wave–Particle Dualism*. Cambridge: Cambridge University Press, 1990.

C

⊠ Carnot, Nicolas Léonard Sadi
(1796–1832)
French
Theoretical Physicist (Thermodynamics)

Sadi Carnot was a pioneer in the field of thermodynamics, whose search for more efficient steam engines led him to fundamental discoveries about the relationship between work and heat.

He was born in Paris on June 1, 1796, the eldest son of an illustrious, erudite family, and named after the medieval Persian poet and philosopher Sa'di of Shiraz. This was a time of political upheaval, and Lazare, his father, was at the center of events as a member of the Directory, the French revolutionary government. During Sadi's early childhood Lazare's political career underwent dramatic reversals, as he was forced into exile and later commanded to return and become Napoleon's minister of war. When shortly afterward he was made to resign, he took charge of his son's education, giving him a solid grounding in mathematics and science, music, and languages. The young Sadi attended the École Polytechnique in Paris, whose faculty boasted such scientific luminaries as Siméon Denis Poisson, ANDRÉ-MARIE AMPÈRE, and François Arago, who exposed him to the newest ideas in physics and chemistry, between 1812 and 1814. By this time, Napoleon's armies were in retreat, foreign armies had entered France, and Sadi and his classmates engaged in skirmishes on the outskirts of Paris in an unsuccessful attempt to turn back the invaders. In 1815, Lazare fled to Germany, where he would remain for the rest of his life, while his son remained in France, preparing himself for a career as a military engineer at the École Genie in Metz.

When he graduated in 1816, Carnot was first assigned to inspect fortifications. In 1819, he managed to transfer to the office of the general staff in Paris, and soon afterward he retired on half-pay, in order to pursue his varied interests, which included industrial development, mathematics, and the fine arts. As he became increasingly fascinated with the design of steam engines, he was led to explore basic questions about the nature and dynamics of heat. Carnot embarked on his most fertile period, which culminated, in 1824, in the publication of his classic work, *Reflections on the Motive Power of Heat*. Even in this most fundamental scientific inquiry, patriotic considerations played a role. It was an Englishman, James Watt, who had invented the steam engine, and England continued to outdistance France in its development of this versatile, if inefficient (it had an efficiency of only 6 percent) workhorse. Carnot's desire to create the optimal steam engine design was at least partly motivated by his belief that France's failure to make adequate use of steam had contributed to its political downfall. When he presented this work formally to the Académie des Sciences, it was given an enthusiastic reception but was ignored until a decade later, in 1834, when Emile Clapeyron, a railroad engineer, quoted and extended his results.

In *Reflections* Carnot strove to determine whether there was a fixed limit to the work a steam engine could produce. He was in search of a theory applicable to all kinds of heat engines. Carnot noted that a steam engine produces motive power when heat flow "falls" from the higher temperature of the boiler to the lower temperature of the condenser, analogously to the way water, when falling, provides power in a water wheel. On this basis, he developed an equation, now known as Carnot's theorem, which showed that the maximal amount of work an engine using a hot gas can produce depends only on the absolute temperature difference of the gas as it flows through the engine from the boiler to the condenser. He was also able to show that the maximal amount of work obtained is independent of (1) whether the temperature drops rapidly or slowly or in a number of stages and (2) the nature of the gas used in the engine. Carnot's work established a scientific basis for steam engine design. His experimental data led him to recommend that steam be used over a large temperature interval and with minimal losses due to conduction or friction.

Carnot arrived at his theorem by considering the case of an ideal heat engine engaged in a reversible cyclic process now known as the Carnot cycle. For a cyclic process to be completely reversible (an idealization), no work is done in overcoming friction at any stage and no heat is lost to the surroundings. In the Carnot cycle a quantity of gas undergoes a combination of two basic processes that produce work and consume heat: (1) an isothermal process that takes place at a constant temperature and (2) an adiabatic process in which heat can neither enter nor leave the system, as any work done changes the internal energy and hence the temperature of the system. In the Carnot cycle an ideal gas (i.e., a gas whose internal energy depends only on the absolute temperature) is first allowed to expand isothermally to do work, absorbing heat in the process, and then adiabatically expanded again without transfer of heat but with a temperature drop. The gas is then compressed isothermally, heat being given off, and finally it is returned to its original condition by another adiabatic compression, accompanied by a rise in temperature.

In terms of this series of operations Carnot was able to show that even an ideal engine cannot convert into mechanical energy all the heat supplied to it since some of the heat energy must be expelled in the process. Hence, Carnot's law states that no real, physical engine can be more efficient than an ideal reversible engine working within the same range of temperature.

During the time that he developed his theorem, Carnot still believed in the caloric theory of heat, which held that heat was a form of fluid. However, unpublished notes, discovered in 1878, showed that Carnot was beginning to believe heat was energy that could change into work. He had calculated a conversion constant for heat and work, indicating that he believed that the total quantity of energy in the universe was constant. In essence, he had thought out the basics of the first law of thermodynamics, conservation of energy. He proposed experiments for observing the temperature effects of friction in fluids, some of which were identical with those performed by JAMES PRESCOTT JOULE 20 years later.

Sadly, Carnot had little time to continue his research; he died suddenly in a cholera epidemic in Paris on August 24, 1832, at the age of 36. According to the custom of the times, his personal belongings, including most of his notes, were burned. He would be largely forgotten until LORD KELVIN (WILLIAM THOMSON) confirmed his conclusions and published them in his *Account of Carnot's Theorem* in 1849. Carnot's work formed the basis for Kelvin's and RUDOLF JULIUS EMMANUEL CLAUSIUS's derivations of the second law of thermodynamics, or entropy. In his short but brilliant career Carnot made seminal contributions to both pure and applied physics, laying the foundations on which the modern science of thermodynamics would be built.

Further Reading

Carnot, Sadi. *Reflections on the Motive Power of Heat.* R. H. Thurston, ed. Translation of 1897 original. New York: John Wiley, 1990.

Van Ness, H. S. *Understanding Thermodynamics.* New York: Dover, 1983.

⊠   Cavendish, Henry
(1731–1810)
British
Experimentalist (Gravitation, Electromagnetism), Physical Chemist

Henry Cavendish devised an ingenious method for determining the gravitational constant, which allowed him to make the first accurate measurements of the mass and density of the Earth. He did pioneering work in electricity and was the first to recognize hydrogen as a distinct substance and to show that water is composed of hydrogen and oxygen. Because of his exacting standards and refusal to make public any results that failed to meet them, Cavendish published very little of his prodigious output. For this reason, his influence on subsequent researchers was far less than it might have been.

He was born in Nice, France, where his mother had traveled to improve her health, on October 10, 1731. Descended from the dukes of Devonshire and Kent, he was the first son of Lord Charles Cavendish, a well-known experimental scientist. His mother died when he was

only two, after giving birth to his brother, Frederick. After receiving his early education at home, Cavendish entered the Hackney Seminary at age 11. In 1749, he began his studies at Peterhouse College, Cambridge University; he left four years later without a degree, a common enough practice in his day. He took the customary "grand tour" of Europe with his brother before settling down in the family house in Soho, London, with his father.

Encouraged by his father to pursue scientific research, Cavendish became a vital member of the scientific community and was elected to the Royal Society in 1760. Despite his immense, inherited wealth, he lived simply and chose to socialize only with other scientists. He appeared to be terrified of women and instructed his female servants to stay out of sight if they wanted to keep their jobs. His long bachelor life would be devoted wholly to science. In 1783, his father died and left him an inheritance that made him one of the wealthiest men in Europe. He then moved to a villa in Clapham Common, where he established a well-equipped laboratory and library. Apart from such expenses, however, he maintained his frugal way of life, spending nothing on himself and neglecting his appearance.

In 1776, Cavendish published his first paper, which demonstrated the existence of hydrogen, a discovery that earned him the Royal Society's Copley Medal. He had arrived at his discovery by using a quantitative approach, which characterized most of his research. By calculating the densities of hydrogen gas and ordinary air, he showed that they are entirely different substances.

For the rest of his life, however, he published rarely. His most significant publications were a theoretical study of electricity (1771), a demonstration of the synthesis of water (1784), and the determination of the gravitational constant (1798). In his 1771 paper on the nature of electricity, he assumed that electricity was an elec-trio fluid. This was the beginning of a decade of work on the subject, in which he strove to do nothing less than explain all electrical phenomena and produce a sequel to SIR ISAAC NEWTON's great work, *The Principia*. Cavendish's accomplishments were substantial: the discovery that electrical fields obey the inverse square law as well as important work on conductivity, in which he compared the electrical conductivities of equivalent solutions of electrolytes and formulated a variant of GEORG SIMON OHM's law. His ideas on electricity predated those of CHARLES AUGUSTIN COULOMB and MICHAEL FARADAY, although his experiments would not be known until a century later, when JAMES CLERK MAXWELL published them. Had Cavendish published them, he doubtless would have accelerated progress in that field.

In 1798, he announced that he had measured Newton's gravitational constant, the constant of proportionality in the equation expressing Newton's law of universal gravitation. Since this law contained two unknowns—the gravitational constant and the mass of the Earth—determining one would determine the other. In what has become known as the Cavendish experiment he used an apparatus consisting of a delicate suspended rod with two small lead spheres attached to each end. He placed two large immobile spheres in a line at an angle to the rod. Because of the gravitational attraction of the large spheres, the small spheres twisted the rod toward them. By observing the period of oscillation set up in the rod, Cavendish determined the gravitational constant and consequently was able to measure the density of the Earth (about five and one-half times that of water) and its mass (6×10^{24} kg). So sensitive was his apparatus, no one was able to produce more accurate measurements for over a century.

Cavendish died alone in his London home on February 24, 1810, at the age of 79, after announcing to his valet, "Mind what I say. I am going to die," and giving him precise instruc-

tions on how to deal with the event. Only after his death did the cornucopia of scientific manuscripts he left behind reveal the immense scope of the research he had done over a span of 60 years and his stature as one of the greatest of 18th-century physicists. The illustrious Cavendish Laboratory at Cambridge University was named in his honor.

Further Reading

Jungnickel, Christa, and Russell K. McCormach, eds. *Cavendish: The Experimental Life*. Bucknell, Pa.: Bucknell University Press, 1999.

Maxwell, James Clerk, ed. *The Electrical Researches of the Honourable Henry Cavendish*. London: Frank Cass, 1967.

McCormach, Russell K. *Cavendish*. Newark, Del.: American Philosophical Society, 1996.

⊠   **Chadwick, Sir James**
(1891–1974)
British
Experimentalist, Particle Theorist, Nuclear Physicist

Sir James Chadwick was a British experimentalist who won the 1935 Nobel Prize in physics for his discovery, three years earlier, of the neutron, a particle with no charge made up of a proton and an electron. This discovery, which has been hailed as the beginning of nuclear physics, led directly to nuclear fission and the development of the atomic bomb.

Born in Cheshire, England, on October 20, 1891, James Chadwick was the son of John Joseph Chadwick and Anne Mary Knowles. He received his secondary education in Manchester and entered Manchester University in 1908. Chadwick is said to have intended to major in mathematics, but when he accidentally entered the line of registering physics majors was too shy to admit the error. On receiving his degree in physics in 1911, he continued at Manchester, doing research with ERNEST RUTHERFORD on the emission of gamma rays from radioactive materials. Intent on gaining more research experience, he left for Berlin in 1913, to work with HANS WILHELM GEIGER. During this apprenticeship, he became the first to demonstrate that beta particles (particles emitted in a type of radioactive decay known as beta decay) possess a range of energies up to a characteristic maximal value of the nucleus. Chadwick soon found himself a hostage of World War I; considered an enemy alien, he was kept at a stable, which had been transformed into an internment camp, for the duration of the war. While there he somehow managed to conduct research on the ionization present during the oxidation of phosphorus and the photochemical reaction between chlorine and carbon monoxide.

After his release in 1919, Chadwick accepted Rutherford's invitation to join him at the Cavendish Laboratory in Cambridge, where he would remain until 1935. While in Liverpool, he met and married, in 1925, Aileen Stewart-Brown and had twin daughters with her. That year Rutherford had produced the first artificial nuclear transformation: by bombarding nitrogen with alpha particles (particles emitted in a type of radioactive decay known as alpha decay) he had produced atom disintegration. Working together the two men observed the transmutation of other light elements by means of alpha particle bombardment and studied the properties and structure of atomic nuclei. On the basis of this work Chadwick established the equivalence of atomic number and charge.

Chadwick's most significant achievement, in the words of the Nobel Prize Committee, "by a combination of intuition, logical thought, and experimental research [proving] the existence of the neutron and establishing its properties," occurred in 1932. Rutherford had hypothesized the existence of a neutral particle with the same weight as a proton as early as 1920. Lacking electric charge, such particles would be difficult to

Sir James Chadwick discovered the neutron, a nuclear particle with zero electric charge. His discovery led directly to nuclear fission and ultimately to the atomic bomb. *(AIP Emilio Segrè Visual Archives, William G. Meyers Collection)*

detect. Chadwick attacked the problem but made little progress until 1930, when the researchers Walter Bothe and Herbert Becker described an unusually penetrating type of gamma ray (a form of electromagnetic radiation emitted by atomic nuclei) produced by bombarding the metal beryllium with alpha particles. Because of the properties of this radiation, Chadwick guessed that it represented not gamma rays, but Rutherford's neutral particle, made up of a proton and an electron and therefore having a mass slightly greater than that of a proton. Since the mass of the beryllium nucleus had not yet been measured, he devised an experiment in which boron was bombarded with alpha particles. This bombardment produced neutrons. Chadwick confirmed that the neutron's mass was slightly greater than the proton's, on the basis of the mass of the boron nucleus and other elements and the properties involved. The neutron provided physicists with an incomparable tool for investigating the atom. Since neutrons are devoid of electric charge, they need not overcome the considerable electric forces present in the nuclei of heavy atoms and therefore are capable of penetrating and splitting the nuclei of even the heaviest elements. Thus, Chadwick's work with neutrons paved the way to the fission of uranium 235 and the creation of the atomic bomb.

The year of his great discovery Chadwick became professor of physics at the University of Liverpool, where he ordered the construction of a cyclotron that he would use to investigate the nuclear disintegration of the light elements. When World War II began, he was one of the first in Great Britain to urge the possibility of developing an atomic bomb. He directed the British atomic bomb effort at Liverpool and in 1943 moved to Los Alamos to head the British team participating in the Manhattan Project. He was knighted in 1945 after his return to Liverpool, where he developed a research school in nuclear physics. In 1948 he became Master of Gonville and Caius College at Cambridge and remained there until his retirement a decade later. He died on July 24, 1974.

Chadwick's discovery of the neutron signaled the beginning of nuclear physics and inspired ENRICO FERMI to study nuclear reactions produced by neutrons, leading to the discovery of nuclear fission.

Further Reading

Brown, Andrew P. *The Neutron and the Bomb: A Biography of Sir James Chadwick*. Oxford: Oxford University Press, 1997.

Chandrasekhar, Subramanyan
(1910–1995)
Indian
Theoretical Physicist, Astrophysicist

Subramanyan Chandrasekhar laid the basis for modern astrophysics with his theories about the evolution of stars, which led to the concept of black holes. He was part of the pioneering generation that melded physics and astronomy into a dynamic, unified discipline. His career illustrates the formidable barriers faced by any physicist whose work represents a paradigm shift—a fundamental change in the way we view physical reality. Despite the public ridicule that greeted the theories he began developing in the 1930s, while still a student, Chandrasekhar's work was eventually recognized as fundamental to the understanding of how stars are born, live, and die. When he died in 1995, he had been awarded the 1983 Nobel Prize in physics for this work and was widely hailed as an astrophysicist who had forever changed the way we look at the universe.

The man known as Chandra to his friends and colleagues (the name means "moon" or "luminous" in Sanskrit) was born on October 13, 1910, in Lahore, in colonial India, now a part of Pakistan. His was a highly educated South Indian family, which would boast the only other Indian Nobel Laureate in physics, his uncle, SIR CHANDRASEKHARA VENKATA RAMAN. His father, C. S. Ayyar, was employed by the Indian railways, which transferred the family to Madras when Chandrasekhar was eight. The oldest boy in a family of three boys and five girls, he was home schooled by his parents and private tutors until the age of 12. Once enrolled in a regular school, he developed a passionate interest in mathematics. He was a precocious student, who entered Presidency College in India at the age of 15. There he met the English physicist ARNOLD JOHANNES WILHELM SOMMERFELD, who exposed him to the new quantum mechanics. Reading all he could find on this revolutionary theory, he came across a 1926 article, "On Dense Matter," by R. H. Fowler, a professor at Cambridge University, which would lead him to begin developing his first original ideas and to leave India in 1930, at 19, to study in England.

Fowler's article removed a daunting roadblock to scientific understanding of so-called white dwarfs or collapsed stars, that is, stars nearing the end of their lives. In the 1920s, the work of Arthur Eddington, the most eminent astrophysicist of the time, had led to a quandary known as the Eddington paradox. Eddington believed that as a star continued gradually to cool, by radiating heat into interstellar space, it must gradually shrink. The star must ultimately turn cold and then support itself not by thermal pressure, but rather by the only other type of pressure known in 1925: that found in solid objects such as rocks, which is due to repulsion between adjacent atoms. Such "rock pressure" was possible only if the star's matter had a density like that of a rock: a few grams per cubic centimeter. This, however, was 10,000 times less than the density of white dwarf stars like Sirius B! In order to reexpand to the lesser density of rock and thereby be able to support itself when it turned cold, a white dwarf star would have to do enormous work against its own gravity. Physicists knew of no energy supply inside the star adequate for such work.

Fowler resolved this paradox by replacing the physical laws Eddington had used with the new quantum mechanics. He ascribed the pressure inside Sirius B and other white dwarf, not to heat, but to a quantum mechanical phenomenon known as the degenerate motion of electrons or electron degeneracy. This degenerate motion is a consequence of a feature of matter that Newtonian physicists never dreamed of: the wave–particle duality. An electron inside the very dense matter of a white dwarf star has a short wavelength and accompanying high energy, which implies rapid motion. This means that the electron must fly around inside its cell,

behaving as an erratic, high-speed mutant: half-particle, half-wave. Physicists say that the electron is degenerate, and they call the pressure that its erratic high-speed motion produces electron degeneracy pressure. Fowler concluded that when a white dwarf like Sirius B cools off, it need not reexpand to the density of rock in order to support itself; rather, it continues to be supported quite satisfactorily by quantum degeneracy pressure at its own density of 4 million grams per cubic centimeter.

On his long sea voyage to England, Chandrasekhar applied the effects of ALBERT EINSTEIN's special relativity and the new quantum mechanics to Fowler's work. He calculated the limiting mass for collapsing stars to become white dwarfs as less than 1.44 solar masses. This value became known as the Chandrasekhar limit. If the mass of the star exceeds this limit, Chandrasekhar concluded, its gravity will overcome pressure inside the star, which will continue to collapse into a very dense object, which would later become known as a black hole. In Cambridge, Chandrasekhar refined his discovery. Yet, when he presented it in 1935 at a meeting of the Royal Astronomical Society, Eddington, who had been highly supportive of Chandrasekhar's research, criticized him in devastating terms. The older scientist's life work had demonstrated that all stars, regardless of their mass, had stable configurations and, in their final life stage, became white dwarfs. Chandrasekhar's contention that there was a limit to the mass of a star in its old age was anathema to him. Unfortunately, Eddington's credibility was far greater than that of Chandrasekhar, a young unknown and a foreigner to boot. Humiliated but still believing in his work, Chandrasekhar succeeded in having a number of famous physicists confirm his calculations. Nonetheless, decades would pass before the physics community would accept the Chandrasekhar limit and make it the basis for the theory of black holes.

After failing to find a position in England, where Eddington's influence prevailed, or in India, where he was the victim of academic infighting, Chandrasekhar was invited to work at the University of Chicago's Yerkes Observatory in 1937. Chandrasekhar gladly accepted but first returned to India to marry Lalitha Doraiswamy, who had been a fellow physics student at Presidency College. She would share his 58 years at the University of Chicago as a beloved teacher and researcher.

Eventually, the Chandrasekhar limit was universally accepted, even by Eddington, with whom he made peace, but the pain of that conflict with his former mentor was sufficient for him to abandon research on the black hole fates of massive stars for nearly 40 years. During this time he laid many of the foundations of modern astrophysics: the theories of stars and their pulsations, of galaxies, and of interstellar gas clouds, to name but a few. But his enduring fascination with the fates of massive stars led him to build upon the efforts of a younger generation of astrophysicists that, from 1964 to 1975, had created the "golden age" of black hole research. Black holes were found to be dynamic objects with enormous energies, whose gravitational and other properties, according to general relativity, could be predicted from just three numbers: the hole's mass, its rate of spin, and its electric charge. Only a few of these properties were known in 1975, when Chandrasekhar took up the exhilarating challenge of computing all the remaining ones, a task to which his formidable mathematical skills proved to be equal. Eight years later, at age 73, he completed his task and published *The Mathematical Theory of Black Holes—A Treatise,* which will be a mathematical handbook for black hole researchers for decades to come, enabling them to extract methods for solving black hole problems in general. That same year Chandrasekhar was honored for "his theoretical studies of the physical processes of importance to the structure and evolution of the stars" by the Nobel Prize in physics.

Four years after his death of a heart attack in 1995, at the age of 84, the National Aeronautics and Space Administration's (NASA's) Advanced X-Ray Astrophysics Facility (AXAF), which was launched and deployed in July 1999 by the space shuttle *Columbia,* was renamed the Chandra X-Ray Observatory in his honor. Referred to by astrophysicists simply as "Chandra," it combines high resolution, a large collecting area, and sensitivity to higher energy X rays, to study extremely faint sources, sometimes strongly absorbed, in crowded fields.

See also BROGLIE, LOUIS-VICTOR-PIERRE, PRINCE DE; WHEELER, JOHN ARCHIBALD.

Further Reading

Chandrasekhar, S. *An Introduction to the Study of Stellar Structure.* New York, Dover, 1989.

Srinivasan, G. ed. *Chandrasekhar: The Man behind the Legend—Chandra Remembered.* Chicago: University of Chicago Press, 2000.

Thorne, Kip S. "The Mystery of the White Dwarfs" and "The Golden Age," in *Black Holes and Time Warps: Einstein's Outrageous Legacy.* New York and London: W. W. Norton, 1994, pp. 140–163 and pp. 258–299.

Wali, Kamewashar C. *Chandra: A Biography of S. Chandrasekhar.* Chicago: University of Chicago Press, 1992.

———, ed. *From White Dwarfs to Black Holes: The Legacy of S. Chandrasekhar.* Singapore: World Scientific, 2000.

⊠   **Cherenkov, Pavel Alekseyevich**
(1904–1990)
Russian
Experimentalist, Particle Physicist

Pavel Alekseyevich Cherenkov discovered what came to be known as the Cherenkov effect or Cherenkov radiation, associated with the characteristic electromagnetic radiation that is emitted by charged atomic particles moving at velocities higher than the speed of light in a transparent medium. His discovery earned him a share of the 1958 Nobel Prize in physics and led to the invention of the Cherenkov detector, which proved to be of great importance for experimentation in nuclear physics and the study of cosmic rays.

Cherenkov was born on July 18, 1904, in the Voronezh region in the northern part of the Russian Empire. His parents, Aleksey and Maria Cherenkov, were peasants. After a youth marked by war and revolution, he studied physics and mathematics at Voronezh University and graduated in 1928. Two years later, he was appointed a senior scientist at the renowned Lebedev Institute of Physics in Moscow, in what had become the Soviet Union. That year, he married Marya Putintseva, the daughter of a Russian literature professor, with whom he would have two children, a son, Aleksey, and a daughter, Elena.

In 1934, the eminent physicist Sergey Vavilov assigned as his thesis work the study of what happens when the radiation from a radium source penetrates into and is absorbed in different fluids. Others had done this before Cherenkov and had observed what he observed: the emission of blue light from a bottle of water subjected to radioactive bombardment. In the past, however, physicists who had seen the bluish glow thought it was a manifestation of fluorescence (i.e., the absorption of energy by atoms, followed by a short-lived emission of electromagnetic radiation, as the particles move to lower energy states).

Cherenkov, however, was not persuaded that what he was looking at was fluorescence. To begin with, using experiments with distilled water, he found that the radiation continued after the exciting source had been removed; had he been looking at fluorescence, the radiation would have stopped. Investigating further, he showed that the light was caused by fast secondary electrons produced by the radiation.

Cherenkov was able to create the effect by irradiating the liquid with the electrons alone. He had discovered a new phenomenon, later dubbed the Cherenkov effect, in which a charged particle that is emitted in radioactive decay with velocity greater than the speed of light in the medium (which is smaller than the speed of light in a vacuum) generates a "shock wave" of photons in the blue light range. It is often compared to the sonic boom of a jet flying faster than the speed of sound.

Cherenkov published his findings in Russian periodicals between 1934 and 1937. Although his papers established the general properties of the newly discovered radiation, they did not provide a mathematical description of the effect. This was done by two of Cherenkov's colleagues, Igor Yevgenyevich Tamm and Il'ya Milhailovich Frank, with whom he would later share the 1958 Nobel Prize in physics. Tamm and Frank's mathematically rigorous formulation explained how a fast electron passing through a liquid could give rise to the type of radiation Cherenkov had observed. As does the bow wave of a ship that moves through the water with a velocity exceeding that of the waves, they explained, a charged particle passing through a medium with a velocity greater than that of the light in the medium creates the subatomic equivalent of a "bow wave," causing the medium to glow as the movement of the electrons outdistances the light. None of this contradicts ALBERT EINSTEIN's theory that nothing can exceed the speed of light, since this applies only to the speed of light in a vacuum or in empty space. In a medium such as water or a transparent solid, the speed of light is less than in a vacuum and varies with the wavelength.

Cherenkov radiation propagates as a cone whose opening angle depends on the particle speed. When this cone of radiation hits a flat surface, a characteristic circular ring of light is seen. In the 1950s, Cherenkov circular rings of light were photographed by Valentin Zrelov using proton beams at the Joint Institute for Nuclear Research (JINR), Dubna, near Moscow.

This work led to the development of the Cherenkov detector, an instrument based on the Cherenkov effect, capable of registering the passage of single particles through a transparent medium, if the particles have a velocity sufficiently high to exceed the speed of light in the medium. As a particle detector, the Cherenkov counter had the advantage of suitability for use as a device for detecting fast particles and determining their speeds, as well as distinguishing between particles of different speeds. It worked in the following fashion: for a given transparent medium, the light associated with Cherenkov radiation is emitted only in directions inclined at a certain angle to the direction of the particles' momentum. Thus, for a given transparent medium, by simply measuring the angle between the radiation and the path of the particles, physicists can measure the particles' speed. This new type of radiation detector became one of the most important instruments used in particle accelerators. It has become a standard piece of equipment in atomic research for observing the existence and velocity of high-speed particles, including cosmic rays.

Cherenkov was eventually made a section leader at the Lebedev Institute and received a doctoral degree from the institute in 1940. He was named professor of experimental physics in 1953 and taught at several institutes of higher learning for many years. In 1959, he was put in charge of the photomeson processes laboratory.

In the latter part of his career, Cherenkov worked on cosmic rays and on the design of large particle accelerators. He shared in the work of development and construction of electron accelerators and in investigations of photonuclear and photomeson reactions. He died on January 6, 1990.

Cherenkov's insightful discovery strikingly demonstrates how a relatively simple observa-

tion, thoroughly investigated and traced to its fundamental dynamics, can lead to important findings and open up new paths to research. The Cherenkov effect played a key role in the 1955 discovery of the antiproton. Cherenkov detectors are used in nuclear medicine and continue to be used extensively in modern high-energy particle accelerator experiments to determine the nature of unified particle theories and the structure of the early universe.

See also REINES, FREDERICK; RUBBIA, CARLOS.

Further Reading
Jelley, J. V. *Cherenkov Radiation and Its Applications.* Oxford, N.Y.: Pergamon Press, 1958.

⊠   **Chu, Steven**
(1948–   )
American
Experimentalist, Atomic Physicist, Biophysicist

Steven Chu is a brilliant and versatile experimentalist, who was awarded the 1997 Nobel Prize in physics, with Claude Cohen-Tannoudji and William D. Phillips, for his pioneering research in cooling and trapping atoms by using laser light.

He was born on February 29, 1948, in Saint Louis, Maryland, the middle son of Ju Chin Chu, an American-educated chemical engineer, and Ching Chen Li, an economist. With China in a state of political turmoil, in 1945 the Chus decided to start their family in the United States and eventually settled in Garden City, New York, near the Brooklyn Polytechnic Institute, where Ju Chin Chu taught. Although there were only two other Chinese families in the town, the Chus chose to live there because of the high quality of the public schools.

As a child he found joy in building plastic model airplanes and warships, eventually graduating to "constructing devices of unknown purpose where the main design criterion was to maximize the number of moving parts and overall size." Developing an interest in various sports, he taught himself to play tennis by reading a book.

Chu describes himself as the "academic black sheep" of the family, whose "mediocre" performance was no match for his older brother's. He found relief from the tedium of memorization in the problem solving approach of a geometry course and, during his senior year, in two well-taught classes in advanced placement physics and calculus. Encouraged to experiment, he capitalized on his years of experience in building devices and constructed a physical pendulum, designed to measure gravity. Chu has remarked, "Ironically, twenty-five years later, I was to develop a refined version of this measurement, using laser cooled atoms in an atomic fountain interferometer."

Rejected by the Ivy League schools, "because of my relatively lackluster A– average," he attended the University of Rochester, New York. There, he was inspired by a course that used *The Feynman Lectures in Physics*, which proved the decisive factor in his choice of physics over math as a major. When he graduated in 1970 with a B.S. in physics and an A.B. in math, his ambition was to become a theoretical physicist. But at the University of California, Berkeley, where he pursued his graduate studies, he soon realized he would be happiest as an experimentalist. Working with Eugene Commins, he defined an exciting experimental thesis topic: the search for a parity-violating effect in atomic transitions associated with the Z^0 particle, a neutral mediator of the weak force predicted by the electroweak theory, a unified theory of quantum electrodynamics and the weak interactions developed by STEVEN WEINBERG, ABDUS SALAM, and SHELDON LEE GLASHOW. Although Chu and his grad student colleagues were eventually scooped by a group at the Stanford Linear Collider, their observation of parity nonconservation in atomic transitions was nonetheless one of the earliest confirmations of the electroweak theory. After receiving a doctorate in 1976 and working for two years as a postdoctoral fellow at Berkeley, he was invited to join the physics faculty as an assistant professor. Fearful of inbreeding, however, the department allowed him to take a leave of absence before starting his own group at Berkeley. Chu saw this "as a wonderful opportunity to broaden myself."

In 1978, he joined the staff of Bell Laboratories in Murray Hill, New Jersey, where he was given the freedom to devote himself exclusively to research in an atmosphere permeated with "the joy and excitement of doing science":

> The cramped labs and office cubicles forced us to interact with each other and follow each other's progress. The animated discussions were common during and after seminars and at lunch and continued on the tennis courts and at parties. The atmosphere was too electric to abandon and I never returned to Berkeley.

At Bell, in 1982, Chu and Allen Mills did the first laser spectroscopy of positronium, an atom made up of an electron and its antiparticle (a positron). Physicists had long been trying to obtain a precise measurement of this most basic of atoms in order to test the predictions of quantum electrodynamics. Chu and Mills measured the 1s-2s energy level splitting of positronium to an accuracy of a few parts per billion. They went on to measure the corresponding transition in muonium, an atom consisting of a muon and an electron.

In 1983, when he became head of the quantum electronics research department at AT&T Bell Laboratories, Holmdel, New Jersey, Chu's interests had broadened considerably. During this period he accidentally discovered what he later called "a counterintuitive" pulse-propagation effect while using picosecond laser techniques to examine the possibility of using exci tons (quasi particles associated with solid state phenomena) as a means for observing metal-insulator transitions.

Then, in 1985, spurred by conversations with his colleague Art Ashkin, Chu led his group to the discovery of how first to cool and then to trap gaseous atoms with laser light. Since temperature is defined by the average random speed of the gaseous atoms, cooling the atoms reduces their average random speed in this gas phase. Employing a method called Doppler cooling, Chu's team used an array of intersecting laser beams to create an effect they called "optical molasses," in which the average random speed of target atoms was reduced from about 4000 kilometers per hour to about one kilometer per hour, as if the atoms were moving through thick molasses. When this method was used, the temperature of the slowed atoms approached –273.15 °C, the absolute zero of temperature. Chu considers the underlying dynamics of this process "quite simple":

> When an atom absorbs a photon from the laser, it absorbs the momentum of the photon. And every time it absorbs a photon, the velocity slows down by three centimeters a second. It then re-radiates the photon with no preferred direction; what this means is if you average over many photon absorptions, the atom constantly gets momentum kicks in the direction of the laser beam. . . . [T]he atom can absorb a photon and re-radiate it in about 30 nanoseconds. So, in the course of one second, you can get a pretty substantial force on the atom, something like 100,000 times the force of gravity. This means that over the course of a millisecond you can slow an atom to an ant's crawl.

However, since gravity will cause atoms to fall out of the optical molasses in about one sec-

ond, Chu and his coworkers had to find a way of actually trapping them. Once more using lasers and magnetic coils, they developed their magnetooptical trap (MOT) by creating a force greater than gravity, which drew the atoms into the middle of the trap to enable them to capture and study the chilled atoms. Phillips and Cohen-Tanoudji expanded Chu's work, devising ways to use lasers to trap atoms at temperatures even closer to absolute zero.

Atomic traps make it possible to study atoms with great accuracy and to determine their inner structure. They allow scientists to improve the accuracy of atomic clocks used in space navigation, to construct atomic interferometers that can precisely measure gravitational forces, and to design atomic lasers that can be used to manipulate electronic circuits at an extremely fine scale. In the future they may well play a key role in the construction of quantum computers.

After doing this major work, Chu felt the need to "spawn scientific progeny" and left his "cozy ivory tower" at Bell to join the faculty of Stanford University in 1987. Since then he has enjoyed working with a large number of "extraordinary" students on a wide range of problems. His group has demonstrated the first atomic fountain clock, which has exceeded the short-term stability of atomic clocks maintained by standards laboratories, and developed an atom interferometer, which has exceeded the accuracy of the most accurate commercial inertial sensors.

A year after his arrival, he developed the optical tweezer method of manipulating individual deoxyribonuclei acid (DNA) molecules. On the basis of Ashkin's demonstration that the atomic trap also works on tiny glass spheres embedded in water, Chu attached micrometer-sized polystyrene spheres to the ends of the DNA molecule. Using this technique of simultaneously visualizing and manipulating single bio molecules, his group has used single DNA molecules to examine various problems in polymer science. They discovered molecular individualism, which unexpectedly showed that identical molecules in the same initial state will choose distinct pathways to a new equilibrium state.

He is currently the Theodore and Frances Geballe Professor of Physics at Stanford University, where he pursues a wide variety of research interests in experimental atomic physics, quantum electronics, laser physics, biophysics, and polymer physics.

See also SCHAWLOW, ARTHUR LEONARD; TOWNES, CHARLES HARD.

Further Reading

Chu, Steven. "Laser Trapping of Neutral Particles," *Scientific American*, February 1992, p. 71.

Metcalf, Harold J., et al., eds. *Laser Cooling*. New York: Springer-Verlag, 1999.

⊠   **Clausius, Rudolf Julius Emmanuel**
(1822–1888)
Prussian
Theoretical Physicist (Thermodynamics)

Rudolf Clausius discovered the second law of thermodynamics, otherwise known as the law of entropy, which states that heat cannot flow spontaneously from a cooler body to a hotter one. His espousal of the mechanical theory of heat and his penetrating insights into the relationship between heat and work created the foundations of modern thermodynamics.

He was born in Koslin in Prussia, now Koszalin in Poland, on January 2, 1822, the sixth son in a large family. As a boy he attended a small local school run by his father, who was also a minister, before going on to the gymnasium in Stettin. He attended the University of Berlin, concentrating in mathematics and physics, and received his degree in 1844; after graduating, he taught advanced classes in those subjects at a gymnasium. He was awarded a Ph.D. from the University of Halle in 1848.

When Clausius began his research on heat and thermodynamics, the caloric theory was generally accepted. It was based on two axioms: (1) heat in the universe is conserved and (2) heat in a substance is a function of the state of the substance. Using JAMES PRESCOTT JOULE's experimental evidence, Clausius showed that both axioms were false and replaced them with two laws of thermodynamics. The first law, conservation of energy, states that the total energy of a system and its surroundings remains constant, even if it changes from one energy form to another. This means that heat and work are equivalent, since whenever work is done by heat, an equivalent amount of heat energy is consumed. Clausius then went on to formulate the second law of thermodynamics, in an 1850 paper, published in *Annalen der Physik*, that established the foundations of modern thermodynamics. Drawing on NICOLAS LÉONARD SADI CARNOT's and LORD KELVIN's (WILLIAM THOMSON) concept of the continuous dissipation of energy, he developed the concept of entropy, which is a measure of disorder and of the extent to which energy can be converted into work. The greater the entropy, the less energy is available for work. Clausius identified two sources of entropy: the conversion of heat to work and the transfer of heat from high to low temperatures, which is the normal behavior of heat and produces positive entropy. He rejected the opposite process—the spontaneous flow of heat from low to high temperatures (which would produce negative entropy)—as contrary to the normal behavior of heat. Since the change in entropy could be only zero, in a reversible process, or positive, in an irreversible process, he concluded that entropy inevitably increases in the universe.

After publishing the 1850 paper that established his reputation, Clausius took a teaching position at the Royal Artillery and Engineering School in Berlin. He would stay there for five years, before leaving his beloved Germany for Switzerland, where he had been offered a double post—as professor of physics at the Zurich Polytechnic and at the University of Zurich. There he pursued his work on the laws of thermodynamics, over the next 15 years, publishing eight more papers, in which he refined his ideas and mathematical formulations. In his 1865 paper he elegantly defined the first and second laws as follows: (1) the energy of the universe is constant; (2) the entropy of the universe tends toward a maximum. He improved the mathematical treatment of HERMANN LUDWIG FERDINAND VON HELMHOLTZ's law on the conservation of energy and jointly formulated the Clausius–Clapeyron equations, which describe the relationship between pressure and temperature for matter undergoing a change of state.

While in Zurich, Clausius began to branch out in other directions as well. He studied the relationship between thermodynamics and the kinetic theory of gases, which JAMES CLERK MAXWELL and LUDWIG BOLTZMANN had developed, and made numerous contributions to that theory. At the same time, he delved into the theory of electrolysis, correctly proposing that an electric current could induce the dissociation of materials.

In 1859, he married a German woman, Adelheid Rimpan, who would bear him six children. Enjoying the stimulation of Zurich's distinguished scientific community, Clausius more than once refused positions in Germany. Not until 1867 did he succumb to homesickness and accept a position as professor of physics at the University of Warzburg. Two years later, he would exchange this position for the physics chair in Bonn.

The serenity of his academic life would soon be shattered by the outbreak of the Franco-Prussian War in 1870. A German patriot, Clausius was not one to shirk his duties. He organized a volunteer ambulance service, working side by side with his students on the major battlefields of the war. While carrying out his mission of mercy, which earned him an Iron Cross, Clausius received a serious leg wound, which would cause him pain for the rest of his life. During this period, his wife died in childbirth, leaving him sole responsibility for their large brood. Clausius appears to have coped bravely with these setbacks, taking up riding to strengthen his leg, raising his children warmly and conscientiously, and eventually remarrying. If he was understandably less productive in the later half of his career, he nonetheless continued to work until his final illness, holding his physics chair at Bonn until his death on August 24, 1888. His many honors included the British Royal Society's Copley Medal in 1879.

A brilliant theorist, Clausius was instrumental in establishing theoretical physics as a recognized discipline. His seminal work led to important developments in thermodynamics, statistical mechanics, and the kinetic theory of gases.

See also HELMHOLTZ, HERMANN LUDWIG FERDINAND VON.

Further Reading

Cardwell, D. S. L. *From Watt to Clausius: The Rise of Thermodynamics in the Early Industrial Age.* Cornell, N.Y.: Cornell University Press, 1971.

Dugdale, J. S. *Entropy and Its Physical Meaning.* London: Taylor & Francis, 1996.

Von Baeyer, Hans Christian. *Warmth Disperses and Time Passes: A History of Heat.* New York: Modern Library, 1999.

⊠ **Cockcroft, John Douglas**
(1897–1967)
British
Experimentalist, Nuclear Physicist

In collaboration with Ernest Walton, John Douglas Cockcroft built the first particle accelerator, which produced the first nuclear transformations by means of artificially accelerated particles, in 1932. For this achievement, which led to new insights into the properties of atomic nuclei, they shared the 1951 Nobel Prize in physics.

He was born in Todmorden, Yorkshire, on May 27, 1897, into a family that had been involved in cotton manufacturing for several generations. In 1914, he entered Manchester University, planning to study mathematics, when England became embroiled in World War I. Cockcroft promptly volunteered for war service and did not resume his studies until the end of hostilities, in 1918. After serving as a signals officer in the Royal Field Artillery, Cockcroft found himself drawn to the study of electrical engineering; he enrolled in Manchester's College of Technology, where he earned a master's degree in technical science in 1922. He served a two-year apprenticeship with an electrical company before enrolling at Saint John's College, at Cambridge University, where he returned to his study of mathematics, receiving a B.A. in 1924. The following year he married Eunice Elizabeth Crabtree, with whom he would have four daughters and a son.

Cockcroft's career blossomed at Cambridge when he went to work at the Cavendish Laboratory, assisting the great nuclear physicist ERNEST RUTHERFORD, who had attracted a group of bright young scientists to work with him. Among them was the visiting Russian physicist PYOTR LEONIDOVICH KAPITSA, who collaborated with Cockcroft on the development of intense magnetic fields and low temperatures. Cockcroft's most important collaboration, however, would be with Walton. By the late 1920s, Rutherford had performed the first artificial transformation of one element into another, by bombarding the nuclei of certain light elements with alpha particles and tracking the results in a cloud chamber. He found, however, that bombardment with alpha particles had its limits, because large nuclei repelled them without disintegrating. Cockcroft's electrical engineering expertise, unavailable to most nuclear physicists, helped him to find a way to overcome this barrier. Working with Walton, he constructed a voltage multiplier that built up a charge of 710,000 volts and accelerated protons in a beam through a tube containing a high vacuum. With this instrument they produced their first artificial transformation in 1932. Beginning with the transformation of lithium into helium, they confirmed the production of helium nuclei by observing their tracks in a cloud chamber. They went on to disintegrate other elements such as boron.

In 1929, Cockcroft was elected to a fellowship at Saint John's College. He later served in various capacities at Cambridge, including Jacksonian Professor of Natural Philosophy. When global war erupted once again, in 1939, Cockcroft went to work on the application of radar to coast and air defense problems, becoming head of the Air Defense Research and Development Establishment. In 1944, he took over the construction of the first nuclear reactor in Canada. When the war ended he returned to England and became the nation's first director of the Atomic Energy Research Establishment in Hartwell. He was knighted in 1948 and in 1959 became master at Churchill College, Cambridge. From 1954 to 1959, he was a scientific research member of the U. K. Atomic Energy Authority.

He received the Atoms for Peace Award in 1961 and became president of the Pugwash Conference, a gathering of distinguished scientists from around the world to express concern over nuclear developments. He died in Cambridge on September 18, 1967.

The voltage multiplier invented by Cockcroft and Walton was the prototype for the more advanced particle accelerators, such as the cyclotron, by means of which the world of subatomic particles would be revealed, in ever greater detail, over the course of the next century.

See also VAN DE GRAAFF, ROBERT JEMISON; WILSON, CHARLES THOMSON REES.

Further Reading
Wilson, Edmund J. *An Introduction to Particle Accelerators.* Oxford: Oxford University Press, 2001.

Compton, Arthur Holly
(1892–1962)
American
Experimentalist (Electromagnetism),
Particle Physicist

Arthur Holly Compton won the 1927 Nobel Prize in physics for his discovery of what came to be known as the Compton effect, a quantum mechanical explanation of the phenomenon in which high-energy electromagnetic waves such as X rays undergo an increase in wavelength after being scattered by electrons. This work, which was recognized as experimental proof that electromagnetic radiation is quantized, possessing both classical wavelike properties and particlelike photon properties, was a foundation of the new field of quantum electrodynamics. As a member of the Manhattan Project, Compton also played a major role in the development of the atomic bomb.

He was born on September 10, 1892, in Wooster, Ohio, into an educated, pious family. His father was a Presbyterian minister and professor of philosophy at the College of Wooster, his mother a Mennonite, dedicated to missionary causes, who had been the 1939 Mother of the Year. While choosing a science career, like his older brother, Karl, Arthur maintained the family tradition of religious dedication. He attended the College of Wooster, where he earned a bachelor's degree in 1913. He then went on to Princeton University for postgraduate study, earning a master's degree in 1914 and a Ph.D. in 1916. Here he devised an elegant method for demonstrating the Earth's rotation. In 1916, he married Betty Charity McCloskey, with whom he would have two sons. Then he spent a year as an instructor of physics at the University of Minnesota. His next position was as a research engineer with the Westinghouse Lamp Company in Pittsburgh. In 1919, he left for Cambridge University, England, as a National Research Council Fellow; there he worked with ERNEST RUTHERFORD at the Cavendish Laboratory on the properties of scattered gamma rays.

In these experiments he discovered that scattering X rays on a graphite block lowered their energy, thereby increasing their wavelength. This phenomenon was clearly unlike the scattering of classical waves, in which the wavelength remains the same when it bounces off a surface. Since the groundbreaking work of JAMES CLERK MAXWELL, the classical wave nature of electromagnetic radiation had been firmly established. Then, in 1905, ALBERT EINSTEIN announced his theory of the photoelectric effect: that light is simultaneously a wave *and* a particle he called a photon. In this context Compton conjectured that the X rays must be acting as light particles or photons, that is, transferring their energy to the graphite electrons with which they collide. Theoretically, Compton treated the X-ray scattering as being due to a collision between a quantum of the electromagnetic field (i.e., a photon) and an electron, the latter regarded as a free particle initially at rest. Then, conservation of energy and momentum required the photon to transfer energy and momentum to the electron, causing the wavelength of the photon to increase. Compton's discovery confirmed the particle nature of the photon and placed the quantum prediction of wave–particle duality on a firm experimental foundation. In the same set of experiments he also discovered the phenomenon of complete reflection of X rays and their total polarization, which led to a more accurate determination of the number of electrons in an atom.

In 1923, he moved to the University of Chicago as professor of physics, and four years later he shared the 1927 Nobel Prize with CHARLES THOMSOM REES WILSON, the discoverer of the cloud chamber.

In the 1930s, the emphasis of his research shifted to cosmic rays, whose nature physicists were then debating: are they a form of electromagnetic radiation that passes in undeflected straight lines from outer space through the Earth's magnetic field, as ROBERT ANDREWS MILLIKAN suggested? Or are they streams of charged particles, which *would*, therefore, be deflected along curved paths as they traversed the Earth's magnetic field, as the German physicist Walther Bothe believed? To attempt to settle the question, Compton carried out measurements at thousands of locations around the world and found conclusive evidence that cosmic rays must consist of charged particles.

In 1941, finding himself unable to embrace the pacifism of his mother's Mennonite creed, he agreed to serve as chairman of the National Academy of Sciences Committee to Evaluate Use of Atomic Energy in War, organized at the University of Chicago, which included ENRICO FERMI, Leo Szilard, and EUGENE PAUL WIGNER. Compton designed methods of isolating fissionable plutonium and, with Fermi, produced a self-sustaining nuclear chain reaction, which led to the development of the atomic bomb. He also played a role in the government's decision to use the bomb. He was responsible for the appointment of J. ROBERT OPPENHEIMER to head the atomic bomb project, and afterward he defended Oppenheimer's loyalty to the United States, when, during the postwar McCarthy era, he was investigated by the House Un-American Activities Committee.

In 1945, at the war's conclusion, Compton returned to Saint Louis as chancellor, a position he held until 1954, when he became professor of natural philosophy. He retired in 1961, when he became Distinguished Service Professor of Natural Philosophy at Washington University. He died on March 15, 1962, in Berkeley, California.

A prolific writer, he wrote several books on social and moral issues, including *The Freedom of Man* (1935), *The Human Meaning of Science* (1940), and *Atomic Quest—a Personal Narrative* (1956), in addition to his numerous technical publications.

Discovery of the Compton effect played an important role in the development of quantum electrodynamics, in which the electromagnetic field itself behaves according to the rules of quantum mechanics. The orbiting Compton Gamma Ray Observatory, named in his honor, was launched on April 5, 1991, and returned to Earth on June 5, 2000. It conducted experiments that relied on the intricate understanding of the interaction of matter and high-energy gamma rays that Compton discovered in his groundbreaking experiments.

See also BARKLA, CHARLES GLOVER; RÖNTGEN, WILHELM CONRAD.

Arthur Holly Compton discovered the Compton effect, a quantum mechanical effect in which high-energy X rays increase in wavelength after scattering by electrons. *(AIP Emilio Segrè Visual Archives)*

Further Reading

Allison, Samuel K. "Arthur Holly Compton." *Biographical Memoirs Nat. Ac. Sci*, 38: 81, 1965.

Compton, Arthur Holly. *Atomic Quest—a Personal Narrative*. Oxford: Oxford University Press, 1956.

———. *The Cosmos of Arthur Holly Compton*. New York: Knopf, 1967.

⊠   **Coulomb, Charles-Augustin de**
(1736–1806)
French
*Theoretician and Experimentalist
(Electromagnetism, Electrostatics,
Mechanics)*

Charles Augustin Coulomb pioneered the science of electrostatics and established the laws governing electric charges. In recognition of his achievement the unit of electric charge is named the coulomb.

He was born on June 14, 1736, in Angoulême, in southwestern France, into a wealthy family, active in legal and governmental circles. When Charles was a boy, the family moved to Paris, where he received a solid grounding in both the sciences and the humanities at the Collège Mazarin. He then entered the Engineering School in Mezières and graduated in 1761 as a military engineer with the rank of first lieutenant in the engineering corps.

Over the next 20 years he traveled extensively with the corps, working on projects involving engineering, structural design, fortifications, and soil mechanics. He spent eight years (1764–1772) on Martinique in the West Indies, when the island was attacked by both British and Dutch fleets, and took charge of building a new fort.

In 1773, he returned to France, to Bouchain, where he began to write important work on applied mechanics. In the first paper he delivered before the Académie des Sciences in Paris, he discussed the influence of friction and cohesion in problems of statics. At his next posting, in Cherbourg, he wrote a famous paper on the magnetic compass, which contained his first work on the torsion balance, a device he invented for measuring very small forces by the torsion (twist) they cause in a fiber or a wire. He was the first to show physicists how the torsion suspension could provide a method of accurately measuring extremely small forces. Sent to Rochefort in 1779, he continued his study of mechanics, using Rochefort's shipyards as laboratories for his experiments. He wrote a major work on friction, *The Theory of Simple Machines*, which won him the Grand Prix from the Académie des Sciences in 1781. In this classic work he virtually created the science of friction, extending knowledge of the effects of friction caused by factors such as lubrication and differences in materials and loads.

The recognition Coulomb received for this work enabled him to leave engineering projects behind him and to devote himself to physics. He was elected to the mechanics section of the Académie and moved to Paris, where he obtained a permanent position. Between 1785 and 1791, Coulomb wrote seven important treatises on electricity and magnetism and submitted them to the academie. In these treatises, influenced by the work of Joseph Priestley on electrical repulsion, he developed a theory of attraction and repulsion: that bodies with the same electrical charge repel each other, whereas bodies with opposite electrical charge attract each other. He examined perfect conductors and insulators (dielectrics) and concluded that a perfect dielectric does not exist in nature, since, above a certain limit, every substance conducts electricity. He treated the electrical force in a manner similar to the way SIR ISAAC NEWTON treated gravitational force, that is, as an action at a distance. Coulomb's major contribution was in electrostatics, the study of time-independent electric fields, in which he made extensive use of an adapted version of his torsion balance. Coulomb performed his experiment by using certain reasonable assumptions, namely, (1) that the electrical forces behave as if concentrated on a point and (2) that the dimensions of the bodies are small compared with the distance between them. Under these conditions he found that the force between two electric charges on these bodies was proportional to the product of the charges and inversely proportional to the square of the distance between their centers; this relationship became known as Coulomb's law. Coulomb also investigated the distribution of electric charge over a body and found that it is located only on the surface of the charged body and not in its interior.

In other work, Coulomb demonstrated the inverse square law of repulsion and attraction in like and unlike magnetic poles. He wrote a total 25 papers between 1781 and 1806, working closely with other eminent scientists of the time. His life was also filled with the preparation of hundreds of committee reports for the academie, with his engineering consulting, and with his multifaceted service to the French government. At different times, Coulomb found himself in charge of the education in public schools, hospital reform, care of the royal fountains, and administration of the water supply of Paris. From 1802 to 1806, as inspector general of public education, he was mainly responsible for setting up lycées (secondary schools) throughout France.

When, in 1789, the French Revolution led to the reorganization of many institutions, Coulomb retired from the engineering corps, and he withdrew to his country home to continue his research in 1791. He would return to Paris a few years later, when the abolished Académie des Sciences was replaced by the Institut de France in 1795. Coulomb became its president in 1801. The following year, he married Louise Françoise LeProust Desormais, who had already borne him two sons. Four years

later, on August 23, 1806, at the age of 70, he died in Paris.

Coulomb's work established electrostatics as an exact science. A century later Coulomb's law would become an important component of the unified theory of electrodynamics developed by JAMES CLERK MAXWELL.

Further Reading

Stewart, C., ed., *Coulomb and the Evolution of Physics and Engineering in Eighteenth Century France*. Princeton, N.J.: Princeton University Press, 1971.

⊠ **Curie, Marie**
(1867–1934)
Polish/French
Experimentalist (Radioactivity), Physical Chemist

Marie Curie was a brilliant and dedicated experimentalist, who, with her husband, Pierre, investigated the atomic process first observed by ANTOINE-HENRI BECQUEREL, which she named *radioactivity*, and discovered two new chemical elements, polonium and radium. She was not only the first woman to win a Nobel Prize—she won two, in physics and in chemistry.

She was born Marya Sklodowska on November 7, 1867, in Warsaw, Poland, the fifth and youngest child of Vladislav, a professor of mathematics and physics, and Bronislava, a pianist, singer, and schoolteacher. Poland was then under Russian dominance and the Sklodowskis were obliged to hide their strong feelings of Polish patriotism. Marya's childhood was marked by the death of her mother of tuberculosis, when Marya was 10. She graduated at the top of her high school class at the age of 15 and then worked for eight years as governess for the children of wealthy families. During this period, she never lost sight of her goals, studying mathematics and physics in her spare time and attending a clandestine university run by Polish professors in defiance of the Russian occupiers. With her salary, she helped her older sister, Bronya, pay her tuition at the Sorbonne (the University of Paris), where she was studying to be a physician. When Bronya in turn, after obtaining her medical degree in 1891, sent for her, the 24-year-old Marya promptly changed her name to its French form, *Marie*, and launched her studies of science and math at the Sorbonne. Although Bronya helped out, Marie led a Spartan existence in an unheated attic, where her diet consisted primarily of bread, butter, and tea. She managed to graduate at the top of her class in the spring of 1893, with the equivalent of a master's degree in physics. When granted a generous fellowship, she was able to continue her studies, earning a master's degree in mathematics the following year.

In 1894, she went to work for a French industrial society and, while looking for adequate laboratory conditions, met Pierre Curie, the laboratory director of the Municipal School of Industrial Physics and Chemistry in Paris. Eight years her senior and already established as a physicist, Pierre persuaded her to remain with him in Paris instead of returning to Poland. They married in 1895, just after Pierre had earned his doctorate and become a full professor. Their union would be fruitful on the personal level—their first daughter, Irène (later the scientist Irène Joliot-Curie), was born in 1897 and a second, Eve, in 1904—and spectacular on the scientific.

Marie was determined to pursue a Ph.D. and, for her doctoral research, decided to follow up on Becquerel's 1896 observation that the element uranium spontaneously emitted radiation. She discovered that the intensity of the radiation was in direct proportion to the amount of the uranium in her sample, and nothing she did to alter the uranium (such as combining it with other elements or subjecting it to light, heat, or cold) affected the rays. This led her to hypothe-

size that the rays were the result of something happening within the atom itself, which was due to a process she called radioactivity.

Next she tested minerals that contained uranium or thorium and found that pitchblende (a mineral that contains uranium) gave off four times as much radiation as would be expected from the amount of uranium it contained. This led her to believe that the mineral must contain other elements that also give off radiation. In April 1898, she published a paper announcing the radioactivity of thorium and speculating that an even more strongly radioactive element existed.

At this point Pierre abandoned his own physics research to collaborate with Marie on hers. The Curies embarked on what Marie would later describe as the happiest time in their life together, doing rigorous work, at their own expense, in a makeshift lab, lacking heat or ventilation. They focused their investigations on pitchblende because it emitted the strongest rays. Using a painstaking refining method in which tons of the material had to be refined to obtain a tiny sample of radioactive material, they quickly succeeded in isolating a substance from pitchblende that was 400 times more active than uranium. Marie called it polonium, in honor of her native Poland. They soon found a second, even more radioactive element, and called it radium.

Although the Curies announced their discovery on December 26, 1898, it was not until September 1902 that they finally produced 0.0035 ounce (0.1 g) of pure radium chloride—enough to confirm the existence of radium. When, in June 1903, Marie described this research to her doctoral committee, she had the pleasure of hearing that hers was the greatest contribution ever to be made by a dissertation. That fall she won the 1903 Nobel Prize in physics, which she shared with her husband and Becquerel. With the public imagination captured by their discovery, the Curies now enjoyed international renown and enough money to ease some of their financial burdens. The Sorbonne created a new professorship for Pierre in 1904 and promised to build an excellent laboratory for him and Marie.

This spiral of good fortune ended tragically when Pierre was killed on April 19, 1906, as he absentmindedly stepped in front of a horse-drawn wagon. With two small daughters to support, Marie found the strength to master her grief and persuaded the Sorbonne to hire her as its first woman professor. Two years later she was promoted to full professor. She independently continued the research she and Pierre had done together, setting out to refute the critics' claim that radium was not really an element, by producing pure radium and pure polonium. In 1911, after four years of exacting work, she succeeded in producing radium as a pure metal. That year she won a second Nobel Prize, this time in chemistry, for her discovery and isolation of radium and polonium.

By 1914, she was the head of two laboratories, one in her native Warsaw and one at the Sorbonne called the Radium Institute. Unable to continue her research during World War I, she supported the French war effort by organizing a fleet of wagons, which were called "little Curies," to carry portable X-ray equipment to battle sites. With her characteristic energy and dedication, she opened 200 X-ray stations that examined over a million soldiers. When the war ended, she campaigned during a 1920 tour of the United States and again in 1929, to raise money for a hospital and laboratory devoted to radiology, the branch of medicine that uses X rays and radium to diagnose and treat disease. She used her celebrity status to campaign for the Radium Institute and other causes she believed in, serving on the council of the League of Nations and on its international committee on intellectual cooperation.

As the 1920s drew to a close, a number of debilitating symptoms, including fatigue, dizzi-

Marie Curie discovered the phenomenon of radioactivity and two new chemical elements, polonium and radium. *(AIP Emilio Segrè Visual Archives, W. F. Meggars Collection)*

ness, low-grade fever, humming in her ears, and a progressive loss of eyesight, became Curie's constant companions. Although she was aware that many colleagues had suffered similarly and died, for a long time, she refused to attribute their deaths to the element she and Pierre had discovered—radium. When she did finally admit radium's role, she continued to work with it. When doctors discovered she had leukemia, they concealed the news from the public and from her. She succumbed to the disease on July 4, 1934, at the mountain sanatorium where she had gone to recuperate. Ironically, one of the enduring applications of her work has been in the treatment of cancer with radiation.

A devoted scientist, whom fame could not distract from the profound pleasures of her laboratory, she once said:

A scientist in his laboratory is not a mere technician: he is also a child confronting natural phenomena that impress him as though they were fairy tales.

Marie Curie lived to see her work give rise to the field of atomic physics. In 1995, her remains, together with her husband's, were enshrined in the Pantheon, the memorial to the nation's "great men," in Paris. She was the first woman to be so honored.

Further Reading

Curie, Eve. *Madame Curie*. New York: Doubleday, 1937.

Pasachoff, Naomi. *Marie Curie and the Science of Radioactivity*. Oxford: Oxford University Press, 1997.

Quinn, Susan. *Marie Curie: A Life*. New York: Simon & Schuster, 1995.

D

Davisson, Clinton Joseph
(1881–1958)
American
Experimentalist, Particle Physicist

Clinton Joseph Davisson shared the 1937 Nobel Prize in physics with George Thomson as the first to observe experimentally the wave nature of the electron, thereby confirming LOUIS-VICTOR-PIERRE, PRINCE DE BROGLIE's theory of the wave–particle duality of subatomic particles.

He was born in Bloomington, Illinois, on October 22, 1881, to Joseph Davisson, an artisan, and Mary Calvert, a schoolteacher. He attended Bloomington public schools and, in 1902, won a scholarship on the basis of his mastery of mathematics and physics to the University of Chicago. There he studied under ROBERT ANDREWS MILLIKAN, the first physicist to measure the charge of the electron, who was impressed with his abilities. For financial reasons, however, Davisson had to leave school and return to his hometown, where he began working for the telephone company. In 1904, Millikan came to his rescue by recommending him for an assistantship in physics at Purdue University. In June of that year, he was able to return to Chicago, where he remained until, in 1905, once more thanks to Millikan, he became a part-time instructor of physics at Princeton University. Continuing to study for his degree, he earned a B.Sc. in 1908, from Chicago. In 1911, a fellowship in physics at Princeton enabled him to complete a Ph. D. from Chicago, for his thesis "The Thermal Emission of Positive Ions from Alkaline Earth Salts." That same year, he married Charlotte Sara Richardson, the sister of his thesis adviser; they would have three sons and one daughter together.

From 1912 to 1917, Davisson was an instructor in the physics department at the Carnegie Institute of Technology in Pittsburgh. During this period he spent the summer of 1913 at the prestigious Cavendish Laboratory, Cambridge University, working under JOSEPH JOHN (J. J.) THOMSON, the discoverer of the electron. When the United States entered World War I in 1917, Davisson tried to enlist in the army but was rejected. He worked instead for the engineering department of the Western Electric Company, which later became Bell Telephone Laboratories, in New York City. Initially this was a war assignment, but Davisson decided to stay on, resigning his position as assistant professor at Carnegie.

While he was working on an experiment involving the reflection of electrons from metal surfaces under electron bombardment in April 1925, an accidental explosion caused a nickel target he was studying to become heavily oxidized. He removed the coating of oxide by heating the nickel. As he continued working, he made an intriguing observation: a change had taken place in the angle of reflection of electrons from the nickel surface. Davisson and his assistant Lester Germer sought a possible cause and finally attributed the change to the recrystallization of the nickel; it was probable, they conjectured, that in the reheating process many small crystals in the nickel surface had converted into several large crystals. The following year, however, while enjoying a "second honeymoon" with Charlotte in London, Davisson attended a conference that dealt with the new relativistic particle–wave theory of the electron developed by de Broglie. In his theory, the French physicist had found a simple relationship between the velocity of the particle and the wavelength of the "wave-packet" associated with this particle. The greater the velocity of the particle, the shorter the wavelength. If the velocity of the particle is known, it is then possible, by using de Broglie's formula, to calculate the wavelength, and vice versa. Once aware of de Broglie's revolutionary new theory that electrons have wave properties, Davisson no longer found the recrystallization hypothesis so persuasive. He knew that X-ray diffraction had already been observed in crystals. Was it not likely, he reasoned, that the effects he had observed were due to the diffraction of electron waves in the planes of atoms in the nickel crystals?

Upon his return to the United States, he and Germer devised an experiment using a simple nickel crystal. The atoms were in a cubic lattice with atoms at the apex of cubes, and the electrons were directed at the plane of atoms at 45 degrees to the regular end plane. Electrons of a known velocity were directed at this plane, and those emitted were recorded by an electron detection apparatus. In January 1927, the results they obtained indicated that for incident electrons of a certain velocity, electron diffraction occurred, producing outgoing beams that could be related to the interplanar distance. The wavelength of the beams could then be determined and used, together with the known velocity of the electrons, to confirm de Broglie's hypothesis.

Four months later, George Thomson, the son of J. J. Thomson, working independently at Aberdeen University, Scotland, with different experimental apparatus, made the same discovery. For their definitive confirmation of the particle–wave duality of subatomic particles, Davisson and Thomsom would share the 1937 Nobel Prize in physics.

From 1930 to 1937, Davisson focused on the theory of electron optics, seeking ways to apply it to engineering problems. He also studied the scattering and reflection of very slow electrons by metals. During World War II, he investigated the theory of electronic devices, as well as a series of problems in the field of crystal physics.

Davisson remained with Bell Telephone Laboratories for 29 years, retiring in 1946, when he became visiting professor of physics at the University of Virginia, Charlottesville. He retired in 1949 and died in Charlottesville, on February 1, 1958, at the age of 76.

Davisson's groundbreaking experiments, by confirming the de Broglie hypothesis that a quantum wave–particle duality is inherent in matter, opened the door to the new world of quantum mechanics and elementary particle physics.

See also LAUE, MAX-THEODOR FELIX VON.

Further Reading

Madey, Theodore E., and William C. Brown. *History of Vacuum Science and Technology: A Special Volume Commemorating the 30th Anniversary of the American Vacuum Society.* Philadelphia: ISI Press, 1984.

⊠ **Dirac, Paul Adrien Maurice**
(1902–1984)
British
Theoretical Physicist, Quantum Theorist, Particle Physicist

Paul Dirac ranks among the most original and creative thinkers in the history of physics. His most important contribution was the Dirac equation, which reconciled quantum mechanics and special relativity, expanded the quantum mechanical model of the atom to include spin and magnetism, and led to the prediction of antimatter. For these seminal achievements in quantum mechanics and particle theory he was awarded the 1933 Nobel Prize in physics.

Paul Adrien Maurice Dirac was born on August 8, 1902, in Bristol, England, the middle of three children born to Charles Dirac, who had emigrated from Switzerland, and Florence Hannah Holten, an Englishwoman from Cornwall. They did not make a particularly happy family. Charles Dirac, who taught French at a secondary school, insisted that only French be spoken at dinner, so that Paul was the only one to dine with him. When his older brother later committed suicide, Paul's estrangement from the father he held responsible was complete; he invited only his mother to his anointment as a Nobel Prize laureate. Paul suffered throughout his life from shyness in social situations and a verbal reticence that gave rise to a body of "Dirac stories." One of the most famous of these tells how he intended to turn down the Nobel Prize in 1933 in order to avoid the publicity. He changed his mind only when ERNEST RUTHERFORD assured him there would be more publicity if he refused the prize.

Dirac attended Merchant Venturer's Secondary School in Bristol, before enrolling at Bristol University, where he studied electrical engineering. He received a B.Sc. degree in 1921 and was accepted for graduate study at Cambridge. Financial problems forced him to postpone matriculation for two more years, during which he studied mathematics at Bristol. At Cambridge, he worked with Cambridge's leading theoretician, Ralph Fowler, who introduced him to Rutherford and NIELS HENRIK DAVID BOHR's work on the structure of the atom and supervised his work on statistical mechanics. He was an unusually prolific graduate student, publishing no fewer than 11 papers on statistical

mechanics and quantum theory. He earned a Ph.D. in 1926 for his thesis "Quantum Mechanics," in which he formulated a mathematically consistent general theory of quantum mechanics in correspondence with Hamiltonian mechanics. The following year he became a Fellow of Saint John's College. As Paul Dirac was establishing himself as a physicist of genius, his brother, Reginald, took his life. The tragedy seems to have only intensified Dirac's tendency toward social isolation.

He began to travel extensively, making numerous trips to the Soviet Union, as well as to Copenhagen and Göttingen, where he worked with the leading lights of quantum mechanics: Bohr, WOLFGANG PAULI, MAX BORN, and WERNER HEISENBERG. Both Heisenberg and ERWIN SCHRÖDINGER had just published rival theories, later to be proved mathematically equivalent, which provided a conceptual foundation for Bohr's quantum model of the atom. Dirac was disturbed by the nonrelativistic form of the Schrödinger–Heisenberg quantum theory and wanted to integrate it with Einstein's special relativity. In 1928, he succeeded in doing so with his renowned Dirac equation, which allowed him to formulate the relativistic theory of the electron. The Dirac equation, which is still used widely today, was both mathematically elegant and breathtakingly productive. It led to a more complex model of the electron by endowing it with an intrinsic quantized rotation property, known as spin, which generated an intrinsic magnetic field around the electron. Until Dirac, the unit of quantization was integer multiples of Planck's constant. The Dirac equation implied half-integers of Planck's constant for the intrinsic spin of the electron. It gave a quantitative explanation of the Compton effect, which describes the way photons and electrons collide and scatter off each other. It also accounted perfectly for certain anomalies in the spectrum of atomic hydrogen.

Two years later, in 1930, Dirac published the first edition of his classic book, *Principles of*

Quantum Mechanics. In its final chapters, he first applied the rules of quantum mechanics to the electromagnetic field. His rudimentary calculations would become the basis for what later became known as quantum electrodynamics (QED), the study of the quantum mechanical interaction of electrons and photons.

The following year his relativistic theory of the atom would yield an even more astonishing prediction: the existence of a positron or positive electron. Dirac's prediction was motivated by his discovery that the mathematics describing the electron contained twice as many states as were expected. He proposed that the positive energy states described the electron, and that the negative energy states described a particle

Paul Dirac discovered the wave equation for a spinning electron, which reconciled quantum mechanics and special relativity and ultimately led to the prediction of antimatter. *(AIP Emilio Segrè Visual Archives, Ramsey and Musprat Collection)*

with a mass equal to that of an electron but with an opposite (positive) charge of equal strength: that is, an antiparticle of the electron. When independent experiments by CARL DAVID ANDERSON and LORD PATRICK MAYNARD STUART BLACKETT, in 1932 and 1933, confirmed that a positron could be produced by a photon, Dirac was invited to share the 1933 Nobel Prize with Schrödinger. His discovery of the positron led to the prediction of other antiparticles, such as the antiproton discovered by EMILIO GINO SEGRÈ in 1955. Today the existence of antimatter—an antiparticle for all particles—is universally accepted.

With the recognition of his work, Dirac was invited, in 1932, to become Lucasian Professor of Mathematics at Cambridge, a post last held by SIR ISAAC NEWTON. He would retain this position until 1969. In 1934, he visited the Institute for Advanced Study at Princeton, where he met the eminent Hungarian physicist EUGENE PAUL WIGNER at a time when Margit Balasz, Wigner's sister, who lived in Budapest, happened to be visiting him. An acquaintance was struck up, and, in 1937, Paul and Margit were married. They had two daughters together. In addition, Paul adopted Margit's daughter and son from a previous marriage, one of whom, Gabriel Andrew Dirac, became a famous pure mathematician.

Dirac's work continued to yield important insights. In 1938, he published his famous paper on classical electron theory, which presented an elegant mathematical formulation of two issues not well understood at the time: mass renormalization and radiative reaction. He was led to yet another significant discovery when he noticed that particles with half-integral spins (i.e., the electron) that obeyed the Pauli exclusion principle also obeyed statistical rules different from those of particles with integer spin, such as the photon. Dirac worked out the statistics for these particles (now called fermions), only to learn that ENRICO FERMI had already done so. History has credited both men: Fermi–Dirac statistics continue to be useful in nuclear and solid state physics (in determining the distribution of electrons at different energy levels).

During World War II, as a consultant to a group in Birmingham working on atomic energy, Dirac worked on uranium separation and nuclear weapons. He retained his academic post in Cambridge until 1969, when he moved to the United States; in 1971, he became professor of physics at Florida State University.

In 1973 and 1975, Dirac gave a series of notable lectures in Leningrad on the large number hypothesis. He had published his first paper on this subject in 1937; now, in his later years, it once more absorbed his attention. The theory deals with pure, dimensionless numbers such as the ratio of the electrical and gravitational forces between an electron and a proton, which is 10^{39}. This ratio happens to represent the age of the universe expressed in terms of atomic units. If there is some meaningful connection between these two values, there must be a connection between the age of the universe and either the electric force or the gravitational force. This implies that the gravitational force may not be a constant but is decreasing at a rate proportional to the rate of aging of the universe. The large number hypothesis also points to the intriguing related fact that the number of particles in the universe, which is 10^{78}, is equal to the square of the age of the universe. This suggests that matter may be being continuously created.

In addition to his numerous, fundamental contributions to quantum mechanics, particle theory, and cosmology, Dirac bequeathed to physics his unique aesthetics: a conviction that beauty or mathematical elegance is inseparable from scientific truth. Subsequent generations of physicists have been inspired and challenged by his 1963 assertion in *Scientific American* that

> it is more important to have beauty in one's equations than to have them fit

experiment. . . . If one is working from the point of view of getting beauty in one's equations, and if one has really a sound insight, one is on a sure line of progress. If there is not complete agreement between the results of one's work and experiment, one should not allow oneself to be too discouraged, because the discrepancy may well be due to minor features that are not properly taken into account and that will get cleared up with further development of the theory.

Paul Dirac died in Tallahassee, Florida, on October 20, 1984, after a long illness.

See also COMPTON, ARTHUR HOLLY; FEYNMAN, RICHARD PHILLIPS; PLANCK, MAX ERNEST LUDWIG; SCHWINGER, JULIAN SEYMOUR.

Further Reading

Dirac, Paul A. M. *The Principles of Quantum Mechanics.* Vol. 27. Oxford: Oxford University Press, 1982.

Fraser, Gordon. *Antimatter: The Ultimate Mirror.* Cambridge: Cambridge University Press, 2000.

Kursunoglu, Behram N., and Eugene Wigner, eds. *Paul Adrien Maurice Dirac: Reminiscences About a Great Physicist.* Cambridge: Cambridge University Press, 1987.

Pais, Abraham, et al., eds. *Paul Dirac, the Man and His Work.* Cambridge: Cambridge University Press, 1998.

⊠ Doppler, Christian
(1803–1853)
Austrian
Theoretical Physicist (Acoustics, Optics)

Christian Doppler discovered the so-called Doppler effect, an apparent change in the frequency of a wave motion caused by relative motion between the source and the observer. It was quickly and easily verified with respect to sound waves, for which the frequency increases as source and observer are in motion toward one another and decreases as they are in motion away from one another. The Doppler effect would have its greatest impact, however, when applied to light waves by astronomers, who found in it an indispensable tool for measuring the velocities and distances of celestial bodies.

Christian Doppler was born in Salzburg, Austria, on November 29, 1803, into a family that had operated a successful stone-masonry business since 1674. He was heir to a family tradition that decreed the son would take over the family business, but for the frail Christian, a less physically arduous path seemed advisable. He progressed from primary school in Salzburg to secondary school in Linz, where his talent for mathematics was increasingly evident. Upon graduating, he attended the Polytechnic Institute in Vienna from 1822 to 1825; there he continued to excel in mathematics and other studies. He then returned to his native city, where he attended lectures at the Salzburg Lyceum and continued his studies privately while tutoring in physics and math.

By 1819, Doppler was back in Vienna, where he studied higher mathematics, mechanics, and astronomy at the university and graduated in 1833. Remaining in Vienna to work as assistant to the professor of higher mathematics and mechanics, A. Burg, he wrote his first papers on math and electricity. Doppler, who, at 30, was somewhat older than most assistants, began seeking a permanent academic post. In the Austro-Hungarian Empire of that time aspiring academics had to compete for vacant professorships (a term applied to secondary as well as university teaching positions) by taking state exams and giving demonstration lectures. Doppler entered several of these competitions, on both the secondary school and university levels, while supporting himself as a bookkeeper at a cotton-spinning factory.

When success continued to elude him, he decided to emigrate to the United States in 1835. But he did an abrupt about-face, just before setting out, when a job teaching mathematics at the Technical Secondary School in Prague was offered to him, two years after he had competed for it. Doppler accepted, glad to have any teaching position, even if it was not the one he dreamed of. He competed for a job teaching advanced mathematics in Prague, but the closest he came to realizing his ambitions was a four-hour-a-week teaching assignment at a technical college. He had married in 1836 and was grateful for the extra income. Not until 1841 would he obtain a professorship in mathematics at the State Technical Academy in Prague. At the same time, his research was producing exciting results.

On May 25, 1842, at a meeting of the Royal Bohemian Society of Sciences, Doppler read his paper "On the Colored Light of the Double Stars and Certain Other Stars of the Heavens." Here he first presented what came to be called the Doppler effect, the relation of the frequency of a source to its velocity relative to an observer. Doppler derived his principle by treating both light and sound as longitudinal waves in the ether and matter, respectively. He pointed out that sound waves from a source moving toward an observer reach the observer at a greater frequency than if the source is stationary, thus increasing the observed frequency and raising the pitch of the sound. Similarly, sound waves from a source moving away from the observer reach the observer more slowly, resulting in a decreased frequency and a lowering of pitch. In 1845, the first experimental test of Doppler's principle was made at Utrecht in the Netherlands. A locomotive was used to carry a group of trumpeters in an open carriage back and forth past a second group of musicians, who jotted down the pitch of the notes being played. The variation of pitch produced by the motion of the trumpeters verified Doppler's equations for the case of sound waves. He would later publish an improved version of his principle, which took into account both the motion of the source *and* the motion of the observer.

In the 1842 paper, Doppler hypothesized that his principle would apply to any wave motion, including motion of light waves, and predicted that this would be of particular value to astronomers.

Time would prove Doppler eminently correct. In 1848 ARMAND-HIPPOLYTE-LOUIS FIZEAU suggested that applying the Doppler principle to observed shifts in the spectral lines of stars would enable astronomers to determine their motion. Twenty years later, William Huggins applied this idea when he determined that Sirius was moving away from our solar system by detecting a small red shift, that is, a displacement of lines toward the red end of the visible spectrum, where wavelengths are longer. This is caused by the Doppler effect and indicates that the observed body is moving away from the Earth. Later observations found that the spectra of some astronomical objects show a blue shift, indicating movement toward the observer. Both red and blue shifts became known as Doppler shifts.

While making his important discovery, Doppler was finding his long-desired professorship, involving the examination of hundreds of students in different scientific and mathematical areas, more than he could handle. In 1844, with his health failing, he requested a leave. He would not return to teaching until 1846. The following year he eagerly accepted a professorship of mathematics, physics, and mechanics at the Academy of Mines and Forests in Banska Stiavnica. Only a year later political unrest throughout the Austro-Hungarian Empire forced him to uproot himself once more. That year he was elected to membership in the Academy of Sciences in Vienna and was awarded an honorary doctorate from the University of Prague. With a solid reputation as a researcher established, he then returned to Vienna and in 1850 became director of the new Physical Institute and professor of experimental physics at the Royal Imperial University of Vienna. This was the apogee of his career. His lung problems steadily worsened until, in November 1852, he traveled to Venice, seeking the healing influence of the warmer climate. His devoted wife, who had remained in Vienna with their three sons and two daughters, rushed to his side when it became clear that he was failing. She was with him when he died in Venice on March 17, 1853.

Doppler's legacy is most readily reflected in its impact on astronomy, in which, applied to the case of light waves, measuring the Doppler shift of celestial bodies remains a fundamental method of estimating their relative velocities and distances. When, in 1929, Edwin Hubble linked the velocity of a galaxy to its distance from Earth, it became possible to use the red shift to determine, not only the distances of galaxies, but, ultimately, the size and structure of the universe.

Further Reading
Eden, Alan. *The Search for Christian Doppler.* New York: Springer-Verlag, 1992.

⊠ **Dyson, Freeman**
(1923–)
British/American
Theoretician, Quantum Field Theorist, Mathematical Physicist

Freeman Dyson was a brilliant theorist and mathematician, who discovered the "Rosetta Stone," capable of harmonizing the relativistic quantum field theoretic language developed by JULIAN SEYMOUR SCHWINGER and SIN-ITIRO TOMANAGA with the space time diagrammatic language of RICHARD PHILLIPS FEYNMAN into a coherent theory of quantum electrodynamics (QED). In the latter part of his career, he has become famous for his speculative work on the possibility of life on other planets.

Dyson was born on December 15, 1923, in Crowthorne, Berkshire, England, the second child of Sir George Dyson, a gifted composer and conductor, who eventually directed the British Royal College of Music in London, and Mildred Lucy Dyson, a highly educated woman who had been trained as a lawyer. Dyson spoke of his upbringing by parents who began raising their family in their 40s as "more like being with grandparents in their own fashion. It was more intellectual than physical." Encouraged to explore the arts, Dyson wrote a futuristic novel, "Sir Phillip Roberts's Erolunar Collision," inspired by Jules Verne, when he was nine; its tale of a mission to the Moon that is aborted for lack of funding is a blend of science fiction and social satire. He was sent to Twyford, a boarding school, when he was nine and, by age 12, he had won first place in the scholarship exams to Winchester, which had the reputation of being the best mathematical school in England's public school system. He read popular books about ALBERT EINSTEIN and relativity and became increasingly obsessed with mathematics, teaching himself calculus and most of complex function theory from the *Encyclopaedia Britannica*.

After graduating from Winchester in the summer of 1941, with England already at war, he enrolled at Trinity College, Cambridge University, where England's greatest mathematicians were then to be found and where PAUL ADRIEN MAURICE DIRAC was the leading light in physics. In 1943, however, his education was interrupted by the British war effort. He was a pacifist in the Gandhi tradition and considered declaring himself a conscientious objector. But the courageous example of the people of his country in their struggle to survive inspired him to do his part. He allowed himself to be recruited into the Royal Air Force Bomber Command at High Wycombe, where he spent his time performing futile statistical studies on the safety and efficacy of the British strategic bombing campaign, which were ignored by the military bureaucracy.

His frustration and sense of impotence at his inability to minimize bomber losses would remain with him for many years.

At the war's conclusion, Dyson accepted a job as a demonstrator in mathematics at Imperial College, London, but subsequently left mathematics, which he had come to view as an intriguing game, for the "reality" of physics, in which he felt the true challenges lay. When he returned to his studies at Cambridge in 1946, he immersed himself in physics and earned a B.A. Feeling that the United States was the only place to pursue his new field, he decided to do his graduate studies at Cornell University, in order to work with HANS ALBRECHT BETHE. When he arrived at Cornell in 1947, he quickly became involved in a moment of high drama in the world of physics.

Bethe had just returned from the Shelter Island conference at which WILLIS EUGENE LAMB JR. had announced the observation of a highly significant experimental discrepancy from the predictions of Dirac's long-accepted theory, which physicists used to calculate the energy levels of the atom. In his experiment Lamb shone a beam of microwaves onto a hot wisp of hydrogen gas blowing from an oven. He found that two fine structure levels in the next lowest group, which should have coincided with the Dirac theory, were in reality shifted relative to each other by a certain amount (the Lamb shift). He measured it with great accuracy and later made similar measurements on heavy hydrogen. On the basis of this experimental discovery Bethe and other quantum theorists such as Schwinger, Feynman, and Tomonaga began to realize that what was missing from Dirac's theory was a proper interpretation of the unwieldy concept of the self-interaction of the electron, which by its very nature contained infinities, thus preventing a straightforward physical interpretation.

When the electromagnetic field is quantized, according to the rules of quantum mechanics, particles of light called photons are generated. At the heart of the quantum electro-dynamic process is the quantum exchange force through which different electrons interact by exchanging photons with each other; in this context an electron can also exchange a photon with itself. How were physicists to deal with this self-interaction? QED, as it was formulated in the mid-1940s, was not considered to be a rela-tivistically covariant formalism (i.e., it was not formally compatible with the rules of special rel-ativity). This lack of relativistic covariance pre-vented a unique mathematical interpretation of the physical effects of self-interaction.

Schwinger changed all this when he discov-ered a relativistically covariant form for QED. This enabled him to introduce the concept of renormalization, which allowed a consistent math-ematical interpretation of the self-energy infinities. On the physical level, renormalization implied that physical particles are surrounded by a cloud of "virtual particles," that is, ghostly particles that exist within the context of the uncertainty princi-ple, whose energy, momentum, and charge modify the physical appearance of the bare original parti-cle. In applying the method of renormalization Schwinger found that the self-energy infinities could be subtracted out. This led to a fully consis-tent relativistic theory of quantum electrodynam-ics, which explained the Lamb shift as due to the virtual particle modification of the Coulomb force between the electron and the proton in the hydro-gen atom. Using his new relativistically covariant QED formalism with renormalization, Schwinger was also able to calculate the anomalous magnetic moment of the electron.

In the midst of this revolutionary turmoil, Dyson arrived at Cornell as a graduate student with a glowing reputation as a mathematician. Since Bethe had been the first to calculate the Lamb shift theoretically, on the basis of a non-relativistic approximation to electron theory, he gave Dyson the problem of developing a more rigorous version of the Lamb shift in the context of a relativistic electron theory, which ignored

Freeman Dyson proposed the S-matrix theory of quantum electrodynamics (QED), which unified the relativistic quantum field theory of Julian S. Schwinger and Sin-Itiro Tomanaga with the space-time diagram theory of Richard P. Feynman. *(AIP Emilio Segrè Visual Archives)*

the electron's spin. Dyson found that an exact relativistic calculation could be carried out with-out impossible complication and gave a finite answer in agreement with Bethe's earlier approx-imate Lamb shift calculation.

While at Cornell, Dyson learned about Schwinger's new version of QED secondhand, from Victor Weisskopf. He was already forming a friendship with the young Richard Feynman, who had come up with an alternate approach to the problems besetting QED, radically different from Schwinger's but equally effective. He used space-time diagrams, easily visualized spacetime analogs of the complicated mathematical expressions needed to describe the quantum probabilities of the behavior of electrons, positrons, and photons. His idea of a diagrammatic approach to QED resulted in a highly effective computational scheme. Instead of quantum field operators, his fundamental building blocks were particle processes in space-time. Feynman's diagrams visualized the construction of the quantum mechanical probabilities associated with fundamental quantum processes in terms of the space-time trajectories of real and virtual particles. More importantly, this space-time diagrammatic formulation of QED had the great advantage of simplifying all of the intricate calculations needed by Schwinger and Tomonaga to predict such interactions in their formulation of QED.

Feynman's space-time diagrammatic approach to QED intrigued Dyson but seemed magical to him. It troubled him that Feynman was merely writing down answers instead of solving equations in the usual way. Dyson began to conceive his mission as synthesizing these rival theories of QED. In the summer of 1948, Dyson and Feynman drove cross-country together, becoming intimate friends. Dyson then spent the summer in Ann Arbor, listening to Schwinger lecture, and came away feeling that Schwinger's theory was "unbelievably complicated." At summer's end, he left Ann Arbor to continue his graduate studies at the Institute of Advanced Study at Princeton, where J. ROBERT OPPENHEIMER was director. On the 48-hour bus ride from Ann Arbor to Princeton, he had an epiphany—the answer to the problem he had been pondering all year. He wrote to his family that his work

> consisted of a unification of radiation theory, combining the advantageous features of the two theories put forth by Schwinger and Feynman. Now it happened that Schwinger and Feynman talk such completely different languages, that neither of them is able to understand properly what the other is

doing. It also happened that I was almost the only young man in the world who had worked with the Schwinger theory from the beginning and had also had long personal contact with Feynman at Cornell, so I had a unique opportunity to put the two together.

Without consulting Oppenheimer, he sent off a paper to *Physical Review* in which he synthesized Feynman's and Schwinger's work. This seminal 1949 paper, "The S-Matrix in Quantum Electrodynamics," formed the basis on which future physicists would devote themselves to problems of renormalization, doing calculations of staggering complexity. Dyson had found the mathematical common ground between Schwinger and Feynman by focusing his attention on the so-called scattering matrix, or S matrix, which described the probability of all of the possible quantum electrodynamic scattering interactions that could occur in spacetime. By using the S-matrix, Dyson derived Feynman's diagrams from Schwinger's more complex quantum field operator language. He did this by devising a graphical technique, which enabled him to show that there was a one-to-one correspondence between the S-matrix elements of his graphs and those of the Feynman spacetime diagrams. Thus, the Dyson graphs provided a means of accurately and unambiguously cataloging the arrays of probabilities corresponding to various Feynman spacetime diagrams. Dyson's formulation was more reliable than Feynman's and more usable than Schwinger's. Despite initial opposition by Oppenheimer and others, in January 1949, at the American Physical Society conference in New York city, the Feynman-Dyson method, as it came to be called, was enthusiastically endorsed. Dyson became an overnight celebrity and was offered half a dozen jobs.

In 1949, he returned to England, where he was awarded a prestigious Royal Society Warren Research Fellowship at the University of Birmingham, where he obtained the Ph.D. he had neglected to earn at Cornell. Returning for a year at Princeton, Dyson met and fell in love with Verena Haefeli. Then, after a brief period in Europe, in May 1950, he agreed to succeed Feynman as professor of physics at Cornell. Later that year he and Verena were married. In 1952, finding that he did not like the professorial life, he moved to the Princeton institute as a permanent member. He became a U.S. citizen in 1957, the same year his wife, with whom he had begun a large family, left him. With this cataclysm in his personal life, an era in his scientific life ended, as well. Dyson would never again work on QED or devote himself to the exploration of fundamental physics problems.

In the late 1950s, he took leave from Princeton in order to join the Orion Project research team, which was attempting to build a crewed spacecraft and send it to Mars. Dyson describes his Orion period as one of the happiest times of his life. The project, conceived at the General Dynamics Corporation by former Manhattan Project scientists who were eager to find peaceful uses for nuclear power, aimed to create a propulsion system that would allow human beings to explore the entire solar system. The proposed vehicle, which would have been propelled into space by several repeated nuclear explosions, never made it to the launch pad and was declared defunct in 1965. (The Nuclear Test Ban Treaty of 1963 outlawed it.) Dyson attributed its demise to scientific conservatism.

In the early 1960s, he became a member of the U.S. Arms Control and Disarmament Agency (ACDA) and took part in test ban negotiations. Later in the 1960s, he chaired the Federation of American Scientists, an organization founded in 1945 as the Federation of Atomic Scientists by Oppenheimer and Los Alamos colleagues to address the dangers of the nuclear age.

In addition he has long advocated the exploration and colonization by earthlings of the solar system and beyond and has studied ways of searching for intelligent life. He is well known for his theories about advanced civilizations being able to take a planet like Jupiter apart to build a star-bound biosphere (known in science fiction as a Dyson sphere).

Dyson has said that his real life began at age 45, when he began publishing a series of books interpreting science to the general public. These include *Disturbing the Universe*, an autobiographical account (1979); *Weapons and Hope*, reflections on nuclear disarmament (1984); *Origins of Life* (1985); *Infinite in All Directions* (1988); and *The Sun, the Genome, and the Internet*, his exploration of the most important technologies for the 21st century.

Dyson's great achievement in synthesizing the work of Schwinger and Feynman gave the physics community easy access to the calculational techniques of QED and thus was the key to innumerable future breakthroughs. A complex man and thinker, Dyson has defined his relationship to science differently at different times of his life. Calling himself "an artist with mathematical tools," he has likened the pleasures of doing mathematical physics with those of writing a novel, "where you as author have complete control over the characters . . . a self-contained world where you understand everything, the parts and the whole."

See also KUSCH, POLYKARP.

Further Reading

Dyson, Freeman. *Disturbing the Universe*. Cambridge, Mass.: Perseus Books, 2001.
———. *Imagined Worlds*. Cambridge, Mass.: Harvard University Press, 1998.
———. *The Sun, the Genome, and the Internet: Tools of Scientific Revolutions*. Oxford: Oxford University Press, 2000.
Schweber, Silvan S. *QED and the Men Who Made It: Dyson, Feynman, Schwinger, and Tomonaga*. Princeton, N.J.: Princeton University Press, 1994.

E

Einstein, Albert
(1879–1955)
German/American
Theoretical Physicist, Relativist,
Quantum Theorist, Philosopher of
Science

Albert Einstein's extraordinary life in physics was a quest for nothing less than to "know how God created this world," to uncover the fundamental, unifying laws of the universe. In this he succeeded more than any scientist before him, with the possible exception of SIR ISAAC NEWTON. His theories of special and general relativity grew out of the paradoxes facing physicists as the 20th century began: the cracks in the seemingly solid walls of the house of classical mechanics that Newton and his successors had built. Revolutionary ideas about the nature of space, time, and matter were in the air that turn-of-the-century physicists breathed, but only a daring leap of intuition and imagination would unify and transform them into a new vision of physical reality. This astounding insight into the workings of nature was the essence of Einstein's genius. Early in his career, it led him to discoveries that confirmed the existence of atoms, launched quantum mechanics, demolished classical Newtonian notions of absolute space and time, redefined gravitation, and revolutionized

cosmology. In Einstein's later years, however, the same sense of inner rightness that guided him to these discoveries led him to reject quantum mechanics and to search futilely for a theory that would unify electromagnetic and gravitational forces.

Little in Einstein's early years presaged his future greatness. Born in Ulm, in Wurttemberg, Germany, on March 14, 1879, into a Jewish family, he moved to Munich six weeks after his birth, when his father's business ventures failed. Late in speaking, he talked slowly, pausing to consider what he would say. When he was four or five, his father showed him a magnetic compass, which convinced him that there was "something behind things, something deeply hidden." As a boy, he sang hymns praising God, which he had composed himself, on his way to school. But at 12, reading a book on Euclidean plane geometry, his religious impulse took the form it would retain all his life: a sense of profound wonder before the natural universe. Young Albert did not extend this sense of awe to secular authorities, however. He was a rebellious, disruptive student at his Munich gymnasium, where successful students learned by memorizing and doing what they were told. Albert preferred to study at home and, at 15, quit school and rejoined his family, who had moved once again, in Italy. The following year he failed the entrance examination to the Swiss Federal Polytechnic in Zurich, Switzerland, and was obliged to beef up his math knowledge at a preparatory school in Arrau, Switzerland.

It was in Arrau, walking along the river, that Albert began the relentless questioning of nature's laws that he would pursue all his life. He asked himself an almost childlike question, What would he see if he were to chase a beam of light at the velocity of light? He knew that Newtonian physics would say that he *could* catch up with the beam of light and would then observe it as a spatially oscillatory electromagnetic field at rest. But experience told him that no such thing could ever be observed, an assertion verified by the great JAMES CLERK MAXWELL, whose famous equations for a unified electromagnetic field indicated that velocity is inherent in light. Albert's *gedanken experiment* (thought experiment) led him to a fork in the road: he must give up either Maxwell's equations or Newton's laws of motion. For the moment, he would live with this paradox.

The next year, he passed his entrance exams and began his studies at the polytechnic. When his physics professor, Heinrich Weber, an old-fashioned classical physicist, ignored the giants of electromagnetic field theory, Maxwell and MICHAEL FARADAY, Albert began skipping Weber's classes and reading physics on his own. When he was not hanging out in coffee houses, playing his violin, or spending time with one of the few women physics students, Mileva Maric, he conducted experiments in the polytechnic's laboratories, one of which ended in an explosion that almost cost him a hand. He managed to pass his exams and graduate in 1900 with unexceptional grades.

Having failed to impress his professors, who might have eased his way into the university system, he floundered at first, taking low-paying teaching jobs. When he was offered a well-paid post as a patent clerk in the Swiss Patent Office in Bern, in 1901, he grabbed it and became a Swiss citizen. By 1903, he was in a position to marry Mileva, who had, meanwhile, given birth to an illegitimate daughter, Lieserl, and entrusted her to the care of relatives in Serbia. Albert never knew his daughter, whose fate has remained a mystery. Over the next few years, Mileva and Albert would have two sons, Hans Albert and Edward. Albert would work as a technical expert at the patent office until 1909, years he would later remember as "my best time of all." In his friend and coworker Michael Besso he had an ideal sounding board for the physics theories he dreamed up in his spare time. It was Besso who steered him to the work of ERNST MACH, one of the few leading scientists to ques-

Albert Einstein's theories of special and general relativity demolished classical Newtonian notions of absolute space and time, redefined gravitation, and revolutionized cosmology. *(AIP Emilio Segrè Visual Archives, Bundy Library)*

tion the Newtonian paradigm that underlay the belief in an ether, the mysterious fluid pervading all of space, in which electromagnetic waves were said to propagate. To Mach, the ether was a "metaphysical obscurity." Observation, he insisted, was the only way for scientists to *know*. As a corollary, he held that space was no abstract *thing*, but an expression of interrelationships among events: "*All* masses and *all* velocities, and consequently *all* forces, are relative." Although Mach would later deny his role as the progenitor of relativity, his influence on the young Einstein was profound. By 1905, Albert's thoughts began to crystallize. That year he obtained a doctoral degree from the University of Zurich and published three papers—on Brownian motion, the photoelectric effect, and special relativity. Each unraveled a mystery that had stumped the best scientific minds.

Einstein's first paper had a major impact on the running debate between physicists of the atomist and energeticist schools. In 1827, Robert Brown had observed through a microscope the random motion of small particles in a fluid: the motion of the particles increases when the temperature increases but decreases if larger particles are used. Since then physicists had been trying to explain the phenomenon. To the energeticists, who rejected the concept of atoms and thought of all matter as continuous, Brownian motion, with its discrete bumps, was disturbing. Einstein only increased their consternation when he explained the phenomenon as the effect of large numbers of molecules bombarding the particles. His assumptions allowed him to predict the movement and size of the particles, values that were later verified experimentally by the French physicist JEAN-BAPTISTE PERRIN. Experiments based on this work were used to obtain an accurate value of the Avogadro number, which is the number of atoms in one mole of a substance, and the first accurate values of atomic size. Einstein had struck a decisive blow in favor of the theory that matter is composed of atoms.

In his second classic paper, Einstein addressed himself to a puzzle surrounding the so-called photoelectric effect, the ejection of electrons from the surface of a substance by radiation. In the classical theory of electromagnetism, light was viewed as a wave. Maxwell's equations for the electromagnetic field predicted that when light waves fall on a metal surface, the energy of the electrons that are ejected depends on the intensity as well as the frequency of the light. But the experiment that produced the photoelectric effect showed that the energy of the electrons ejected is quantized according to the frequency and not the intensity of the light.

In 1900 MAX ERNEST LUDWIG PLANCK, while studying blackbodies (objects that do not reflect surface light and are thus perfect emitters and absorbers of radiation at all frequencies), had discovered a formula that related the energy of the radiation to its frequency. This was his famous blackbody radiation law, which predicted that $E = hv$, where E is the energy, h is a number known as Planck's constant, and v is the frequency of radiation. He was led to the realization that a sound derivation of his law could only be based on the postulate that the energy of radiation is emitted and absorbed, not continuously, but in discrete packets, which he called quanta. Planck postulated that the material oscillators in the walls of the blackbody had units of energy that were quantized in terms of the frequency of light, but he did not quantize the light itself.

It was Einstein who took that step, generalizing from Planck's quantum postulate. Suppose, he said, the light itself is quantized according to its frequency. Light then would consist of particles, which he called light quanta or photons. He used Planck's constant as a way to determine the energy of these light particles, suggesting that the kinetic energy of each electron is equal to the difference in the incident energy and the light energy needed to overcome the threshold of emission. The equation expressing this is known as Einstein's photoelectric law. This work would earn him the Nobel Prize in physics 16 years later, in 1921. It would also mark the beginning of his long friendship with Planck, who disliked the quantum idea he himself had fathered as much as Einstein had. For years Einstein would struggle in vain to understand how the classical Maxwell's equations could consistently produce his light quanta. The key to this enigma would be found 30 years later by RICHARD PHILLIPS FEYNMAN, JULIAN SEYMOUR SCHWINGER, and others, in the form of quantum electrodynamics, a theory that Einstein never accepted.

Although solving the enigma of Brownian motion and discovering that light has a particle property were no small achievements, they paled next to the sweeping discoveries enunciated in Einstein's third 1905 paper, "On the Electrodynamics of Moving Bodies," which unveiled the theory of special relativity. Nineteenth-century physicists had amassed a growing body of knowledge indicating that the behavior of light and other electromagnetic radiation regularly contradicts classical Newtonian physics. Since 1881, they had been living with the results of the Michelson–Morley experiment, which conclusively demonstrated that the velocity of light is constant and does not vary with the motion of either the source or the observer. Designed to measure the effect of the ether, the mysterious medium pervading the universe, in which light waves were thought to be propagated, the Michelson–Morley experiment appeared to disprove the ether's very existence. A number of theories arose, attempting to rescue this pillar of the Newtonian universe, including the brash idea, arrived at independently by HENDRIK ANTON LORENTZ and GEORGE FRANCIS FITZGERALD, that objects moving through the ether contract slightly in the direction of their motion and thereby hide the effect of the change in the velocity of light.

As for Einstein, the lack of an ether was not a problem but an opportunity to propose the central tenet of special relativity: since there is no frame of reference against which absolute motion can be measured, all motion can only be measured as relative to the observer. He further proposed that the velocity of light is constant and does not depend on the motion of the observer. From these two postulates and some elementary algebra, he concluded that when an object is in uniform motion relative to an observer, length decreases and time slows by the amount postulated by Lorentz, while the inertial mass of particles increases. At ordinary velocities, the magnitude of these effects is negligible and Newton's laws still apply. But at velocities approaching that of light, they become substantial. To quantify this strange state of affairs, Einstein found that he could use the Lorentz transformations, a group of equations that mathematically predict the increase of mass, shortening of length, and dilation of time for an object traveling at near the speed of light, while the velocity of light is always the same. Many years later, Einstein's conclusions on time dilation, length contraction, and mass increase would be confirmed by observations of fast-moving subatomic particles and cosmic rays.

Since an object moving at the velocity of light would appear to an observer to have zero length and infinite mass, while time would stand still, Einstein concluded that an object cannot move at a velocity equal to or greater than the velocity of light. Answering the question he had asked himself in Arrau, at age 16, he concluded that one could never catch up with a beam of light. Maxwell's equations, which stated that the velocity of light was constant, were correct, whereas the ether and Newton's absolute space were superfluous. There was no need for any preferred universal frame of reference. What matter are observable events, and no event can be observed until the light that communicates it reaches the observer. If space and time had been the absolutes in Newton's universe, the speed of light was the only absolute in Einstein's. Indeed, no velocity greater than that of light has ever been detected.

In the process of exploring the further implications of special relativity, Einstein was led, in 1907, to the discovery that mass is equivalent to energy, which he expressed in what is undoubtedly the most famous equation ever written: $E = mc^2$, or energy equals mass times the square of the speed of light. Given the magnitude of the speed of light, Einstein's equation revealed that an enormous amount of energy is stored as mass. It would provide the key to both nuclear power and nuclear weapons, although such applications were far from Einstein's mind at the time.

Special relativity would undergo a vital evolution in 1908 when Einstein's former professor at the polytechnic, the brilliant mathematician Herbert Minkowski, expressed Einstein's ideas in terms of a geometric form: a four-dimensional continuum made up of three dimensions of space and one of time: *Minkowski space* or *spacetime*. If this interpretation helped make relativity acceptable to most physicists, it led Einstein himself to remark, "Since the mathematicians have invaded the theory of relativity, I do not understand it myself anymore."

With the lay public, not to mention a good many scientists, confounded by special relativity, recognition was not immediate. Einstein was brusquely rejected when he applied for an academic job at the University of Bern. But by 1909 his discoveries were gaining appreciation in the scientific community. Einstein was offered a junior professorship at the University of Zurich. As his reputation grew, in 1911 he became a full professor, first in Prague, and, in 1912, in Zurich. Then in 1914, just as war was breaking out in Europe, Einstein moved his family to Berlin, where he had been offered a post, without teaching duties, as director of the Kaiser Wilhelm Institute.

By this time, Einstein was nearing the end of his eight-year struggle to make the theory of relativity generally applicable by considering not only systems that are in uniform motion but also those that are accelerating. The special theory dealt with inertial mass, which is the resistance objects offer to change in their state of motion. Inertial mass is what you feel when you glide a bowling ball along the floor. Gravitational mass is what you feel when you lift it. Yet, for some reason, the inertial and gravitational masses of any given object are equivalent. Moreover, because they are equivalent, they cancel each other out. Since GALILEO GALILEI's discovery that an apple and a cannonball fall at the same velocity, scientists had been aware of this. But what had struck Galileo and Newton as mere coincidence appeared to Einstein in a different light.

He had begun his relentless pursuit of a relativistic theory of gravity in 1907. In his famous elevator *gedanken experiment*, he had shown that people in a sealed elevator would be unable to determine whether they were in a real gravitational field or just feeling the inertial forces due to acceleration. Einstein would later call this idea, known as the principle of equivalence, "the happiest thought of my life." Einstein reasoned that if the effects of gravity are identical to those of acceleration, gravitation itself might be regarded as locally equivalent.

In this context he investigated the effect of gravitation on light and in 1911 concluded that light rays would be bent in a gravitational field. He realized, however, that if gravitation were a form of acceleration, the new theory would require something other than the four-dimensional Euclidean geometry that underlay the spacetime continuum of special relativity. Moreover, to prevent reintroducing the concept of the ether, the mathematical formalism had to be expressed in a covariant manner, that is, one in which the physical laws seen by observers are not tied to a preferred frame of reference. In 1912, Einstein enlisted his friend Marcel Grossman to find a way of expressing the new theory in terms of a new mathematical language called covariant tensor calculus.

Despite many setbacks, by 1915, Einstein had developed these ideas into the general theory of relativity, which states that masses distort the structure of spacetime and that this distortion produces the effects of gravitation. Matter curves space, and what we call gravitation is only the acceleration of objects as they fall along their trajectories in curved spacetime. The ability of this new formulation to account for a phenomenon that Newton's theory could not—the anomalous part of the perihelion in Mercury's orbit—convinced Einstein himself that his theory was correct. He had only a handful of supporters, however, during the chaotic years of World War I. When not involved in scientific research, Einstein was embroiled in pacifist activities and a bitter separation from Mileva, who returned to Zurich with their two sons. When his divorce was finalized in 1919, Einstein married his cousin and longtime mistress Elsa Einstein.

That same year, his theory of general relativity would achieve spectacular recognition. The English astronomer Arthur Eddington traveled to Brazil to observe a solar eclipse, hoping to confirm general relativity's prediction that the apparent position of stars would shift when they are seen near the Sun because its intense gravity would bend the light rays from the stars as they pass the Sun. When Eddington announced that the apparent bending of the light rays seen from the stars was offset to just the degree predicted by general relativity, Einstein became an overnight international celebrity. General relativity was further confirmed in 1925, when its prediction that a red shift is produced if light passes through a strong gravitational field was observed.

General relativity gave an exciting new answer to the age-old question, What exists beyond the edge of the universe? Since gravity is linked to the geometry of a curved four-dimensional space-time, the universe can be both infinite in four dimensions and bounded in three, to those who observe it from a three-dimensional universe. Since matter warps three-dimensional space, the total of the mass in all the galaxies in the universe may be sufficient to cause three-dimensional space to close around them.

Despite burgeoning recognition of his relativity theory, the conservative Nobel Prize committee awarded Einstein the 1921 prize for his work on the photoelectric effect. Einstein would spend the next decade traveling widely—within Europe, to the United States, to Palestine, and to South America—explaining his theories to mostly rapt audiences. The 20s was also the time when Einstein would commence his lifelong debate with the adherents of the new quantum mechanics. As early as 1909 Einstein had pointed to the need to reconcile the particle and wave theories of light. It would be LOUIS-VICTOR-PIERRE, PRINCE DE BROGLIE, who in 1923, using Einstein's mass–energy equation and Planck's quantum theory, would find a way to describe the wave nature of a particle. Whereas Einstein was supportive of de Broglie's work, he rejected its further development by ERWIN SCHRÖDINGER, WERNER HEISENBERG, NIELS HENRICK DAVID BOHR, and others, into a theory expressed in terms of probabilities, which essentially banished microscopic causality in spacetime. For Einstein the revolutionary, this was one revolution too many, signaling the end of physics itself. His meeting with Bohr in 1927 at the Solvay Conference marked the beginning of a famous series of debates between the two giants, in which Einstein would present a *gedanken experiment* designed to debunk quantum theory, which Bohr would then proceed to demolish. For Einstein, the great stumbling block was the principle of indeterminacy: "God does not play dice," he told Bohr, who replied, "Einstein, stop telling God what to do." To this day the conceptual and mathematical incompatibility of relativity and quantum mechanics has not been resolved.

With Hitler's ascendance to power in 1933, the heady period of discovery and unfettered debate in European physics came to an end. That year Einstein emigrated to the United States, accepting a position at Princeton University's Institute for Advanced Study, where he would remain for the rest of his life. His second wife, Elsa, died in 1936, and he never remarried. In 1939, with the Hungarian physicist Leo Szilard, he wrote to President Roosevelt, informing him that Hitler had the ability to build an atomic bomb, an admonition that led to the establishment of the Manhattan Project. When World War II ended, Einstein was tormented by his role in the development of the American atomic bomb and campaigned actively to abolish nuclear weapons. In 1952, Einstein, who had helped establish the Hebrew University of Jerusalem, was offered the presidency of Israel. He refused, believing he was temperamentally unsuited for the job. He eloquently described the two poles of his nature:

> My passionate sense of social justice and social responsibility has always contrasted oddly with my pronounced lack of need for direct contact with other human beings and human communities. I am truly a "lone traveler" and have never belonged to my country, my home, my friends, or even my immediate family, with my whole heart; in the face of all these ties, I have never lost a sense of distance and a need for solitude.

In the solitude of his later years, believing that "God is subtle, but he is not malicious," Einstein labored tirelessly to plumb that subtlety, seeking a theory that would unify gravity and electromagnetism. This time the mystery did not yield to him.

Albert Einstein died in Princeton on April 18, 1955, when an aneurysm in his abdominal aorta burst. He was cremated that day at 4 P.M. in Trenton, New Jersey. His ashes were scattered at a nearby river.

After Einstein, the mathematical language of relativity became the stuff of which all the laws of physics would have to be constructed. From then on, one of the key tests of any new physical law would be whether it could be written in a "relativistic form," capable of satisfying the fundamental structure of space and time that Einstein had revealed.

See also BOLTZMANN, LUDWIG; DOPPLER, CHRISTIAN; MICHELSON, ALBERT ABRAHAM; OPPENHEIMER, J. ROBERT.

Further Reading

Aczel, Amir D. God's Equation: Einstein, Relativity, and the Expanding Universe. New York: Dell, 2000.

Balibar, Françoise. Einstein: Decoding the Universe. New York: Harry N. Abrams, 2001.

Bernstein, Jeremy. Albert Einstein and the Frontiers of Physics. Oxford: Oxford University Press, 1996.

Bodanus, David. E = mc²: A Biography of the World's Most Famous Equation. New York: Walker, 2000.

Brian, Denis. Einstein: A Life. New York: John Wiley & Sons, 1997.

Calder, Nigel. Einstein's Universe. New York: Random House, 1990.

Einstein, Albert. The Expanded Quotable Einstein, Alice Calaprice, ed. Princeton, N.J.: Princeton University Press, 2000.

———. Ideas and Opinions. New York: Random House, 1988.

———. Out of My Later Years. Secaucus, N.J.: Carol Publishing Group, 1972.

Howard, Donald R., and John Stachel, eds. Einstein: The Formative Years 1879–1909. Boston: Birkhauser, 1986.

Overbye, John. Einstein in Love. New York: Viking, 2000.

Pais, Abraham. Subtle Is the Lord . . . the Science and the Life of Albert Einstein. Oxford: Oxford University Press, 1982.

F

Faraday, Michael
(1791–1867)
British
*Theoretical and Experimental Physicist
(Electromagnetism), Physical Chemist*

Michael Faraday was among the greatest 19th-century physicists to engage in the quest to understand electromagnetism and to harness electricity for human ends. His experimental work led him to discover electromagnetic induction and invent the electric motor, the electric generator, and the transformer. As a theorist, Faraday introduced the concept of force fields to explain the relationship of electricity and magnetism, replacing the time-honored idea of action at a distance. This new paradigm would dominate every aspect of modern physics.

Nothing in the circumstances of Michael Faraday's birth, on September 22, 1791, in Newington, Surrey, England, pointed to his illustrious future. His father, a poor blacksmith, would leave for London that same year to seek work. As a boy, Michael learned to read and write but was given little formal education and had a weak grounding in mathematics. When he was 14 he was apprenticed to a bookbinder and bookseller in London. It was there that his eclectic scientific training began. Encouraged to read, Michael devoured every book on science he could find in the shop and was soon performing his own crude experiments on electricity. Reading the works of the eminent French chemist Antoine Lavoisier sparked his fascination with chemistry. His bookbinding training helped him develop the manual dexterity that he would use to good purpose as an experimenter.

The young Faraday also took advantage of the educational opportunities London had to offer. In 1810, he was introduced to the City Philosophical Society, in which he received a basic education in science. He also began to attend the lecture series sponsored by the Royal Institution. In 1812, after listening to lectures presented by Sir Humphry Davy, the British chemist, Faraday managed to get himself hired as the great man's laboratory assistant. The following year, he accompanied him to France and Italy, where he was exposed to the latest advances in scientific research and attended lectures by some of Europe's most renowned researchers, including the pioneer of current electricity, ALESSANDRO GUISEPPE ANTONIO ANASTASIO VOLTA.

When he returned to London Faraday embarked on a 20-year period at the Royal Institution when he would make most of his pioneering discoveries in electricity and chemistry. At the beginning he concentrated on chemistry, developing a method for liquefying chlorine and isolating benzene, a compound now widely used in chemical products. He combined his interests in chemistry and electricity in the study of electrolysis, the production of chemical change through certain conducting liquids, that is, electrolytes, and in 1834 formulated what became known as Faraday's laws of electrolysis, which state that (1) the amount of chemical change produced is proportional to the charge passed, and (2) the amount of chemical change produced by a given charge depends on the ion concerned.

His great work in electromagnetism began in 1820, when the Danish physicist HANS CHRISTIAN ØRSTED discovered that a current in a wire deflects a magnetic needle. When Faraday repeated this experiment he found that a magnet also exerts a force on a wire carrying an electric current. Faraday explained the orientation of Ørsted's compass by stating that circular lines of magnetic force are produced around the wire, a hypothesis he demonstrated by devising an apparatus that would cause a magnet to revolve around an electric current. In 1821, Faraday showed how electric and magnetic forces could be converted into mechanical motion by positioning a current-carrying wire between the north and south poles of a horseshoe magnet. The interaction of forces made the wire rotate, thus creating a simple electric motor. He demonstrated the following dynamics: (1) If a current flows in a wire close to a magnet that is not fastened down, the magnet moves. (2) If a current flows in a wire close to a magnet that *is* fastened down, the wire moves. By showing that either the conductor or the magnet could be made to move, Faraday had demonstrated with stunning simplicity that electrical energy could be converted into motive force.

Faraday's fame grew, eventually eclipsing that of his mentor Davy, who felt that his own role in Faraday's discoveries had been ignored. In 1824, he was elected as a member of the Royal Society, England's premier scientific organization. The following year he began to popularize science through his lectures at the Royal Institution, an activity he would engage in throughout his career.

In 1831, Faraday embarked on the groundbreaking work that would lead him to the discovery of electromagnetic induction, that is, the production of electricity by means of varying magnetic intensity. He was not the first to demonstrate the phenomenon, but the first to understand its meaning. Seven years earlier, François Arago had found that a rotating nonmagnetic copper disk caused the deflection of a magnetic needle placed above it. Nobody at the time could explain what was happening. Then Faraday performed the following experiment: using a ring of soft iron wrapped in two windings of insulated wire, he connected one wire to a galvanometer (which measures electric current) and the other to a battery. At first it appeared that nothing had happened. But when the circuit was broken and reconnected, the galvanometer recorded the pulse of an electric current. Faraday realized that an induced current was produced while the intensity of the magnetized iron ring was rising or falling: that is, a changing magnetic field can induce a current. (At the same time another American physicist, JOSEPH HENRY, made the same discovery but could not spare time from his teaching to publish his findings.) The device Faraday constructed to demonstrate induction was the first transformer, or mechanism for changing the voltage supplied by an electrical current. Faraday made another great discovery when he realized that the motion of the copper wheel relative to the magnet in Arago's experiment caused an electric current to flow in the disk, which in turn set up a magnetic field and deflected the magnet. He built a similar device in which the current produced could be led off; this was the first electric generator.

Throughout the 1830s Faraday's string of discoveries showed no signs of abating. In 1832, he showed that the flow of electrostatic charge gives rise to the same effects as electric currents, thus proving that there is no basic difference between them. Then, in 1837, he investigated electrostatic force and demonstrated that it consists of a field of curved lines of force, and that different substances take up different amounts of electric charge when subjected to an electric field. This led Faraday to conceive of specific inductive capacity.

Faraday explained magnetism in terms of lines of force, which stretch out in all directions, flowing out of the north pole and into the south pole of a magnet. He knew little mathematics and found this concrete approach to the description of electricity and magnetism much more useful than the equations for the force between charges or currents in terms of an action at a distance. The field concept allowed the electric and magnetic forces to be clearly visualized and formed the basis for the mathematical description of electricity and magnetism, known today as electromagnetic theory.

Faraday suggested that the propagation of light through space consisted of vibrations in the field lines of electromagnetic force. He also found that when light passes through a medium, a magnetic field rotates the plane of polarization of the light. This is now known as the Faraday effect. His concepts of electric and magnetic fields were put into rigorous mathematical form a generation later by JAMES CLERK MAXWELL, who showed that light is, in fact, an oscillatory disturbance in the electromagnetic field lines, just as Faraday had predicted.

Another aspect of Faraday's work, which illustrates his prescience, was his suggestion that a link between gravitation and electromagnetism may exist. Such a link was observed 70 years later in a test of ALBERT EINSTEIN's general theory of relativity, when light rays passing near the Sun were found to be deflected.

By 1850, Faraday's years of brilliant scientific invention had ended. He abandoned research and then lecturing, retiring in 1858 on a small pension. Faraday was a deeply religious man, whose principles would not allow him to take part in preparation of poison gas for use in the Crimean War. Despite his fame, he remained modest, refusing to be knighted or receive honorary degrees. He died on August 25, 1867, in the Hampton Court apartment that had been given to him by Queen Victoria.

He is honored by the use of his name in the farad, the Système International d'Unités (SI) unit of capacitance, and the Faraday constant, the quantity of electricity needed to liberate a standard amount of substance in electrolysis.

Faraday's extraordinary ability to visualize and pictorially represent physical processes underlay his discoveries of basic laws, which led directly to the age of electricity. His special genius enabled him to visualize not only his experimental apparatus (the coil and the magnet), but also the invisible field that surrounds it and conveys the electromagnetic force. His introduction of the concept of a field of force may well be his greatest contribution to modern science.

Further Reading

Cantor, Geoffrey, David Gooding, and Frank A. James. *Michael Faraday*. Amherst, N.Y.: Prometheus Books, 1996.

Day, Peter, compiler. *The Philosopher's Tree: A Selection of Michael Faraday's Writings*. Bristol, England: Institute of Physics, 1999.

Keithley, Joseph F. *The Story of Electrical and Magnetic Measurements: From Early Days to the Beginnings of the 20th Century*. New York: Wiley-IEEE Press. 1998.

Ludwig, Charles. *Michael Faraday: Father of Electronics*. Scottdale, Pa.: Herald Press, 1978.

Russell, Colin Archibald. *Michael Faraday: Physics and Faith*. Oxford: Oxford University Press, 2000.

⊠ **Fermi, Enrico**
(1901–1954)
Italian/American
*Theoretician and Experimentalist,
Particle Theorist, Nuclear Physicist*

Enrico Fermi was the last great physicist to wear the double hat of theorist and experimentalist

Enrico Fermi developed the theory of radioactivity known as beta decay, which is based on a weak force acting within the nucleus of the atom. He also created the first known chain reaction in a nuclear reactor. *(AIP Emilio Segrè Visual Archives, Bainbridge Collection)*

before the explosion of knowledge after 1900 forced physicists to specialize. His early research in using slow neutrons to produce new radioactive elements won him a Nobel Prize in physics in 1938. His theory of radioactivity, known as beta decay, introduced the last of the four basic forces known in nature: gravity, electromagnetism, and, operating within the nucleus of the atom, the strong force and Fermi's weak force. His most famous achievement was the creation of the first known chain reaction in a nuclear reactor, which led to both the atomic bomb and the production of nuclear power for peaceful purposes.

Enrico Fermi was born in Rome on September 19, 1901, the youngest of the three children of Alberto Fermi, an administrator with the Ital-

ian national railroad, and Ida de Gattis. He discovered physics at age 14, when, stunned by the death of his older brother, he took refuge in reading from cover to cover two old volumes of elementary physics that he had chanced to find. He was a prodigy in secondary school, composing an essay for university admission that was deemed worthy of a doctoral examination. In 1918, he entered the elite Reale Scuola Normale Superior, associated with the University of Pisa. Fermi received his Ph.D. four years later, at age 21, for a thesis on X rays. His subsequent postdoctoral study at Göttingen University, where he worked under MAX BORN among such leaders of the quantum revolution as WOLFGANG PAULI and WERNER HEISENBERG, was perhaps the only time in his career when his own talents were eclipsed.

In 1924, Fermi accepted a position as lecturer in mathematics at the University of Florence. There he published a paper on the behavior of perfect gases, which spurred the physics department at the University of Rome to make him an offer; Fermi accepted, becoming a full professor of theoretical physics at age 25. Intent on reviving Italian physics, he set about attracting the best young minds to Rome.

He soon developed a statistical method for predicting the characteristics of electrons according to Pauli's exclusion principle, which states that no two subatomic particles can be described with the same quantum numbers. Fermi–Dirac statistics, as they came to be known, continue to be useful in nuclear and solid state physics, for example, to determine the distribution of electrons at different energy levels. Particles with half-integer spins, such as the electron, that obey Fermi–Dirac statistics, were later named *fermions*.

Fermi gained national recognition for his 1928 book *Introduction to Atomic Physics*, the first textbook on modern physics to appear in Italy, and the following year became the youngest member of the prestigious Royal Academy of Italy.

In the 1930s he engaged in experimental work that yielded some of his most important results and earned him an international reputation. In 1930, Pauli had proposed that the emission of an electron in beta decay is accompanied by the production of an unknown particle. Pauli's particle had neither charge nor mass; as a result, it had never been detected. In 1934, Fermi experimentally confirmed Pauli's hypothesis and dubbed the new particle the *neutrino*. That same year Frederik and Irène Joliot-Curie discovered artificial radioactivity, using alpha particle bombardment. Learning of their work, Fermi began producing new radioactive isotopes by neutron bombardment. He found that a block of paraffin wax or a jacket of water around the neutron source produced slow, or "thermal," neutrons, which are more effective in producing such elements. This work garnered him the 1938 Nobel Prize in physics.

Receiving the Nobel Prize proved to be more than an honor for Fermi; it saved his life. In 1928, he had married a Jewish woman, Laura, with whom he had two children. As Italian anti-Semitism grew during the 1930s, under Benito Mussolini's fascist regime, the Fermis' life became increasingly precarious. Taking the opportunity to leave Italy for the Nobel Prize ceremonies, they emigrated to the United States.

Their new home was in New York, where Fermi received a position at Columbia University. Together with two eminent Hungarian émigrés, Leo Szilard and EUGENE PAUL WIGNER, Fermi wrote a famous letter to Franklin Delano Roosevelt, warning him of the danger of Hitler's scientists' applying the principle of the nuclear chain reaction to the production of an atomic bomb. Roosevelt acted on the warning, and Fermi and Szilard soon found themselves working on the development of a nuclear reactor at Columbia.

Continuing his work on the fission of uranium induced by neutrons, Fermi constructed an elegantly simple machine that he called an "atomic pile." It had a moderator consisting of a pile of purified graphite blocks, to slow the neutrons, with holes drilled in them to take rods of enriched uranium. Other neutron-absorbing rods of cadmium, called control rods, could be lowered into or withdrawn from the pile to limit the number of slow neutrons available to initiate the fission of uranium.

In 1941, when the Manhattan Project consolidated its operations at the University of Chicago, Fermi and his team moved there and began building a nuclear reactor on the university's squash court. On December 2, 1942, the control rods were withdrawn for the first time and the reactor began to work, powered by the first self-sustaining nuclear chain reaction. It was the forerunner of the modern nuclear reactor, which releases the basic binding energy of matter.

Fermi moved to Los Alamos, New Mexico, and, with ARTHUR HOLLY COMPTON, led the team that constructed an atomic bomb, which used the same nuclear reaction process, but now without control mechanisms. This bomb produced a nuclear explosion in July 1945 at Alamogordo Air Base, New Mexico. Fermi used a simple experiment to estimate its explosive yield: he dropped scraps of paper both before the explosion and afterward, when the blast wind arrived, and compared their displacement. A few weeks later atom bombs were dropped on Hiroshima and Nagasaki, in Japan.

After the war Fermi, who had become a U.S. citizen in 1944, returned to the University of Chicago as Distinguished-Service Professor of Nuclear Studies. There he continued his research on the basic properties of nuclear particles, focusing on mesons, which are the quantized form of the force that holds the nucleus together. He also played a key role in the construction of the university's synchrocyclotron particle accelerator.

In 1949, he argued against U.S. development of the hydrogen bomb, calling it "a weapon, which in practical effect is almost one of genocide." The warning went unheeded. Fermi died prematurely in Chicago of stomach cancer on November 28, 1954, but his name and his achievements have been honored in many ways: element number 100, discovered in 1955, was named *fermium*. The Fermi is a unit of length used in atomic and nuclear physics, and the Fermi level is the energy level in a solid at which a quantum state would be equally likely to be empty or to contain an electron. Finally, the highly prestigious Fermi Award, which he himself received shortly before his death, was established in his honor.

See also DIRAC, PAUL ADRIEN MAURICE.

Further Reading

Cooper, Dan. *Enrico Fermi and the Revolutions in Modern Physics.* Oxford: Oxford University Press, 1998.

Fermi, Laura. *Atoms in the Family: My Life with Enrico Fermi.* Chicago: University of Chicago Press, 1994.

Segrè, Emilio. *Enrico Fermi, Physicist.* Chicago: University of Chicago Press, 1995.

⊠ **Feynman, Richard Phillips**
(1918–1988)
American
Theoretician, Quantum Field Theorist, Particle Physicist

Richard P. Feynman was the brilliant, charismatic architect of the modern space-time diagram formulation of quantum electrodynamics (QED), the theory that describes the quantum interaction of electrons with electromagnetic fields. For this work he shared the 1965 Nobel Prize in physics with JULIAN SEYMOUR SCHWINGER and SINITIRO TOMONAGA.

He was born on May 11, 1918, in New York City, the descendant of Russian and Polish Jews who had immigrated to the United States in the late 19th century. His father, Melville, had attended a homeopathic medicine institute but chose not to practice, engaging instead in a series of precarious business ventures. He strongly influenced his young son's approach to problem-solving activities, stressing that what mattered was not facts per se, but the process of finding things out. His mother, Lucille Phillips, was born in New York, into a family of prosperous, assimilated Jews, and attended the broadly humanistic Ethical Culture School. Feynman credited her for teaching him that "the highest forms of understanding we can achieve are laughter and human compassion." When Feynman was four or five, a baby brother died shortly after birth. His sister, Jean, was born when he was nine; she, too, would become a physicist. Feynman grew up in Far Rockaway, Queens, where he had excellent instruction in chemistry and mathematics in the public schools. In addition to impressive mathematical skills, he manifested an early mechanical aptitude: he started collecting old radios for use in his "personal laboratory" at age 10 and by age 12 was repairing his neighbors' radios.

As an undergraduate at the Massachusetts Institute of Technology in Cambridge, he wrote a thesis proposing an original and enduring approach to calculating forces in molecules. After graduating in 1939, he continued his studies at Princeton University, earning a Ph.D. in 1942 for a thesis directed by JOHN ARCHIBALD WHEELER, in which he put forth a theory of advanced and retarded electromagnetic interactions that travel backward and forward in time. This relativistic "action-at-a-distance" approach to electromagnetism attempted to replace the wave-oriented electromagnetic picture of JAMES CLERK MAXWELL with one based entirely on particle interactions mapped in space and time. He developed an approach to quantizing the theory governed by the principle of least action, a technique that minimized a quantity that was equal to the energy multiplied by the time associated with the physical processes. Feynman's method calculated all the probabilities of all the possible paths a particle could take in moving from one space-time point to another. His first lecture on this subject at Princeton was intriguing enough to attract ALBERT EINSTEIN, WOLFGANG PAULI, and John von Neumann to the audience.

Shortly after receiving his doctorate, over his parents' objections, he married his first love, Arlene Greenbaum, who was ill with tuberculosis. She was with him, in nearby sanitoriums, when, during World War II, he joined the Manhattan Project to build the atomic bomb, working first at Princeton, New Jersey, then at Los Alamos (1943–1945). When Arlene's health deteriorated rapidly and she died, Feynman buried his sorrow in research. He became the youngest group leader in the theoretical division, headed by HANS ALBRECHT BETHE, with whom he devised the formula for predicting the energy yield of a nuclear explosive. He was also in charge of the project's primitive computing effort, using a hybrid of new calculating machines and human workers to try to process the vast amounts of numerical computation required by the project. At Los Alamos, he flouted military discipline with a series of eccentric practical jokes, exposing the inadequate security by breaking into safes where atom bomb plans were stored. Present at the first test of an atomic bomb on July 16, 1945, at Alamogordo, New Mexico, he was initially euphoric but later anxious about the deadly force that he and his colleagues had unleashed.

From 1945 to 1950, Feynman was at Cornell University as professor of theoretical physics; there he returned to his research on formulating a consistent theory of QED. He was present at the Shelter Island conference in 1947 when the foundations of QED were shaken by two key experimental results. In 1947, WILLIS EUGENE LAMB, working in ISIDOR ISAAC RABI's laboratory at Columbia University, began experimental investigations that would have a profound impact on the formulation of QED. Applying the art of spectroscopy with unprecedented precision, he shone a beam of microwaves onto a hot wisp of hydrogen gas blowing from an oven. He found that two fine structure levels in the next lowest group, which should have coincided with the Dirac theory, were in reality shifted relative to each other by a certain amount (the Lamb shift). A second flaw in the Dirac theory was found that same year in the same Columbia lab, by POLYKARP KUSCH, who discovered a tiny discrepancy from what the theory predicted when he made highly accurate measurements of the magnetic moment of the electron. On the last day of the Shelter Island meeting, which he described 20 years later as the most important conference in his life, Feynman lectured on his spacetime approach to quantum mechanics and became stimulated once more to address the problems of QED.

After Shelter Island Schwinger, Tomonaga, and other quantum field theorists such as Bethe began to realize that what was missing from Dirac's theory was a proper interpretation of the unwieldy concept of the self-interaction of the electron, which by its very nature contained

Richard P. Feynman created the modern space-time diagram formulation of quantum electrodynamics (QED), the theory that describes the quantum interaction of electrons with electromagnetic fields. *(AIP Emilio Segrè Visual Archives, Photo by Christopher Sykes)*

infinities, thus preventing a straightforward physical interpretation. When the electromagnetic field is quantized, according to the rules of quantum mechanics, particles of light called photons are generated. At the heart of the quantum electrodynamic process is the quantum exchange force through which different electrons interact by exchanging photons with each other; an electron can also exchange a photon with itself.

How were physicists to deal with this self-interaction? QED, as it was formulated in the mid-1940s, was not considered to be a relativistically covariant formalism (that is, it was not formally compatible with the rules of special relativity). This lack of relativistic covariance precluded a unique mathematical interpretation of the physical effects of self-interaction. Schwinger changed all this when he discovered a relativistically covariant form for QED. In doing so, he introduced the concept of renormalization, which allowed a consistent mathematical interpretation of the self-energy infinities. On the physical level, renormalization implied that physical particles are surrounded by a cloud of "virtual particles," ghostly particles existing within the context of the uncertainty principle, whose energy, momentum, and charge modify the physical appearance of the bare original particle. In applying the method of renormalization Schwinger found that the self-energy infinities could be subtracted out. This led to a fully consistent relativistic theory of quantum electrodynamics, which explained the Lamb shift as due to the virtual particle modification of the Coulomb force between the electron and the proton in the hydrogen atom.

Feynman's intuitive space-time diagram formulation of QED was radically different from Schwinger's approach. By 1948, he had completed this construction. In the process he introduced the idea of Feynman diagrams, easily visualized spacetime analogs of the complicated mathematical expressions needed to describe the quantum probabilities of the behavior of electrons, positrons, and photons. His idea of a diagrammatic approach to QED, which stemmed from his earlier work with Wheeler on an action-at-a-distance formulation of electrodynamics, resulted in a highly effective computational scheme. Instead of quantum field operators, his fundamental building blocks were particle processes in spacetime. Feynman's diagrams visualized the construction of the quantum mechanical probabilities associated with fundamental quantum processes in terms of the spacetime trajectories of real and virtual particles. More importantly, Feynman's formulation of QED had the great advantage of simplifying all of the intricate calculations needed by Schwinger and Tomonaga to predict such interactions in their formulation of QED.

Feynman first presented his space-time diagrams at the 1948 Pocono conference attended by the elite of theoretical physics, who had just listened to Schwinger's elegant mathematical arguments. Prominent physicists such as NIELS HENRIK DAVID BOHR, EDWARD TELLER, and PAUL ADRIEN MAURICE DIRAC were mystified and unconvinced by Feynman's original and intuitive presentation. Later a new generation of physicists would see matters differently. FREEMAN DYSON, who had befriended Feynman at Cornell, called Feynman diagrams "this wonderful vision of the world as a woven texture of world lines in space and time, with everything moving freely . . . a unifying principle that would either explain everything or explain nothing." In 1949, Feynman sent off a paper with the full version of the diagrams, which discussed the "fundamental interaction": a ladderlike diagram that showed two electrons moving forward in spacetime interacting by exchanging a single "virtual" photon. By 1950, the relative simplicity of the Feynman diagrams began to overshadow the more intimidating operator QED methods of Schwinger and a flood of papers began to cite Feynman.

At first, Feynman diagrams seemed to be unrelated to the relativistic quantum operator field formulation of QED developed by Schwinger and Tomonaga. This perception changed dramatically when Dyson was able to demonstrate that Feynman's diagrammatic results and insights were directly derivable from the quantum probability matrix (called the S matrix) formulation of Schwinger and Tomonaga. Dyson was also able to use the Feynman diagrams to show that mass and charge renormalization removed all the infinities from the S matrix of QED to all orders of approximation. Further, he demonstrated how Feynman formalism made it easy to make predictions about observable phenomena associated with the anomalous magnetic moment of the electron and the Lamb shift. The Feynman diagrammatic method allowed Dyson to prove that renormalized QED was a consistent quantum field theory.

In 1950, Feynman accepted a professorship of theoretical physics at the California Institute of Technology, where he remained for the rest of his life. In the early 1950s, he provided a quantum mechanical explanation for the Soviet physicist LEV DAVIDOVICH LANDAU's theory of superfluidity: the bizarre, frictionless behavior of liquid helium at temperatures near absolute zero. In 1952, Feynman was a visiting professor at the University of Rio de Janeiro, Brazil, where he lectured on electromagnetism for 10 months, while preparing to parade in the carnival of a samba school in Copacabana. After this exhilarating period, he returned to the United States, and in June 1952 he married Mary Louise Bell of Neodesha, Kansas, whom he had met at Cornell. The marriage lasted only four years.

In 1958, Feynman contributed to the theory of nuclear interactions with MURRAY GELL-MANN. They devised a theory that accounted for most of the phenomena associated with the weak force, which is the force at work in radioactive decay. Their theory, which is based on the asymmetrical "handedness" of particle spin, has proved extremely productive in modern particle physics.

On September 24, 1960, Feynman married Gweneth Howarth, a native of Yorkshire, England, with whom he would have two children, Carl (born in 1962) and Michelle (adopted in 1968). In 1968, while working with experimenters at the Stanford Linear Accelerator on the scattering of high-energy electrons by photons, he invented a theory of *partons*, or hypothetical hard particles inside the nucleus of the atom, that helped lead to the theory of quarks.

Feynman's lectures at Caltech evolved into the books *Quantum Electrodynamics* (1961) and *The Theory of Fundamental Processes* (1961). Then, between 1963 and 1965, he published his classic textbook based on his introductory physics course, *The Feynman Lectures on Physics*. Two additional books, also distilled from his lectures, *The Character of Physical Law* (1965) and *QED: The Strange Theory of Light and Matter* (1985), contain his views on a range of subjects, including quantum mechanics, the scientific method, the relationship between science and religion, and the role of beauty and uncertainty in scientific knowledge.

After the explosion of the National Aeronautics and Space Administration's (NASA's) space shuttle *Challenger* in 1988, he was appointed to the council investigating the cause of the disaster. In typical Feynman fashion, he swept away bureaucratic constraints and identified the cause as the failure of an o ring in the unusually cold launch-pad temperatures by dunking a similar O ring in a glass of ice water in front of other committee members to emphasize his conclusion.

After a five-year battle against abdominal cancer, he died in 1988, at the age of 69. Colorful and iconoclastic, brilliant and unconventional in his thinking, Richard Feynman enjoyed a level of esteem within the physics community that was unique, even in a discipline given to mythologizing the great. His space-time diagram formulation of QED has remained valid, leading to far-reaching consequences. His invention of Feynman diagrams, a pictorial representation of the quantum probability for particle interactions in spacetime, adopted by the world of science as the basic language of quantum physics, continues to permeate many areas of theoretical physics in the 21st century.

Further Reading

Feynman, Richard. *The Meaning of It All: Thoughts of a Citizen Scientist*. Reading, Mass.: Addison Wesley, 1995.

———. *QED: The Strange Theory of Light and Matter*. Princeton, N.J.: Princeton University Press, 1985.

Feynman, Richard, and R. Leighton. *Surely You're Joking, Mr. Feynman: Adventures of a Curious Character*. New York: W. W. Norton, 1997.

Gleick, James. *Genius: The Life and Science of Richard Feynman*. New York: Pantheon Books, Random House, 1992.

Leighton, R. *Tuva or Bust! Richard Feynman's Last Journey*. New York: W. W. Norton, 1991.

Mehra, J. *The Beat of a Different Drum: The Life and Science of Richard P. Feynman*. Oxford: Oxford University Press, 1994.

⊠   Fitch, Val Logsdon
(1923–   )
American
Experimentalist, Particle Physicist

Val Logsdon Fitch won the 1980 Nobel Prize in physics, together with James W. Cronin, for the 1964 discovery of charge parity (CP) violations in the decay of neutral K mesons, one of the approximately 100 particles turned up by giant particle accelerators in the 1950s and 1960s. Fitch's experiment disproved the long-held theory that particle interactions should always be indifferent to the direction of time.

He was born on March 10, 1923, on a cattle ranch in sparsely populated Cherry County, Nebraska, not far from a Sioux reservation and the site of the battle of Wounded Knee. He was the youngest of three children born to Fred Fitch, who had acquired a ranch of four square miles by the time he was 20, and Frances Logsdon, a local schoolteacher. Fred Fitch could speak the Sioux language and was made an honorary chieftain in recognition of "his friendly interests on their behalf." Injured by a fall from a horse shortly after Val's birth, he was forced to leave the management of the ranch to others and move his family to nearby Gordon, where he became an insurance broker. His son would remember less the romance of ranching, than the hard, mundane work of keeping things in repair.

After attending public schools in Gordon, Fitch served in the United States Army during World War II and was sent to Los Alamos, New Mexico, to work on the Manhattan Project for the development of the atomic bomb. The experience proved a turning point, deflecting him from his original interest in chemistry. In his three years of working in a laboratory there, he had the chance to interact with such giants of physics as NIELS HENRIK DAVID BOHR, SIR JAMES CHADWICK, ENRICO FERMI, ISIDOR ISAAC RABI, and Richard Tolman, and to learn the basic techniques of experimental physics.

After the war, he completed the work for a bachelor's degree in electrical engineering at McGill University in Montreal in 1948. He then went to Columbia University for his graduate studies, at the time when its Nevis Laboratory Cyclotron was beginning operation. Working under Jim Rainwater, for his thesis he did pioneering experiments conducted at the Cyclotron on μ-mesic atoms and received a Ph.D. in 1954.

That year, he joined the faculty of Princeton University, where he pursued his latest interest, the decay properties of subatomic particles, concentrating on strange particles and K mesons.

For the next 20 years, most often working with a small group of students, he studied K mesons.

When Fitch began his work, there were three known symmetry principles in particle physics: (1) the principle of charge symmetry (C), which states that the laws of nature are exactly alike for both antimatter and matter; (2) the principle of parity symmetry (P), which states that fundamental laws have exact mirror symmetry; and (3) the principle of time-reversal symmetry (T), which states that fundamental laws do not change when all motions are reversed. T-symmetry is equivalent to the combination of charge and parity (CP) symmetries. The neutral K mesons are most suitable for testing the validity of the combination of charge and parity. All three symmetries separately exist in electromagnetic, strong, and gravitational interactions.

In 1957, the theorists TSUNG-DAO LEE and CHEN NING YANG and the experimentalist CHIEN-SHIUNG WU showed that left hand–right hand symmetry is violated in the weak interactions and hence that exact mirror symmetry (P) is violated in the weak interactions. The weak interactions also violated the symmetry principle associated with (C): that laws of nature are exactly alike for matter and antimatter. However, in the weak interactions, the two effects of C and P symmetry violations canceled each other out, thus preserving the CP symmetry, which was equivalent to the T symmetry. Time-reversal symmetry is valid for all processes governed by electromagnetic forces and is thus a cornerstone of chemistry. Similarly, T = CP symmetry is also presumed to be valid for the strong force, the weak force, and gravity.

Earlier, in 1955, MURRAY GELL-MANN and Abraham Pais had analyzed the neutral K-mesons and found that they had a new physical property called strangeness, which can make their decay properties ambivalent with respect to matter and antimatter. If perfect symmetry were to prevail, a K-meson would have to decay into antimatter in exactly half the cases and into matter in the other half. Lee, Yang, and Wu's discovery of P violation left open the possibility of CP = T violations.

In 1964, Fitch conducted experimental studies with Cronin at the Brookhaven National Laboratory to test whether the decay of K mesons could violate CP symmetry. In order to find out whether CP violating decay of K mesons occurred, they used the Alternating Gradient Synchrotron proton accelerator, to produce a beam of neutral K mesons. Then, using a specially designed large and complicated detector array, they recorded and measured with great precision the radioactive decay of the neutral K mesons in flight. The type of K meson they studied decayed into one-half ordinary matter and one-half antimatter. They found that two of a thousand of these K meson decays did in fact violate the CP symmetry.

Fitch and Cronin's experiment demonstrated that left hand–right hand asymmetry is not always completely compensated for by transforming matter to antimatter while maintaining left–right symmetry. They interpreted the results of their experiment, which were confirmed by a long series of subsequent experiments, as a small but clear lack of CP = T symmetry.

A few years after Fitch had done his groundbreaking work, his life was marred with tragedy: his wife, Elise Cunningham, with whom he had two sons, died in 1972. Four years after her death, he married Daisy Harper, who added three stepchildren to his life. That same year, he was made head of the physics department at Princeton. Fitch has stayed involved in elementary particle research and continues to express his belief that "the delights and challenges of unexpected discovery will continue always."

The results of Fitch's experiments forced physicists to abandon the long-held principle of time-reversal T invariance and explained why matter and antimatter did not immediately annihilate one another during the big bang origin of the universe. It is now believed that a net positive amount of matter was created in the big bang only because of a CP = T symmetry violation in the particle decay processes in the early universe. This allowed matter to gain an edge over antimatter.

See also DIRAC, PAUL ADRIEN MAURICE.

Further Reading
Wilson, Jane, ed. *All in Our Time: The Reminiscences of 12 Nuclear Physicists*. Chicago Bulletin of the Atomic Scientists, 1975.

⊠ **FitzGerald, George Francis**
(1851–1901)
Irish
Theoretical Physicist
(Electromagnetism)

George Francis FitzGerald was a gifted mathematical physicist best known for what is now called the Lorentz–FitzGerald contraction hypothesis. Developed independently by FitzGerald and HENDRIK ANTOON LORENTZ, the hypothesis was an ingenious attempt to explain away the negative result of the Michelson–Morley experiment, which was inconsistent with the existence of the ether.

FitzGerald was born on August 3, 1851, in Dublin, into a family of clerics and intellectuals. His father, William FitzGerald, was a minister in the Irish Protestant Church and rector at Saint Ann's in Dublin when George was born and would later become a bishop. The boy was tutored at home by the sister of the mathematician George Boole, who noted his strong mathematical aptitude, great inventiveness for mechanical constructions, and skills as an observer. When he was only 16, he entered Trinity College in Dublin, where he concentrated on mathematics and experimental science, as well as pursuing athletics and social activities. When he graduated four years later, in 1871, he was recognized

as the outstanding student in mathematics and experimental science.

For the next six years, he devoted himself to the study of mathematics and physics, in preparation for the competition for a fellowship from Trinity College. It was during this period, in 1873, that JAMES CLERK MAXWELL published his *Treatise on Electricity and Magnetism*. In this summary work Maxwell presented the four partial differential equations, now known as Maxwell's equations, that demonstrated that electricity and magnetism are aspects of a single electromagnetic field, and that light itself is a variety of this field. FitzGerald, as did Lorentz and HEINRICH RUDOLF HERTZ, with whom he exchanged ideas over the years, insightfully perceived that Maxwell's work provided a jumping-off point for further discoveries and began exploring directions for developing the theory.

FitzGerald published a paper on the electromagnetic theory of the reflection and refraction of light and sent it to Maxwell, who commented that Lorentz was developing his ideas in the same general direction. In 1883, on the basis of his own research, he predicted that a rapidly oscillating (alternating) electric current would produce electromagnetic waves, adding that he himself "did not see any feasible way of detecting the induced resonance." This would be left to Hertz, who in the late 1880s conducted his early experiments with radio. FitzGerald drew Hertz's work to the attention of the scientific community in Britain, announcing that Hertz had "observed the interference of electromagnetic waves, quite analogous to those of light."

By this time FitzGerald had won his fellowship, on his second try in 1877, and had become Erasmus Smith Professor of Natural and Experimental Philosophy at Trinity College in 1881. He would remain at Trinity for the rest of his life and do his most important work there. The path to this contribution, once more, originated with Maxwell, who, in the last year of his life, had suggested an experiment that would be carried out in the 1880s by ALBERT ABRAHAM MICHEL-SON and Edward Morley. Using an ingenious optical system capable of making extremely precise measurements, Michelson and Morley attempted to measure the ether, a hypothetical medium in which light waves were thought to propagate. The idea was to measure changes in the speed of light, on the basis of the interference patterns created when a light beam was split in two and the separated beams guided along perpendicular paths and then recombined. When no changes in the speed of light were observed, physicists scrambled to find a plausible explanation, that would not involve relinquishing their long-held belief in the existence of the ether.

FitzGerald proposed that the null result of the Michelson–Morley experiment could be explained by assuming that the length of an object moving through the ether contracts in the direction of the motion. He further suggested that light emitted by such an object does have a different velocity but travels over a shorter path and thus seems to have a constant velocity regardless of the direction of motion. Only an observer outside the moving system would be aware of the reduction in light velocity; within the system the contraction would also affect the measuring instruments and result in no change in the perceived velocity. He worked out a simple mathematical relationship to show how velocity affects physical dimensions. Lorentz independently arrived at the same idea in 1895, giving a more detailed description of what became known as the Lorentz–FitzGerald contraction hypothesis. Each man readily acknowledged the contribution of the other. In 1905, four years after FitzGerald's death, the contraction hypothesis was given a different interpretation and incorporated into ALBERT EINSTEIN's theory of relativity.

In addition to his research, FitzGerald was deeply involved in educational issues in Ireland. At Trinity, he waged a long-running battle to increase the amount of teaching of experimental physics and advocated expanding the role of the university to "teach mankind," rather than the elite few. He served as a commissioner of national education, working to reform primary education in Ireland by introducing more opportunities to do experimental work into the curriculum.

In 1883, he married Harriette Mary Jellett, daughter of the Reverend J. H. Jellett, provost of Trinity College. During the eight years of their marriage, the couple had eight children, three sons and five daughters. In 1900, FitzGerald experienced intestinal problems and, despite surgery, died, in Dublin, on February 22, 1901, at the age of 49.

FitzGerald was honored in his lifetime by election in 1883 as a Fellow of the Royal Society, which awarded him its Royal Medal in 1899 for his contributions to optics and electrodynamics. He was an important transitional figure, whose bold idea that objects moving at very high speeds contract, conceived in the context of SIR ISAAC NEWTON's universe, would provide a stepping-stone to Einstein's.

Further Reading
Whittaker, E. "G. F. FitzGerald." *Scientific American* 185(5): 93–98, 1953.

Fizeau, Armand-Hippolyte-Louis
(1819–1896)
French
Experimentalist (Optics), Astronomer

Louis Fizeau was a brilliant experimental physicist who made the first measurements of the speed of light on the Earth's surface and, concurrently with JEAN BERNARD LÉON FOUCAULT, showed that light travels faster in air than in water, thus confirming the wave theory of light. He also discovered the fact that the relative motion of a star affects the position of the lines in its spectrum.

Fizeau was born in Paris on September 23, 1819, at Nanteuil, Seine-et-Marne, into a wealthy family. As a young man, he hoped to follow the career of his father, an illustrious physician and professor of pathology at the Paris Faculty of Medicine. When poor health prevented him from pursuing this course, he turned instead to physics, studying optics at the Collège de France and working with François Arago at the Paris Observatory.

Arago would inspire Fizeau to study the new science of photography, which led him to invent a method of increasing the permanence of daguerreotypes in 1839. Then, in 1845, in collaboration with Foucault, he made the first detailed photographs of the Sun. Fizeau and Foucault went on to investigate the interference of light rays and showed that the spacing of the interference fringes was proportional to the wavelength of the light. In 1847, they discovered that heat rays from the Sun undergo interference, thus showing that radiant heat also has a wave nature.

That same year, he and Foucault went their separate ways; each took up Arago's suggestion for an experiment to determine whether light travels faster in air than in water—a question that would settle the heated debate between supporters of AUGUSTIN JEAN FRESNEL's and THOMAS YOUNG's position that light is a wave and the majority of physicists, who advocated the particle theory of light. Whereas Foucault adopted Arago's suggestion that a rotating mirror method be used, Fizeau tried a less complex method, which featured a rotating toothed wheel, with over a hundred teeth, rotating hundreds of times a second, and two mirrors situated over five miles apart. He shone a light through a gap between the teeth of a rapidly rotating toothed wheel. A mirror reflected the light through the same gap to the second mirror. Fizeau varied the speed of the wheel, thereby determining the speed at which the wheel was spinning too fast for the light to pass back and

forth through the same gap. Since he already knew the distance the light traveled, and since the speed of rotation of the wheel determined the time the light took to travel back and forth through the gap, he was able to divide the distance by time to get the speed of light. In this way, in 1849, he made the first measurement of the speed of light, putting it at 315,000 kilometers per second (about 5% larger than the currently accepted modern value). The following year, working with Louis Breguet and using a rotating-mirror method, he measured the comparative speed of light in air and water. As did Foucault, who had just done a similar experiment, he discovered that light moves faster in air than in water, thus coming down on the side of the wave theory of light.

While performing these historical experiments, Fizeau also made a landmark contribution to astronomy when in 1848 he proposed (independently of the work of CHRISTIAN DOPPLER) that a moving light source such as a star undergoes a change in observed frequency that can be detected by a shift in its spectral lines. This meant that stars moving away from the Earth would show a red shift in their spectral lines, whereas those moving toward the Earth would show a blue shift. Fizeau's method is used extensively today to determine the distances of galaxies and other astrophysical phenomena

In 1851, returning to his investigation of light, he conducted an experiment that measured the amount of ether drift of light waves in a moving, transparent medium. He exactly reproduced Fresnel's formula for the velocity of ether drift of waves in a moving, medium. This effect was later explained, without the concept of the ether, in the context of the theory of relativity.

Fizeau died in Venteuil, France, on September 18, 1896. His most far-reaching contribution to physics was the development of experimental methods to determine both the speed and the wave nature of light, which subsequently led to the development of special relativity.

Further Reading
Buchwald, Jed Z. *The Rise of the Wave Theory of Light: Optical Theory and Experiment in the Early Nineteenth Century*. Chicago: University of Chicago Press, 1989.

⊠   Foucault, Jean-Bernard-Léon
(1819–1868)
French
Experimentalist (Classical Electrodynamics, Optics), Astronomer

Jean Foucault is famous for inventing the gyroscope and developing the Foucault pendulum, with which he demonstrated the rotation of the Earth. He made the first accurate measurement of the speed of light and validated the wave theory of light by showing that light moves faster in air than in water.

Born in Paris on September 19, 1819, he was a sickly child who was educated at home. As a young man, he began to study medicine, with the intention of becoming a surgeon, but soon found himself more strongly drawn toward science. He began his scientific career as what we would today call a free-lancer, writing scientific textbooks and newspaper articles on scientific matters, while performing his experiments at home.

Foucault's first piece of original research would prove to be a gateway to a more far-reaching achievement than he anticipated. Collaborating with ARMAND-HIPPOLYTE-LOUIS FIZEAU, he used the photographic techniques of the time to take the first detailed pictures of the surface of the Sun in 1845. In Foucault's time photography was in its infancy. For the long exposures then required he used a clockwork pendulum device to turn the camera slowly so that it would follow the Sun. He noticed that the pendulum device was behaving strangely,

in that it tended to remain in the same swing plane when rotated. This observation led him to demonstrate experimentally the Earth's rotation: he showed that a pendulum maintains the same swing plane relative to the Earth's axis, and as a result the swing plane appears to rotate slowly as the Earth turns beneath it. He first performed this experiment at home in 1851, before making a spectacular demonstration by suspending a 200-foot pendulum that traced its path in sand on the floor from the dome of the Panthéon in Paris. The pendulum continued to swing in a single plane as the Earth rotated beneath it and left a series of traces in the sand that revealed its motion. The slow clockwise veering of the swing plane of the pendulum demonstrated that the Earth was slowly turning counterclockwise beneath it.

Although the fact of the Earth's rotation was universally accepted by then, Foucault offered the first actual demonstration of this phenomenon, ending a quest that had begun in GALILEO GALILEI's time two centuries earlier. Since the back and forth movement of a pendulum is actually a form of rotation, Foucault's experiment led him to realize that a body rotating in one direction would behave in a similar manner as a pendulum. In 1852, this realization led him to invent the gyroscope, a body rotating in one direction only, in one plane of rotation. His discovery of the nature of the movements of the pendulum and gyroscope demonstrated the effects that conservation of angular momentum had on rotating bodies in motion over the Earth's surface.

At the same time Foucault devised experiments to determine accurately the value of the speed of light and to discover whether light consisted of waves or particles. Both he and Fizeau, following a suggestion of François Arago, attempted to find the comparative speed of light in air and water. According to the physics of the time, if light traveled faster in water, the particle theory of light would be vindicated. Conversely, if light traveled faster in air, the wave theory of light would be confirmed. Foucault made his measurements by reflecting a beam of light from a rotating mirror to a stationary mirror and back again to the rotating mirror. The time taken by the light to travel this path caused a deflection of the image. The deflection would be greater if the light traveled through a medium that slowed its velocity. The more the medium slowed the velocity, the greater the deflection. In 1849, in competition with Foucault, Fizeau developed a slightly different method, involving a rotating toothed wheel, which yielded a moderately accurate value for the speed of light. In 1850, using the rotating-mirror method, Foucault showed that light travels faster in air than in water, just ahead of Fizeau's making the same discovery. In 1862, Foucault refined his rotating mirror technique and made the first highly accurate measurement of the speed of light. He performed the first terrestrial measurement of the speed of light in absolute units: kilometers per second. The value he obtained in this way, 298,000 kilometers per second, remains within 1 percent of the results obtained today by more advanced methods.

With a solid scientific reputation established by his pendulum exposition, in 1855, Foucault was able to transfer his experiments from his home to the Paris Observatory, where he would make several important contributions to practical astronomy. He died of a brain disease in Paris on February 11, 1868, at age 49.

Foucault's original and exacting experiments resulted in enduring contributions to several areas of theoretical and applied physics. His precise measurement of the speed of light in air and water confirmed the wave nature of light. His invention of the gyroscope was applied in the navigational gyrocompass and automatic stabilizers in ships and aircraft. Every major science museum has a Foucault pendulum on display.

Further Reading

Tobin, W. "Léon Foucault." *Scientific American*, July 1998, pp. 52–59.

Tobin, W., and A. B. Pippard. "Foucault, His Pendulum and the Rotation of the Earth." *Interdisciplinary Science Reviews*, 19 (1994): 326–337.

⊠ **Franck, James**
(1882–1964)
German/American
Atomic Physicist, Quantum Theorist,
Physical Chemist, Biophysicist

James Franck and his collaborator, Gustav Hertz, performed experiments that provided convincing evidence for the quantum theory and the quantum model of the atom, developed, respectively, by MAX ERNEST LUDWIG PLANCK and NIELS HENDRIK DAVID BOHR. This work earned them the 1925 Nobel Prize in physics. Franck participated in the development of the atomic bomb but was a leading advocating for demonstrating it in an uninhabited area, rather than dropping it on Japan, without warning.

Franck was born on August 26, 1882, in Hamburg, Germany, into a well-to-do Jewish banking family. His father, who denigrated Franck's pursuing a scientific career, sent him to the University of Heidelberg in 1901 to prepare himself to join the family firm by studying law and economics. Franck was saved by his friendship with another wealthy Jewish youth, MAX BORN, whose parents fully approved of his decision to pursue physics. Franck's father was gradually converted to their point of view. Initially a chemistry student, he went on to study physics at the University of Berlin, where he earned a Ph.D. in 1906 for a dissertation on the kinetics of electrons, atoms, and molecules and the mobility of ions in gases.

In 1911, Franck married Ingrid Jefferson of Göteborg, Sweden, with whom he would have two daughters, Dagmar and Lisa. He lectured at the University of Berlin from 1911 to 1918, taking time out to participate in World War I, and was awarded an Iron Cross, first class, for his service. After the war he became the head of the physical chemistry division at the Kaiser Wilhelm Institute, where he worked on the theory of gases. In 1920, he became professor of experimental physics and director of the Second Institute for Experimental Physics at the University of Göttingen. Franck emerged as an inspiring teacher of such brilliant students as the future Nobel Prize winner PATRICK MAYNARD STUART, LORD BLACKETT and the man who would spearhead the development of the U.S. atomic bomb, J. ROBERT OPPENHEIMER. Between 1920 and 1933, Göttingen became an international center for the new quantum physics and Franck's old friend Max Born was then head of Göttingen's Institute of Theoretical Physics. The two collaborated on molecular processes, developing their classic potential energy diagrams found in physical chemistry textbooks. Franck went on to calculate the dissociation energy of molecules by extrapolating data on the vibration of molecules obtained from spectra. Later on Edward Condon interpreted this method in terms of quantum mechanical wave mechanics, formulating what became known as the Franck–Condon principle.

At Göttingen, Franck did his most important work in collaboration with Gustav Ludwig Hertz, studying collisions of free electrons in various noble gases. In these experiments they discovered that for certain inelastic collisions between electrons and atoms energy can be transferred from the electrons to the atoms only in discrete amounts. In particular for the mercury atom, they found that electrons accept energy only in quantum units of 4.9 electron volts (an electron volt is the kinetic energy gained by an electron that is moving through a difference in electric potential of one volt). For an inelastic collision (a collision in which an electron transfers energy to the mercury atom), the electron needs kinetic energy in excess of 4.9 electron volts. The inelastic transfer of the electron's energy to the mercury atoms was then revealed by the fact that the atoms became excited, and that state caused them to emit light at a specific wavelength of 2537 Å. This result offered the first experimental proof of Planck's quantum hypothesis that $E = h\nu$, where E is the change in energy, h is Planck's constant, and ν is the frequency of light emitted. These experiments also confirmed the existence of the discrete energy levels postulated by Bohr in his theory of the atom. In 1925, Franck shared the Nobel Prize in physics with Hertz for this work.

When the Nazis came to power in 1933, Franck, though Jewish, was allowed to keep his post because of his outstanding military service in World War I. He was, however, instructed to fire his Jewish colleagues. Franck refused and resigned in open protest against the regime's racist policies. He left Germany with his family and settled in Baltimore, Maryland, where he became a professor at Johns Hopkins University. He remained there until 1935, when he went to the Niels Bohr Institute in Copenhagen, Denmark, for a year as visiting professor. He then returned to the United States and, in 1938, moved to the University of Chicago. During World War II, Franck was director of the Chemistry Division of the Metallurgical Laboratory at the University of Chicago, which was the center of the Manhattan Project to develop an atomic bomb. He headed a group of atomic scientists in preparing the "Franck Report" for the U.S. War Department; it urged an open demonstration of the atomic bomb in an uninhabited place, instead of dropping it without warning on Japan. Although the report was not heeded, it remains an important testimony to the social conscience of scientists attempting to circumscribe the destructive impact of their wartime work.

In 1946, several years after the death of his first wife, he married Hertha Sponer, who was a professor of physics at Duke University in Durham, North Carolina. In 1947, at age 65, he became professor emeritus at the University of Chicago, but he continued his work as head of the university's photosynthesis research group until 1956.

He died in Germany on a visit to friends in Göttingen, where he and Born had been made honorary citizens, on May 21, 1964, at the age of 81.

Franck is remembered for both his contributions to the development of quantum mechanics and his personal courage in taking difficult moral stands that he believed to be compatible with his responsibility as a physicist.

Further Reading

Gamow, George. *Thirty Years That Shook Physics: The Story of Quantum Theory.* New York: Dover, 1985.

Walker, J. Samuel. *Prompt and Utter Destruction: Truman and the Use of Atomic Bombs Against Japan.* Chapel Hill: University of North Carolina Press, 1997.

⊠ **Fraunhofer, Joseph von**
(1787–1826)
German
Theoretician (Optics), Spectroscopist, Astronomer

A self-taught theoretician and a master of applied optics, Joseph von Fraunhofer invented the first spectroscope, an instrument for producing and analyzing spectra, and established Germany as the world leader in the production of scientific and optical instruments. His name entered the lexicon of physics when the dark lines in the spectra of the Sun and stars, which he was the first to investigate, were named Fraunhofer lines.

He was born on March 6, 1787, in Straubing, Bavaria, the 11th child of Franz Xaver Fraunhofer, a glazier. From the beginning, his life was directed toward the skillful making and refining of objects. His formal schooling was brief. At age 10, after losing his mother, he began helping out in his father's glazing work-

shop. A year later his father died and he became an apprentice to a glazier in Munich, then the capital of Bavaria. His master trained him well but forbade him to engage in the reading and studies he craved. A happy accident, however, changed his life. When the house he was working in collapsed around him but left him uninjured, he came to the notice of both the future king, Maximilian I, who helped him financially, and an influential politician and businessman named Joseph Utzschneider, who gave him books and exposed him to ideas in physics and optics. At the age of 19, he found employment in the optical shop of the Munich Philosophical Instrument Company, producers of scientific instruments. Working under a master glazier, he perfected his skills and became an expert in both the theory of optics and its practical applications. He was a great success within the company, which made him a director in 1811.

In the area of lens making, Fraunhofer applied his knowledge of optical theory to the construction of lenses with the smallest possible degree of aberration. Rather than using the trial-and-error approach that was typical at that time, he attacked the problem conceptually, working out the dispersion power and refractive index of different kinds of optical glass. Using glass prisms he invented the first spectroscope capable of scientific study of spectra and was able to make very accurate measurements of the dispersion and refractive properties of various kinds of glass.

In 1814, while trying to obtain more accurate values of the dispersion power and refractive index of various kinds of optical glass, he used a monochromatic light source consisting of a flame that contained two bright yellow spectral lines. By comparing the effect of the light from the flame to that of the light from the Sun he found that the solar spectrum contained 574 dark spectral lines between the red and violet ends of the spectrum. Later, in 1859, GUSTAV ROBERT KIRCHHOFF would explain that these dark spectral lines, called Fraunhofer lines, were due to the absorption of light by sodium and other elements present in the Sun's atmosphere. Using this interpretation, Kirchhoff was able to identify other elements in the Sun's spectrum and developed the law that states that the ratio of the emission line and absorption line power of all material bodies is the same at a given temperature and a given wavelength of radiation produced. He went on to demonstrate the existence in the Sun of many chemical elements isolated on Earth and to argue that the Sun primarily comprises a hot, incandescent liquid.

In 1821, Fraunhofer perfected another important invention: a diffraction grating that was able to decompose white light into its spectral components. He was able to do this by first looking at the patterns produced by light when it is diffracted through a single slit and then determining the relationship between the dispersion angles of the light and the width of the slit. He then built a diffraction grating, using a large number of slits consisting of 260 parallel wires, and made the first study of spectra produced by the diffraction grating method. Using the diffraction grating to study the dark spectral lines from the Sun, he noted that the dispersion of the spectra is greater with a diffraction grating than with a prism. Then by examining the relationship between the spectral dispersion and the separation of the wires in the grating, he concluded that the dispersion is inversely related to the distance between successive slits in the grating. By measuring the dispersion, he was able to determine the wavelengths of light of specific colors and the dark lines. In addition to diffraction gratings he constructed reflection gratings in order to study the effect of diffraction on oblique rays. Finally, by using the wave theory of light, Fraunhofer was able to derive a general form of the diffraction grating equation that is still in use today.

For this groundbreaking work Fraunhofer was recognized as a leader in the field of optics,

and in 1823 he became director of the Physics Museum of the Bavarian Academy of Sciences in Munich. The following year he was knighted and elected as a member of the Civil Order of the Bavarian Crown. Tragically only two years later, at the age of 39, he contracted tuberculosis and he died in Munich on June 7, 1826.

Fraunhofer left a rich legacy to many branches of physics, from astronomy to the structure of the atom. Spectroscopy became a vital tool used by modern physicists to understand atomic structure.

Further Reading

Jackson, Myles W. *Spectrum of Belief: Joseph von Fraunhofer and the Craft of Precision Optics.* Transformations: Studies in the History of Science and Technology. Cambridge, Mass.: MIT Press, 2000.

⊠ **Fresnel, Augustin-Jean**
(1788–1827)
French
Theoretical and Experimental Physicist (Optics)

The deft mathematical analysis of Augustin-Jean Fresnel persuaded his contemporaries to abandon the prevailing Newtonian theory that light consists of particles in favor of his own transverse wave theory of light. By applying his new ideas on light, he developed a revolutionary lens design that came to be known as the Fresnel lens, which still has applications today, from lighthouses to stage spotlights to the solar panels on spacecraft.

He was born on May 10, 1788, in Broglie, Normandy, in France, into a religious family and received home schooling until the age of 12. He then studied at the École Centrale in Caen, where he was introduced to science. In 1804, he entered the École Polytechnique in Paris; there he received a solid grounding in mathematics and, two years later, went on to the École des Pont et Chaussées, where he studied for three years, in preparation for an engineering career. Upon graduating, he worked as a civil engineer for the government. It was during this period that he developed an interest in optics and, thanks to unforeseen political circumstances, found he had plenty of time to pursue research in this field. In 1815, when Napoleon returned to France from exile in Elba, Fresnel, calling the former emperor's homecoming "an attack against civilization," protested vigorously, despite his precarious health. Placed under house arrest in Normandy, he embarked upon the work that would persuade his peers to reconsider totally their point of view about optical phenomena.

At the end of the 19th century physics was dominated by SIR ISAAC NEWTON's particle theory of light, which the great 17th century physicist had extrapolated from his particle laws for matter. Fresnel disliked Newton's theory of light, both because of its failure to explain such basic optical phenomena as the interference effect and because of its lack of the simplicity he valued in a theory. Using his mathematical prowess, intuition, and skills as an experimentalist, he undertook an ambitious project to "create a revolution in science."

One of the major challenges for optics in Fresnel's time was to interpret diffraction, the tendency of light images to become fuzzy around the edges, within the context of a mathematically solid wave theory. In 1815, Fresnel demonstrated mathematically that the dimensions of light and dark bands produced by diffraction could be related to the wavelength of the light, if one assumed that light consisted of waves. In 1816, the new theory enabled him to predict the intensity of the "fuzziness" or fringes.

His second major challenge was to interpret polarization, the tendency of light waves to vibrate in directions perpendicular to the direction of wave propagation. Earlier, the

Dutch physicist CHRISTIAAN HUYGENS had proposed that light consists of longitudinal waves analogous to sound waves. Huygens's light wave theory made it possible to interpret refraction and reflection. It also allowed him to explain, as the light particle theory could not, how two light rays can intersect without causing each other to deviate from their paths. It failed, however, to explain light polarization. And, like the particle theory, the longitudinal wave theory could not explain double refraction, the splitting of incident transmitted light into two beams, each polarized perpendicularly to the other. Working with the French physicist François Arago, Fresnel showed that polarized light could be refracted through two different angles because one ray would consist of waves oscillating in one plane and the other ray would consist of waves oscillating in a plane perpendicular to the plane of the first. His mathematical analysis led him to conclude in 1821 that polarization could occur only if light consists of transverse waves. Fresnel's explanation of polarized light was later carried beyond the visible spectrum in JAMES CLERK MAXWELL's theory of electromagnetism.

By this time, Napoleon had departed French shores once more and Fresnel was back on the job as a civil engineer. Despite having to relegate scientific research to his spare time, he continued to do important work. He designed what came to be called the Fresnel lens, which has one surface cut in steps so that transmitted light is refracted just as it would be if a much thicker, heavier, and more expensive conventional lens were used. Designed to replace mirrors in lighthouses, Fresnel lenses have myriad uses today.

Fresnel went on to unify the mysterious fluids Newton had postulated to account for the phenomena of heat, electricity, magnetism, and light into a single universal fluid, that is, an *ether*, whose modifications explained the different observed physical phenomena. At the suggestion of Arago, Fresnel undertook research on the aberration of starlight, an effect tied to the annual movement of the Earth in its orbit, which makes starlight seem to slant, altering the apparent position of the stars hour by hour. The aberration of starlight causes the light from stars to describe an ellipse over the course of a year. This aberration effect, which is constant for all stars, led Fresnel to conjecture that the movement of light emitted by all stars is uniform, an idea contrary to Newton's theory, in which particles of light emitted by different stars have different speeds. Fresnel succeeded in explaining this phenomenon by proposing that the ether was partially dragged by the Earth's movement.

Although Fresnel's light wave theory gained many adherents, the question of whether light consisted of particles or waves awaited experimental demonstration. Fresnel never learned the answer since, continually plagued by ill health, he died of tuberculosis in Ville-d'Avray, near Paris, on July 14, 1827, at the age of 39, before a decisive experiment could be performed. He had been elected to the Académie des Sciences in 1823 and awarded the Rumford Medal of the Royal Society just before his death.

More than two decades after Fresnel's death, at the request of Arago, ARMAND-HIPPOLYTE-LOUIS FIZEAU and JEAN-BERNARD-LÉON FOUCAULT independently conducted experiments that determined whether a light wave moves more rapidly in air than in water, as the wave theory predicted, or, on the contrary, more rapidly in water than in air, as the particle theory predicted. The former proved to be the case and Fresnel's wave theory was validated.

By developing his ideas on the wave nature of light into a comprehensive mathematical theory, Fresnel determined properties that every future theory of light would have to satisfy and created the groundwork for the later work of Maxwell.

Further Reading

Buchwald, Jed Z. *The Rise of the Wave Theory of Light: Optical Theory and Experiment in the Early Nineteenth Century.* Chicago: University of Chicago Press, 1989.

Silverman, Mark P. *Waves and Grains.* Princeton, N.J.: Princeton University Press, 1998.

G

Gabor, Dennis
(1900–1979)
Hungarian/British
Theoretician and Experimentalist
(Holography, Information Theory)

Dennis Gabor is known as the father of holography, a method of producing three-dimensional images by using laser light. Although lasers, which are necessary to producing a hologram, would not be invented until 1960, Gabor conceived and fully elucidated the nature of holograms in 1948. He received the 1971 Nobel Prize in physics for this work.

He was born in Budapest on January 5, 1900, the oldest son of Bertalan Gabor, director of a mining company, and his wife, Adrienne. He fell in love with physics at the age of 15 and, with his brother George, set up a small laboratory at home, where they duplicated experiments on X rays and radioactivity. However, he chose to pursue engineering instead of physics, which was not yet a profession in Hungary, studying first at the Technical University in Budapest and then at the Technische Hochschule of Berlin in Charlottenberg, where he received his diploma in 1924. Nonetheless, as often as he could, he would "sneak over" to the University of Berlin, where ALBERT EINSTEIN, MAX ERNEST LUDWIG PLANCK, WALTHER

NERNST, and MAX THEODOR FELIX VON LAUE were then in residence. Although Gabor was officially an engineer, most of his work was in applied physics. In 1927, he was awarded a Ph.D. for developing one of the first high-speed cathode ray oscillographs, devices that monitor oscillating currents in high-voltage transmission lines. While doing this work, he also developed the first iron-shrouded magnetic electron lens. He then joined the firm of Seimens and Halshe in Berlin, where he invented a high-pressure quartz mercury lamp with superheated vapor and a molybdenum tape seal, which is now extensively used in street lamps.

When the Nazis came to power in 1933, Gabor fled Germany and, after a short time in Hungary, left for England, where he remained for the rest of his life, becoming a British citizen. He first worked at the British Thomson-Houston Company in Rugby as a research engineer and in 1936 married Marjorie Louise Butler. He remained at Thomson-Houston until 1948. During the postwar years he wrote his first papers on communication theory and developed a system of stereoscopic cinematography.

In 1948, Gabor carried out the basic experiments in holography, which at the time was called wavefront reconstruction. He would later call this "an exercise in serendipity," the art of looking for something and finding something

110

else. His original goal was to develop an improved electron microscope that used electrons instead of light rays, one capable of resolving atomic lattices and seeing single atoms. He studied the idea of coherent beams of light (i.e., light that contains a single frequency of oscillation), in which all the waves are exactly in phase with each other. Gabor noted that a coherent beam reflected from an object should travel a varying number of wavelengths to an observer, depending on the distance to the surface of the object. In this context he realized that a beam of coherent light reflected from an object should contain information on the shape of the object.

In Gabor's holographic method the beam of coherent light reflected from the object and the beam of coherent light emitted directly from the coherent light source are simultaneously recorded on a photographic film plate. The superposition of the two coherent beams produces an interference pattern on the photographic film plate.

The interference pattern produced depends on the amplitude and phase of the light waves in the reflected beam at the surface of the plate, which in turn depend on the distance of each part of the plate from each part of the object. The interference pattern thus contains information on the shape of the object. The interference pattern, called a hologram, appears on the film as an apparently meaningless pattern of whirls when the plate is developed. However, when the hologram is placed in a beam of coherent light and the viewer looks through the hologram, the interference pattern of the two beams is deconstructed and a three-dimensional image of the object is seen (i.e., if the viewer moves, a different perspective of the object is obtained). Gabor invented this method before coherent light sources became available, calling it holography from the Greek *holos,* meaning "whole." In three papers published between 1948 and 1952, he laid out an exact analysis of the method. Holography made its breakthrough when the first laser was invented in 1960. This is because the laser generates continuous coherent light wave trains of such length that one can reconstruct the depth in the holographic image. Twenty years after doing his pioneering work in the development of holography Gabor would be awarded the 1971 Nobel Prize in physics.

In 1949, he joined the faculty of the Imperial College of Science and Technology in London. With young doctoral students as collaborators, he worked on many interesting and challenging problems: he constructed an advanced version of the Wilson cloud chamber and made a number of inventions, including a holographic microscope; an electron-velocity spectroscope; an analog computer that was a universal, nonlinear "learning" predictor, recognizer, and stimulator of time series; a flat, thin color television tube; and a new type of thermionic converter. He also did work on communication theory, plasma theory, and magnetron theory.

Finally, in 1958, Gabor became professor of applied electron physics at Imperial College, at which point his interest shifted to a study of the future of industrial civilization. He wrote:

> I became more and more convinced that a serious mismatch has developed between technology and our social institutions, and that inventive minds ought to consider social inventions as their first priority.

He developed these ideas in three books: *Inventing the Future,* 1963; *Innovations,* 1970; and *The Mature Society,* 1972.

Gabor retired in 1967 and became a senior research fellow and a staff scientist of CBS Laboratories, Stamford, Connecticut, where he worked on many new schemes of communication and display. From 1967 until his death in London on February 8, 1979, he was Professor Emeritus of applied electron physics at the University of London.

Gabor's invention of holography continues to have a profound impact on the development of technology. With its ability to determine an object's point in space to a fraction of a light wavelength, the hologram has enriched optical measurement techniques. Many physics and engineering applications have been found for holography, and in the future we may see the development of holographic computers and holographic three-dimensional television or motion pictures.

See also TOWNES, CHARLES HARD.

Further Reading
Gabor, Dennis. *Innovations: Scientific, Technological and Social*. Oxford: Oxford University Press, 1970.

Kock, Winston E. *Lasers and Holography: An Introduction to Coherent Optics*. New York: Dover, 1981.

⊠   **Galilei, Galileo**
(1564–1642)
Italian
Theoretical and Experimental Physicist, Astronomer, Mathematician

Galileo Galilei was called "the father of modern physics—indeed of modern science altogether" by ALBERT EINSTEIN. Over the course of his long and stormy career, he established the modern scientific method of deducing mathematical laws to explain the results of experiment and observation. His discovery of the properties of the pendulum, his invention of the first crude thermometer, and his formulation of the laws governing the motion of falling bodies all flowed from these principles. In astronomy, the enhanced telescopes he constructed enabled him to observe the heavens in unprecedented detail and led him to espouse the Copernican theory that the Earth revolves around the Sun. His subsequent prosecution by the Inquisition enthroned him in the popular imagination as a defender of scientific truth against the dogmas of religious authority. Galileo

himself, however, was a devout Catholic who saw no conflict in the truths of religion and science.

He was born in Pisa on February 15, 1564, and given, as was not uncommon at the time, a first name based on the family name. His father, Vincenzio Galilei, was a musician who performed experiments to test mathematical formulations on the properties of music and an advocate of using reasoned argument rather than authority to prove one's point. He had his son tutored privately in Pisa and, in 1575, when the family moved to Florence, sent him to study at a nearby monastery. Galileo remained there until 1581; at the age of 17, he returned to Pisa, where, at his father's urging, he took up the study of medicine at the university. When it became clear that Galileo was more attracted to mathematics, Vincenzio allowed him to be tutored by the Tuscan court mathematician. Galileo the son made his first important discovery: by using his pulse to time the swing of a lamp in the Cathedral of Pisa, he observed that a pendulum always swings back and forth in the same period of time, regardless of the amplitude of the swing.

When Galileo returned home to Florence in 1585, he had not earned a formal academic degree. The following year he extended ARCHIMEDES' work in hydrostatics by inventing an improved version of the hydrostatic balance for measuring specific gravity. In 1589, he was back at the University of Pisa, a professor of mathematics at age 25. He lost no time in attacking Aristotle's belief that the speed of a falling object is determined by its weight. According to legend, Galileo dropped two cannon balls of different weights from the Leaning Tower of Pisa; they hit the ground together, demonstrating that unequal weights fall at the same speed. He published his first ideas on motion in *De Motu* (On motion) in 1590. In Pisa he also invented one of the first scientific measuring instruments: a primitive thermometer consisting of a bulb of air that expanded or contracted as the temperature changed, causing the level of a column of water to rise and fall. Despite these achievements, his attacks on Aristotle had antagonized his colleagues and his contract was not renewed.

In 1592, friends found him a better position as professor of mathematics at the University of Padua, where he would flourish over the next 17 years. In 1599, he met a 21-year-old courtesan, Marina Gamba, who would bear him a son and two daughters. Galileo, who never married, would end his liaison with Marina when he left Padua but remain a devoted father all his life. His family responsibilities led him to market a military compass he had invented. He also managed to continue his motion studies, determining the law of falling bodies: the distance fallen by a body is proportional to the square of the time of descent. From this he was able to deduce that neglecting air resistance, all bodies fall with the same acceleration in a gravitational field. It was this ability to separate out the significant components of an experiment to determine what was fundamental to the investigation that made him the first modern scientist.

In 1609, on learning that a Dutchman named Hans Lippershey had invented the telescope the year before, Galileo immediately set to work improving the instrument. Using a telescope that magnified up to 20 times, he made the celestial observations that would revolutionize astronomy: he discovered the mountains of the Moon, sunspots, the phases of Venus, and the four largest (Galilean) satellites of Jupiter. He also noted that Saturn had an oval shape, an observation that CHRISTIAAN HUYGENS later showed to be caused by its rings, and that the Milky Way is composed of stars. When he published his findings in *Sidereus nuncius* (The starry messenger) in 1610, the book caused a sensation throughout Europe. Galileo was offered a lifelong position in Padua but returned to his native Florence instead, as mathematician and philosopher to the grand duke of Tuscany.

Galileo Galilei established the modern scientific method and was the first to provide experimental support for the Copernican theory that the Earth revolves around the Sun. *(AIP Emilio Segrè Visual Archives)*

In Florence, in 1612, he returned to his earlier work on hydrostatics. He published a study of the behavior of bodies in water, *Bodies That Stay Atop Water or Move Within It,* in which he championed Archimedes' law of buoyancy, which states that the buoyant force on a body in water is equal to the weight of the water displaced. In 1613, he published his treatise on sunspots, *Sunspot Letters,* in which he argued that the spots were on or near the Sun's surface and not, as the German Jesuit Christoph Scheiner proposed, the Sun's satellites.

Meanwhile, his work in both mechanics and astronomy was yielding new arguments for Copernicus' heliocentric theory, which Galileo had always favored. Anti-Copernicans argued that the Earth must be stationary, since birds and clouds would fall off a turning or moving Earth, just as, unanchored objects would fall off a moving carousel. But, on the basis of his motion studies, Galileo conjectured that a horizontal component of motion provided by the Earth always exists to keep such objects in position, even though they are not attached to the ground. Furthermore, Galileo had found that when seen through a telescope, Venus showed phases like those of the Moon and therefore must orbit the Sun, not the Earth. This left astronomers with a choice of either the Copernican system or the Earth-centered model proposed by Tycho Brahe, in which everything but the Earth and Moon revolves around the Sun, which in turn revolves around the Earth. Most astronomers favored the Earth-centered model, and Galileo increased the ranks of his enemies by using all his formidable mathematical and verbal skills to ridicule them. In 1616, however, the Catholic Church banned all work on the Copernican theory, forcing Galileo to make a tactical retreat. That year he wrote his *Theory on the Tides*, in which he used the two Copernican motions—the Earth's daily rotation on its own axis and its annual revolution around the Sun—to explain the tides. Throughout his life he would disdain the hypothesis that the Moon caused the tides, considering it tainted with occultism and astrology.

In 1623, Galileo published *Il saggiatore* (The assayer), in which he set forth his philosophy of the nature of physical reality and modern scientific inquiry:

> The grand book of the universe . . . cannot be read until we have learnt the language and become familiar with the characters in which it is written. It is written in mathematical language, and the letters are triangles, circles and other geometric figures, without which means it is humanly impossible to comprehend a single word.

Il saggiatore was about to be printed when a longtime admirer and patron, Mafeo Cardinal Barberini, was chosen to become Pope Urban VIII. Galileo had time to dedicate the work to the new pope; he then set out for Rome and succeeded in obtaining Urban's permission to present arguments for the rival geocentric and heliocentric positions, so long as Copernicus' theory was treated as a mere hypothesis. It was not until 1632 that Galileo published his *Dialogue Concerning the Two Chief World Systems: Ptolemaic and Copernican*. The book was banned and Galileo was taken to Rome to stand trial for heresy in April 1633. Although Galileo may have believed his own claim that he had been even-handed in his presentation, most readers would agree with the Inquisition's view that he had used observations and experiments to favor Copernicus. Legend has it that when forced to abjure his belief that the Earth moves around the Sun, he muttered, "Eppur si muove" (Yet it does move). On the margins of his copy of the *Dialogue* he wrote: "In questions of science, the authority of a thousand is not worth the humble reasoning of a single individual."

Galileo's sentence of life imprisonment was commuted to house arrest at his villa at Arcetri near Florence for the rest of his life. Although his final years were saddened by the death of his beloved daughter, Sister Maria Celeste, in 1634, and the loss of his eyesight in 1637, he continued to work, with the help of an assistant, and published his most important book, *Discourses and Mathematical Discoveries Concerning Two New Sciences*, in Leiden, Holland, in 1638. In it he summed up his life's work in motion studies. He rejected the idea that a force is necessary to sustain motion and showed that a body in uniform horizontal motion will continue moving. He realized that this will occur only in a vacuum, and that air resistance always causes a uniform terminal velocity to be reached. He demonstrated that gravity not only causes a body to fall, but also determines the motion of rising bodies, and furthermore that gravity extends to the center of the Earth. He then showed that the motion of a projectile is made up of two components: uniform motion in a horizontal direction and vertical motion under the downward acceleration due to gravity. He also deduced, by combining horizontal and vertical motion, that the trajectory of a projectile is a parabola.

Galileo's second "new science" was engineering, particularly the science of structures, in which he pointed out that the dimensions of a structure are important to its stability: a small structure will stand, whereas a larger structure of the same relative dimensions may collapse. Using the law of levers, he went on to examine the strengths of the materials necessary to support structures.

In his last years, Galileo designed a pendulum clock. He died at Arcetri on January 8, 1642. Galileo's work formed the basis for the subsequent development of classical mechanics and observational astronomy. His work in motion studies would be developed by SIR ISAAC NEWTON into a full understanding of inertia, the laws of motion, and gravitation.

Further Reading

Drake, Stillman. *Galileo at Work: His Scientific Biography*. Chicago: University of Chicago Press, 1978.

Galileo. *Discoveries and Opinions of Galileo: Including the Starry Messenger (1610), Letter to the Grand Duchess Christina (1615), and Excerpts from Letters on Sunspots (1613), The Assayer (1623)*. Stillman Drake, Translator. Based on work by Galileo. New York: Doubleday & Company, 1989.

Reston, James. *Galileo: A Life*. New York: Harper-Collins, 1994.

Sobel, Dava. *Galileo's Daughter: A Historical Memoir of Science, Faith, and Love*. New York: Walker, 1999.

⊠   **Gamow, George**
(1904–1968)
Russian/American
Theoretical Physicist, Nuclear Physicist, Cosmologist

George Gamow is famous for the theoretical work that led to discovery of the first evidence supporting the big bang theory of the origin of the universe. In 1948, he predicted that if the universe had begun with a fiery explosion, then a cosmic microwave background radiation would remain. Seventeen years later, Gamow's prediction was confirmed, when ARNO ALLAN PENZIAS and Robert Wilson, using a horn-shaped radio antenna, detected an unchanging noise that pervaded all of space.

Born in Odessa, Ukraine, on March 4, 1904, Gamow attended local schools. His father's gift of a telescope when George was 13 sparked his interest in astronomy. In 1922, he entered Novorossiya University in Odessa; he then transferred to Leningrad State University, where he studied optics and cosmology, earning a Ph.D. in 1928.

In 1926, Gamow traveled to Germany to attend summer school at the University of Göttingen, where he did his first important research in nuclear physics. Attempting to formulate a theory of alpha particle decay, he constructed a model in which the new quantum mechanics was applied to the study of nuclear structure for the first time. He explained why uranium nuclei could not be penetrated by alpha particles (ionized atoms of helium) that have twice the energy of the alpha particles emitted by the nuclei. His hypothesis was that this was due to a "Coulomb potential barrier" generated by the repulsion of the electrostatic Coulomb forces that acted between the uranium nuclei and the alpha particles. In classical mechanics, charged subatomic particles like protons, when hurled against the Coulomb barrier, had no chance of penetrating it. In quantum mechanics, however, because of the uncertainty principle, protons can sometimes penetrate the Coulomb barrier by tunneling through it. In 1928, Gamow applied quantum probabilities to the question of nuclear fusion in stars and found that alpha particles could surmount the Coulomb barrier by quantum tunneling, which now stood revealed as the mechanism by which the stars shine.

That year, on the basis of this work, NIELS HENRIK DAVID BOHR, the guiding spirit of quantum mechanics, invited Gamow to his Institute of Theoretical Physics in Copenhagen, where he continued his work on nuclear physics and began to study nuclear reactions in stars. There Gamow worked on the liquid drop model of nuclear structure, in which the nucleus was regarded as a collection of alpha particles that interacted via strong nuclear and Coulomb forces. Ultimately, the liquid drop model of the nucleus was used to develop the present theory of nuclear fission and fusion.

In 1929, as a Rockefeller Fellow, he worked with ERNEST RUTHERFORD at the Cavendish

George Gamow is famous for his theoretical work that led to discovery of the first evidence supporting the big bang theory of the origin of the universe. *(AIP Emilio Segrè Visual Archives)*

Laboratory, Cambridge, England, calculating the energy required to split a nucleus by using artificially accelerated protons. This work led directly to the 1932 construction, by JOHN DOUGLAS COCKCROFT and Ernest Watson, of a linear accelerator, in which protons were used to disintegrate boron and lithium, resulting in the production of helium. This was the first experimentally produced transmutation generated without the use of radioactive materials.

In 1930, Gamow engaged in further research in Copenhagen. Then in 1931 he returned to the Soviet Union and married Lyubov Vokhminzeva, with whom he would have a son, Rustem Igor. His career as a Soviet physicist was thriving: in 1931, he became professor of physics at Leningrad State University as well as head of research at the Academy of Sciences in Leningrad. But that same year, he was made aware of the limits on his freedom when the Soviet Union denied him permission to attend a conference on nuclear physics in Rome. In 1933, he was allowed to travel to the Solvay Conference in Brussels, where he defected to the West. After spending some time in Europe, as a fellow at the Pierre Curie Institute in Paris and visiting professor at the University of London, he traveled to the United States.

His first American home was in Washington, D.C., where he assumed the chair of physics at George Washington University, in 1934. He was able to offer a position at George Washington to another distinguished émigré, EDWARD TELLER, thus taking him to the United States before the war. A fruitful collaboration began between the two physicists, who, in 1936, announced their discovery of the Gamow–Teller selection rule for beta decay, the process in which radioactive transmutations of matter from one element to another occur.

Much of Gamow's later work centered around the immense question that had long intrigued him: the origin of the universe. In 1938, he contributed to the interpretation of the Hertzsprung–Russell diagram, which plots the spectral classes (or colors) of stars against their brightness. Using this technique, he examined the origin of nebulae and the production of energy occurring in the internal structure of red giant stars and, in the early 1940s, published his neutrino theory of supernovae in which he examined neutrino emission during stellar explosions.

Gamow made his most fundamental contribution, however, in 1948, when he presented evidence in support of the big bang model of the expanding universe. At the time, the two competing cosmological theories were the steady state theory, which held that the universe was as it had always been; homogenous in space and time, and the big bang theory, which arose from Edwin Hubble's 1929 finding that the galaxies are rapidly moving away from each other. Supporters of the big bang theory argued that galaxies must have been closer to one another in the past and that, in the furthest reaches of the past—the cataclysmic moment in which the universe was created some 15 billion years ago—they were merged in a single, infinitely dense, hot mass. In Gamow's view these conditions must have existed in order for atomic nuclei to fuse into different combinations, thereby creating chemical elements. He reasoned that if the universe had been expanding and cooling since the big bang, it would not have cooled completely. For this reason, the radiation emitted by the initial cosmic explosion should still be present in the universe at a lower but nonetheless observable frequency. In particular the photons carrying the energy of the big bang, having originated in the wavelengths of light, ought to have been redshifted by the subsequent expansion of the universe into the lower frequencies of electromagnetic energy called microwave radio radiation. On the basis of this idea, Gamow and his colleagues estimated that the universe today should be permeated by an ocean of photons with an ambient absolute temperature of about three

kelvin (3 K). At the time, this prediction was impossible to verify, since radio astronomy was in its infancy. But in 1965, Penzias and Wilson were having trouble accounting for a persistent hiss in a microwave horn Bell Laboratories had built for satellite communication experiments. When the Princeton physicist Robert Dicke made them aware of Gamow's work, they realized that Gamow's refined estimate of 2.7 K was just the temperature of this unwanted noise. They had stumbled across confirmation of Gamow's cosmic background radiation hypothesis, which, in turn, provided evidence for the big bang. Penzias and Wilson, rather than Gamow, received the Nobel Prize for the work, but Gamow's seminal role in this historical discovery is universally recognized.

In 1948, the year of his great work in cosmology, Gamow's career took a radically different direction. After gaining a top secret security clearance, he went to work on the hydrogen bomb project at Los Alamos, New Mexico, where he collaborated closely with Teller on the design of the weapon.

Another major turning point in his life occurred in the mid-1950s, when he became interested in molecular biochemistry and contributed to the solution of the genetic code. Gamow's theory was found to be correct, and in 1961 the genetic code was understood. In 1956, he divorced his first wife and left George Washington University to become professor of physics at the University of Colorado. Two years later he married Barbara Perkins. He remained at Colorado until his death in Boulder on August 20, 1968.

Later in life, Gamow became famous as the author of popular science books, especially his Mr. Tompkins series. Others included *One, Two, Three . . . Infinity* (1947), *The Creation of the Universe* (1951), *The Atom and Its Nucleus* (1961), and *Gravity* (1962).

George Gamow was a major figure in 20th-century physics, whose personality and intellect left an indelible impression on the physicists of his time. Unconventional and charismatic, he was known for his wit and irreverence toward human affairs, while retaining his sense of awe before nature. His curiosity about how the universe began moved the physics community a giant step closer to unraveling that mystery.

See also COULOMB, CHARLES AUGUSTIN; HEISENBERG, WERNER; MEITNER, LISE.

Further Reading

Gamow, George. *The Creation of the Universe.* New York: Mentor, 1951.

———. *Mr. Tompkins in Paperback.* Canto Series. Cambridge: Press Syndicate of the University of Cambridge, 1968.

———. *My World Line: An Informal Autobiography.* New York: Viking Penguin, 1970.

———. *Thirty Years That Shook Physics.* Garden City, N.Y.: Anchor, 1966.

⊠ **Gauss, Johann Carl Friedrich**
(1777–1855)
German
Mathematical Physicist
(Electromagnetism, Terrestrial
Magnetism), Astronomer

Johann Gauss was a scientific virtuoso, most famous for his seminal work in mathematics. As a physicist, he made substantial contributions in the fields of electromagnetism and terrestrial magnetism, deriving a method for defining magnetic units and formulating Gauss's law, which was later used to simplify the mathematical formulation of JAMES CLERK MAXWELL's equations for the electromagnetic field.

Gauss was born on April 30, 1777, into a very poor and uneducated family in Brunswick, Germany. He was a child prodigy, who taught himself to read and do arithmetic and who was doing calculations before he had learned to talk. Upon entering school, Gauss offered proof of the genius that would rescue him from his humble

beginnings, presenting a masterfully simple solution to the problem of adding up any set of whole numbers. His reputation spread, and, at the age of 14, he was given a stipend by the duke of Brunswick, so that he could devote himself to science. Before his 25th birthday, Gauss would be famous for his work in mathematics and astronomy.

When Gauss entered the Brunswick Collegium Carolinum in 1792, he already possessed a superior scientific and classical education. He spent three years there, during which he independently discovered Bode's law of planetary distances, the binomial theorem for rational exponents, and the arithmetic–geometric mean, as well as the law of quadratic reciprocity and the prime number theorem. In 1795, he went on to the University of Göttingen, where he constructed a 17-sided regular polygon, using only a ruler and compass, the first construction of a regular figure in modern times. Leaving Göttingen in 1798 without a degree, he returned to Brunswick, where he lived on another stipend from the duke. He received a degree in 1799, for a dissertation submitted to the University of Helmsted on the fundamental theory of algebra.

Gauss now entered upon a period of renowned discoveries. Continuing to pursue his mathematical research, in 1801, he published his famous work, *Arithmetic Disquisitions,* most of which was devoted to number theory. That same year, he made a remarkable prediction of the position in which the asteroid Ceres could be found. Ceres was the first asteroid to be located by astronomers, who later lost track of it. When it was rediscovered, it turned out to be exactly where Gauss, using his least squares approximation method, had calculated it would be.

For the next few years Gauss's personal life underwent a series of triumphs and reversals. He married his beloved first wife, Johanna Ostoff, on October 9, 1805. The following year, however, he lost his patron when the duke of

Brunswick was killed fighting Napoleon. Forced to seek a paying job, Gauss became director of the Göttingen Observatory in 1807, a position he would retain for the next 47 years, until his death at age 78. In 1808, Gauss's father died, and the next year his wife died giving birth to their second son, who died soon afterward. Distraught and left with a son to raise, he married Johanna's best friend, Minna, the next year. They had three children, although the match seems to have been a marriage of convenience.

Despite personal tragedy, and the political turmoil of the times—the French Revolution, the Napoleonic period, and democratic revolutions in Germany—external events never seemed to lessen the flow of Gauss's scholarly output. In 1809, he published his second book, the two-volume *Theory of the Motion of a Celestial Body Following an Elliptical Orbit Around the Sun*, in which he discussed conic sections and elliptic orbits and showed how to refine the estimate of a planet's orbit. He published several works on mathematical subjects and worked on the construction of a new observatory, which was completed in 1816. The following year he began important work at the observatory in the field of geodesy. Asked to conduct a geodesic survey of the state of Hanover, he created new and improved surveying methods, inventing the heliotrope in 1821 to reflect an image of the Sun to mark positions over long distances; it worked by reflecting the Sun's rays by using a design of mirrors and a small telescope. Gauss published over 70 papers on geodesy between 1820 and 1830.

In 1825, he experienced heart problems; since surveying work was physically rigorous and mathematical breakthroughs no longer came so easily to him, he sought a new research direction. In 1831, the year that his second wife died after a long illness, he began working in the field of terrestrial magnetism in collaboration with WILHELM EDUARD WEBER, who had just arrived in Göttingen as a physics professor.

The two men carried out a worldwide magnetic survey from which, in 1839, Gauss derived a mathematical formula expressing magnetism at any location.

During their six productive years together, Gauss and Weber also studied electromagnetism. In 1833, they constructed a moving-needle telegraph similar to that later developed by Charles Wheatstone. However, Gauss failed to scour up financial support for his telegraph, and it was abandoned. Their theoretical work was more successful. Believing that all units should be assembled from a few basic or absolute units, that is, length, mass, and time, Gauss and Weber derived units for magnetism; the centimeter–gram–second (CGS) unit for magnetic flux density was called the gauss. (It has now been replaced by the tesla in the Système International d'Unités [SI] system, which does, however, derive a full set of units from seven basic units in accordance with Gauss's ideas.)

During this period, he formulated what came to be known as Gauss's law, as follows: if the electric flux through an area is defined as the value of the electric field multiplied by the area of a surface perpendicular to the field, then the total of the electric flux out of a closed surface is proportional to the electric charge enclosed. This is a general law applying to any closed surface, for example, a sphere, and represents a language that can be used to simplify the formulation of Maxwell's equations for the electromagnetic field.

After a political dispute forced Weber to leave Göttingen, Gauss's productivity decreased. He did no more magnetic research but continued astronomical observations and mathematical investigations. His students included the great mathematician G. F. B. Riemann. In his later years he made a fortune through shrewd investments in bonds issued by private companies. Gauss died in his sleep, of heart disease, in Göttingen on February 23, 1855.

A versatile genius, he once described the restless process of discovery that fueled his long, productive career:

> When I have clarified and exhausted a subject, then I turn away from it, in order to go into darkness again. The never satisfied man is so strange, if he has completed a structure, then it is not in order to dwell in it peacefully, but in order to begin another.

See also MAXWELL, JAMES CLERK.

Further Reading

Bühler, W. K. *Gauss: A Biographical Study.* New York: Springer-Verlag, 1987.

⊠ **Geiger, Hans Wilhelm**
(1882–1945)
German
Experimentalist, Nuclear Physicist

Hans Wilhelm Geiger helped lay the foundations of nuclear physics by inventing the Geiger counter, the first instrument capable of detecting and determining the quantity of radioactivity.

Geiger was born in Neustadt, Rheinland-Pfalz, on September 30, 1882, the son of a philologist. He studied physics at the Universities of Munich and Erlangen and received a Ph.D. from Erlangen in 1906 for a dissertation on electrical discharges in gases. That year he went to England and became a research assistant at the University of Manchester, where he worked under ERNEST RUTHERFORD from 1907 to 1912. His early postdoctoral work involved the application of his expertise in electrical charges in gases to radioactive decay.

While working with Rutherford, in 1908, Geiger developed the first version of his groundbreaking invention: a device that counted alpha particles, which are emitted in the type of radioactive decay known as alpha decay. The instrument consisted of a metal tube containing a gas (usually argon) at low pressure and a thin wire in the center of the tube. A high voltage was applied between the wire and the tube. When an alpha particle arrived through a window at one end of the tube, it caused the gas to ionize and produced a momentary flow of current. The electrical signals that resulted from this process corresponded to the number of alpha particles entering the tube. The counter enabled Geiger to show that approximately 3.4×10^{10} alpha particles are emitted per second by a gram of radium, and that each alpha particle has a positive charge, with twice the value of the electron charge. Rutherford later demonstrated that the doubly charged alpha particle was in fact a helium nucleus.

Later, Ernest Marsden joined the group as a research assistant. In 1909, he and Geiger studied the interactions of alpha particles from radium with metal reflectors. Although a small number of alpha particles were deflected at wide angles to the beam, most alpha particles passed straight through the metal. They found that the deflection of the alpha particles was a function of the material of which the reflector was made: as the atomic weight of that material decreased, so did the amount of deflection. To show that the deflection was unrelated to the thickness of metal reflectors, they experimented with deflectors of varying thickness. This research led Rutherford to propose, in 1911, that the atom consists of a central positively charged nucleus surrounded by electron shells. In Rutherford's model of the atom, the deflection of the alpha particles occurs when the positively charged particle interacts with the positively charged nuclei in the metal and is deflected by repulsive forces. The mathematical relationship between the amount of alpha scattering and atomic weight was later formulated by Geiger and Marsden.

In 1910 and 1911, Geiger and Rutherford studied the relationship between the range of an alpha particle and its velocity and examined the various disintegration products of uranium. They determined the relationship between the range of an alpha particle and the radioactive constant, together with John Nuttall.

Geiger returned to Germany in 1912 to become head of the Radioactivity Laboratories of the Physikalische Technische Reichsanstalt in Berlin. There he established a successful research group, which included SIR JAMES CHADWICK, who would later discover the neutron. He embarked upon the task of refining his radiation counter and created an instrument capable of detecting beta particles and other kinds of radiation. When World War I broke out, Geiger served in the German artillery; he returned to his position in Berlin at the war's conclusion. He remained there until 1925, when he decided to accept a position as professor of physics at the University of Kiel. In 1928, collaborating with Walther Müller, he developed a more advanced version of the Geiger counter, known as the Geiger–Müller counter, which has the advantage of emitting audible clicks, in response to radioactivity, that are recorded automatically. He used this instrument to confirm the Compton effect, that is, photon–electron scattering, in 1925. Four years later, he moved again, this time to the University of Tübingen. From 1931 on, he devoted himself to the study of cosmic rays. In 1936, he became head of the physics department at the Technical University of Charlottenberg-Berlin and editor of the *Zeitschrift für Physik*.

Geiger's fortunes declined during World War II, as he was debilitated by illness and thereby forced to slow the pace of his research. After the war, when the Allies occupied Germany, he lost his home and possessions. He died on September 24, 1945, in Potsdam.

By discovering an accurate way of detecting radiation, Geiger made an invaluable contribution to the development of nuclear physics.

See also COMPTON, ARTHUR HOLLY.

Further Reading
Stoner, Donald R., and Buford Guy. *Geiger Counter*. New York: American Institute of Physics/American Association of Physics Teachers, 1975.

⊠ Gell-Mann, Murray
(1929–)
American
Theoretician, Particle Physicist

Murray Gell-Mann revolutionized our understanding of the subatomic world by his bold reconstruction of the fundamental assumptions of elementary particle physics. His "eightfold way" theory described the nuclear force in terms of a new fundamental quantity called strangeness, which obeyed a new kind of symmetry principle (called Unitary Symmetry SU[3]). This new theory presupposed a more basic elementary particle, which Gell-Mann called a quark, hidden inside the protons and neutrons of the nucleus. On the basis of this insight, he was able to impose order on the chaos of the particle zoo that was created by the discovery, in high-energy particle accelerator experiments, of some 100 excited states of particles associated with the atomic nucleus. The scope of his achievement was recognized in his selection as the sole winner of the 1969 Nobel Prize in physics.

He was born in New York City on September 15, 1929. His father, Arthur, was a Jewish immigrant from Austria, who learned English so well that he opened an English language school for immigrants in the early 1920s. But he was plagued by business failures and imposed his disappointment, perfectionism, and frustration on his precocious younger son. Gell-Mann's older brother taught him to read and awakened his love of language, science, and art. Known as "Wonder Boy" and "Most Studious" to his classmates at Columbia Grammar, a private school on the Upper West Side

Murray Gell-Mann described the nuclear force in terms of a new fundamental quantity called strangeness, which acts in accordance with a symmetry principle called Unitary Symmetry SU3, which presupposes elementary particles called quarks hidden inside the protons and neutrons of the nucleus. *(AIP Emilio Segrè Visual Archives, Physics Today Collection)*

of New York, Gell-Mann entered Yale University at age 15 and received a B.Sc. in 1948. With an innate talent for many subjects, he did not immediately gravitate to physics. Shortly before his 19th birthday, he was accepted for graduate studies at the Massachusetts Institute of Technology (MIT) in Cambridge by Victor Weisskopf and arrived just in time to watch the competition in quantum electrodynamics between the theorists

JULIAN SEYMOUR SCHWINGER at Harvard and RICHARD PHILLIPS FEYNMAN at Cornell.

After receiving a Ph.D. from MIT in 1951, he spent a year at Princeton University's Institute for Advanced Study. He then moved on to the University of Chicago, where he joined ENRICO FERMI's group and over the next four years rose in the ranks from instructor to associate professor. It was during this period that he was led to his great work by focusing on a puzzling phenomenon in particle physics: the large number of new excited nuclear states (new particles) generated by high-energy particle accelerators appeared quickly but disappeared slowly. This suggested that their creation relied on the strong forces inside the atomic nucleus and that their demise was controlled by the weak forces associated with the slow processes of radioactive decay.

In his quest to understand how this process worked, Gell-Mann proposed a new physical attribute, which he called "strangeness," a quantity analogous to electric charge, which obeyed the unitary symmetry principle called SU(3). Whereas in electromagnetic particle collisions, electric charge is conserved—the total going in equals the total going out—strangeness is conserved in strong and electromagnetic interactions, but *not* in weak interactions. In this theory, strong interactions create a pair of particles with equal and opposite values of strangeness, which cancel each other out. The separated members of such a pair cannot spontaneously decay through a strong interaction, because that would violate conservation of strangeness. Thus, the slower weak interaction, in which the violation of conservation of strangeness *is* allowed to occur, takes over and causes radioactive decay of the particles.

The idea of a theory of nuclear force based on a limited conservation of strangeness proved to be an organizing principle. The unitary symmetry (SU[3]) inherent in Gell-Mann's concept of strangeness led him directly to his eightfold way, a method of sorting all known particles, according to certain general characteristics, into eight "families." In calling this grouping the eightfold way, he was alluding tongue in cheek to Buddhism's teaching that there are eight attributes of right living. The logic of his system was so tight that it revealed the obvious missing family members. In 1953, Gell-Mann published a series of papers predicting specific, as yet undiscovered new particles, as well as other particles

he insisted could not be discovered. His timing could not have been better. Successive experiments confirmed each of his positive predictions and did not contradict the negative ones.

In 1955, Gell-Mann married the archaeologist J. Margaret Dow, of Birmingham, England, with whom he would share a marriage of over 25 years and raise two children, Elizabeth Sarah and Nicholas Webster. The following year he moved to the California Institute of Technology, where his most intense collaboration would be with his new colleague, Richard Feynman. The personal differences between these two giants of theoretical physics, Feynman's earthiness and disdain of formality versus Gell-Mann's high-brow, intellectually polished manner, as well as their competing research styles and theoretical ideas, became legendary. Despite their rivalries, they managed to work together and, in 1958, devised a theory that accounted for many of the phenomena associated with the weak force—the force at work in radioactive decay. In a joint paper, they extended the underlying principles beyond beta decay to other classes of particle interaction, a proposal that experiments would validate many years later. They also suggested the idea that a new kind of current (analogous to electric current, a flow of charge) should be conserved. This concept was later developed and became a basic tool of high-energy physics.

Picking up the thread of his earlier breakthroughs in particle physics, that is, the success of his SU(3) symmetry structure, to explain the nuclear force and produce the eightfold way, Gell-Mann now addressed himself to the question, What is the basic building block (called the irreducible representation) of this symmetry? His answer, given in 1964, when George Zweig came up with a similar solution, was an as yet undiscovered particle that had three different manifestations (hadrons), each holding a fraction of a charge. Gell-Mann gave them the whimsical name *quarks* (rhymes with *corks*) and was later pleased to associate this nomenclature with a line in James Joyce's *Finnegan's Wake:* "Three quarks for Muster Mark!" Quarks were the building blocks of all of the new excited particle states, including the more stable neutron and the proton. Initially, Gell-Mann suggested that hadrons were composed of three quarks or antiquarks, called up, down, and strange (u, d, s), with spin 1/2 and electric charges 2/3, −1/3, −1/3, respectively. (That year, SHELDON LEE GLASHOW proposed that a fourth, charm quark was needed to complete the theory. Later, beauty and truth quarks were added.)

Physicists debated, Were quarks real or mere mathematical quirks? Fractional charge seemed an outrageous suggestion at first—it had never been observed. For this reason, the introduction of quarks was treated more as a mathematical idea than as a proposal for an actual physical object. Gell-Mann himself, who had little use for philosophy, refused to participate in this debate. Whatever the nature of his abstract invention, its predictions continued to be confirmed by particle accelerator experiments.

In terms of the new SU(3) symmetry, the quarks inside the protons and neutrons in the nucleus were permanently confined by forces resulting from the quantum exchange of massive vector particles called gluons. Following an analogy with quantum electrodynamics (QED) theory, Gell-Mann and others later constructed the quantum field theory of quarks and gluons, called quantum chromodynamics (QCD), which can account for all the nuclear particles and their strong interactions.

In 1992, 11 years after losing his wife, Margaret, to cancer, Gell-Mann married Marcia Southwick of Boston. Shortly afterward, he became a professor and Distinguished Fellow at the Santa Fe Institute, where he pursues his interests in natural history, biological evolution, the history of language, and the study of creative thinking. Over the succeeding decade, after discovering the simplicity underlying nature's complexity, Gell-Mann turned to the mystery of

complex adaptive systems, which learn or evolve by utilizing acquired information. He deals with these issues in his book *The Quark and the Jaguar: Adventures in the Simple and the Complex*, in which he relates the basic laws of physics to the complex diversity of the natural world.

He is also concerned about policy matters related to world environmental quality (including conservation of biological diversity), restraint in population growth, sustainable economic development, and stability of the world political system.

By developing the quark theory of the strong interactions, Gell-Mann redefined the concept of an elementary particle and opened the door for physicists to create what is known today as the Standard Model, a global unification of the electroweak theory with the quantum chromodynamic SU(3) strong interaction theory. Even this theory, however, is not complete; the search for a unified particle "theory of everything" continues today.

See also LEDERMAN, LEON M.; SALAM, ABDUS; WEINBERG, STEVEN.

Further Reading

Gell-Mann, Murray. *The Quark and the Jaguar: Adventures in the Simple and the Complex.* New York: W. H. Freeman, 1994.

Johnson, George. *Strange Beauty: Murray Gell-Mann and the Revolution in Twentieth-Century Physics.* New York: Alfred A. Knopf, 1999.

Kane, Gordon. *The Particle Garden: Our Universe as Understood by Particle Physicists.* Cambridge, Mass.: Perseus, 1995.

⊠   **Glaser, Donald Arthur**
(1926–   )
American
Experimentalist, Nuclear Physicist, Particle Physicist, Biophysicist

Donald Arthur Glaser received the 1960 Nobel Prize in physics for his invention of the bubble chamber, a device that gave nuclear physicists a powerful new tool for investigating elementary particles.

He was born on September 21, 1926, in Cleveland, Ohio, to Lena Glaser, a homemaker, and William Glaser, a businessman. After attending public schools in Cleveland Heights, Ohio, he entered the Case Institute of Technology; he graduated with a bachelor's degree in physics and mathematics in 1946. He spent the spring semester of that year teaching math at Case before beginning his graduate studies at the California Institute of Technology. His doctoral thesis was an experimental study of the momentum spectrum of high-energy cosmic rays and mesons at sea level, which he completed in 1949.

After receiving a Ph.D. in 1950, he joined the physics faculty at the University of Michigan, in Ann Arbor, where he remained until 1957. During this period, his research focused on elementary particles. He was especially interested in the strange particles, the transient particles generated in the high-energy collisions of nuclei, and sought new methods of detecting and recording their presence in elementary particle processes. Initially, he experimented with the cloud chamber, which had been developed in the 1920s by the British physicist CHARLES THOMSON REES WILSON. It contained a saturated vapor that condensed as a series of liquid droplets along the ionized path left by a particle crossing the chamber. The cloud chamber played an enormous role in the "golden age of nuclear physics" during the 1930s, when the nuclear particles being investigated had energy ranges on the order of a few million electron volts. With advances in particle accelerators, such as the one built at the European Nuclear Research Center (CERN) in Geneva, in the early 1950s, nuclear physicists began to deal with fast high-energy particles, with energies more than 1000 times larger than those earlier obtainable, which would be missed by the cloud chamber.

In 1952, Glaser solved this problem when he built his first bubble chamber, in which particle tracks are composed of small gas bubbles in a liquid. It consisted of a vessel only a few centimeters across containing superheated liquid ether under pressure. When the pressure was released suddenly, particles traversing the chamber left tracks consisting of streams of small bubbles formed when the liquid boiled locally. The tracks were photographed by using a high-speed camera. In later bubble chambers, supercooled liquid hydrogen was substituted for ether, and chambers two meters across are now in use. Glaser worked on the development of various types of bubble chambers for experiments in high-energy nuclear physics.

He also did experiments on elementary particles at two giant accelerators, the Cosmotron of the Brookhaven National Laboratory in New York and the Bevatron at the Lawrence Radiation Laboratory in California, which yielded information on the lifetime, decay modes, and spins of strange particles, as well as differential cross sections for the production of these particles by pions.

In 1959, Glaser accepted an appointment at the University of California, where he switched his research focus from physics to molecular biology. In 1960, he won the Nobel Prize and married Ruth Bonnie Thompson. He later became professor of physics and biology in 1964.

Because it allowed the physics community to exploit the gigantic atomic accelerators in the United States, Western Europe, and Russia, Glaser's invention of the bubble chamber had an enormous impact on nuclear and particle physics. The bubble chamber has allowed large research teams to make rapid progress in investigating the strange new particles that are formed, transformed, and annihilated when the beam from these machines is directed into the chamber. This has led to a greater understanding of the structure of nuclear particles and the internal quantum forces that allow them to be created and destroyed inside atomic nuclei.

See also RICHTER, BURTON; RUBBIA, CARLO; TING, SAMUEL; CHAO CHUNG.

Further Reading

Wilson, E. J. N. *An Introduction to Particle Accelerators.* Oxford: Oxford University Press, 2001.

⊠   **Glashow, Sheldon Lee**
        (1932–   )
        American
        Theoretician, Particle Physicist,
        Quantum Field Theorist

Sheldon Lee Glashow shared the 1976 Nobel Prize in physics with ABDUS SALAM and STEVEN WEINBERG for his contributions to the creation of the electroweak theory, a quantum field theory that unifies the electromagnetic and weak forces. In particular, he predicted the existence of the fourth, "charmed" quark, which was necessary to generalize the electroweak theory into what is now known as the Standard Model of electromagnetic, weak, and strong interactions. He has remained active in the quest for a Grand Unified Theory (GUT) capable of synthesizing all four interactions—the weak, electromagnetic, strong, and gravitational—within the context of quantum field theory.

He was born on December 5, 1932, in New York City to Lewis Glashow and Bella Rubin Glashow, who had immigrated to the United States in the early years of the century to escape anti-Semitism in tsarist Russia. Lewis Glashow became a successful plumber and secured a comfortable, middle-class life for his family. Having had no opportunity to obtain a higher education, Lewis and Bella were determined that their children would. "In comfort and in love," Sheldon would later recall, "we were taught the joys of knowledge and of work well-done." His brothers chose medicine and dentistry; Sheldon knew from an early age that he would be a scientist, and his parents encouraged him in this direc-

tion. His father built a basement lab for him, where he explored aspects of chemistry and biology and played with frogs.

He attended the highly competitive Bronx High School of Science, where his talented schoolmates included Steven Weinberg, who became a close friend. His interests soon shifted from biology to physics. Glashow, Weinberg, and Gary Feinberg, who also became a physicist, taught each other relativity and quantum mechanics on the subway ride to school. In his Nobel Prize speech, Glashow would thank these friends "for making me learn too much too soon of what I might otherwise not have learned at all." Gregarious, easy-going, anything but a "grind," he seemed to learn physics effortlessly. In his senior year, Glashow was named a finalist in the nationwide Westinghouse Science Talent Search.

Graduating in 1950, Glashow and Weinberg went on to Cornell University together; their class included the mathematical physicist Daniel Kleitman, who would become Glashow's brother-in-law. Glashow was disappointed with the lackluster faculty and turned in a mediocre performance in his classes; he received a bachelor's degree in 1954. Matters improved considerably when he enrolled at Harvard University for his graduate studies. At Harvard, he absorbed the belief that the weak and electromagnetic forces could be explained by a single, unified gauge theory from his thesis adviser, JULIAN SEYMOUR SCHWINGER. In his thesis "The Vector Meson in Elementary Particle Decay," Glashow first explored the possibility of an electroweak synthesis. In 1958, when Glashow completed his work, he and Schwinger planned to write a paper on the topic, but one of them lost the first draft and somehow the project was dropped.

After finishing the work for which he would receive a Ph.D. in 1959, Glashow won a National Science Foundation Post-Doctoral Fellowship, which he hoped would take him to Moscow, to work with Igor Tamm at the Lebedev Institute. These were the cold war years, however, and his visa never materialized. Instead, working at the Niels Bohr Institute in Copenhagen, he was able to construct a unified Yang–Mills theory (a generalization of quantum electrodynamics [QED]) of the weak and electromagnetic interactions, which contained the $SU(2) \times U(1)$ symmetry structure of what would later be called the electroweak theory. However, the theory had the drawback of containing infinities. Glashow tried to renormalize his equations to get rid of the infinities and, once convinced he had done so, presented his results at a conference in London in the spring of 1959. Abdus Salam, an expert on renormal-

ization, who with John Ward had been trying unsuccessfully to do the same thing, was in the audience; he lost little time in invalidating Glashow's results. Undeterred, Glashow continued to look for links between the weak and electromagnetic forces.

During his stay in Europe, he met MURRAY GELL-MANN, who became a champion of his ideas. Gell-Mann presented his theory on the algebraic structure of weak interactions at a 1960 conference in Rochester, New York, and arranged for his protégé to become a research fellow at the California Institute of Technology. Soon afterward, Gell-Mann invented the "eightfold way"

Sheldon Lee Glashow predicted the existence of the fourth, "charmed" quark, necessary for what is now known as the Standard Model of electromagnetic, weak, and strong interactions. *(AIP Emilio Segrè Visual Archives, Physics Today Collection)*

method of grouping elementary particles, which Glashow studied for several years.

In 1961, he became an assistant professor at Stanford University and, independent of Weinberg and Salam, published a paper, "Partial Symmetries of Weak Interactions." In it, he pointed out "remarkable parallels" between electromagnetism and the weak force, depicting them as linked by a broken symmetry, and predicted the existence of the W and Z force-carrying particles. These previously undetected particles would play a significant role in experimental tests of the unified electroweak theory, but Glashow could not predict their masses, with the result that the experimenters were in the dark as to what to look for.

The next year, he moved on to the University of California, Berkeley, as an associate professor; he remained there until 1966. During this time, he continued to develop the phenomenological successes of the SU(3) symmetry applied to the various species of quarks, which were labeled in terms of their flavors (up quark, down quark, strange quark, etc.), and attempted to understand the departures from exact symmetry as a consequence of spontaneous symmetry breaking. Then, in 1964, he became aware of the work of Peter Higgs, demonstrating that the spontaneous symmetry breaking effects of scalar fields, coupled to Yang–Mills gauge theory, could create new kinds of force-carrying particles, some of them massive. This led Glashow to speculate that if the virtual particles that carry the electromagnetic and weak forces (known collectively as the intermediate vector boson W and Z particles) were related by a broken symmetry through a coupling to a scalar Higgs field, it might be possible to estimate their masses in terms of the unified, more symmetrical force from which the two forces were thought to have arisen.

During this same period, Weinberg and Salam, independently of Glashow and of one another, devised similar electroweak theories.

All three theories, however, had the limitation that they could only be applied to leptons, a class of particles that includes electrons and neutrinos, but not to protons and neutrons. Glashow, who in 1966 became Eugene Higgins Professor of Physics at Harvard, studied various means of extending the electroweak theory to include the hadronic class of particles, which, according to Gell-Mann's SU(3) symmetry, was made up of quarks such as protons, neutrons, and mesons. In order to do so, he found it necessary to postulate the existence of a flavor for Gell-Mann's quarks, which he called charm. In 1969, with John Iliopolos and Luciano Maiani he developed the arguments that predicted the existence of charmed hadrons built out of charmed quarks. The prediction would not be verified until 1974, when SAMUEL CHAO CHUNG TING and BURTON RICHTER independently discovered the J/psi particle.

Over the next two years, this extended version of the unified electroweak theory attracted scant attention since, unlike QED, it had not yet been shown to be renormalizable (i.e., capable of being altered by a mathematical procedure that cancels the unwanted infinities in a quantum field theory by introducing the appropriate renormalization constants). All this changed in 1971, when the Dutch physicist GERARD 'T HOOFT used computer algebra techniques to prove that the extended electroweak theory coupled to a Higgs scalar field with spontaneous symmetry breaking was indeed renormalizable. This technique showed that the theory was viable and immediately attracted the attention of the particle physics community. Although experimental verification of the electroweak theory would not occur until 1983, 't Hooft's proof was the final step that led the Nobel Committee to award Weinberg, Salam, and Glashow the 1979 prize in physics.

In the intervening years between the formulation of the electroweak theory and its international recognition, Glashow pushed his ideas further. In 1974, he and his Harvard colleague Howard Georgi correctly predicted that the effects of charmed quarks would be discovered in neutrino physics and in particle production processes generated by high-energy electron positron annihilation. Continuing their fruitful collaboration, they suggested that a connection analogous to that between the electromagnetic and weak forces might be forged with the strong force. Their proposed "grand unification" of three of the four forces in the universe differed in one essential way from that of the electroweak theory; whereas the electroweak theory argued that the electromagnetic and weak forces crystallized out of their symmetric union when the temperature of the universe dropped to about a 10^{15} K above absolute zero, Georgi and Glashow showed that the union of the electroweak force with the strong force would have been apparent only at a temperature some 10^{12} times higher. From the point of view of energy, this is about 10^{15} times the mass of the proton or about four orders of magnitude less than the Planck mass (the mass energy at which ALBERT EINSTEIN's gravitation theory must be quantized). Thus, their idea of a GUT boldly took theoretical physics into an energy realm beyond that which anyone had dared explore, namely, that of the early universe, when the quantization of gravity became important.

In his Nobel Prize speech, Glashow commented on the relationship of the extended electroweak theory (which by then had become known as the Standard Model) and GUTs:

While keeping an open mind with respect to the source of the next big breakthrough, Glashow has pursued his own vigorous path as a researcher and educator. He has written a number of technical and popular science books, including *Lie Algebras in Particle Physics*, with Howard Georgi (1999); *The Charm of Physics* (1995); *From Alchemy to Quarks: The Study of Physics as a Liberal Art* (1994); and *Interactions: A Journey Through the Mind of a Particle Physicist and the Matter of This World*, with Ben Bova (1988). His most recent book project deals with a favorite activity of his: solving physics problems while engaged in a game of billiards.

He has had visiting professorships to CERN, the University of Marseilles, the Massachusetts Institute of Technology, Texas A&M University, and Boston University. He has also been a consultant to the Brookhaven National Laboratory and affiliated senior scientist at the University of Houston.

He lives with his wife, the former Joan Shirley Alexander, whom he married in 1972 and with whom he has had four children. In recent years, he has moved to Boston University, where he is Arthur G. B. Metcalf Professor of Physics.

Glashow's current work involves grand unified theories and the search for antimatter in the universe, focusing on the theoretical consequences of such a discovery. He continues to believe that there will always be another fundamental discovery just around the corner:

Can we really believe that nature's bag of tricks has run out? Have we finally reached the point where there is no longer a new particle, a "fifth" force, or a bewildering new phenomena to observe? Of course not. Let the show go on!

See also NE'EMAN, YUVAL; RUBBIA, CARLO; VELTMAN, MARTINUS J.G.

Further Reading

Crease, Robert P. *The Second Creation: Makers of the Revolution in Twentieth-Century Physics.* Rutgers, N.J.: Rutgers University Press, 1995.

Glashow, Sheldon L. *The Charm of Physics.* New York: Springer-Verlag, 1991.

———. *Interactions: A Journey Through the Mind of a Particle Physicist and the Matter of This World.* New York: Warner Books, 1988.

⊠   Goeppert-Mayer, Maria Gertrude
(1906–1972)
German/American
*Theoretical Physicist, Nuclear Physicist,
Physical Chemist*

Maria Goeppert-Mayer was a brilliant theoretical physicist who developed a shell model theory that explained the arrangement of the protons and neutrons in the atomic nucleus. After a lifetime of underrecognition, she was awarded the Nobel Prize in physics in 1963 for this work, the second woman to be honored in this way.

She was born on June 28, 1906, in Katowice, Germany (now part of Poland), into a family of academics: seven generations of university professors. Her father, Friedrich, a pediatrician, moved the family to Göttingen, when Maria was four, on his appointment to the medical faculty at the University of Göttingen. Maria inherited her scientific bent from her father, whom she idolized, and a love of music and social life from her mother. In 1924, after attending a small women's preparatory school, she was admitted to the University of Göttingen, which had very few women students in those years. She began studying mathematics, but Göttingen was then a renowned world physics center. Intrigued by the exciting developments of atomic physics and the new quantum mechanics, she changed her major in 1927. Working under the great quantum theorist MAX BORN, in 1930, she completed her Ph.D. thesis, in which she calculated the probability that an electron orbiting an atomic nucleus would emit a two-photon state of light as it jumped to an orbit closer to the nucleus. This bold result would be experimentally confirmed 30 years later.

While Maria was a graduate student, her father died and her mother began to take in boarders, one of whom was an American chemistry student, Joseph Edward Mayer, whom Maria married in 1930. The young couple moved to the United States, where Joseph became an assistant professor of chemistry at Johns Hopkins University in Baltimore. Because of the antinepotism laws common at the time, which forbade the hiring of both husband and wife, as well as the economic hardship of the Great Depression, Johns Hopkins could not offer her a paying academic job. Her only income was from helping a professor with his German correspondence; but she kept doing physics, for the sheer love of it. Encouraged by her husband, she began to explore physical chemistry and did significant research in that area.

In 1933, the year she became a naturalized U.S. citizen, her daughter Maria Anne was born. When her child was four, she began working with her husband on a textbook on statistical mechanics describing the behavior of molecules, which was eventually published in 1940. In 1938, just before her husband took a job at Columbia University, she gave birth to her second child, Peter.

As World War II loomed, Goeppert-Mayer and her husband joined a group of expatriate scientists who urged the United States to develop an atomic bomb. She was offered her first paid work as a scientist to search for ways to separate the bomb's potential fuel, the radioactive form of uranium, from the more common, nonradioactive form. However, her work had little direct effect on the bomb project.

In early 1946, she and her husband were invited to join the University of Chicago's new Institute for Nuclear Studies (later the Enrico Fermi Institute), where the faculty included EDWARD TELLER and ENRICO FERMI. Goeppert-Mayer was made an unsalaried associate professor and earned a part-time salary as a senior researcher at the Argonne National Laboratory. The Chicago group was one of the most brilliant gatherings of scientists in the 20th century. A frequent topic of their weekly seminars was the arrangement of protons and neutrons in the atomic nucleus. At the time, the most commonly accepted model of the atomic nucleus pictured it as something like a drop of water, in which protons and neutrons moved randomly. However, physicists also knew that electrons orbited the atomic nucleus in distinct layers called shells, and by analogy some physicists had suggested that the protons and neutrons in the atomic nucleus might also be arranged in shells. They based their supposition on findings made in the period between 1920 and 1930 that protons and neutrons each formed particularly stable systems in an atomic nucleus when the number of either kind of nucleon was one of the so-called magic numbers: 2, 8, 20, 50, 82, and 126. Several physicists tried to interpret these magic numbers by analogy with NIELS HENRIK DAVID BOHR's successful explanation of the periodic table of the elements. In this picture it was then assumed that the nucleons move in orbits in a common field of force and that these orbits are arranged in so-called shells that are energetically well separated. The magic numbers should correspond to completed shells of nucleons. Although this interpretation was successful for light nuclei, it could only explain the first three magic numbers (2, 8, 20), and for many years no further physical or mathematical evidence was found to support the shell model of the nucleus.

In 1946, while working with Edward Teller on a theory of the origin of chemical elements, Goeppert-Mayer noticed that elements whose nuclei contained certain numbers of protons or neutrons (i.e., magic numbers again) were unusually abundant and stable, never undergoing radioactive decay. Searching for the reason for these magic numbers, she hypothesized that they had some relation to the arrangement of protons and neutrons in shells. She proposed that just as filled electron shells prevented elements from reacting chemically, the magic number of protons and neutrons represented the filled shells in the nucleus that prevented atoms from breaking down radioactively. In this picture the nucleus was like an onion with the protons and neutrons revolving around each other in

Maria Gertrude Goeppert-Mayer developed the nuclear shell model theory, which explained the arrangement of the protons and neutrons in the atomic nucleus. *(AIP Emilio Segrè Visual Archives, Physics Today Collection)*

layers. The magic numbers represented the most abundant elements because in these elements the nuclear particles were very tightly bound together, moving in a clockwise–counterclockwise pattern. These tightly fitting nuclei would not change properties with any other elements, and that was why such elements as tin and lead were so abundant.

The way to prove her idea occurred to Goeppert-Mayer in a discussion with Enrico Fermi when he asked her whether there were any possibility that spin-orbit coupling might play an important role in the shell model of the nucleus. Spin-orbit coupling is a process whereby the direction in which a particle spins helps determine which orbit or shell it will occupy. Whereas spin-orbit coupling among electrons is weak, Goeppert-Mayer realized that if it were powerful in the nucleus, requiring much more energy for particles to spin in one direction than in the other, it could explain her magic numbers and prove that nuclear particles are arranged in shells. Her idea was that a nucleon's energy should differ when it "spins" in the same direction from that when it spins in the opposite direction as it revolves around the nucleus. Fermi's question was the key to unlocking this brilliant insight. Ten minutes after he asked it, she was bombarding him with the mathematical proof of her theory.

Her landmark paper on this topic, published in 1948, marked the beginning of a new era in the acceptance of the shell model. Goeppert-Mayer had given convincing evidence for the existence of the higher magic numbers, noting, moreover, that experiments strongly supported the existence of the closed nuclear shells. At about the same time, Hans D. Jensen, a German physicist, developed and published a similar idea. The two physicists met in 1950 and five years later published their important book *Elementary Theory of Nuclear Shell Structure*. Her achievements at last opened academic doors, and, in 1959, both Goeppert-Mayer and her husband were invited to join the faculty of the University of California at La Jolla in San Diego, California.

Although a few months later, she suffered a stroke, she was able to continue her research, refining her shell model theory of the nucleus.

In 1963, she shared the Nobel Prize in physics with Jensen and EUGENE PAUL WIGNER, the second woman, after MARIE CURIE, to attain that distinction. She died of heart disease on February 20, 1972, at the age of 65, in La Jolla.

Goeppert-Mayer's work created new insight into the nature of nuclear structure, the process of radioactive transformation of the elements, and the properties of the forces that can hold nuclei together in closed nuclear shells. The existence of such closed shell nuclei, which cannot radioactively transform into other elements, explains why there is an abundance of stable elements in nature.

Further Reading
Dash, Joan. *A Life of One's Own*. New York: Harper & Row, 1973, pp. 231–346.

McGrayne, Sharon Bertsch. *Nobel Prize Women in Science: Their Lives, Struggles, and Momentous Discoveries*. New York: Birch Lane Press, 1993.

Sachs, Robert. *Maria Goeppert-Mayer, 1906–1972, a Biographical Memoir*. Washington, D.C.: National Academy of Sciences, 1979.

H

Hawking, Stephen
(1942–   )
British
Theoretician, Relativist, Astrophysicist,
Cosmologist

One of the best known physicists alive today, Stephen Hawking developed groundbreaking mathematical theorems delineating the physical requirements dictated by general relativity for the astrophysical formation of black holes. He later shocked the physics community with his discovery of a physical mechanism that required that black holes must always emit radiation (known as Hawking radiation), leading to their eventual evaporation. His ideas on cosmology have deeply influenced our current understanding of space and time, black holes, and the origin of the universe.

Hawking was born on January 8, 1942, in Oxford, England, where his mother had retreated from the family's London home to escape the German bombing during World War II. Soon afterward, they returned to London, where Stephen's father worked as a medical researcher. When Stephen was eight, his father took a job at the Institute for Medical Studies in Mill Hill and the family moved to Saint Albans, about 20 miles north of London, to shorten his commute. Stephen first attended the Saint Albans School for Girls (where boys up to the age of 10 were accepted) then transferred to Saint Albans School, at age 11.

He wanted to study mathematics, but his father was determined he should attend his alma mater, University College, at Oxford, which did not have a mathematics fellow. So it was that, on winning a scholarship, he studied physics and natural science instead. At Oxford, where the prevailing attitude was that "you were supposed to be brilliant without effort," he coasted for three years, without working very hard, and just managed to obtain a first-class honors degree in natural science.

He continued on to Cambridge University to do research in cosmology and general relativity under Dennis Sciama. But during his first semester, the clumsiness and unexplained falls he had begun to experience during his last year at Oxford grew more worrisome. Persuaded by his mother to see a doctor, he underwent a series of diagnostic tests. The verdict was amyotrophic lateral sclerosis (ALS), popularly known as Lou Gehrig's disease, a degenerative, incurable neuromuscular illness. Hawking had just turned 21. Although the doctors offered him no hope and no treatment, he found his own source of strength:

My dreams at that time were very disturbed. Before my condition had been

133

Stephen Hawking developed theorems delineating the physical requirements for the astrophysical formation of black holes. His ideas have deeply influenced our current understanding of space and time and the origin of the universe. *(AIP Emilio Segrè Visual Archives, Physics Today Collection)*

diagnosed, I had been very bored with life. There had not seemed to be anything worth doing. But shortly after I came out of hospital, I dreamt that I was going to be executed. I suddenly realized there were a lot of things I could do if I were reprieved. . . . I found to my surprise that I was enjoying life in the present more than before.

Hawking fell in love with Jane Wilde, whom he had met just before learning his diagnosis and soon became engaged to her. With

something to work for, he returned to Cambridge, continued his research, and earned a doctorate in 1966. He became a research fellow and later a professorial fellow at Cambridge's Gonville and Caius College. In 1973, he left the Institute of Astronomy and joined the Department of Applied Mathematics and Theoretical Physics, where, since 1979, he has held the highly prestigious post of Lucasian Professor of Mathematics, once filled by SIR ISAAC NEWTON.

Between 1965 and 1970, Hawking worked on the thorny issue of singularities in classical gravitational theory. A singularity is a hypothetical region in space in which gravitational forces cause matter to be infinitely compressed and space and time to become infinitely distorted. Singularities seemed to be associated with the gravitational collapse process described by ALBERT EINSTEIN's general theory of relativity. In collaboration with Roger Penrose, Hawking developed a series of reasonable mathematical physics theorems that showed that the general theory of relativity contained singularities, which implied that space and time would have a beginning in the big bang and an end in black holes. The development of these Penrose–Hawking theorems led relativistic astrophysicists to devise new mathematical techniques for studying this area of cosmology.

From 1972 to 1975, stimulated by a series of meetings and discussions with Kip Thorne and the eminent Russian physicist Yakov Zeldovich, Hawking made a startling discovery. By effecting a "partial marriage" between quantum field theory and general relativity, he showed that black holes must always emit radiation. This was perhaps his most spectacular achievement: the discovery that black holes are not really black; rather, they emit every possible type of elementary particle within the thermal spectrum of a hot body, ultimately ending their lives in a general explosion.

During this same period Hawking studied the creation of the universe and postulated that after the big bang, many objects as heavy as 10^9 tons but only the size of a proton were created.

Because of their large gravitational attraction, these primordial mini black holes would be governed by general relativity, but their subatomic size would require them to obey the laws of quantum mechanics. To date, however, observational searches for the characteristic Hawking radiation emitted by these primordial black holes have not been successful.

Hawking's next major work was a logical continuation of this partial marriage of quantum field theory and general relativity. In 1983, with Jim Hartle of Santa Barbara, he made the remarkable proposal that our universe and all it contains are in a unique quantum state having no boundary or edge. The no-boundary universe is one in which the universe does not start with a singularity. The hypothesis uses the American physicist RICHARD PHILLIPS FEYNMAN's idea of treating quantum mechanics as a "sum over histories," meaning that a particle does not have one history in spacetime but instead follows every possible path to reach its current state. By summing these histories—a difficult process that must be done by treating time as imaginary—you can find the probability that the particle passes through a particular point.

Hawking and Hartle then wedded this idea to general relativity's view that gravity is just a consequence of curved spacetime. Under classical general relativity, the universe either has to be infinitely old or has to have started at a singularity. But Hawking and Hartle's proposal raises a third possibility—that the universe is finite but had no initial singularity to produce a boundary. The geometry of the no-boundary universe would be similar to the geometry of the surface of a sphere, except that it would have four dimensions instead of two. You can travel completely around Earth's surface, for instance, without ever encountering an edge. In this analogy, unfolding in imaginary time, Earth's North Pole represents the big bang, marking the start of the universe. However, just as the North Pole is not a singularity, neither is the big bang.

As Hawking writes:

Both time and space are finite in extent, but they don't have any boundary or edge . . . there would be no singularities, and the laws of science would hold everywhere, including at the beginning of the universe.

In the mid-1970s, when he did this important work, despite his physical handicaps, Hawking was managing to lead a rich professional and personal life. He and Jane had three children, and until 1974, with Jane's help, he was relatively independent. However, as his condition progressed, the couple employed outside help, including a live-in research assistant and part-time nurses. In 1984, Hawking completed the first draft of what would be his phenomenally selling popular science book, *A Brief History of Time*.

But the following year, while visiting the giant particle accelerator at the European Center for Nuclear Research (CERN) in Geneva, Hawking contracted pneumonia, underwent a tracheotomy, and lost his voice. His speech had become increasingly slurred over the years, but now it was gone altogether. David Mason, of Cambridge Adaptive Communication, fitted a small portable computer and a speech synthesizer to his wheelchair. Using this system, he has written books and papers and given a large number of scientific and technical talks. After 25 years of marriage, Stephen and Jane divorced. He subsequently married Elaine Mason, one of his nurses.

His technical books include *The Large Scale Structure of Spacetime*, with G. F. R. Ellis (1989); *General Relativity: An Einstein Centenary Survey*, with W. Israel (1980); *300 Years of Gravitation*, with W. Israel (1990), and *The Nature of Space and Time*, with Ruger Penrose (1996). His popular books include *A Brief History of Time* (1998), *Black Holes and Baby Universes and Other Essays* (1994), *Stephen Hawking's Universe: The Cosmos Explained* (1998), *The Universe in a Nutshell*

(2001), *The Theory of Everything*, (2002), *The Future of Spacetime* (2002), and *On the Shoulders of Giants: The Great Works of Physics and Astronomy* (2002).

Hawking continues to do research, write, and travel all around the world to lecture. In January 2002, physicists gathered at Cambridge University to honor him on his 60th birthday.

Although Hawking modestly dismisses talk of himself as a "modern day Einstein" as media hype, there is no doubt about the substantial contributions he has made, both to astrophysics and to cosmology, and to the popular fascination with these frontiers of knowledge. With clarity and incisive humor, he has consistently raised issues that lie on the boundary between physics and philosophy.

See also CHANDRASEKHAR, SUBRAMANYAN; GAMOW, GEORGE; WHEELER, JOHN ARCHIBALD.

Further Reading

Hawking, Steven. *A Brief History of Time: From the Big Bang to Black Holes*. New York: Bantam Books, 1998.

———. *The Future of Spacetime*. New York: W. W. Norton, 2002.

———. *Stephen Hawking's Universe: The Cosmos Explained*. New York: Basic Books, 1998.

———. *The Universe in a Nutshell*. New York: Bantam Books, 2002.

⊠   Heisenberg, Werner
(1901–1976)
German
Theoretical Physicist, Quantum Theorist

One of the greatest of 20th-century physicists, Werner Heisenberg is recognized as the founder of quantum mechanics. His most famous discovery was the uncertainty principle, which introduced the notion of probability into the complex of startling ideas on which the structure of a quantum mechanical view of nature was being built. For this dubious feat—depriving physics of modern science's most hallowed tenet, the deterministic idea of cause and effect—he was awarded the 1932 Nobel Prize in physics. If Heisenberg was fortunate to begin his career at a time of exciting breakthroughs in physics, the historical circumstances of his middle years posed agonizing choices. Choosing to remain in his beloved Germany through the Nazi era, he managed to avoid internment as a "white Jew" and eventually played a major role in the failed attempt to develop a German atomic bomb. The price he may have paid for his survival remains controversial to this day.

Werner Karl Heisenberg was born on December 5, 1901, in Würzberg, in the southern German state of Bavaria. His father, August Heisenberg, was professor of Greek at the University of Munich; his mother, Anna, who converted from Catholicism to her husband's Lutheranism, was the daughter of the principal of the gymnasium Werner and his older brother, Erwin, would attend. Erwin later became a chemist. As a prestigious academic family, the Heisenbergs participated in the cultural and social life of the upper middle class.

In 1911, Werner entered the Maximilians Gymnasium in Munich; nine years later he would graduate at the top of his class. But these schoolboy years were anything but idyllic: the outbreak of World War I produced danger and physical deprivation. Heisenberg's generation, which was indoctrinated with German nationalism, felt betrayed by their elders when Germany was defeated in 1918 and the monarchy collapsed. Heisenberg appears to have found refuge from political and social turmoil in two parallel, but quite different, directions. The first was his fascination with the mathematics of the number system, because, he said, "It's clear, everything is so that you can understand it to the bottom." The second was his role as leader of a small group of younger boys, associated with the New Boy

Scouts, who sought the essence of German culture in nature, music, and poetry. His "boys" would be the companions of his leisure, hiking and camping in and around Germany with him, until Hitler banned all independent youth groups in 1933.

Upon entering the University of Munich in 1920, Heisenberg decided to study theoretical physics under ARNOLD JOHANNES WILHELM SOMMERFELD and soon produced a publishable contribution to the early quantum mechanics of the atom. He formed a lasting friendship with WOLFGANG PAULI, who, in 1922, began to cast doubt on NIELS HENRIK DAVID BOHR's results. Completing his doctoral dissertation on turbulence in fluid streams in 1923, Heisenberg became MAX BORN's assistant at Göttingen. From 1924 to 1925, he took a leave of absence to work with Bohr in Copenhagen. In later years, he described his apprenticeship by saying, "I learned optimism from Sommerfeld, mathematics at Göttingen, and physics from Bohr."

During these dynamic years in Munich, Göttingen, and Copenhagen, Heisenberg attacked the three areas of research in which the failure of theory to explain experimental results had led to talk of a "crisis" in quantum theory: the study of light emitted and absorbed by atoms (spectroscopy), the predicted properties of atoms and molecules, and the question of whether light behaved as a wave or a particle. Heisenberg was less interested in creating a picture of what happens inside the atom than in finding a mathematical system that explained the features of the atom—in this case the position of the spectral lines of hydrogen, the simplest atom. Since the electron orbits in atoms could not be observed, Heisenberg tried to develop a quantum mechanics without them. He relied instead on what can be observed, namely, the light emitted and absorbed by atoms. He used a modification of the quantum rules to explain the anomalous Zeeman effect, in which single spectral lines split into groups of closely spaced lines in a strong magnetic field. Information about atomic

structure can be deduced from the separation of these lines. His collaboration with Born and Born's other assistant, Pascual Jordan, resulted in a famous "three-man paper" in 1925, introducing a system called matrix mechanics. By mathematical treatment of values within matrices or arrays, the frequencies of the lines in the hydrogen spectrum were obtained. Heisenberg had formulated the first precise mathematical description of the workings of the atom, thus becoming the founder of quantum mechanics.

The next year, however, Heisenberg's formulation was challenged when ERWIN SCHRÖDINGER presented a system of wave mechanics that

Werner Heisenberg's uncertainty principle introduced the notion of probability into the complex ideas on which the structure of a quantum mechanical view of nature was being built. *(AIP Emilio Segrè Visual Archives. Photo by A.B. Truckeri)*

accounted mathematically for the 1923 discovery of LOUIS VICTOR-PIERRE, PRINCE DE BROGLIE, that electrons do not occupy orbits but exist in standing waves around the nucleus. Wave mechanics, with its more familiar concepts and equations and its ability to visualize the atom, rapidly became the theory of choice. In May 1926, Schrödinger published an article proving that wave and matrix mechanics were equivalent mathematically, while claiming the superiority of wave mechanics. Heisenberg's opinion, expressed in a letter to Pauli, was unambiguous:

> The more I think about the physical portion of Schrödinger's theory, the more repulsive I find it. . . . What Schrödinger writes about the visibility of his theory "is probably not quite right," in other words it's crap.

With the addition of Born's statistical interpretation of the wave function and the unified equations known as transformation theory, created by Jordan in Göttingen and PAUL ADRIEN MAURICE DIRAC in Cambridge, England, the mathematical foundation of the new quantum mechanics seemed complete. The search was on for an "interpretation" of the mathematics capable of linking observations in the macroscopic world, in which classical physics prevails, to events and processes in the quantum world of the atom.

In February 1927, Heisenberg, by uncovering a problem in the way physicists could measure basic physical variables appearing in quantum equations, formulated his famous uncertainty principle:

> The more precisely the position (of an atom) is determined, the less precisely the momentum is known in this instant, and vice versa.

Heisenberg showed that this uncertainty, sometimes referred to as the principle of indeterminacy, is not the fault of the experimentalist, but inherent in quantum mechanics. Hence, the uncertainty principle negates cause and effect, maintaining that the result of an action can only be expressed in terms of the probability that a certain effect will occur. Heisenberg's revolutionary idea, together with Bohr's complementarity principle and Schrödinger's wave function, became the cornerstones of the Copenhagen interpretation of quantum mechanics. Over the protests of so prominent a holdout as ALBERT EINSTEIN, who could not accept the notion of probability, the Copenhagen interpretation was accepted by the greater part of the international physics community.

At the historic Solvay Physics Conference held in Brussels in October 1927, Heisenberg declared the new quantum mechanics to be complete and irrevocable. At the end of his life, looking back at this time, he would say that the next five years

> looked so wonderful that we often spoke of them as the golden age of atomic physics. The great obstacles that had occupied all our efforts in the preceding years had been cleared out of the way; the gate to an entirely new field, the quantum mechanics of the atomic shells, stood wide open, and fresh fruits seemed ready for the picking.

Heisenberg's personal prospects could hardly have been better: Only 25, he had just been offered the Chair of Theoretical Physics at the University of Leipzig. That very year, he made another major breakthrough by solving the mystery of ferromagnetism. By using the Pauli exclusion principle, which states that no two electrons can have four identical quantum numbers, he proved that ferromagnetism is caused by electrostatic interaction between the electrons. Heisenberg would remain at Leipzig until 1942 and make it a first-rate international research center. Collaborating with Pauli and others, he made important progress in joining quantum mechanics and relativity theory into a relativistic quantum theory of "fields" (such as electromagnetic or material fields). At a time when high-energy particle accelerators did not yet exist, he and Dirac pioneered high-energy physics research by examining the highly energetic particles entering the Earth's atmosphere from outer space (cosmic rays).

When Hitler gained power in 1933 and attacked relativity theory and quantum mechanics as "Jewish physics," Heisenberg, who had made the decision to remain in his homeland, defended the new physics and tried to preserve what was left of German science. But in 1937, Heisenberg found himself the target of attack. Branded by an official newspaper as a "representative of Judaism in German spiritual life who must be eliminated just as the Jews themselves," he endured a frightening year-long investigation, which ended in his exoneration. In that year, he met and married Elisabeth Schumacher, a student of German literature, who was to bear him seven children.

Heisenberg went on to serve as director of the Kaiser Wilhelm Institute for Physics in Berlin from 1942 to 1945 and led Germany's atomic research effort. Given its preeminence in nuclear fission research, Germany might well have developed an effective nuclear weapons program. The fact that it did not may have been due simply to lack of resources. But Heisenberg's defenders suggest that he was reluctant to hand the Nazis nuclear weapons and sabotaged the effort by diverting research to peaceful uses of nuclear energy. Historians continue to debate the issue hotly.

After the war, he was briefly interned in Britain along with other leading German scientists. From 1946 to 1970, he directed Max Planck Institute for Physics and Astrophysics in Munich. Even though his collaboration with Pauli to find a consistent unified quantum theory of elementary particles failed, he was successful in making his institute an internationally renowned research center. In his later years, he wrote two books on the philosophy of physics: *Physics and Philosophy* (1962) and *Physics and Beyond* (1971). He died of cancer at his home in Munich on February 1, 1976.

Further Reading

Cassidy, David. "Heisenberg, Uncertainty and the Quantum Revolution." *Scientific American* 266: 106–112.

———. *Uncertainty: The Life and Science of Werner Heisenberg.* New York: W. H. Freeman, 1992.

Heisenberg, Werner. *The Physicist's Conception of Nature.* Westport, Conn.: Greenwood, 1970.

Rose, Paul Lawrence. *Heisenberg and the Nazi Atomic Bomb Project: A Study in German Culture.* Berkeley: University of California Press, 1998.

⊠ **Helmholtz, Hermann Ludwig Ferdinand von**
(1821–1894)
German
Theoretician (Thermodynamics, Hydrodynamics, Optics, Acoustics), Physical Chemist

Hermann Helmholtz's major contribution to physics is generally considered to be the first precise formulation of the law of conservation of energy. However, his impact on 19th-century science transcends the value of any specific discovery. A polymath of great intellectual range, deeply concerned with the implications of science for philosophy and culture, Helmholtz was the dominant figure in German science in the mid-19th century.

He began his life in Potsdam on August 31, 1821, the eldest of four children born to a mother who descended from William Penn, the founder of Pennsylvania, and a father who taught philosophy and literature at the local gymnasium. A delicate child, Hermann eagerly absorbed the erudition showered on him by his father. In addition to imbuing in him a love of the fine arts, his father taught him Latin, Greek, Hebrew, French,

Italian, and Arabic. As a student at his father's gymnasium, Hermann showed a talent for physics. The family was poor, however, and, since scholarships were awarded only in popular fields, Hermann decided to study medicine. In exchange for financial aid, he obligated himself to 10 years of service as an army doctor. In 1838, he entered the Friedrich Wilhelm Institute of Medicine and Surgery in Berlin. In his spare time, he attended courses at the university on chemistry, studied advanced mathematics on his own, and became an expert pianist.

His research career began in 1841 with work on a dissertation on the relationship between nerve fibers and nerve cells of invertebrates. In this work, he rejected the accepted view that living things possess an innate vital force, arguing that life processes obey the same principles that govern nonliving systems. Physiology, he proposed, should be based wholly on the principles of physics and chemistry. After receiving an M.D. in 1842, he became a surgeon with his regiment in Potsdam, where he spent all his time conducting experiments in the laboratory he had set up in the barracks.

In 1847, Helmholtz published an important paper, presenting the mathematical principles behind the principle of conservation of energy. He derived a general equation in which the kinetic energy of a moving body is equal to the product of the force and distance through which the force moves to bring about the energy change. This equation could be applied in many fields to show that energy is always conserved. It led to formulation of the first law of thermodynamics, which states that the total energy of a system and its surroundings remains constant even if it changes from one form of energy to another. He demonstrated that in various situations—collisions, expansion of gases, muscle contraction—in which energy appears to be lost, it is in fact converted into heat energy. Helmholtz examined the applications of this work in such fields as electrostatics, galvanic phenomena, and electrodynamics.

The quick recognition of Helmholtz's valuable work led to his early release from his military duties, in 1848. The following year he married Olga Von Velten and settled down to an academic career as associate professor of physiology at Konigsberg University. While there, he developed the three-color theory of vision first proposed, in 1801, by THOMAS YOUNG, demonstrating that a single primary color must also affect retinal structures sensitive to the other primary colors. This hypothesis successfully explained the color of after-images and the effects of color blindness. He also developed the ophthalmoscope, which is used to examine the retina, and the ophthalmometer, which measures the curvature of the eye.

In 1855, Helmholtz moved to Bonn to become professor of anatomy and physiology; there he made important discoveries in the physiology of vision and hearing by studying nerve impulses. He would move yet again, in 1858, to become professor of physiology at the University of Heidelberg. That same year, he wrote an important paper on hydrodynamics, in which he established the mathematical principles that define motion in a vortex.

Helmholtz's first years at Heidelberg were beset with personal losses. First, his father died in 1858, and then, at the end of 1859, his ailing wife died. Left to raise two young children on his own, Helmholtz remarried within 18 months. His second wife, Anna von Mohl, the daughter of another professor at Heidelberg, was much younger than he. Worldly and handsome, she would bear him three children as well as expanding his social world.

His next major work was an 1862 study on acoustics, examining musical theory and the perception of sound. He explained the origin of music on the basis of fundamental physiological principles. He also formulated a theory of hearing, correctly suggesting that structures within the inner ear resonate at particular frequencies so that both pitch and tone can be perceived.

Around 1866, he began to move away from physiology and more toward physics, engaging in the heated discussions on the properties of non-Euclidean space that preoccupied many scientists in the late 19th century.

In 1871, Helmholtz accepted the chair of physics at the University of Berlin, where he worked on thermodynamics in physical chemistry and attempted to establish a mechanical foundation for thermodynamics. Throughout the 1870s he was embroiled in debates with WILHELM EDUARD WEBER on the compatibility of Weber's electrodynamics with the principle of conservation of energy. The question became moot when, in the 1880s, JAMES CLERK MAXWELL's theory of electrodynamics was widely accepted. Helmholtz then derived Maxwell's equations from the least action principle, a mathematical technique for deriving field equations.

In 1882, he derived an expression that relates the total energy of a system to its free energy (which is the portion that can be converted to forms other than heat) and to its temperature and entropy. His findings enabled chemists to determine the direction of a chemical reaction. In 1887, he became director of the new Physical-Technical Institute of Berlin. This was the last post he would hold. After a period of poor health, he died in Berlin on September 8, 1894.

In his staggeringly diverse inquiries, Hermann Helmholtz pursued his lifelong ambition to understand the unifying principles in nature, as well as the subjective sources of knowledge—the sense organs—that mediate experience. He was the brilliant representative of a generation who emphatically rejected metaphysics in favor of mathematics and mechanics. Among the countless students he influenced, not the least was HEINRICH RUDOLF HERTZ, whom he set upon the path that would lead him to the discovery of radio waves.

A classical physicist himself, Helmholtz would inspire the generation who would create the revolutions of relativity and quantum mechanics, imbuing them with his high mathematical and experimental standards.

Further Reading

Cahan, David, ed. *Hermann Von Helmholtz and the Foundations of Nineteenth-Century Science.* California Studies in the History of Science, No. 12. Berkeley: University of California Press, 1993.

Helmholtz, Hermann Von, and David Cahan, eds. *Science and Culture: Popular and Philosophical Essays.* Chicago: University of Chicago Press, 1995.

⊠ Henry, Joseph
(1797–1878)
American
*Experimental Physicist
(Electromagnetism)*

Joseph Henry was the greatest experimental physicist in America in the mid-19th century. His most enduring work was in the field of electromagnetism; he built extremely large electromagnets and independently discovered the dynamics of electromagnetic induction at the same time as MICHAEL FARADAY was making similar discoveries in Great Britain. Henry's work with electromagnetism played a crucial role in Samuel F. B. Morse's invention of the telegraph. Later in his career, as the first secretary of the Smithsonian Institution, he played a pivotal role in the organization and development of American science.

Joseph Henry's life exemplified the American ideal of the self-made person who rises to eminence from modest origins. Born on December 17, 1797, in Albany, New York to Ann Alexander and William Henry, a day laborer of Scottish descent, he received little schooling and was obliged to work in a general store after school hours. At age 13, he was apprenticed to a watchmaker, learning skills that would later serve him well in the physics laboratory. In 1819, he was able to enroll in the Albany Academy, a school for boys from the elementary grades up to the college level, which charged no tuition. It was here that his lifelong fascination with science was born. By 1822, Henry had already become a teaching assistant in science courses. He left the Albany Academy in 1822 and worked as a schoolteacher for a time, before returning, in 1826, as professor of mathematics and natural philosophy. In spite of spending up to seven hours a day in the lecture hall, he found time to begin his research on electromagnetism.

When Henry set to work, the Danish physicist HANS CHRISTIAN ØRSTED had already announced his seminal experiment in which an electric current generated a magnetic force. Ørsted's work launched the great age of electromagnetic research in Europe, spearheaded by ANDRÉ-MARIE AMPÈRE in France and Faraday in Great Britain. Henry began his own research by using the skills he had learned as a watchmaker's apprentice to construct electromagnets of unprecedented strength. By insulating the wire instead of the iron core, he was able to wrap a large number of turns of wire around the core; the resulting magnet, built for Yale University, was capable of lifting more than 2000 pounds, while requiring relatively small electric currents.

While experimenting with such magnets, Henry discovered mutual induction—the induction of electric currents by a moving or changing magnetic field—the principle behind the transformer and the generator. He was unaware of the work of Faraday, who independently discovered mutual induction and published his results before Henry, with his heavy teaching load, managed to do so. If Faraday is credited with discovering induction, Henry nonetheless made an entirely unique discovery: noticing that a large spark was generated when an electric circuit was broken, he deduced the property known as self-induction, a phenomenon that acts to prevent a current from changing. If a current is flowing, self-induction keeps it flowing; if an electromotive force is applied, self-induction prevents

current from increasing. He found that self-induction depends on the configuration of the circuit, especially the coiling of the wire. He also discovered how to make non-self-inductive windings by folding the wire back on itself.

The early 1830s were years of personal and professional growth. In 1830, he married Harriet Alexander. The following year he published seminal papers on his experiments on electromagnetism and his invention of an oscillating electromagnetic motor, the first demonstration of continuous motion produced by magnetic attraction and repulsion. In 1832, Henry became professor of natural philosophy at the College of New Jersey, later Princeton University, a post he would retain until 1848. He was a popular instructor, who taught not only physics, but also chemistry, geology, mineralogy, astronomy, and architecture.

At Princeton, lighter teaching responsibilities and the company of stimulating colleagues nurtured his research. Continuing to work on magnets, Henry built one for Princeton that could lift 3500 pounds. He also invented the first primitive magnetic relay by rigging a wire to send signals from his laboratory to his home on campus, a distance of one mile. This system, whereby Henry sent signals to his wife, used a remote electromagnet to close a switch in order to generate a stronger local circuit. It was similar to the arrangement Morse used in his telegraph and may thus be considered a prototype of that history-making invention.

In the following years, as Henry became unwillingly involved as a witness in the legal challenges of other inventors to Morse's rights to the patent for the telegraph, his own role in creating Morse's machinery became an issue. Henry never said he had invented the telegraph, but he correctly claimed credit for discovering its basic underlying principles and for demonstrating, in his laboratory and lecture hall, that an electromagnetic telegraph was possible. The two men would argue bitterly about issues of scientific and technological priority for the rest of their lives.

In 1837, Henry traveled to Europe, where he met Faraday, Wheatstone, and other British scientists, and he returned with new equipment for his experiments. The following year he discovered that by rearranging his electrical coils, he could either step-up or step-down the voltage of a current. This innovation perfected the magnetic induction mechanism, paving the way for what is now known as a transformer. Four years later, he discovered that electrical induction could be detected over long distances, an observation that would lead to radio transmission.

Henry was a remarkably versatile and prolific investigator, publishing papers in astrophysics and terrestrial magnetism, optics and acoustics, and founding the field of applied acoustics in the United States. But he truncated his life as a researcher prematurely in 1846, when he agreed to serve as the first secretary of the Smithsonian Institution, created by Congress for "the increase and diffusion of knowledge." He made the decision reluctantly, commenting, "If I go, I will probably exchange permanent fame for transient reputation." Under his leadership, the Smithsonian became the premier American institution for organizing and developing science.

As a "public scientist," Henry undertook many services to the nation. In 1848, he organized and supported a corps of volunteer weather observers, a highly successful venture that led to the creation of the U.S. Weather Bureau. From 1849 to 1850, he was president of the American Association for the Advancement of Science. From 1852 to 1878, he took charge of experiments and testing for the U.S. Light-House Board, responsible for the construction and repair of lighthouses, making it into a center for applied research in optics, thermodynamics, and acoustics. He served as President Lincoln's chief technical adviser during the Civil War. From 1868 to 1878, Henry was president of the National Academy of Sciences. When Joseph Henry died on May 13, 1878, in Washington, D.C., his funeral was attended by the U.S. president and numerous illustrious government officials and scientists.

By demonstrating the ability of electricity to do useful work through an electromagnet, Henry laid the foundation of the 19th-century communications revolution. His name has been honored in many ways, most notably in 1893, when the standard electrical unit of inductive resistance was named the *henry*.

Further Reading

Coulson, Thomas. *Joseph Henry: His Life and Work.* Princeton, N.J.: Princeton University Press, 1950.

Moyer, Albert E. *Joseph Henry: The Rise of an American Scientist.* Washington, D.C.: Smithsonian Institution Press, 1997.

⊠   **Hertz, Heinrich Rudolf**
(1857–1894)
German
Experimentalist (Classical Electromagnetism)

Heinrich Hertz experimentally confirmed the electromagnetic theory of JAMES CLERK MAXWELL, who in 1873 had predicted the existence of electromagnetic waves that traveled at the speed of light. In the process, he discovered radio waves and paved the way to the invention of the radio and the wireless telegraph.

He was born in Hamburg, on February 22, 1857, the son of a prominent lawyer and legislator. He showed an early interest in understanding how things work and by the age of twelve had equipped his own workshop. At age 15, he entered the Johanneum Gymnasium. Three years later, having decided on an engineering career, he moved to Frankfurt to gain some practical experience in that profession. In 1876 he moved to Dresden to prepare for the state examinations that would officially qualify him as an engineer. But his studies were interrupted by a year of compulsory military service, during which time he decided that his true calling was in pure science. When he was discharged he entered Munich University with the intention of studying mathematics but soon switched his field of concentration to physics.

In 1878, he transferred to the University of Berlin and studied under GUSTAV ROBERT KIRCHHOFF and HERMANN LUDWIG FERDINAND VON HELMHOLTZ. The latter quickly recognized his abilities and became his mentor. In 1880, after receiving a Ph.D. magna cum laude for a theoretical study of electromagnetic induction in rotating conductors, Hertz remained in Berlin to work as Helmholtz's assistant. The next step in his academic career was to gain experience as a lecturer; with this in mind, in 1881, Hertz moved again, this time to the University of Kiel. Four years later, he accepted a position as professor of physics at the Polytechnic in Karlsruhe and the following year married Elizabeth Doll, daughter of a Karlsruhe professor, who would bear him two daughters. It was in Karlsruhe in 1888, that Hertz would perform his history-making experiments designed to test the validity of Maxwell's theory of electromagnetism.

Maxwell's equations had demonstrated that electricity and magnetism are aspects of a single electromagnetic field, and that light itself is a variety of this field. In his 1873 *Treatise on Electricity and Magnetism* he established that light has a radiation pressure and suggested that a whole family of electromagnetic radiations must exist, of which light is only one. According to Maxwell's theory, waves exist in space and travel with a finite velocity, as opposed to the prevailing view that electric effects are examples of action at a distance: that electricity travels infinitely fast.

To test Maxwell's predictions, Hertz used an oscillator made of polished brass knobs, each connected to an induction coil and separated by a tiny gap over which sparks could leap. He reasoned that if Maxwell's predictions were correct, electromagnetic waves would be transmitted during each series of sparks. To confirm this, he made a simple receiver of looped wire. At the ends of the loop were small knobs separated by a tiny gap. The receiver was placed several yards from the oscillator. According to Maxwell's theory, if electromagnetic waves were spreading from the oscillator sparks, they would induce a current in the loop that would send sparks across the gap. This occurred when Hertz turned on the oscillator, producing the first transmission and reception of electromagnetic waves.

Hertz's oscillator experiments demonstrated what Maxwell had only theorized: that the velocity of radio waves is equal to the velocity of light. This proved that radio waves are a form of light. The experiments also demonstrated how to make electric and magnetic fields detach themselves from wires and move through space freely. When Hertz described his experiments in 13 papers and published them in 1893, they were successfully repeated around the world. As a result, Maxwell's theory was universally accepted.

Sadly, in the very year that he did his groundbreaking experimental work, Hertz began to suffer from a toothache, which proved to be the first symptom of the degenerative bone disease that would cut short his brilliant career. In 1889, he accepted an appointment as professor of physics in Bonn, succeeding the great RUDOLF JULIUS EMMANUEL CLAUSIUS. His health steadily worsened and on January 1, 1894, at age 36, in Bonn, he died of blood poisoning.

Hertz's discovery of radio waves would lead to a revolution in communication, triggering the invention of the wireless telegraph and of radio, as well as paving the way for radar and television. In his honor the unit of frequency of an electromagnetic wave, which is equal to one complete vibration or cycle per second, is called the hertz.

Further Reading

Baird, David, Alfred Nordman, and R. I. Hughes, eds. *Heinrich Hertz: Classical Physicist, Modern Philosopher*. Boston: Kluwer Academic, 1997.

Buchwald, Jed Z. *The Creation of Scientific Effects: Heinrich Hertz and Electric Waves*. Chicago: Chicago University Press, 1994.

Susskind, Charles. *Heinrich Hertz: A Short Life*. San Francisco: San Francisco Press, 1995.

Susskind, Charles, Johanna Hertz, and Mathilde Hertz. *Heinrich Hertz: Memoirs, Letters, Diaries*. San Francisco: San Francisco Press, 1977.

⊠ 't Hooft, Gerard
(1946–)
Dutch
Theoretician, Quantum Field Theorist

Gerard 't Hooft shared the 1999 Nobel Prize in physics with MARTINUS J. G. VELTMAN for proving that the quantum structure of electroweak interactions is renormalizable. In so doing, he placed particle physics on a firmer mathematical foundation, showing how the theory could be applied in precise calculations of finite physical quantities that could be used to predict the properties of new particles. Particle accelerator experiments in Europe and the United States have recently confirmed many of the calculated results. This work, which used algebraic computer algorithms, also stimulated the development of superfast quantum computers.

Gerard 't Hooft was born on July 5, 1946, in Den Helder, the Netherlands, into a family boasting a number of eminent scientists: his grand-uncle, Fritz Zernike, won the 1953 Nobel Prize in biology for his invention of the phase contact microscope; his grandfather, Pieter Nicolaas von Kampen, was a famous zoologist at the University of Leiden; and his uncle, Nicolaas Godfried van Kampren, was a professor of theoretical physics at the State University of

Utrecht. From early childhood, Gerard had a natural inclination toward pure science, to the dismay of his father, a naval engineer, who hoped his son would study engineering. 'T Hooft grew up in the Hague, with his father; his mother, who was a French teacher; and his two sisters.

'T Hooft received his secondary education at the Dalton Lyceum in the Hague, where he followed the classical track, which required Greek and Latin, because he was attracted by the challenge of it. Already drawn to physics, he was inspired by his high school physics teacher, who would spur his students to deeper thinking by saying, "If there were any real geniuses in this class, then they could have argued as follows. . . ." When he was 16, he competed in the Dutch National Math Olympiad and, to his own considerable amazement (he was aware of having made several errors), won second prize.

After passing his final high school exams in 1964, 't Hooft enrolled at the State University of Utrecht. At his father's urging, he joined the most elite student organization, enduring the humiliating initiation rituals. He was also a member of the university's renowned rowing team and its science club, as well as an organizer of a national congress for science students. Although all of this was time-consuming, it failed to lessen his preoccupation with physics. His uncle–professor invited him into the Theoretical Physics Institute, an informal arrangement of three houses next to a canal, where, he recalls, "I adored the discussions and the laughter."

Despite the reservations of his uncle, he was determined to study elementary particles, which he perceived as "the heart of physics." Fortunately, a new professor of theoretical physics, Martinus Veltman, specialized in particle theory. When "Tini," as everyone called Veltman, took 't Hooft on as his student, the first material he gave him to study was a paper by CHEN NING

Yang and Robert Mills, telling him, "This stuff you must know." 'T Hooft found the work "beautiful, elegant, and unique." In 1950, Yang and Mills had shown that the quantum electrodynamic (QED) formalism, developed earlier by JULIAN SEYMOUR SCHWINGER, RICHARD PHILLIPS FEYNMAN, and SIN-ITIRO TOMONAGA, could be generalized to include internal dynamic symmetries that were more general than the standard spacetime C (charge), P (parity), and T (time-reversal) symmetries.

In the early 1960s, STEVEN WEINBERG, SHELDON LEE GLASHOW, and ABDUS SALAM used this new generalization of QED to unify the electroweak forces into one quantum formalism. At the heart of their quest was the fascinating phenomenon of so-called broken symmetries, which seemed to permeate matter—asymmetric relations that have spontaneously arisen from the functioning of symmetric laws (for instance, the asymmetric crystal structure of ice that freezes out from the symmetric liquid structure of water when the temperature becomes low enough).

They were aware of the fact that in the early 1960s Peter Higgs had published papers demonstrating that spontaneous symmetry breaking events could create new kinds of force-carrying particles, some of them massive. This led them to speculate that if the virtual particles that carry the electromagnetic and weak forces (known collectively as the intermediate vector boson W and Z particles) were related by a broken symmetry, it might be possible to estimate their masses in terms of the unified, more symmetrical force from which the two forces were thought to have arisen. Working independently, each constructed a unified quantum field theory of electromagnetic and weak interactions (a quantum electroweak theory) that could make a verifiable prediction of the approximate masses of the required triplet W and the singlet Z particles needed to describe the weak interactions.

However, in the beginning the physics community found it difficult to accept the electroweak theory. This was because when they tried to use the theory to calculate the properties of the W and Z particles (which were the carriers of the weak force associated with radioactivity) and many other physical quantities, the theory predicted nonsensical infinite results. The situation resembled that of the 1930s, before Tomonaga, Schwinger, and Feynman "renormalized" QED, that is, found a mathematical way to eliminate the infinities.

Meanwhile, in the Netherlands, Veltman had not given up hope of renormalizing theories like the electroweak theory. At the end of the 1960s, he had developed a computer program called "Schoonschip," which, using symbols, performed algebraic simplifications of the complicated expressions that all quantum field theories result in when quantitative calculations are performed. Twenty years earlier, Feynman had systematized the calculation problem with his diagrams. Veltman believed in the possibility of finding a way of renormalizing the theory, and his computer program was the cornerstone of the comprehensive work of testing different ideas.

When 't Hooft chose a topic for his Ph.D. thesis in 1969, nothing appealed more to his imagination than the problem Veltman was working on: renormalization of the so-called Yang–Mills non-Abelian gauge field theories, which at the time were being applied to the study of the weak interactions. In 1971, 't Hooft succeeded in this task and published two articles describing his breakthrough. However, it was the second paper, in which he renormalized massless Yang–Mills fields for theories using the Higgs spontaneous symmetry breaking mechanism to generate the masses of the fields, that attracted world attention. First, 't Hooft verified his results, using Veltman's computer program; when this was done, the two men were able to work out a detailed calculation of the renormalization method. The renormalized non-Abelian gauge theory of electroweak interaction was now a functioning theoretical machinery capable of performing precise calculations. They presented their results at a 1971 international particle physics conference in Amsterdam.

'T Hooft received his Ph.D. for this work in 1972 and that year married Albertha (Betteke) A. Schik, who had studied medicine at Utrecht University. The couple began their married life at the European Center for Nuclear Research (CERN), Geneva, where Gerard had a fellowship and Betteke began training to be a specialist in anesthesia. 'T Hooft recalls this as an exciting time:

> Veltman also came to CERN, and together we refined our method for Yang–Mills theories. We were delighted with the great impact that our theories had. From 1971 onwards, all theories for the weak interactions that were proposed were Yang–Mills theories. Experiments were set up aimed at selecting out which of these Yang–Mills theories were correct.

The reaction of the physics community is nicely summed up by the physicist Kenneth Lane:

> For a decade, people didn't know how to calculate beyond the most elementary level with the Glashow–Salam–Weinberg model. It was impossible. And so the model was more or less forgotten. Then Gerard and his thesis advisor, Tini Veltman, showed how to do it. Their findings came out in 1971, and it was a bombshell. Everybody immediately jumped onto this theory, and over the next decade it became clear that the Glashow, Weinberg, and Salam model was correct. Experiments verified it in every detail.

The decisive experiments were first made by CARLO RUBBIA and his team at CERN in 1981. Two years later, in 1983, they announced the discovery of the triplet W and singlet Z particles, which had been based on signals from detectors specially designed for this purpose.

As his work made possible a new physical understanding of the subatomic world, 't Hooft's personal and professional life flourished. In 1974, he returned to Utrecht, where he was made assistant professor. Two years later, he was a visiting professor at Harvard and Stanford, where he worked on problems related to quantum chromodynamics (QCD), including the issue of quark confinement. In the meantime, his family was growing: his daughter, Saskia Anne, was born in Boston; his second daughter, Ellen Marga, was born in 1978, in Utrecht, where 't Hooft became a full professor.

He and Veltman went on to calculate the mass of the top quark, the heavier of the two quarks in the third family of what became known as the Standard Model. Many years later, in 1995, the top quark was observed directly for the first time at the Fermi Laboratory in Batavia, Illinois. Another highly important element of the model is the existence of a massive Higgs particle, which has not yet been observed. Physicists hope that the Large Hadron Collider (LHC), to be completed at CERN around 2005, will do the job.

'T Hooft's books include *Under the Spell of the Gauge Principle* (1994), *In Search of the Ultimate Building Blocks* (1997), and *Introduction to General Relativity* (2001).

'T Hooft is currently professor of theoretical physics at both the Spinoza Institute and the Institute for Theoretical Physics at the University of Utrecht, where his research focuses on magnetic monopoles, gauge theory in elementary particle physics, quantum gravity and microscopic black holes, and other fundamental aspects of quantum physics.

As do other physicists who have greatly contributed to the Standard Model of elementary particles, 't Hooft believes that, despite its immense productivity, it must eventually be replaced by a deeper understanding, involving a new paradigm about the nature of space and time, and incorporation of the gravitational force. He writes, "To me, Nature is a big jigsaw puzzle, and I see it as my task to try to fit the pieces together."

See also BETHE, HANS ALBRECHT; DYSON, FREEMAN; GELL-MANN, MURRAY.

Further Reading

'T Hooft, Gerard. *In Search of the Ultimate Building Blocks*. Cambridge: Cambridge University Press, 1997.

———. *Under the Spell of the Gauge Principle*. Singapore: World Scientific, 1994.

⊠ Hooke, Robert
(1635–1703)
British
*Theoretician and Experimentalist
(Classical Mechanics, Gravitation,
Optics), Astronomer*

Robert Hooke was a brilliant and versatile 17th-century scientist, who is best known for the derivation of Hooke's law of elasticity, which states that the stress placed on an elastic body is proportional to the strain produced.

He was born on July 18, 1635, in Freshwater, Isle of Wight. Poor health prevented him from preparing for the ecclesiastical career his father, who was a church curate, had planned for him. When his father died in 1648, Hooke went to London and was educated at Westminster School, where he studied ancient languages, learned to play the organ, and mastered the first six books of Euclid's *Elements* in a week.

At age 18 he went on to Oxford University, where he joined an outstanding group of young scientists. In 1658, while working for Robert Boyle, he constructed an improved version of the air pump of Otto Guericke, which led, four years later, to Boyle's formulation of his famous law, defining the relationships among the temperature, pressure, and volume of a gas. By the time Hooke

received an M.A. in 1663, he and most of his Oxford colleagues had moved to London, where, in 1662, they established what would become England's most prestigious organization of scholars, the Royal Society. At first, Hooke served as curator of experiments at the society, a paid post that involved the presentation of several new experiments at every weekly meeting. This caused him to know a little bit about a lot of things. He would later become a fellow of the society and one of its secretaries. In the mid-1660s he accepted two posts he would hold for the rest of his life: lecturer in mechanics for the Royal Society and professor of geometry at Gresham College in London. A fine architect, he also served as London's city surveyor and was chief assistant to Christopher Wren in rebuilding London after the Great Fire of 1666.

Hooke enjoyed a prolific 40-year career in London as a natural philosopher, as physicists were called then. In 1665, he published *Micrographia*, the book that would establish his international reputation. The first important work on microscopy, it contained observations made with magnifying lenses, some of very small objects, others of astronomical bodies, including a series of observations of lunar craters with speculations about their origin. It was in this work, which contained a number of discoveries in biology, that Hooke used the term *cell* to describe the empty spaces in the structure of cork; modern biology would adopt the term to denote a living unit of protoplasm. In *Micrographia* Hooke also put forth the controversial idea, which had emerged from his observations of spectral colors and patterns in thin films, that light might have a wave nature. He would develop this idea in 1672, by suggesting what AUGUSTIN-JEAN FRESNEL would prove almost 150 years later, in 1821, namely, that the vibrations in light might be perpendicular to the direction of propagation.

Another major focus of Hooke's research was gravitation, a phenomenon that absorbed him for more than 20 years. In 1664, he made the suggestion that astronomical bodies are pulled into orbits around larger bodies by a gravitational force directed toward the center of the larger body. He developed this idea into a theory of planetary motion in his 1674 publication *Attempt to Prove the Motion of the Earth*, which correctly proposed that planets are held in their orbits through a balance between an outward centrifugal force and an inward gravitational attraction to the Sun. In 1679, he wrote to SIR ISAAC NEWTON, suggesting that this attraction would vary inversely as the square of the distance from the Sun. When, in 1687, Newton published his law of universal gravitation, which built upon these ideas, Hooke, who seems to have been constantly embroiled in disputes over who should be credited with a discovery, was angry and embittered.

The most important discovery, indisputably his, was what came to be known as Hooke's law. Enunciated in 1678, the principle, which grew from his longtime interest in the physical properties of springs, states that the stress placed on an elastic body is proportional to the strain produced. Hooke's law has important practical applications in both physics and the physical design of mechanical devices of machines.

Hooke himself was no mean designer of instruments. Over the course of his career, he invented the compound microscope; a wheel barometer, which registered air pressure with a moving pointer; a weather clock, which recorded such factors as air pressure and temperature on a revolving drum; the universal, or Hooke's, joint, found in all cars; and the balance wheel in watches.

In 1696, Hooke's health began to deteriorate. He died in London on March 3, 1703.

Hooke was an outstanding example of the "Renaissance physicists" of his time: he made substantial original contributions to a stunning variety of fields and stimulated the thinking of his colleagues, most prominently, Newton's ideas on gravitation.

See also CARNOT, NICOLAS LÉONARD SADI.

Further Reading

Drake, Ellen Tan. *Restless Genius: Robert Hooke and His Earthly Thoughts*. Oxford: Oxford University Press, 1996.

Nichols, Richard. *Robert Hooke and the Royal Society*. Philadelphia: Trans-Atlantic, 1999.

Huygens, Christiaan
(1629–1695)
Dutch
Mathematical Physicist (Optics, Classical Mechanics), Astronomer

Christiaan Huygens was a brilliant physicist and mathematician who is best known for developing the first wave theory of light, which would be further developed by 19th-century physicists from AUGUSTIN-JEAN FRESNEL to JAMES CLERK MAXWELL. An extraordinarily versatile scientist, Huygens also explained the mechanics of the pendulum and invented the pendulum clock, created improved telescopes, and made important advances in pure and applied mathematics and in mechanics. He was the first astronomer to recognize the rings of Saturn and discover its satellite Titan.

Huygens was born on April 14, 1629, at the Hague, in the Netherlands, into a cultured family prominent in diplomatic circles. He was educated at home by his erudite, multitalented father, Constantin. In 1645, he entered the University of Leiden, where he studied mathematics and law until 1647. Intending to pursue a diplomatic career, he then went on to study law for another two years at the College of Orange in Breda. But in 1649, after returning from a diplomatic mission in Denmark, he decided not to enter his father's profession after all. Huygens senior provided him with an ample allowance, which enabled him for the next 16 years to devote himself to his scientific pursuits.

During this period Huygens applied new mathematical techniques to physical problems.

Motivated by disbelief in René Descartes's laws of impact, he carried out studies on impact and collisions, using the idea of relative frames of reference and considering the motion of one body relative to that of another. He discovered the law of conservation of momentum, stating it in terms of the conservation of the center of gravity of a system of bodies under impact. He then was able to prove by experiment that the momentum in a fixed direction before the collision of two bodies is equal to the momentum after the collision. In addition, he showed that the kinetic energy of motion of a system of particles is conserved in an elastic collision.

In 1655, Huygens and his brother began to make telescopes of high technical quality, devising a better way of grinding and polishing lenses. That year he discovered Titan, the largest satellite of Saturn, and determined its period of rotation. The next year he solved the problem of what had appeared to be the "arms of Saturn," ingeniously deducing that they must be the profile view of a ring surrounding the planet. As telescopes improved, Huygens's observations were confirmed, and by 1665 his theory was accepted by the important astronomers of the day. In the early 1660s, he traveled to Paris and London, where he met the leading scientists and showed the English his telescopes, which proved superior to their own. He was elected to the newly formed Royal Society of London in 1663.

In 1657, Huygens developed and patented a clock regulated by a pendulum, which greatly increased the accuracy of time measurements. By 1658, major towns in the Netherlands had pendulum tower clocks. He had been drawn to this work by both the need in astronomy for accurate timekeeping and in navigation for determination of longitude at sea. For the latter purpose, he built several pendulum clocks, which underwent sea trials in 1662 and 1668. He would work on theories related, first, to the simple pendulum and, second, to harmonically oscillating systems, throughout the rest of his life; eventually he solved the problem of the compound pendulum.

In 1659, Huygens studied centrifugal force and showed its similarity to gravitational force. He also derived the law of centrifugal force for uniform circular motion. During the same period, he studied projectiles and gravity, developing GALILEO GALILEI's ideas. Later, in the 1670s, studying motion in resisting media, he became convinced by experiment that the resistance in such media as air is proportional to the square of the velocity. In London, at the Royal Society, he met SIR ISAAC NEWTON. Although admiring Newton, Huygens was unconvinced by his theory of universal gravitation, since he did not believe that two different masses could attract one another if there were nothing between them.

In 1666, when the Académie Royale des Sciences was founded, Huygens was invited to Paris, to join; he assumed leadership of the group, on the basis of his knowledge of how the Royal Society operated in England. Supported by a generous pension by King Louis XIV, he lived and worked in luxurious quarters at the Bibliothèque Royale for 15 years, frequently traveling abroad. In 1671, when Louis XIV invaded the Netherlands, Huygens found himself in a country at war with his own. Like many scientists of this era, Huygens considered himself above political wars; with the help of his friends, he stayed on in Paris and continued his work.

Huygens is most famous for his wave theory of light, which he published in 1678. In it he presented what is now known as Huygens's principle: an expanding sphere of light behaves as if each point on the wave front were a new source of radiation of the same frequency and phase. He used this principle to explain reflection and refraction, showing that refraction is related to differing velocities of light in media. He then used this argument as a counter to Newton's particle theory of light. The essence of Huygens's theory was that light is transmitted as a pulse through the ether, which sets up a whole train of vibrations in the ether. However, as successful as this approach was, he could not use his ideas to explain the phenomenon of polarization of light.

In the final years of his life, he composed one of the earliest discussions of extraterrestrial life, published posthumously in 1698. He was frequently ill and returned home to the Hague, where he died on July 8, 1695.

Huygens was one of the foremost scientists and mathematicians of his day. In physics, his wave theory of light would be recognized as the forerunner of theories challenging Newton's particle theory of light, which, in turn, led the way to modern theories of electromagnetism.

Further Reading

Struik, Dirk Jan. *The Land of Stevin and Huygens: A Sketch of Science and Technology in the Dutch Republic During the Golden Century*. Dordrecht, Netherlands: D. Reidel, 1982.

Yoder, Joella G. *Unrolling Time: Christiaan Huygens and the Mathematization of Nature*. Cambridge: Cambridge University Press, 1988.

J

Josephson, Brian David
(1940–)
Welsh
*Theoretician, Solid State Physicist,
Condensed Matter Theorist*

Brian David Josephson discovered the phenomenon of superconducting quantum tunneling, which later became known as the Josephson effect, when he was a 22-year-old graduate student. Apart from its important role in the theory of superconductors, the discovery of the Josephson effect led to major scientific breakthroughs in the development of computer technology and earned Josephson a share in the 1973 Nobel Prize in physics.

He was born on January 4, 1940, in Cardiff, Glamorgan, in Wales, where he attended the public schools. After graduating from Cardiff High School, he entered Trinity College, Cambridge University; while still an undergraduate, he published an important paper, which dealt with certain aspects of the special theory of relativity and the Mössbauer effect. He earned a bachelor's degree in 1960 and two years later was elected a fellow at Trinity. A brilliant and exacting student, he was a source of discomfort to his professors, who would politely inform them, after class, of any errors they had made.

In graduate school Josephson's embryonic interest in superconductivity (the absence of electric resistance of certain materials at low temperatures) took wing when he noticed some unique connections between solid state theory and his own experimental work on superconductivity. He was led to calculate the current due to quantum mechanical tunneling across a thin strip of insulator between two superconductors. He began to explore the properties of a junction between two superconductors, which later became known as a Josephson junction. In this work he extended earlier work by Leo Esaki and Ivar Giaever on quantum tunneling, the phenomenon by which electron quantum wave functions can penetrate solids. He showed theoretically that an electron current between two superconductors could flow across an insulating layer without the application of a voltage as a result of quantum tunneling. On the other hand, when a voltage *was* applied, the current stopped flowing and oscillated at high frequency. The frequency of the oscillating quantum tunneling current is very precisely related to fundamental constants of physics and can be used to determine most accurately the ratio of Planck's constant to the charge on the electron and to establish a highly accurate voltage standard. This phenomenon, now called the Josephson effect, can also be used as a generator of radiation, particularly in the microwave and far infrared regions.

When Josephson published his discovery, it was disputed by no less an authority than JOHN BARDEEN, the most renowned solid state physicist of his time, who had invented the transistor and developed the comprehensive microscopic theory of superconductivity, the Bardeen–Cooper–Schrieffer (BCS) theory, announced in 1957. This led to a face-to-face debate by the two men in London in 1962. But the question was eventually resolved by later experimental results in which Josephson's predictions were verified. Ironically the experimental confirmation of Josephson's discovery ultimately reinforced the BCS theory.

Although Josephson makes his permanent academic home at Cambridge, where he was made assistant director of research in 1967 and full professor of physics in 1972 in the United States. In the early 1980s, Josephson was involved in the mathematical modeling of intelligence. By applying the Josephson effect, in 1980, researchers at IBM assembled an experimental computer switch structure that permitted switching speeds from 10 to 100 times faster than those possible with conventional silicon-based chips, vastly increasing data processing capabilities.

Josephson shared the 1973 Nobel Prize in physics with Esaki and Giaever "for their discoveries regarding tunneling phenomena in solids."

A few years before he was awarded the Nobel Prize, Josephson became interested in the possible relevance of Eastern mysticism to scientific understanding. Josephson is currently director of the Mind–Matter Unification Project of the Theory of Condensed Matter Group at the Cavendish Laboratory, Cambridge, which he describes as "concerned primarily with the attempt to understand from the viewpoint of the theoretical physicist, what may loosely be characterized as intelligent processes in nature associated with brain function or with some other natural process." He has published many articles on consciousness, including paranormal phenomena.

Josephson's insight into superconducting quantum tunneling theory has had profound consequences for experimental and theoretical physics. Quantum interference effects are the basis of superconducting quantum interference devices (SQUIDs), which can act as ultrasensitive magnetometers capable of detecting tiny geophysical anomalies in the Earth's magnetic field as well as the anomaly caused by the presence of submarines hidden in the ocean. Finally, the experimental use of the Josephson effect has made the phase of the macroscopic quantum wave function accessible to experimental control and has raised the science of metrology (the science of measuring the values of the fundamental constants of physics) to the extraordinary precision of one part in 10^{19}, thus strengthening the foundations of physics. In the future, Josephson's discovery may also have an important impact on the development of artificial intelligence.

See also DOPPLER, CHRISTIAN; EINSTEIN, ALBERT; MÖSSBAUER, RUDOLF LUDWIG; SCHRIEFFER, JOHN ROBERT.

Further Reading
Josephson, B. D. "The History of the Discovery of Weakly Coupled Supercomputers." In *Physicists Look Back: Studies in the History of Physics,* edited by J. Roche. Bristol, England: Adam Hilger, 1990. Institute of Physics Publishing.
McDonald, Donald G. "The Nobel Laureate Versus the Graduate Student." *Physics Today* July 1991, pp. 46–51.
Utts, Jessica, and Brian D. Josephson. "The Paranormal: The Evidence and Its Implications for Consciousness." *Times Higher Education Supplement* April 5, 1996, p. v.

⊠ **Joule, James Prescott**
(1818–1889)
British
Experimentalist (Thermodynamics)

James Prescott Joule was an ingenious experimentalist, who collaborated with LORD KELVIN

(WILLIAM THOMSON) in developing a thermodynamic theory of gases and discovered what is now known as Joule's law, which defines the relationship between heat and electricity. His highly accurate measurements of the mechanical equivalent of heat were later used by RUDOLF JULIUS EMMANUEL CLAUSIUS to formulate the first law of thermodynamics, conservation of energy.

Joule was born on December 24, 1818, in Salford, England, into a family whose wealth derived from operating a brewery. Described as a shy and delicate child, he received his only formal education, in elementary mathematics, natural philosophy, and chemistry, at home between 1833 and 1837. Although his lack of advanced mathematical training would later limit his role in developing the theory of thermodynamics, it did nothing to dim the luster of his experiments, the first of which he performed in a laboratory at the family brewery.

At the peak of his career, between 1837 and 1847, Joule conducted a series of diverse and highly accurate experiments. The first of these related the chemical and electrical energy expended to the heat produced in metallic conductors and voltaic and electrolytic cells. Published between 1840 and 1843, Joule's results established the relationship between heat and electricity that came to be known as Joule's law. It states that the heat produced in a conductor of resistance r by a current i is equal to i^2r/second.

Joule's next experiments investigated the relationship between heat and mechanical energy. In one of these he built a small electromagnet to show that electrical energy generated by mechanical energy in a dynamo can produce heat, a form of energy. Later, in his famous paddle-wheel experiment, he measured the rise in temperature of water agitated by paddles driven by falling weights. This showed that the kinetic energy of the weights as they fell existed as heat energy in the water. This last experiment has become the best-known method for determining the mechanical equivalent of heat, which veri-

fied that in physical processes energy cannot be created or destroyed, but only changed into different forms. Clausius would use Joule's results in formulating this principle as the first law of thermodynamics, conservation of energy. The currently accepted value of the mechanical equivalent of heat is 4.1868 joules/calorie. (The joule is the unit of all forms of energy and is defined as the energy expended when a force of one newton moves through a distance of one meter. The calorie is the heat energy required to heat one gram of water by one degree Celsius.)

When Joule first presented his results on the mechanical equivalent of heat, they were resisted by many of his peers, who found it difficult to believe he could make such accurate measurements. He had a critically important ally, however, in Lord Kelvin (then William Thomson), who recognized that Joule's experimental prowess was just what he needed to buttress his own theoretical work. In 1847, they embarked on a collaboration, which resulted, in 1852, in the discovery of the Joule–Kelvin effect: the decrease in temperature in a gas when it expands in a vacuum. This is caused by the conversion of heat into work done by the molecules in overcoming attractive forces between them as they move apart. It was to prove vital to techniques in the liquefaction of gases and in low-temperature physics.

The acceptance of Joule's work led to his election in 1850 to the Royal Society, which awarded him its prestigious Copeley Medal in 1866. Throughout his career, Joule's family wealth had allowed him to work as an independent researcher, without taking on an academic position. When his funds did run out in 1878, Queen Victoria rescued him with a state pension. After a long illness he died in Sale, Cheshire, on October 11, 1889.

Joule was a pivotal figure in the development of thermodynamics in the mid-19th century. His major contribution to physics was the demonstration, through a body of precise experiments, that the ratio of equivalence of the different forms of energy is independent of the process in which one form of energy is converted into another and of the materials involved.

See also CARNOT, NICOLAS LÉONARD SADI.

Further Reading

Von Baeyer, Hans Christian. *Warmth Disperses and Time Passes: A History of Heat*. New York: Modern Library, 1999.

K

Kamerlingh Onnes, Heike
(1853–1926)
Dutch
Experimentalist (Low-Temperature Physics, Superconductivity)

Heike Kamerlingh Onnes was a pioneer of low-temperature physics, whose success in producing liquid helium made it possible to study the properties of matter at temperatures close to absolute zero on the Kelvin scale. He received the 1913 Nobel Prize in physics for this work, which then led him to discover superconductivity, the total lack of electrical resistance in some materials when they approach a temperature near absolute zero.

He was born on September 21, 1853, in Groningen, the Netherlands, to Anna Gerdina Coers, the daughter of an architect, and Harm Kamerlingh Onnes, the owner of a brick works. After attending secondary school in his native town, at age 17 he entered the University of Groningen to study physics and mathematics. He won two prizes there for his studies on the nature of the chemical bond. From 1871 to 1873, he studied in Heidelberg, under the German physicists GUSTAV ROBERT KIRCHHOFF and Robert Bunsen. Upon his return to Groningen, he wrote his doctoral dissertation, "New Proofs of the Rotation of the Earth," in which he showed that JEAN-BERNARD-LÉON FOUCAULT's famous pendulum experiment belongs to a larger group of experimental phenomena that can be used to prove that the Earth rotates.

In 1878, Kamerlingh Onnes became an assistant at the Polytechnic in Delft, where Johannes van der Waals aroused his interest in the properties of matter at low temperatures. In 1881, he published a paper, "General Theory of Liquids," that dealt with the kinetic theory of the liquid state. This marked the beginning of his lifelong study of the properties of matter at low temperatures. During this period he formulated his methodological approach to physics, which emphasized the importance of quantitative research. "Knowledge through measurement" became his famous maxim. In the attempt to obtain experimental evidence to support van der Waals's theories on equations of state for gases, he investigated the equations of states of matter of liquids and gases over a wide range of pressures and temperatures and studied their general thermodynamic properties. His goal was to find experimental evidence for the atomic theory of matter and for van der Waals's corpuscular theory of gases at low temperatures.

In 1882, Kamerlingh Onnes became professor of experimental physics at the University of Leiden, which became his academic home, and, in 1887, he married Maria Bijleveld, with whom he had a son, Albert. In 1894, he founded his famous Cryogenic Laboratory in Leiden, which soon became a renowned world center for low-temperature physics. In this laboratory, he first applied the cascade method for cooling gases, developed by James Dewar; finally, in 1908, he succeeded in liquefying helium by using the Joule–Kelvin effect, a process in which the temperature of a gas decreases when it expands in a vacuum. This effect is caused by the conversion of heat into work done by the molecules in overcoming attractive forces between them as they move apart. By lowering the temperature of helium to 0.9 K, he reached the nearest approach to absolute zero then achieved. He also constructed "cold baths" with liquid helium, which enabled him to study the properties of matter at temperatures between 4.3 and 1.15 K, very close to absolute zero.

Then Kamerlingh Onnes surpassed these technical achievements and studied how matter behaves at even lower temperatures. In 1910, he succeeded in lowering the temperature of liquid helium to 0.8 K. LORD KELVIN (WILLIAM THOMSON) had postulated in 1902 that as the temperature approached absolute zero, electrical resistance would increase. In 1911, Kamerlingh Onnes found the reverse to be true: that is, the electrical resistance of a conductor decreases and finally vanishes as temperature approaches absolute zero. This phenomenon, for which he became most famous, was later called superconductivity. His interest in the technical application of low-temperature physics to the industrial and commercial uses of refrigeration led him to study magneto-optical properties at low temperatures of such metals as mercury, lead, nickel, and manganese–iron alloys.

Known for his personal charm and humanity, Kamerlingh Onnes, working with his wife during and after World War I, assisted starving children in war-torn countries. His health had always been delicate. He retired in 1924 and died in Leiden, after a short illness, on February 21, 1926.

The research done by Kamerlingh Onnes essentially created the field of low-temperature physics. Since properties of matter at very low temperatures differ fundamentally from those at normal temperatures, knowledge of these changes was crucial in answering basic questions on the nature of matter at the molecular, atomic, and subatomic levels. His systematic research on superconductivity was also of great importance to the theory of electrical conduction in solids. Superconducting materials have helped scientists make many important advances. For example, superconducting magnets play a crucial role in the particle accelerators physicists use to study subatomic particles. The Kamerlingh Onnes Laboratory in Leiden, named in his honor, is the leading world center for research in low-temperature physics.

See also JOULE, JAMES PRESCOTT; KAPITSA, PYOTR LEONIDOVICH; LANDAU, LEV DAVIDOVICH.

Further Reading
Goudaroulis, Y., and Gavrolu, K., eds. *Through Measurement to Knowledge: The Selected Papers of Heike Kamerling Onnes*. Dordrecht, Netherlands: Kluwer, 1991.

⊠   **Kapitsa, Pyotr Leonidovich**
(1894–1984)
Russian
Experimentalist (Low-Temperature Physics)

Pyotr Leonidovich Kapitsa won the 1978 Nobel Prize in physics for his fundamental discoveries in low-temperature physics, which relates to the properties of materials at temperatures near absolute zero on the Kelvin scale. Best known for his work on the superfluidity of liquid helium, Kapitsa is recognized for both his contributions to the development of physics in England and the Soviet Union and his role as an international advocate for peace and disarmament.

He was born in Kronstadt, near Saint Petersburg, Russia, on July 8, 1894, into an educated, prosperous family. His father, Leonid Petrovich Kapitsa, was a military engineer; his mother, Olga Yeronimovna Stebnitskaya, worked as an educator and researcher in the area of folklore. Kapitsa studied under the eminent Russian physicist A. F. Joffe at the Petrograd Polytechnic Institute and earned a Ph.D. in 1919. During this tumultuous period of revolution and civil war, Kapitsa's wife and two small children died in a flu epidemic. In the aftermath of this tragedy, at Joffe's urging, he left for England, to do his graduate work under ERNEST RUTHERFORD at the Cavendish Laboratory in Cambridge.

Kapitsa arrived at a time of great intellectual excitement at Cambridge. The charming, dedicated young Russian became Rutherford's favorite pupil and mixed with such brilliant young physicists as JOHN DOUGLAS COCKCROFT, SIR JAMES CHADWICK, and LORD PATRICK MAYNARD STUART BLACKETT. Put off by the formality of British academic life, he founded "the Kapitsa Club," in which unfettered discussion was the rule. In his first important piece of work, he developed methods for obtaining strong magnetic fields of transient duration.

By 1924, Kapitsa was deputy director of magnetic research at the Cavendish. That year, using specially constructed accumulator batteries and switching gear, he performed experiments that allowed him to pass very large electric currents through a coil with a very small internal volume for very brief periods. Using this experimental technique, he was able to generate a very large transient magnetic field: 5,000,000 gauss (the centimeter–gram–second unit for magnetic flux density), a record for magnetic field intensity that was not surpassed for 30 years. He kept on improving this technique and, in 1927, by short-circuiting an alternating current generator of special construction, he was able to generate large magnetic field intensities for longer times in larger volumes of space.

Kapitsa's career and personal life were flourishing. In 1927, he was married a second time, to Anna Alekseyevna Krylova, with whom he had two sons, Sergey and Andrey. He was elected a fellow of Trinity College, Cambridge, in 1925, and a fellow of the Royal Society, in 1929. With R. H. Fowler, he founded and edited an international series of physics monographs. In 1930, he became director of the Royal Society Mond Laboratory in Cambridge, which had been built especially for him, and began his most significant research, on liquid helium, one of the most useful means for attaining low temperatures, which was first produced by HEIKE KAMERLINGH ONNES in 1908.

In 1934, while working in his laboratory, Kapitsa invented and designed an original device for liquefying helium, which cooled the gas by periodically controlled expansions. For the first time a machine could produce liquid helium in large quantities without previous cooling with liquid hydrogen. This device was the foundation for the subsequent important advances made in low-temperature physics.

In the autumn of 1934, Kapitsa's life as a distinguished member of the British physics community abruptly ended, when he went to the Soviet Union for a conference and was not allowed to return to Cambridge. On Stalin's orders, his passport was confiscated by a state that considered his services too valuable to be shared with the West. Kapitsa was a patriotic Russian, with a "friendly" attitude toward the revolution. Although depressed at first, he managed to accept his situation and devoted himself to the advancement of Soviet science. In 1936, he was made director of the S. I. Vavilov Institute for Physical Problems of the Soviet Academy of Sciences in Moscow, and, the following year, he persuaded the Soviet physicist LEV DAVIDOVICH LANDAU to head the theory division of his institute. With Rutherford's help, the Soviet Union purchased the Mond Laboratory, which was transported to Kapitsa in

Moscow. His friendship with Cockcroft and other English scientists kept the link between European and Soviet physics alive even during the worst years of Communist oppression.

During this period, Kapitsa developed a new method for the production of liquid oxygen with a low-pressure cycle by using a special high-efficiency turboexpansion device. This work and its application were of great importance to the Soviet Union, particularly in the production of steel. The highly efficient radial compressed gas turboengine he developed still serves as a world model for modern large-scale oxygen production plants. Kapitsa was also one of the first to study the unusual properties of a form of liquid helium that exists below 2.2 K called helium II. Because liquid helium II conducts heat far more rapidly than copper, which is the best heat conductor at room temperature, it flows very easily with a viscosity far less than that of any other liquid or gas. By 1938, Kapitsa was able to show that helium II had such great internal mobility and negligible or vanishing viscosity that it could be characterized as a *superfluid* (a term he coined). During the next few years, his experiments on the properties of the helium II superfluid indicated that it was in a macroscopic or coherent "quantum state" and therefore could be considered to be a "quantum fluid" with zero entropy: that is, that it has a perfect atomic order. Kapitsa published his findings on the superfluidity of helium II in 1941. Meanwhile, Landau successfully worked on a comprehensive theory of superfluidity, earning the 1962 Nobel Prize in physics.

During World War II, Kapitsa headed the government department in charge of oxygen production. However, in 1946, he refused to work on the development of nuclear weapons. He was placed under house arrest until Stalin died, in 1953, and the head of his secret service, Lavrenty Beria, was arrested, in 1955. Kapitsa was not idle during his years of confinement. He worked on high-power electronics, inventing high-power

microwave generators. He also discovered a new kind of continuous high-pressure plasma discharge with electron temperatures greater than 1,000,000 K. In 1955, when he was once more made director of the Vavilov Institute, instead of returning to his work on low temperatures, he continued to study high-power electronics and plasma physics. In the 1950s, he worked on ball lightning, a puzzling phenomenon in which high-energy plasma maintains itself for a much longer period than seems likely. Kapitsa remained head of the Vavilov Institute until his death in 1984. He received the Nobel Prize in 1978, at the age of 84.

Kapitsa became a well-known public figure, highly respected for his personal courage. He was a member of the Soviet National Committee of the Pugwash Conference of scientists devoted to peace and disarmament and received the Frédéric Joliot-Curie Silver Medal of the World Peace Committee in 1959. During the worst years of repression, he defended his colleagues, saving some of them from death in the camps. The writer C. P. Snow, who knew him in his Cambridge days, wrote that only when Kapitsa had returned to the Soviet Union did he and his other English colleagues realize

> how strong a character he was; how brave he was; and fundamentally what a good man. . . . If he hadn't existed, the world would have been worse: that is an epitaph that most of us would like, but don't deserve.

See also KELVIN, LORD (WILLIAM THOMSON).

Further Reading

Badash, Lawrence. *Kapitsa, Rutherford, and the Kremlin*. New Haven, Conn.: Yale University Press, 1985.

Boag, J. W., P. E. Rubinin, and D. Shoenberg, eds. *Kapitsa in Cambridge and Moscow: Life and Letters of a Russian Physicist*. Amsterdam: Elsevier Science, 1990.

⊠ Kelvin, Lord (William Thomson)
(1824–1907)
Irish
Theoretical Physicist (Thermodynamics, Classical Electromagnetism)

William Thomson, better known as Lord Kelvin, was a great 19th-century physicist who first proposed the use of an absolute scale of temperature. The Kelvin temperature scale has the enormous advantage over other temperature scales such as the Celsius scale of defining its zero point independently of the freezing point of a specific substance such as water. It became an invaluable tool in the study of the thermodynamics of gases. The basic unit of thermodynamic temperature was named the *kelvin* in his honor, with absolute zero being equal to zero Kelvin (0 K).

William Thomson was born on July 26, 1824, in Belfast, Ireland. He lost his mother when he was six but had a close relationship with his father, a professor of mathematics at Belfast University. Studying with his father, William became an accomplished mathematician at a very early age. When he was eight, the family moved to Glasgow, Scotland, where his father became a professor of mathematics at Glasgow University. Since it was common practice for universities to compete with secondary schools for the best pupils, William was able to enroll at Glasgow University at age 10. There he studied astronomy, chemistry, and natural philosophy, as physics was then called, which included the study of heat, electricity, and magnetism. At age 17, he went on to Cambridge and graduated four years later at the top of his class.

After a trip to Paris, he returned to Glasgow, where, with his father's help, he was appointed professor of natural science at the university. An innovative pedagogue, he created the first laboratory in a British university. Shortly after his appointment, he began a collaboration with GEORGE GABRIEL STOKES, which would con-

tinue for more than 50 years. Their initial work on hydrodynamics led to Kelvin's discovery, in 1847, that the distribution of electrostatic force in a region and the distribution of heat through a solid are mathematically equivalent. He concluded that electrical and magnetic fields are distributed in the same way that energy moves through an elastic solid. JAMES CLERK MAXWELL would later successfully develop these ideas into his unified electromagnetic theory.

The following year, in 1848, Kelvin would make his most important contribution, which grew out of his early interest in the French mathematical approach to physics. While doing research in Paris he had learned of the research of NICOLAS LÉONARD SADI CARNOT on the ability of heat to do work. By analyzing Carnot's theory, which explains the amount of work produced by an engine using an ideal gas governed only by the temperature at which it operates, Kelvin developed the idea of an absolute temperature scale in which the temperature represents the total energy of a body. Unlike the Celsius scale, in which 0°C is defined as the freezing point of water, Kelvin's scale defined absolute zero as the point at which all vibrational motion associated with heat in the gas vanishes. He did this by considering a perfect gas, that is, a gas of molecules of zero dimension whose pressure depends only on its temperature. Absolute zero was reached when the volume of the perfect gas vanished. This was an idealization, since, in reality, the volume of a gas does not wholly disappear. But real gases approximate perfect gases for some range of temperature and their volume decreases as the temperatures decreases, if the pressure is held constant. In this manner Kelvin determined that absolute zero 0 K occurred at –273°C, showing that Carnot's theory was correct if absolute temperatures were used.

Where Kelvin parted ways with Carnot, however, was on the question of what heat actually *is*. Carnot held the prevalent belief that heat is a fluid, whereas Kelvin believed it to be a form of motion and supported JAMES PRESCOTT JOULE's mechanical theory of heat. In 1851, he announced that Carnot's theory and the mechanical theory of heat were compatible, if one accepted that heat cannot pass spontaneously from a cooler body to a hotter one. In this way Kelvin found his own path to what is now known as the second law of thermodynamics, which RUDOLF JULIUS EMMANUEL CLAUSIUS had independently discovered a year earlier. In 1852, Kelvin suggested that mechanical energy dissipates as heat, an idea that Clausius later developed into the theory of entropy, a measure of the disorder of a system.

In 1852, Kelvin, the brilliant theorist, and Joule, the ingenious experimentalist, began a long, fruitful collaboration. Together they formulated the Joule–Thomson effect, which describes the decrease in temperature in a gas when it expands in a vacuum. The effect, which is caused by the work done as the gas molecules move apart, had great import in the liquefaction of gases. Joule's ideas on heat would change Kelvin's, leading him to believe in a dynamical theory of heat, which he would apply to electricity and magnetism. "Whatever electricity is," Kelvin said in 1856, "it seems quite certain that electricity in motion is heat, and that a certain alignment of the axes of revolution in this motion is magnetism."

Kelvin also expanded the discoveries made by MICHAEL FARADAY into a full theory of electromagnetism. He derived an expression for the energy possessed by a circuit carrying a current and developed a theory of oscillating circuits that was experimentally verified in 1857. This approach was later used to generate and analyze the production of radio waves.

Kelvin's work on electricity and magnetism helped Maxwell develop his theory of electromagnetism. Still, when it was published, Kelvin did not fully support its tenets. In fact, in the latter part of his career, he seemed to be on the los-ing side of many arguments: he refused to accept the concept of atoms, opposed Darwin's theories, speculated incorrectly as to the age of the Sun and the Earth, and opposed ERNEST RUTHERFORD's ideas on radioactivity.

Kelvin joined a group of industrialists in the mid-1850s on a project to lay a submarine cable between Ireland and Newfoundland. He pointed out that a fast rate of signaling could only be achieved by using low voltages, which would require very sensitive detection equipment such as the mirror galvanometer that he had invented. When his advice was finally heeded on the third try, in 1866, the first transatlantic cable worked beautifully. That year he was knighted for this work and became Lord Kelvin. As a further reward for his involvement in the cable project, he amassed a considerable personal fortune. In 1899, he retired from Glasgow; he died at his estate in Largs, Ayrshire, Scotland, on December 17, 1907.

Kelvin's discoveries were crucial to the two main areas of research of 19th-century physics. His absolute scale of temperature became one of the foundations of thermodynamics. His insights into the nature of electricity and magnetism placed stepping-stones along the road leading to Maxwell's comprehensive formulation of an electromagnetic theory.

See also HERTZ, HEINRICH RUDOLF.

Further Reading

Burchfield, Joe D. *Lord Kelvin and the Age of the Earth*. Chicago: University of Chicago Press, 1990.

Shachtman, Tom. *Absolute Zero and the Conquest of Cold*. New York: Mariner Books, 2000.

Smith, Crosbie, and M. Norton Wise. *Energy and Empire: A Biographical Study of Lord Kelvin*. Cambridge: Cambridge University Press, 1989.

Thompson, Silvanus P. *The Life of Lord Kelvin*. New York: Chelsea, 1977.

Tunbridge, P. *Lord Kelvin: His Influence on Electrical Measurements and Units*. IEE History of Technology Series, No. 18. London: Peter Peregrinus, 1991.

⊠   Kilby, Jack St. Clair
(1923–   )
American
Solid State Physicist, Electrical Engineer

Jack St. Clair Kilby had an enormous impact on the world through his invention of the integrated circuit, also known as the microchip, which gave rise to the microelectronics revolution. His work was rewarded with a 2000 Nobel Prize in physics.

The man who would clear a path toward the computer age and space exploration was born on November 8, 1923, in Great Bend, Kansas, and grew up in the midst of what he called "the industrious descendants of the western settlers of the American Great Plains." His father ran a small electric company and had customers throughout the rural western part of Kansas. Kilby traces his decision to pursue electronics to watching his father use an amateur radio to keep in touch with customers who had lost their power and phone service during an ice storm. As a teenager in the 1940s, he was a fan of broadcast radio and particularly enjoyed band music, a love that stayed with him all his life.

After high school graduation, he majored in electrical engineering at the University of Illinois, where he also took courses in vacuum tube engineering physics. He received a B.S. in 1947, a year before Bell Laboratories announced the invention of the transistor. He later observed that this "meant that my vacuum tube classes were about to become obsolete, but it offered great opportunities to put my physics studies to good use." He took a job with Centralab, an electronics manufacturer in Milwaukee, Wisconsin, that made parts for radios, televisions, and hearing aids; there he was responsible for the design of ceramic-base silk screen circuit boards. At the same time, he took evening classes at the University of Wisconsin and earned a master's degree in electrical engineering in 1950. In 1952, Centralab sent him to a transistor symposium at Bell Labs, where he was

able to see firsthand the work of the transistor inventors JOHN BARDEEN, Walter Brattain, and WILLIAM BRADFORD SHOCKLEY.

In 1958, he left Milwaukee and moved with his wife to Dallas, Texas, where he went to work for Texas Instruments. The company put him to work on electronic component miniaturization, looking for ways to produce smaller and more effective electrical components. Electrical engineers were aware of the potential of digital computers but faced the challenge of what was known as the "tyranny of numbers": the exponentially increasing number of components required to design improved circuits and the physical limitations derived from the number of components that could be assembled together. In the summer of 1958, while the rest of the staff was on vacation and he had the laboratory to himself, Kilby worked on this problem. He found his solution in the monolithic (formed from a single crystal) integrated circuit. Rather than design smaller components, he was able to fabricate entire networks of discrete components—resistors, capacitors, distributed capacitors, and transistors—in a single sequence by laying them into a single crystal, or chip of a semiconducting material. On September 12, Kilby was able to demonstrate that an integrated circuit worked.

At the same time, Robert Noyce, a research engineer who had founded his own company, Fairchild Semiconductor Corporation, found the same solution, using a different manufacturing process. Fairchild and Texas Instruments then engaged in more than 10 years of litigation, which was finally resolved in the late 1960s when the firms decided to cross-license their technologies. Both Kilby and Noyce were awarded the National Medal of Science (1969), and both were inducted into the National Inventors Hall of Fame (1982).

When, 42 years after the invention of the integrated circuit, Kilby shared the 2000 Nobel Prize in physics with the semiconductor pioneers

ZHORES IVANOVICH ALFEROV and HERBERT KROEMER, he paid tribute to his colleague:

> While Robert [Noyce] and I followed our own paths, we worked hard together to achieve commercial acceptance for integrated circuits. If he were still living, I have no doubt he would have shared this prize.

Kilby, who was known in the corridors of Texas Instruments as "the humble giant," because of his height (he is six feet six inches tall) and unassuming manner, would later express his astonishment at the fruits of his invention:

> What we didn't realize then was that the integrated circuit would reduce the cost of electronic functions by a factor of a million to one. . . . Nothing had ever done that for anything before.

After proving that integrated circuits were possible, he headed teams that built the first military systems and the first computer incorporating integrated circuits. He also worked on teams that invented the handheld calculator and the thermal printer, which was used in portable data terminals.

In 1970, Kilby took a leave of absence from Texas Instruments and worked on a method to apply silicon technology to help generate electrical power from sunlight. From 1978 to 1984, he was a Distinguished Professor of Electrical Engineering at Texas A&M University. He officially retired from Texas Instruments in the 1980s but has maintained a strong involvement with the company. Texas Instruments named a state-of-the-art digital chip research center in his honor in 1997. He is married and has two daughters and five granddaughters.

Kilby's invention of the integrated circuit is one of the most important discoveries in the history of technology. Today chips that are being made contain nearly a billion bits of memories or logic gates in processors—the brains of computers. Integrated circuit chips containing programmable digital signal processors and analog components are also used in mobile phones, enabling them to place calls at costs per transistor roughly the same as those for simple memory chips.

Further Reading

Queisser, Hans. *The Conquest of the Microchip.* Cambridge, Mass.: Harvard University Press, 1988.

Reid, T. R. *The Chip: How Two Americans Invented the Microchip and Launched a Revolution.* New York: Simon & Schuster, 1984.

Warshofsky, Fred. *The Chip War: The Battle for the World of Tomorrow.* New York: Charles Scribner's Sons, 1984.

⊠ **Kirchhoff, Gustav Robert**
(1824–1887)
German
*Theorist and Experimentalist
(Electromagnetism, Spectroscopy,
Thermodynamics)*

Gustav Kirchhoff's most far-reaching contribution to physics was his discovery of a fundamental law of electromagnetic radiation, which led to his development of the concept of a perfect blackbody, an object that does not reflect any surface light and is therefore a perfect emitter and absorber of radiation at all frequencies. In addition, together with Robert Bunsen, he founded the science of spectroscopy, which led to the discovery of several new chemical elements and to methods of determining the composition of stars and the structure of the atom. He also did important work in electromagnetism, discovering the laws that govern the flow of electricity in electrical networks.

Kirchhoff's distinguished career unfolded within a period of national prosperity and stability that was reflected in a thriving academic community. He was born in Königsberg, Germany, on

March 12, 1824, into a family with a strong tradition of service to the Prussian state. He attended the University of Königsberg, where he was introduced to the new science of electromagnetism and exposed to the ideas and methods of the leading French school of mathematical physics. While still a student, he made an important contribution to electrical circuit theory when he extended Ohm's law, which formulates the relationships of current, electromotive force (voltage), and resistance to networks of conductors. He went on to derive what came to be known as Kirchhoff's laws, a set of rules for calculating unknown currents, resistances, and voltages in an electric circuit. In later work on electricity, Kirchhoff would demonstrate that electrostatic potential is identical to electromotive force, thus unifying the theories of static charges and electric currents. He would also show that an oscillating current is propagated in a conductor of zero resistance at the velocity of light, a discovery that would play an important role in the development, in the 1860s, of the electromagnetic theory of light by JAMES CLERK MAXWELL.

Kirchhoff graduated from the University of Königsberg in 1847 and the same year married Clara Richelot, the daughter of one of his professors. The following year he began his teaching career, as a lecturer at the University of Berlin. A master of mathematical analysis, who insisted on clear-cut logical formulation of physical ideas directly based on observed data, Kirchhoff became a leading figure in the flowering of theoretical physics in Germany. In 1850, he became professor of physics in Breslau, where he met Robert Bunsen, the inventor of the Bunsen burner. The two men began a dynamic collaboration, and, in 1852, Bunsen arranged for Kirchhoff to join him on the physics faculty of the University of Heidelberg. In Heidelberg's stimulating atmosphere, dominated by the great HERMANN LUDWIG FERDINAND VON HELMHOLTZ Kirchhoff, partly in collaboration with Bunsen, made his greatest contributions.

Bunsen had been interested in how chemicals emitted and absorbed light; Kirchhoff had focused on the way substances acted when burned or heated to incandescence. In the 1850s, Bunsen had developed his famous gas burner, which emitted a colorless flame, and used it to investigate the distinctive colors that metals and their salts produce in a flame. He used colored solutions and glass filters to distinguish the colors in a partly successful attempt to identify the substances by the colors they produced. Kirchhoff pointed out to Bunsen that he could accurately identify the colors by using a prism to produce spectra of the colored flames. They developed the spectroscope, a prism-based device that separated light into its primary chromatic components (i.e., its spectrum), and, in 1860, discovered that each of the elements present in a substance has a characteristic set of spectral emission lines. They then began studying the spectral "signature" of various chemical elements in gaseous form. By using this spectral analysis, Bunsen discovered two new elements: cesium in 1860 and rubidium in 1861.

In 1859, Kirchhoff made another important discovery, while investigating spectroscopy as an analytic tool. He noticed that certain dark spots in the Sun's spectrum were intensified if the sunlight passed through a sodium flame. JEAN BERNARD LÉON FOUCAULT had made this discovery 10 years earlier but had not followed up on it. Kirchhoff hypothesized that the sodium flame was absorbing light from the sunlight of the same color that it emitted and explained that the dark lines, known as Fraunhofer lines, were due to the absorption of light by sodium and other elements present in the Sun's atmosphere. He identified other elements in the Sun's spectrum in this way and developed a law, which states that the ratio of the emission line power to absorption line power of all material bodies is the same at a given temperature and a given wavelength of radiation produced. He went on to demonstrate the existence in the Sun of many chemical elements isolated on Earth and to argue that the Sun is mainly composed of a hot, incandescent liquid. Further, he firmly established the hot gaseous nature of the solar atmosphere and produced the first detailed map of the solar system.

From this, Kirchhoff went on in, 1862, to derive the concept of a perfect blackbody, an object that absorbs all the energy that falls upon it, and, because it reflects no light, appears black to an observer. It is also a perfect emitter, and in this context Kirchhoff proved that the energy emitted E depends only on the temperature T and the frequency v of the emitted energy: $E = J(T, v)$. He challenged physicists to find the function J. In 1900, MAX ERNEST LUDWIG PLANCK guessed the correct formula for the J function but in so doing was forced to hypothesize that the energy was emitted, not continuously, but in discrete packets, which he named *quanta*. Thus began the revolution that would result in the quantum theory.

Kirchhoff's Heidelberg life was marred by the death of his wife in 1869. Left with two sons and two daughters, in 1872, he married Luise Brommel. By 1875, he was too ill to continue experimental work and accepted a chair in theoretical physics in Berlin. He was disabled by an accident, which obliged him to use crutches or a wheelchair. In 1886, he was forced by illness to retire, and he died in Berlin on October 17, 1887.

Through his seminal work in electromagnetism, blackbody radiation, and spectroscopy, Kirchhoff made enduring contributions to modern physics. Although he was a classical physicist to the core, his work on blackbodies would lead a new generation of physicists to the discovery of quantum mechanics.

See also FRAUNHOFER, JOSEPH VON; OHM, GEORG SIMON.

Further Reading

Hearnshaw, J. B. *The Analysis of Starlight: One Hundred and Fifty Years of Astronomical Spectroscopy.* Cambridge: Cambridge University Press, 1990.

Kuhn, Thomas. *Black-Body Radiation and the Quantum Discontinuity, 1894–1912.* Oxford: Oxford University Press, 1978.

⊠ **Kroemer, Herbert**
(1928–   )
German
Theoretical Physicist and Experimentalist, Solid State Physicist

Herbert Kroemer is a pioneer in the field of semiconductor theory, who first elucidated the idea of heterostructures and their potential for opening up laser and transistor technology. In recognition of this work, which laid the foundation for today's revolution in information and communication technology, he shared the 2000 Nobel Prize in physics with ZHORES IVANOVICH ALFEROV and JACK ST. CLAIR KILBY.

Kroemer was born on August 25, 1928, in Weimar, Germany. His father was a civil servant, his mother "a classical German housewife." Both were members of families of skilled artisans and had no high school education but wanted the best education for their children and pushed them to achieve. Herbert had no problem doing so: he "breezed through 12 years of schools almost effortlessly" but was often bored and entertained himself by disturbing the class. His high school physics teacher overcame this problem by enlisting him to help teach the subject. Upon graduating in 1947, he entered the University of Jena in East Germany, where he was inspired by the physics lectures of Friedrich Hund. These were precarious years, when political oppression under the Communist regime caused many of his fellow students to flee to the West or to "disappear" into the German branch of Stalin's system of forced labor camps. While spending a summer as a student at the Siemens company, Kroemer took the opportunity to escape to West Berlin via one of the empty return flights of planes participating in the Berlin airlift.

After being grilled by such notable professors as WOLFGANG PAULI, he was admitted to the University of Göttingen, which he described as "intellectually a wonderful place." When the experimentalist he wanted to work with turned out to have a long waiting list, he set the direction of his career by signing on with a theorist, Fritz Sauter, for his diploma thesis. Sauter had a profound influence on Kroemer's development as a physicist, teaching him that

> you had to be able to go back and forth [between the physical concept and the mathematical formulation] with ease. Yet, in the last analysis, concepts took priority over formalism, the latter was simply an (indispensable) means to an end.

Taking NIELS HENRIK DAVID BOHR as his "personal role model," he never developed into a "hard-core Theorist with a capital *T*," but became basically a conceptualist, who remained acutely aware of his limitations as a formalist.

For his diploma thesis, Kroemer chose to extend the 1939 work of WILLIAM BRADFORD SHOCKLEY on the nature of surface states in one-dimensional potentials. While he was engaged in this research, Sauter urged him to write up a talk he had given on hot-electron effects in the newly invented transistor and submit it as his doctoral dissertation. Thus, Kroemer received a Ph.D. in theoretical physics, in 1952, before his 24th birthday, from the University of Göttingen.

Despite this academic success, he had no hopes for a university career. Virtually no new positions for theoretical physicists were being created in Germany, and the prospects in industry were similarly dim. He managed to be hired as a "house theorist" by the small semiconductor research group at the Central Telecommunications Laboratory of the German postal service. There he was given the freedom to choose his own problems and lecture to his experimentalist and

technologist colleagues, an experience that made him into what he termed "an applied theorist."

By focusing on the problem of overcoming the severe frequency limitations of the new transistors, Kroemer was led directly to the idea of heterostructures: two semiconductors whose atomic structures fit one another well but that have different electronic properties. He was the first to point out the significant performance advantages that can be gained in various semiconductor devices by incorporating heterostructures into them. In a 1954 paper, he first outlined the principles of his heterostructure bipolar transistor (HBT). After coming to the United States, where he joined RCA Laboratories in Princeton, New Jersey, in 1954, he returned to his research on heterojunctions. In a seminal 1957 theoretical paper, he set forth the basic design principles for all heterostructures.

In 1963, while working at Varian Associates in Palo Alto, California, Kroemer proposed the concept of the double-heterostructure laser, the central concept in the field of semiconductor lasers, without which that field would not exist. He proposed that the concentration of electrons, holes, and photons would become much higher if they were confined to a thin semiconductor layer between two others—a double heterojunction. His Nobel Prize–winning contributions can be traced directly to these early papers. But his paper was rejected, in 1963, and, when finally published, ignored. Varian refused to fund his work on the new kind of laser, believing that it had no practical applications. For the next decade, Kroemer worked on hot-electron negative resistance effects, playing no role in the technological realization of the laser. It was Alferov and his group at the Ioffe Institute in Leningrad who, in May 1970, first produced a laser that operated continuously, without requiring complex cooling measures.

Kroemer left Varian in 1966 and, two years later, joined the University of Colorado, where he resumed his work on heterostructures. After moving to the University of California in Santa Barbara, he developed a powerful new method to determine physical properties of the heterostructure interface. In 1976, he began assembling a large illustrious team of scientists for the study of the physics and technology of compound semiconductors and devices based on them. During this period, he became a strong advocate for developing the full potential of "the device that started it all," the heterostructure bipolar transistor.

Today the use of heterostructures continues to dominate the design of compound semiconductors—not just lasers and light-emitting diodes (LEDs), but integrated circuits as well—and they are even invading mainstream silicon circuit technology. In addition, Kroemer's work on heterostructures has led to spectacular scientific breakthroughs in which the advanced materials and tools of microelectronics are being used for studies in nanoscience and of quantum effects. The impact of this research on the modern world of electronics and communication has been enormous. Lasers and light-emitting diodes have been further developed in many stages. Without the heterostructure laser, today we would not have had optical broadband links, compact disc (CD) players, laser printers, bar code readers, laser pointers, and numerous scientific instruments. LEDs are used in displays of all kinds, including traffic signals. In recent years, it has been possible to make LEDs and lasers that cover the full visible wavelength range, including blue light. Today, high-speed transistors are found in cellular phones and in their base stations, in satellite dishes and in links. There they are part of devices that amplify weak signals from outer space or from a faraway cellular phone without drowning in the noise of the receiver itself.

Kroemer later turned to experimental work and became one of the pioneers in molecular beam epitaxy, a method used to apply new material surfaces on silicon interfaces. Describing himself as "an opportunist—and not at all ashamed of it," Kroemer continues to work on a wide variety of problems in semiconductor physics, including the superlattice Bloch oscillator, "an exciting combination of heterostructures and hot electron physics," and superconducting weak links.

See also BARDEEN, JOHN; BLOCH, FELIX.

Further Reading

Kroemer, Herbert. "Heterostructures for Everything: Device Principle of the 1980's?" *Japanese Journal of Applied Physics*, 20 (suppl. 20–21): 9–13, 1981.

Riordan, Michael, and Lillian Hoddeson. *The Invention of the Transistor and the Birth of the Information Age*. New York: W. W. Norton & Company, 1998.

⊠   **Kusch, Polykarp**
(1911–1993)
American
Experimentalist, Quantum Physicist

Polykarp Kusch was an ingenious experimentalist who won the 1955 Nobel Prize in physics "for his precise determination of the magnetic moment of the electron." His discovery had a profound impact on the development of the theory of the interaction of electrons and electromagnetic radiation, known as quantum electrodynamics (QED).

Kusch was born on January 16, 1911, in Blankenberg, Germany, the son of a Lutheran clergyman. The family immigrated to the United States the following year and settled in the Midwest, where Kusch received his early education. He became a U.S. citizen in 1922. After entering the Case Institute of Technology, Cleveland, Ohio, intending to major in chemistry, he soon realized that his interests lay elsewhere; he earned a B.S. in physics in 1931. For the next five years, he pursued his graduate studies at the University of Illinois, where he received an M.A. in 1933 and a Ph.D. in 1936, for a thesis on optical molecular spectroscopy directed by F. Wheeler Loomis. He then did postdoctoral work at the

Polykarp Kusch's precise determination of the magnetic moment of the electron had a profound impact on the development of quantum electrodynamics. *(AIP Emilio Segrè Visual Archives, Physics Today Collection)*

University of Minnesota, working with John T. Tate on problems in mass spectroscopy.

In 1937, Kusch began his long career as a member of the physics department at Columbia University, where he became known as a captivating teacher, who delivered his intense, well-thought out lectures in what was known as "the Kusch whisper," a loud, deep, inspirational voice. From the beginning, he worked closely with ISIDOR ISAAC RABI on his investigations of atomic, molecular, and nuclear phenomena via the molecular beam method. Following the work of Rabi, Kusch focused his research on the use of externally applied electric and magnetic fields on atomic and molecular beams as a method to conduct detailed studies of the interactions of particles and molecules.

During World War II, Kusch put his research on atomic and molecular beams aside and studied the use of microwave generators for radar at the Westinghouse Electric Corporation, the Bell Telephone Laboratories, and the Columbia Radiation Laboratories. His work in these laboratories taught him about microwave methods and was crucial to his future applications of vacuum tube technology to a broad span of experiments.

Back at Columbia full time after the war, he became immersed in research on atomic and molecular beams and did his most important work: the experimental observation of the anomalous magnetic moment of the electron and precise determination of its magnitude. The electron had long been known to be capable of acting as a small magnet whose strength, as measured by its magnetic moment, according to PAUL ADRIEN MAURICE DIRAC's theory of the electron, was equal to a quantity known as the Bohr magneton. However, at the beginning of 1947, Rabi and his collaborators found that the hyperfine structure of the electron in external magnetic fields did not entirely conform to the predictions of Dirac's theory. The physicist Gregory Breit hypothesized that this might be due to the fact that the value of the magnetic moment of the electron was larger than the value of one Bohr magneton as predicted by the Dirac equation.

Building on this insight, in 1947, Kusch made a series of scrupulously accurate atomic beam experiments that confirmed Breit's hypothesis: the magnetic moment of the electron was, indeed, larger than the Bohr magneton by about one part in a thousand. Kusch used an extremely refined technique involving atomic beams in order to reveal this tiny difference in the value of the magnetic moment of the electron. Later, he would use molecular beam techniques in experimental studies in chemical physics. As did the Lamb shift, discovered by WILLIS EUGENE LAMB in the Columbia lab that same year, the tiny discrepancy that Kusch detected led the theorists JULIAN SEYMOUR SCHWINGER, RICHARD PHILLIPS FEYNMAN, and FREEMAN DYSON to reformulate QED, the theory of the interaction of electrons and electromagnetic radiation, radically. Lamb and Kusch would share the 1955 Nobel Prize in physics.

In 1949, Kusch was made full professor at Columbia. He later was director of the Radiation Lab, as well as its academic vice president and provost. In the latter part of his career, he became deeply involved with problems of science education. From 1962 to 1965, he served on the Board of the Institute for Scientific Information in Philadelphia.

Kusch was married to Edith Starr McRoberts, with whom he had three children. A year after her death in 1959, he married Betty Pezzoni. In 1971, he retired from Columbia and became a member of the physics department at the University of Dallas, Texas, for the next decade. After a series of strokes, he died on March 20, 1993, at his home in Dallas at the age of 82.

Kusch's experimental discovery of the minuscule anomaly in the magnetic moment of the electron, along with the detection of the Lamb shift, provided the impetus for the theoretical breakthrough that led to the development of modern renormalized quantum electrodynamics.

See also BOHR, NIELS HENRIK DAVID.

Further Reading

Kusch, P. "The Magnetic Moment of the Electron." In *Nobel Lectures on Physics*. New York: Elsevier, 1964, pp. 298–310.

Schweber, Silvan S. *QED and the Men Who Made It: Dyson, Feynman, Schwinger, and Lamb, Willis Eugene Tomonaga*. Princeton, N.J.: Princeton University Press, 1994.

L

Lamb, Willis Eugene Jr.
(1913–)
American
*Theoretical and Experimental Physicist,
Quantum Physicist*

Willis Eugene Lamb Jr. manifested a rare combination of theoretical and experimental prowess in his discovery of what came to be called the Lamb shift, an anomaly in the fine structure of the hydrogen spectrum. For this work, which had profound implications for the development of quantum electrodynamics, he shared the 1955 Nobel Prize in physics with POLYKARP KUSCH.

He was born on July 12, 1913, in Los Angeles, California, the son of Willis Eugene Lamb, a telephone engineer, and Marie Helen Metcalf, originally from Nebraska. After attending public schools in Oakland and Los Angeles, he entered the University of California at Berkeley in 1930; four years later he graduated with a B.S. in chemistry. He then "switched to the other side of the sidewalk," doing his graduate work at Berkeley in theoretical physics. Working under J. ROBERT OPPENHEIMER, he earned a Ph.D. in 1938 for a thesis on the electromagnetic properties of nuclear systems.

Lamb then joined the faculty of Columbia University, New York, and the following year married Ursula Schaefer, a student from Germany. Over the next 10 years, he rose from instructor to full professor. From 1943 to 1951, he was also associated with the Columbia Radiation Laboratory. During World War II, he did extensive work on radar technique, which improved upon ISIDOR ISAAC RABI's magnetic resonance method.

Lamb won his Nobel Prize for the discovery he made, while heading a group at the Columbia lab after the war, of an anomaly in the fine structure of the hydrogen spectrum. The work focused on the hydrogen atom, in which a single electron moves around the nucleus in one of a series of orbits, each having a definite energy. These energy levels exhibit a fine structure, which means that they are arranged in groups of neighboring levels, the groups widely separated. In 1928, PAUL ADRIEN MAURICE DIRAC had offered an explanation of the fine structure, long thought to be correct: a theory of the electron based on relativity and quantum theory. Over the next decade, physicists used optical methods to check Dirac's theory of the fine structure; although these studies suggested some flaw in the theory, no definite results emerged.

In 1947, after a thorough theoretical analysis, Lamb began his experimental investigations, using a modified Rabi resonance method. Applying the art of spectroscopy with unprecedented precision, he shone a beam of

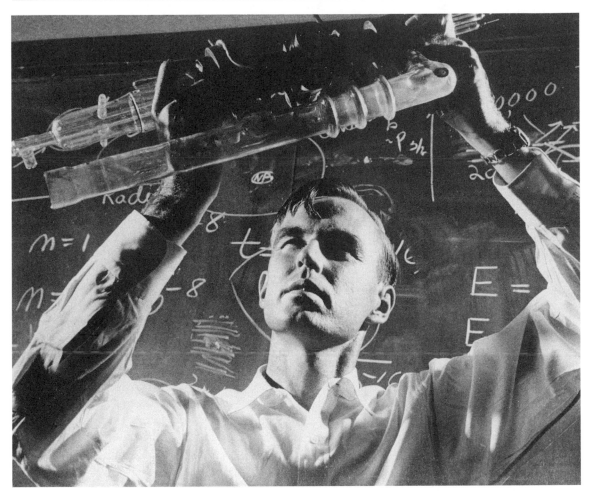

Willis Eugene Lamb Jr. discovered an anomaly in the fine structure of the hydrogen spectrum that had profound implications for the development of quantum electrodynamics. *(AIP Emilio Segrè Visual Archives)*

microwaves onto a hot wisp of hydrogen gas blowing from an oven. He found that two fine structure levels in the next lowest group of fine structure levels, which should have coincided with the Dirac theory, were in reality shifted relative to each other by a certain amount (the Lamb shift). He measured the shift with great accuracy and later made similar measurements on heavy hydrogen. When Lamb announced his news to the participants of the 1947 Shelter Island (New York) Conference, they were unset-tled by this stunning example of the truism that progress in science occurs when experiment contradicts theory. The comment of JULIAN SEYMOUR SCHWINGER, a leading theorist of the time, reveals the level of shock and surprise: "The facts were incredible—to be told that the sacred Dirac theory was breaking down all over the place." Quantum theorists would later realize that what was missing from Dirac's theory was the unwieldy concept of the self-interaction of the electron, which by its very nature contained infinities, thereby preventing a straightforward physical interpretation.

At the heart of the quantum electrodynamic process is the quantum exchange force by which electrons can interact by exchanging photons with each other. However, an electron can also exchange a photon with itself. Encouraged by Lamb's groundbreaking experiment, HANS ALBRECHT BETHE, Schwinger, RICHARD PHILLIPS FEYNMAN, and SIN-ITIRO TOMONAGA were stimulated to develop the method of renormalization, in which the self-energy infinities could be subtracted out. This led to a fully consistent relativistic theory of quantum electrodynamics, in which both the electron and the electromagnetic field were quantized. It explained the Lamb shift as due to the modification of the Coulomb force between the electron and the proton in the hydrogen atom. This modification was caused by the finite part of the quantum electrodynamic self-interaction, which was left after the infinite part was subtracted by the renormalization process.

Lamb has been on the physics faculty of Stanford University, Harvard University, Oxford University, and Yale University. Since 1974 he has been affiliated with the University of Arizona's Optical Science Center and Department of Physics, and from 1983 on with its Arizona Research Laboratories. He became Regents Professor at Arizona in 1989.

Lamb's research has spanned a large number of subjects, including the theory of the interactions of neutrons and matter, field theories of nuclear structure, theories of beta decay, ranges of fission fragments, fluctuations in cosmic ray showers, pair production processes, order–disorder problems, ejection of electrons by metastable atoms, quadrupole interactions in molecules, diamagnetic corrections for nuclear resonance experiments, theory and design of magneton oscillators, theory of a microwave spectroscope, study of the fine structure of hydrogen, deuterium, and helium, and theory of electrodynamic energy level displacements.

At the age of 87, he was still walking a mile each morning to his work at the Optical Sciences Center in Tucson.

Lamb's discovery had a profound impact on the theorists of his time and resulted in a paradigm shift leading to the reshaping of the quantum theory of the interaction of electrons and electromagnetic radiation. The implications of the seemingly tiny effect associated with the Lamb shift probed the depths of the process in which the quantum theory is applied to electromagnetic fields and led to what is now known as quantum electrodynamics.

Further Reading

Lamb, Willis E. "The Fine Structure of Hydrogen," Laurie M. Brown and Lillian Hoddeson, eds. in *The Birth of Particle Physics*. Cambridge: Cambridge University Press, 1983.

Schweber, Silvan S. *QED and the Men Who Made It: Dyson, Feynman, Schwinger, and Tomonaga*. Princeton, N.J.: Princeton University Press, 1994.

⊠   **Landau, Lev Davidovich**
(1908–1968)
Russian
Low-Temperature, Atomic, Nuclear, Plasma, Solid State, and Quantum Physicist; Relativist

Lev Davidovich Landau was a brilliant theorist whose research and teaching raised theoretical physics in the Soviet Union to new levels. The range of his discoveries influenced all branches of theoretical physics, from fluid mechanics to quantum field theory. His theoretical explanation of why liquid helium is a superfluid earned him the 1962 Nobel Prize in physics.

Landau was born in Baku, Azerbaijan, Russian Empire, on January 22, 1908, into a scientifically oriented Jewish family. His father was an engineer, who worked in the Baku oil industry; his mother, a physician, who did research in physiology. Lev was a mathematical prodigy and finished his secondary studies at age 13. Too young to attend university, he studied physics and chemistry at the Baku Economical Technical School for a year before enrolling in Leningrad State University. After graduating in 1927, at the age of 19, he worked at the Leningrad Physico-Technical Institute (known as Phystech). At that time, the most exciting work in physics was being done in Europe, and Landau had the opportunity to interact with it firsthand. Between 1929 and 1931, he traveled to Göttingen, Leipzig, Zurich, and Cambridge. But his most formative stay was in Copenhagen, where he worked at NIELS HENRIK DAVID BOHR's Institute of Theoretical Physics, developing the new quantum mechanics. From then on he became a disciple of Bohr, whose work would have a powerful impact on his subsequent research, which involved applying the quantum paradigm to all realms of theoretical physics.

In 1930, Landau published a quantum theoretical study on the behavior of free electrons in a magnetic field that drew him instant international recognition. It proved essential to an understanding of the properties of metals. Working in collaboration with his students, after he returned to the Soviet Union, he was able to uncover important new theoretical results about the structure of magnetic substances and superconductors and new insights into the quantum theory of phase transformations and thermodynamical fluctuations. In 1932, he became the head of the Theory Division of the Ukrainian Physical-Technical Institute in Kharkov, Ukraine. Under his leadership, it became the center of theoretical physics in the Soviet Union. Three years later, he became head physicist at the Kharkov Gorky State University and began collaboration with E. M. Lifshits on classic texts on theoretical physics. Studying under Landau was no small achievement: to be accepted a student had to master what Landau called "the theoretical

minimum," which meant basic knowledge of all fields of theoretical physics.

In 1937, Landau left Kharkov for Moscow to become head of the Theory Division of PYOTR LEONIDOVICH KAPITSA's Institute of Physical Problems and teach at Moscow University. That year he married K. T. Drobanzeva, with whom he would have a son, Igor, who would become an experimental physicist. The late 1930s, however, was the height of the Stalinist purges, known as the Terror, and Landau soon became one of its victims. Swept up in the dragnet of random, groundless arrests, in April 1938, he was taken into custody and convicted of being a "German spy." After a year in prison, he was seriously ill, and Kapitsa, whose work on behalf of Soviet physics was much valued by Stalin, was

Lev Davidovich Landau was a Russian physicist whose discoveries influenced all branches of theoretical physics, from fluid mechanics to quantum field theory. (*AIP Emilio Segrè Visual Archives, Physics Today Collection*)

determined to save him. He handed the dictator a desperate ultimatum, saying that he, Kapitsa, would resign all his posts if Landau was not released immediately. Kapitsa won his gamble and Stalin assented. Although Landau would later be honored as a Hero of Socialist Labor, his youthful devotion to Communism did not survive this episode.

After Kapitsa's discovery, in 1938, of the superfluidity of liquid helium II, Landau began research that would lead him to a complete theory of the "quantum liquids" at temperatures near absolute zero. Helium gas had previously been liquefied by cooling to about 4 K above absolute zero. However, subsequent experiments by Kapitsa indicated that when cooled another 2 K, liquid helium was transformed into a new state, called liquid helium II, whose high thermal conductivity allowed it to flow without friction through very fine capillaries and slits that almost completely prevent the flow of all other liquids. Kapitsa invented the term *superfluid* to describe the physical flow properties of liquid helium II. On the basis of these experiments Landau set out to explain superfluidity in terms not of single atoms, but of the quantized states of motion of the whole liquid. He began by looking at the fluid in its ground state of absolute zero temperature. He theoretically described the excited states of the superfluid in terms of quantum states that he called quasi particles. He then used Kapitsa's experimental results to deduce the quantum mechanical properties of the quasi particles. Landau's theory, from which the properties of the superfluid could be calculated, was later confirmed in 1957 by studies of the scattering of neutrons in liquid helium II and earned him the Nobel Prize in physics in 1962.

His papers of 1941–1947 are devoted to the theory of the quantum liquids of the Bose type, to which the superfluid liquid helium (the usual isotope ^{4}He) belongs. Between 1956 and 1958, he formulated the theory of the quantum liquids of the Fermi type, among which liquid helium or isotope ^{3}He belongs.

In 1962, the year he won the Nobel Prize, Landau was in a car accident that left him unconscious for six weeks. Several times doctors declared him clinically dead. Although he did regain consciousness and lived for another six years, he was never again capable of creative work. He died in Moscow on April 1, 1968.

Landau is famous for the major discoveries he made in low-temperature, atomic, nuclear, and plasma physics. He will always be remembered for his uncanny ability to see to the core of a physical problem with a unique physical intuition that he was able to apply to almost all areas of theoretical physics.

Further Reading

Adronikashvili, Elevter. *Reflections on Liquid Helium.* Melville, N.Y.: American Institute of Physics, 1998.

Khalatnikov, I. M., ed. *Landau, the Physicist and the Man: Recollections of L. D. Landau.* Oxford: Oxford University Press, 1989.

Livanova, Anna Mikhailovna. *Landau: A Great Physicist and Teacher.* Oxford: Oxford University Press, 1980.

⊠   **Landé, Alfred**
(1888–1976)
German/American
Theoretical Physicist, Quantum Theorist

Alfred Landé was a theoretical physicist best known for discovering the g-formula, which relates the ratio of the magnetic moment of an atom to the angular momentum of the electric current loops that generate it. This fundamental discovery enabled physicists to determine the fine as well as the superfine structure of the optical and X-ray spectra of atomic systems.

Landé was born on December 18, 1888, in Elberfeld, in the Rhine region of Germany. Landé's father served as deputy head of the provincial government in Duesseldorf for a time. The family was a cultured one, imbuing him with a deep love of music. His early fascination with science encompassed cosmology, the study of crystals and minerals, chemistry, and electricity. His superiority in mathematics and physics led his high school teachers to consider him something of a prodigy. He studied at the Universities of Marburg and Göttingen, where he became an assistant to the great mathematician David Hilbert. He went on to earn his Ph.D. at the University of Munich in 1914, working under ARNOLD JOHANNES WILHELM SOMMERFELD, an influential and revered member of the community of European-born physicists who produced the quantum revolution in the first third of the 20th century. Unlike Sommerfeld and other visionary physicists, who regarded the new quantum theory and its wave–particle duality as a radical paradigm shift, however, Landé tried to understand the quantum mystery as a gap in classical mechanics.

When World War I began, Landé enlisted for service with the Red Cross and worked in a hospital for a few years. Later MAX BORN, one of the pioneers of quantum mechanics, managed to have him transferred to one of the few scientific sections of the army, a weapons development center in Berlin, where he worked under Born on sound detection methods. The two physicists did not confine their collaboration to war-related research, but moved into the abstract domain of the new quantum theory. Together they reached the conclusion that NIELS HENRIK DAVID BOHR's 1913 quantum mechanical model of the atom, which was limited to electron orbits moving in the same plane (in analogy to the solar system), was inadequate, since they found evidence that the electron orbits must be inclined toward each other. They published a paper stating these results, which were experimentally justified in 1918, arousing widespread interest and controversy in the physics community.

The following year Landé became a lecturer at the University of Frankfurt, where he would remain until 1922. This would be one of the most important periods of his scientific life. He visited Bohr at his institute in Copenhagen to discuss the Zeeman effect, the splitting of spectral lines in an intense magnetic field, which had been discovered by PIETER ZEEMAN, in 1897. He was able to pursue this issue further, when, in 1922, he was appointed associate professor of theoretical physics at the University of Tübingen, then the most renowned center for atomic spectroscopy in Germany. That year he married Elisabeth Grünewald, with whom he would have two sons, Arnold and Carl.

In 1923, Landé became famous for publishing a formula (known as the g-formula) expressing a factor known as the Landé splitting factor as a function of the quantum numbers of the stationary state of the atom. The Landé splitting factor determined the ratio of the magnetic moment of an atom to its intrinsic angular momentum, measured in quantum units of Bohr magnetons, which equals $eh/4\pi mc$, where e is the charge on an electron, h is the Planck constant, m is the electron rest mass, and c is the speed of light. On the basis of this fundamental discovery physicists were able to determine the fine and superfine structures of the optical and X-ray spectra of atomic systems. Landé went on to collaborate with Louis Paschen and others to analyze in great detail the fine structure of the line spectra and the further splitting of the lines under the action of magnetic fields of increasing strength. He formulated the laws obeyed by the frequencies and intensities of the lines in terms of the sets of spectroscopic quantum numbers, which could take on either integral or half-integral values.

As the Nazi regime was rising to power in 1931, Landé, who had visited Columbus, Ohio, in 1929 and 1930, left Germany for the United States and became a professor of theoretical physics at Columbus State University. He remained there for the rest of his life. His son

Arnold became a surgeon in Minneapolis, and his son Carl became a professor of politics at the University of Kansas.

Despite his many significant contributions to the quantum theory of atomic structure, to the very end of his career Landé remained opposed to the wave–particle duality interpretation of quantum mechanics. In support of this conservative point of view he claimed

> Today, after endless repetition, a dual nature of matter may seem as obvious and indisputable to the experts as the immobility of the earth seemed to Galileo's learned colleagues who refused to look through his telescope because it might make them dizzy.

He died in October 1976 in Columbus, Ohio. *See also* AMPÈRE, ANDRÉ-MARIE; BROGLIE, LOUIS-VICTOR PIERRE, PRINCE DE.

Further Reading

Born, Max. *Atomic Physics*. 8th Ed. New York: Dover Books, 1989, pp. 161–164.

Herzberg, Gerhard. *Atomic Spectra and Atomic Structure*. New York: Dover, 1944.

⊠   **Langevin, Paul**
        (1872–1946)
        French
        *Theoretician, Experimentalist
        (Acoustics, Condensed Matter),
        Mathematical Physicist*

Paul Langevin, the foremost mathematical physicist of his time in France, is most renowned for his invention of a method for generating ultrasonic waves. Langevin's invention became the basis of modern techniques of sonar: a method of finding the range and bearing of an object (target) by transmitting high-frequency sounds and detecting their echoes on their return. He also performed important work on paramagnetic and diamagnetic forces.

Langevin was born in Paris on January 23, 1872. He attended the École Lavoisier and the École de Physiques et de Chimie Industrielles, where Pierre Curie was his laboratory supervisor. He entered the Sorbonne in 1891, and took a one-year leave, in 1893, to serve in the military. In 1894, he entered the École Normale Superieure, where he studied under JEAN-BAPTISTE PERRIN. In 1897, he won an award that allowed him to spend a year at the Cavendish Laboratory in Cambridge, England, then under the direction of the great atomic physicist ERNEST RUTHERFORD; there he worked under JOSEPH JOHN (J. J.) THOMSON, the discoverer of the electron.

Langevin received a Ph.D., in 1902, for work on gaseous ionization done partly at Cambridge and partly under Pierre Curie. In Paris, he spent a lot of time in Perrin's lab and was swept up in the excitement of the early years of the study of radioactivity and ionizing radiation. He joined the faculty of the Collège de France in 1902 and, two years later, was made professor of physics. He remained there until 1909, when the Sorbonne offered him a similar position.

Langevin's early work at the Cavendish and the Sorbonne on the analysis of secondary emission of X rays from metals exposed to radiation resulted in his discovery of secondary electrons from irradiated metals. He was also interested in the dynamics of ionized gases, particularly the mobility of positive and negative ions; in 1903, he published a theory for their recombination at different pressures.

In 1905, ALBERT EINSTEIN published his groundbreaking paper on special relativity. Langevin would become deeply interested in Einstein's work on space and time and a firm believer in the theory of the equivalence of mass and energy. However, it was another paper Einstein published that year, on Brownian motion, the incessant random movement of microscopic particles in a liquid, that would influence Langevin's own seminal work on paramagnetic (weak attractive) and diamagnetic (weak repulsive) phenomena in gases. In 1895, Pierre Curie had shown experimentally that the susceptibility of a paramagnetic substance to an external magnetic field varies inversely with temperature. Ten years later, in 1905, when Langevin produced a model based on statistical mechanics to explain this phenomenon, he was influenced by Einstein's proposal that Brownian motion was due to imbalances in the forces on a particle resulting from molecular impacts from the liquid. Einstein's explanation led Langevin to hypothesize that when an externally applied magnetic field was absent, the alignment of molecular moments in a paramagnetic substance would be random; conversely, when such a field was present, the alignment would be nonrandom. The greater the temperature, however, the greater the thermal motion of the molecules, and thus the greater the disturbance to their alignment by the magnetic field. This theory was extremely useful in describing molecular fluctuations in other systems, including in nonequilibrium thermodynamics.

Langevin went on to propose that the magnetic properties of a substance are determined by the valence electrons, the specific orbital electrons that atomic elements have available to share when forming into molecular compounds. Langevin extended his description of magnetism in terms of electron theory to account for diamagnetism. He showed how a magnetic field would affect the motion of electrons in the molecules to produce a moment that is opposed to the field. This enabled predictions to be made concerning the temperature-independence of this phenomenon and allowed estimates to be made of the size of electron orbits.

In 1911, Langevin, whose work had been instrumental in verifying Einstein's atomic theories, published one of the earliest popular accounts of relativity, *L'evolution de l'espace et du temps* (The evolution of space and time). His scientific career was flourishing, but his personal life was in disarray. Unhappily married and the father of four children, he became romantically involved with MARIE CURIE, who had been widowed in 1906, when Pierre Curie was run over and killed by a carriage in the street. Langevin's wife, Jeanne, arranged for their love letters to be pilfered and published in the French press. Scandal broke out on the eve of the First Solvay Conference in Brussels, which both physicists were attending, nearly overwhelming the burning issues of the new quantum mechanics the conference was called to address. Although his affair with Curie did not last, Langevin obtained a divorce. Later, in 1921, he would become a member of the Solvay International Physics Institute and, in 1928, be elected its president.

Pan-European scientific gatherings such as the Solvay Conference were interrupted with the outbreak of World War I, in 1914. But the war years, when he worked on military technologies, led Langevin to his seminal discoveries on piezoelectricity, the electric current produced by some crystals and ceramic materials when they are subjected to mechanical pressure. Building on the research of other physicists, which showed that the reflection of ultrasonic waves from objects could be used to locate them, he developed an improved technique for accurate detection and location of submarines. His technique was based on the use of high-frequency radio circuitry to oscillate piezoelectric crystals and thus obtain ultrasonic waves at high intensity. Within a few years, this approach led him to a practical system for the echolocation of submarines, which became the basis of modern sonar and is used for scientific as well as military purposes. In 1917, he pioneered the use of the piezoelectric effect, that is, the generation of a small potential difference across certain materials when they are subjected to a stress, as well as vacuum tube amplifiers in underwater sounding equipment, the first use of electronics in this way. In 1918, this new technology enabled him to receive echoes from a submarine as deep as 1800 meters. This work continued after the war, leading to the development of sonar transducers, circuits, systems, and materials.

In 1940, after the German occupation of France, Langevin became director of the École Municipale de Physique et de Chimie Industrielle, where he had been teaching in 1902. However, the Nazis soon arrested him for his outspoken antifascist views. He was first imprisoned in Fresnes and later placed under house arrest in Troyes. After the execution of his son-in-law and deportation of his daughter to Auschwitz (which she survived), he was forced to escape to Switzerland in 1944. Langevin returned to Paris later that year and resumed his directorship of his old school. He died soon after in Paris on December 19, 1946. The Institute Max von Laue–Paul Langevin was established in Grenoble, France, in 1967, in honor of him and German physicist MAX THEODOR FELIX VON LAUE, as a symbol of postwar cooperation between France and Germany.

Langevin's genius as a theoretical physicist was recognized by Einstein, who wrote that Langevin had all the tools for the development of the special theory of relativity at his disposal, and that if he had not proposed the theory himself, Langevin would have done so. Equally talented as an experimentalist, Langevin, through his research on the piezoelectric effect, initiated the modern ultrasonic era.

Further Reading
Graff, K. F. "A History of Ultrasonics," in Warren Mason, ed. *Physical Acoustics*. Vol. 1. New York: Academic Press, 1981.

⊠ **Laue, Max Theodor Felix von**
(1879–1960)
German
*Theoretician (Electromagnetism),
Relativist, Solid State Physicist*

Max von Laue was a brilliant theoretician who played an important role in several crucial developments in the early part of the 20th century. One of ALBERT EINSTEIN's earliest followers, he produced evidence strengthening the theory of relativity, thereby hastening its acceptance. In 1912, through his discovery of X-ray diffraction in crystals, he simultaneously revealed the nature of X rays and crystals and laid the foundations of X-ray crystallography. This work, described by Einstein as one of the most beautiful discoveries in physics, won him the 1914 Nobel Prize in physics.

He was born in Pfaffendorf, near Koblenz, on October 9, 1879, to Julius and Wilhelmine Laue. His father was an official in the German military administration, who was raised to the rank of the hereditary nobility in 1913, thus becoming "von Laue." The nature of his father's work led Max to spend his childhood moving between a number of German towns, including Strasbourg. He studied briefly at the University of Strasbourg, before transferring to the University of Göttingen, where he specialized in theoretical physics. He spent a semester at the University of Munich and, in 1902, continued his studies at the University of Berlin. There, working under the founder of quantum theory, MAX ERNEST LUDWIG PLANCK, he earned his Ph.D. the following year for a dissertation on light interference phenomena between plane-parallel plates. He then returned to Göttingen, where he spent two years studying art and earned a teaching certificate in 1905.

That year Laue became an assistant to Planck at the Institute of Theoretical Physics in Berlin, where the two physicists began what would be a long, productive collaboration. During his four years in Berlin, Laue qualified as a lecturer. He worked on the application of entropy to radiation fields and on the thermodynamic significance of the coherence of light waves. Inspired by Planck's relativity seminar, he began writing to Einstein, with whom he developed a close friendship. When, in 1906, he visited him at his patent office in Bern, Laue emerged feeling he had met the "revolutionary" who had "overturned all of mechanics and electrodynamics." He developed a proof for Einstein's special theory of relativity based on optics. His proof used ARMAND HIPPOLYTE LOUIS FIZEAU's 1851 experimental verification of a theory developed earlier by AUGUSTIN JEAN FRESNEL that hypothesized that the speed of light is affected by the motion of water. This theory seemed to be at odds with the prime assertion of special relativity, that is, that the speed of light is a constant independent of the motion of the observer. However, Laue showed that Fresnel's theory could be derived in a manner consistent with special relativity. In this way, Fizeau's experiment provided confirmation of relativity, contributing greatly to its rapid acceptance. In addition, Laue published the first monograph on Einstein's special theory of relativity, which he would expand in 1919 to include the general theory of relativity; both monographs were reprinted several times.

In 1909, Laue became a lecturer at the Institute of Theoretical Physics at the University of Munich, then under the direction of the renowned ARNOLD JOHANNES WILHELM SOMMERFELD; in the following year he married Magdalena Degen, with whom he would have two children. In Munich, Laue began his research on the nature of X rays, which were only vaguely understood at the time. X rays were thought to be extremely short electromagnetic waves. Because of their short wavelengths, the problem physicists confronted in measuring them was that no grating sufficiently fine to diffract them could be artificially produced. Laue found his measuring instrument in nature. He hypothesized that crystals, known to consist of orderly arrays of atoms, could be used as superfine gratings, providing a medium in which X rays would form some kind of diffraction or interference pattern.

In the spring of 1912, when he had moved to the University of Zurich, Laue's proposal was tested by Sommerfeld's graduate students. They bombarded copper sulfate crystals with X rays and produced a photographic plate with a dark central patch representing X rays that had penetrated straight through the crystal surrounded by a multilayered halo of regularly spaced spots representing diffracted X rays. The results were announced to the Bavarian Academy of Science less than two weeks later along with Laue's mathematical formulation, the "theory of diffraction in a three-dimensional grating." The demonstration of the nature of crystal structure and of X rays by the same experiment was widely acclaimed and promptly led to Laue's selection as the 1914 Nobel laureate in physics.

His fame established, Laue was offered and accepted a full professorship at the University of Frankfurt. By this time, however, Europe was in the grip of World War I; from 1916 on, Laue worked at the University of Würzburg, developing communication equipment for the German army. When hostilities ceased, he exchanged teaching positions with MAX BORN and became professor of theoretical physics at the University of Berlin, where he expanded his original theory of X-ray diffraction. Laue had initially considered only the interaction between the atoms in the crystal and the radiation waves, but he now included a correction for the forces acting between the atoms. This correction accounted for the slight deviations that had already been noticed. Although Laue did not participate directly in the development of quantum theory, he did incorporate the discovery of electron wave particle interference into his diffraction theory.

In 1914, the Kaiser-Wilhelm Institute of Physics in Dahlem, a suburb of Berlin, had been founded with Einstein as first director. Laue now became second in charge, dealing with most of the administrative work of the institute, particularly its financing. In this way he had a strong influence on the direction of German physics in the 1920s and early 1930s, advancing important discoveries in theoretical physics.

Laue's eminence in the scientific community of Germany was recognized in 1932, when he was awarded the Max Planck Medal by the German Physical Society.

However, in 1933, the rise of Hitler and the resulting persecution of Jewish scientists such as Einstein became intolerable to Laue. By virtue of his protest he lost his influence within Germany's scientific establishment. In 1943, he resigned in protest from his position as professor at University of Berlin. When Berlin was bombed in 1944, he moved to Hechingen, along with the Kaiser Wilhelm Institute. When the Allies invaded Germany, his acts of civic courage and refusal to participate in the development of a German atomic bomb did not exempt him from internment in England, along with nine other German atomic physicists. During this dark period, he continued his research and wrote a history of physics, which was translated into many languages and had several editions.

In 1946, Laue returned to Germany and helped rebuild German science, becoming acting director of the Max Planck Institute in Göttingen and professor at the university there. He also did some important work on the effect of magnetic fields on superconductivity, publishing 12 papers and a book on the subject between 1937 and 1947. In 1951, he became director of the Fritz Haber Institute for Physical Chemistry in Berlin, where he focused on research on X-ray optics. In 1958, he retired but continued his research. At the age of 80, he died on April 23, 1960, of injuries sustained in a car accident, when he collided with an inexperienced motorcyclist on the way to his Berlin laboratory.

A deeply religious man, esteemed by his colleagues for both his intellect and his moral integrity, he exerted a strong, positive influence on German physicists during the catastrophic years of World War II and its aftermath. The Institute Max von Laue–Paul Langevin was established in Grenoble, France, in 1967, in honor of him and the French physicist PAUL LANGEVIN, as a symbol of postwar cooperation between France and Germany.

The reverberations of Laue's discovery that X rays were waves of light that obeyed Maxwell's equations were felt in physics, chemistry, and biology. His work gave rise to two entirely new branches of science: X-ray crystallography, which led to the determination of the structure of (DNA), and X-ray spectroscopy, which led to exact measurement of the wavelength of X rays and to advances in atomic theory.

See also MAXWELL, JAMES CLERK; RÖNTGEN, WILHELM CONRAD.

Further Reading

Hammond, Christopher. *The Basics of Crystallography and Diffraction*. Oxford: Oxford University Press, 2001.

Mendelssohn, Kurt. *The World of Walther Nernst: The Rise and Fall of German Science, 1864–1941*. Pittsburgh: University of Pittsburgh Press, 1973.

⊠   **Lederman, Leon M.**
(1922–)
American
Experimentalist, Particle Physicist

Leon M. Lederman shared the 1988 Nobel Prize in physics with Melvin Schwartz and Jack Steinberger for the invention of the neutrino beam method and the demonstration of the doublet structure of the electron and of the muon (both now known as leptons). They arrived at this discovery by observing that the weak interaction decay processes associated with electrons and muons involved electron neutrinos and muon neutrinos, respectively.

He was born on July 15, 1922, in New York City, the son of Jewish immigrants. His father, Morris Lederman, who earned his living by operating a hand laundry, had a profound respect for education. Leon would later trace the beginning of his passion for science to the reading of science books:

> I got hold of a book in big print that was written by Einstein in the 1930s, which presented science as a detective story. And I loved that presentation.

Lederman attended Manhattan public schools, where he was inspired by "absolutely magnificent teachers." He entered the City College of New York as a chemistry major, but, by the time he graduated in 1943 with a degree in chemistry, he had lost his enthusiasm for the subject, "because of all the smells, that was one thing," and found physics to be "simpler somehow, cleaner." With the country embroiled in World War II, he was obliged to put off further studies. For the next three years, he served in the United States Army, rising to the rank of second lieutenant in the Signal Corps. On discharge, he entered Columbia University's graduate program in physics, headed by ISIDOR ISAAC RABI, the inventor of the technique of nuclear magnetic resonance. In 1948, Lederman began working with Eugene T. Booth, director of Columbia's project to construct a 385-million-electron-volt (Mev) synchrocyclotron (a device for accelerating particles to high energies) at its Nevis Laboratory. Lederman's thesis assignment was to build a Wilson cloud chamber, the first instrument to detect the tracks of atomic particles. He was awarded the Ph.D. in 1951 for this work and asked to stay on at Columbia, which remained his academic home for the next 28 years. During this time he directed the work of 50 Ph.D. candidates.

In 1958, after being promoted to full professor, Lederman took his first sabbatical at the European Center for Nuclear Research (CERN), where he organized a group to do the "g-2" experiment, designed to measure the anomalous magnetic moment of elementary particles. He would continue to participate in CERN collaborations through the 1970s. Between 1961 and 1978, he was director of the Nevis Lab. In 1979, he became director of the Fermi National Accelerator Laboratory, where he supervised the construction and utilization of the first superconducting synchrotron, which became the highest-energy accelerator in the world at that time. Although he was a guest scientist at many laboratories, most of his important research was done at Nevis, Brookhaven National Accelerator Laboratory on Long Island, New York, CERN, and Fermilab.

Lederman's Nobel Prize–winning work was carried out in the early 1960s. The experiment was planned when Lederman, Melvin Schwartz, and Jack Steinberger were associated with Columbia University and was carried out with the alternating gradient synchrotron at Brookhaven. The work led to discoveries that opened entirely new opportunities for research into the innermost structure and dynamics of matter. It removed two big obstacles to research into the weak forces (1) by inventing an experimental method for the study of weak forces at high energies and (2) by discovering that the weak interaction of radioactive decay is associated with at least two kinds of neutrinos. One neutrino belongs with the electron, the other neutrino with the muon (a relatively heavy charged elementary particle, discovered in cosmic radiation in the late 1930s by CARL DAVID ANDERSON). The current elementary particle theory of electroweak interactions, which uses a doublet grouping of electron with electron neutrino and muon with muon neutrino, has its roots in this discovery.

Neutrinos are elementary particles with no charge and zero or almost zero rest mass. They are ghostlike particles, which undergo only weak interactions. Classified as leptons with the electron and the muon, they are fermions with spin 1/2. In 1927 the electron neutrino's existence was postulated by WOLFGANG PAULI. Lederman and his colleagues transformed the ghostly neutrino into an active research tool by inventing a method that generated a beam of high-energy neutrinos from the interactions occurring in high-energy proton particle accelerators. Using this method, they found that neutrino beams could reveal the hard inner structure of nucleons.

At Columbia, where TSUNG-DAO LEE and CHEN NING YANG had discovered that the law of parity was violated in the weak interactions, the problem Lederman's group addressed was developing a feasible method of studying the effect of weak forces at high energies. Previously, it had been possible to study only low-energy spontaneous processes of radioactive decay. Hence, beams of electrons, protons, and neutrons had proved unusable. Schwartz proposed using a beam of neutrinos, and for the next two years the three collaborators worked on creating a sufficiently intense beam of high-energy neutrinos free of all other particles, as well as on designing a detector for measuring high-energy neutrino reactions. Their method was to generate a high-energy muon neutrino beam as the by-product of the decay processes associated with high-energy proton collisions and let these muon neutrinos interact with matter. If the muon neutrinos were the same as electron neutrinos, then equal numbers of electrons and muons would be generated by these neutrino interactions. What they found, however, was that their high-energy neutrino beam created only muons. This result proved that the weak interaction of radioactive decay is associated with at least two kinds of neutrinos, an electron neutrino at low energies and a muon neutrino at high energies.

Lederman's first wife was Florence Gordon, with whom he had three children. He is now married to Ellen Lederman. After retiring from Fermilab in 1989, he became professor of physics at the University of Chicago. That year he was appointed science adviser to the governor of Illinois. He helped organize the Teachers' Academy for Mathematics and Science, designed to bolster the skills of teachers in the Chicago public schools. He helped found and is on the board of trustees of the Illinois Mathematics and Science Academy, a residential public school for gifted children. In 1991, he became president of the American Association for the Advancement of Science. Lederman has also developed collaborations with Latin American scientists and been a crusader for science education for gifted children as well as for public understanding of science. He has served on the boards of directors of several museums, schools, science organizations, and government agencies.

Lederman's invention of the high-energy neutrino beam method, which led to the discovery that electrons and muons have their own neutrinos, laid the experimental foundation for theoretical development of the unified theory of electroweak interactions. This theory later became important for quark research that yielded a deeper understanding of the strong nuclear interactions.

In his later years, Lederman, in his characteristically witty, incisive manner, expressed his view of the way science works:

> You get to be a part of the establishment by blowing it up. . . . Because every time you destroy the established dogma . . . the accepted body of knowledge, you are going to get something better to take its place. . . . Although nobody likes his own theory to be overturned . . . at some point deep down inside the scientist will creep in and say, gee, but this gives the possibility of a different theory, which might even be better.

See also GELL-MANN, MURRAY; GLASHOW, SHELDON LEE; SALAM, ABDUS; WEINBERG, STEVEN.

Further Reading

Lederman, Leon M., and David N. Schramm. *From Quarks to the Cosmos: Tools of Discovery.* New York: W. H. Freeman, 1995.

Lederman, Leon M., and Dick Teresi. *The God Particle: If the Universe Is the Answer, What Is the Question?* New York: Houghton Mifflin, 1993.

⊠ Lee, Tsung-Dao
(1926–)
Chinese/American
Theoretical Physicist, Particle Physicist

Tsung-Dao Lee is a world-renowned theoretician who, together with CHEN NING YANG, predicted that conservation of parity is violated in the weak interactions of the atomic nucleus. Their discovery led to significant developments in particle theory and won them the 1957 Nobel Prize in physics. Only 31 years of age at the time, Lee became the second youngest physicist to be awarded the Nobel Prize.

Lee was born on November 24, 1926, in Shanghai, China, the third of six children born to Tsing Kong Lee, a businessman, and Ming Chang Chang. He attended the Kiangsi Middle School in Kanchow, Kiangsi, graduated in 1943, and entered the National Chekiang University in Kweichow province. When the Japanese invasion forced him to flee to Kunming, Yunan, he enrolled at the National Southwest University, where he met his future collaborator, Yang.

Lee entered the United States, in 1946, on a scholarship from the Chinese government. Although he had never formally earned an undergraduate degree, he enrolled in graduate studies in physics at the University of Chicago, where his friend Yang was also enrolled. He earned a Ph.D. in 1950, working under the eminent Indian-born astrophysicist SUBRAMANYAN CHANDRASEKHAR, for his dissertation "Hydrogen Content of White Dwarf Stars." That year he married (Jeanette) Hui Chung Chin, a former university student, with whom he would have two sons, James and Stephen. After spending a few months as a research associate at Yerkes Astronomical Observatory, in Lake Geneva, Wisconsin, from 1950 to 1951, he was a research associate and lecturer at the University of California in Berkeley. He then accepted a fellowship to the Institute of Advanced Study in Princeton and became a member of the staff. He was

appointed assistant professor of physics at Columbia University in 1953, promoted to associate professor in 1955 and to professor in 1956 (at age 29, he was then the youngest professor on the faculty). By this time Lee had a substantial reputation, for his work in statistical mechanics and in nuclear and atomic physics.

In June 1956, Yang collaborated with Lee on a paper that raised the question of whether parity, the assumption that nature makes no distinction between left and right, is conserved in weak interactions. At the time physicists universally believed that physical reactions would be the same (i.e., have parity, or equality) whether the particles involved in them had a right-handed or a left-handed spin (i.e., the quantized rotation property). If the physical process proceeds in exactly the same way when referred to an inverted coordinate system, then parity is said to be conserved. If, on the contrary, the process has definite left- or right-handedness, then parity is not conserved in that physical process. This law of conservation of parity was explicitly formulated in the early 1930s by the Hungarian-born physicist EUGENE PAUL WIGNER and became a component of quantum mechanics.

The strong forces that hold atoms together and the electromagnetic forces that are responsible for chemical reactions obey the law of parity conservation. Since these are the dominant forces in most physical processes, physicists assumed that parity conservation was an inviolable natural law. In the early 1950s, applying the principle of conservation of parity to individual subatomic particles and their interactions had proved highly successful in accounting for the behavior of those particles. By the end of 1955, however, a puzzling contradiction between the parity principle and the other principles employed to order the ever-growing number of subatomic particles had emerged. In particular, questions were raised by results beginning to pour forth from the many high-energy accelerators built in the United

Tsung-Dao Lee together with Chen Ning Yang predicted that conservation of parity is violated in the weak interactions of the atomic nucleus. Only 31 years of age at the time, Lee became the second youngest physicist to be awarded the Nobel Prize. *(AIP Meggars Gallery of Nobel Laureates)*

States after World War II, indicating that one form of radioactive decay appeared to violate the conservation of parity law. One of the newly discovered mesons—the so-called K meson—seemed to exhibit decay modes into configurations with differing parity. Exploring this paradox from every conceivable perspective, Lee and Yang discovered that contrary to what had been assumed, there was no experimental evidence against parity nonconservation in the weak interactions. The experiments that had been done, it turned out, were not relevant to the question. In their landmark 1956 *Physical Review* paper, "Question of Parity Conservation in Weak Interactions," Lee and Yang made

the startling proposal that the universally accepted conservation of parity law might not hold true in weak nuclear interactions, which include radioactive decay.

Their suggestions for experiments capable of deciding the issue were immediately taken up by CHIEN-SHIUNG WU at the cryogenic laboratory of the National Bureau of Standards in Washington, D.C. Testing radioactive cobalt atoms at temperatures approaching absolute zero, Wu found the evidence Lee and Yang were looking for: the law of parity did not apply to weak interactions. Shortly after her announcement in January 1957, other experimentalists confirmed her results. In the wake of this paradigm revolution generated by the three Chinese American physicists, Lee and Yang were awarded the Nobel Prize in physics that very year.

In 1960, Lee became professor of physics at the Institute of Advanced Study, Princeton; he returned to Columbia in 1963 to assume the first Enrico Fermi Professorship in physics. Since 1964 he has also made significant contributions to the explanation of the violations of time-reversal invariance that occur during certain weak interactions.

As did Yang, Lee became deeply involved in the development of Chinese science, and specifically its integration into world science. Working together with the Chinese Academy of Sciences, he founded the China Center of Advanced Science and Technology World Laboratory in 1989. Since then he has traveled to China every year to participate in international academic exchanges.

Through his brilliant insights, Lee has made invaluable contributions to the modern formulation of symmetry principles in particle physics. The renowned American physicist J. ROBERT OPPENHEIMER described Lee's work as characterized by "a remarkable freshness, versatility, and style." As a 77-year-old professor at Columbia University, Lee summarized his relationship to his work in a statement of moving simplicity:

To me, scientific research is as important as breathing, and it equals my life.

Further Reading

Bernstein, Jeremy. *A Comprehensible World: On Modern Science and Its Origins*. New York: Random House, 1967.

Boorse, H. A., and L. Motz, eds. *The World of the Atom*. New York: Basic Books, 1966.

⊠ Lenard, Philipp von
(1862–1947)
Hungarian/German
Experimental and Theoretical Physicist, Atomic Physicist

Philipp von Lenard is most famous for developing an ingenious experiment involving a cathode ray tube with a thin aluminum window that permitted the rays to be studied as they escaped into the open air. The results of this research, which led him to conclude that the volume of an atom is mainly composed of empty space, contributed to the conception of Rutherford's planetary model of the atom and earned Lenard the 1905 Nobel Prize in physics. His pioneering work on the photoelectric effect was influential in ALBERT EINSTEIN's formulation of the theory of light quanta.

He was born in Pressburg, Hungary (now Bratislava, Slovakia), on June 7, 1862. His family was originally from the Tyrol, and Lenard was strongly drawn to German culture. After briefly studying at the University of Budapest, he went to Germany, where he studied physics at the University of Heidelberg under Robert Bunsen, the inventor of the Bunsen burner, and at the University of Berlin under the eminent HERMANN LUDWIG FERDINAND VON HELMHOLTZ. In 1886, he received a Ph.D. from Heidelberg, where he remained for the next three years.

In 1891, Lenard became assistant to HEINRICH RUDOLF HERTZ, the discoverer of radio waves, at the University of Bonn, where he did his first original research in mechanics, on the oscillation of precipitated water drops. He also worked on luminescence and phosphorescence. He then began his work on cathode rays, the phenomenon that produced a fluorescent glow when most of the air was pumped out of a glass tube with wires embedded at either end and a high voltage was sent across it. At the time physicists were asking whether cathode rays were charged particles or some undefined wavelike process in the ether. In 1892, after reading William Crookes's 1879 paper on the movement of cathode rays *inside* discharge tubes, Lenard became interested in designing an experiment that would allow cathode rays to be examined *outside* discharge tubes. Hertz suggested that Lenard construct a cathode ray tube with a thin sheet of aluminum serving as a window, now called the Lenard window, to contain the vacuum, while releasing the cathode rays. Lenard experimented with windows made of aluminum foil of varying thickness and in 1894 published results that revealed the amazing discovery that the cathode rays could move about 8 cm in the air after passing through the thin aluminum window. In this experiment he found that the cathode rays decreased in number as the distance from the tube increased and was able to show that the ability of a material to absorb cathode rays depends on its density. The experimental finding that cathode rays were able to pass through the aluminum foil led him to the hypothesis that the volume that the atoms occupied in the metal consisted of a large amount of empty space. This result proved crucial to JOSEPH JOHN (J. J.) THOMSON in his discovery that cathode rays were really electrons: particles that had to have a mass much smaller than the mass of any atom. Lenard had hoped to make that discovery himself and was embittered when Thomson beat him to it. Subsequent experiments by others measured the charge directly and confirmed Lenard's conclusions.

After his groundbreaking work on cathode rays, Lenard accepted an associate professorship at the University of Breslau and in 1895 became professor of physics in Aachen. In 1896, he became professor of theoretical physics at Heidelberg, and in 1898, he became professor of experimental physics at the University of Kiel.

In 1901, on the basis of his cathode ray experiments, he concluded that the part of the atom where the mass was mostly concentrated consisted of neutral doublets, or "dynamides," of negative and positive electricity, which were very small and separated by wide spaces and had a number equal to the atomic mass. On the basis of this idea he estimated that the solid matter in the atom was about one billionth of the whole atom. This work was influential in HENDRIK ANTOON LORENTZ's formulation of his theory of electrons and, 10 years later, in ERNEST RUTHERFORD's proposal of his planetary model of the atom, which incorporates this basic structure.

From 1902 onward, Lenard studied photoelectricity and worked on extending Hertz's research on the photoelectric effect. In the latter experiments he was able to show that negative electricity can be released from metals by exposure to ultraviolet light. He later found that this electricity was identical in properties to cathode rays (later known as electrons). He also was able to show experimentally that the number of electrons projected is proportional to the energy carried by the incident light, whereas the electron speed or kinetic energy varies inversely with the wavelength of the incident light while remaining independent of its energy. However, it was Einstein who was able to explain the photoelectric effect successfully, in 1905, using the quantum theory rejected by Lenard, by postulating that light consists of particles of energy, which he called photons.

Lenard married Katharina Schlehner and between 1907 and 1931 settled in Heidelberg, as professor of experimental physics. In 1924, he became a dedicated follower of Hitler's National Socialist Party, which rewarded his zeal by making him the "Chief of Aryan [or German] Physics." He spent his later years reviling Jewish scientists such as Einstein, whom he never forgave for discovering and giving his name to the theory of light quanta, which explained Lenard's results.

Antagonistic by nature, Lenard in his book *Great Men of Science* (1934) omitted such figures as WILHELM CONRAD RÖNTGEN, who had used a tube that Lenard had designed to discover X rays but had failed to credit him. Lenard felt underrated by the physics community, despite his Nobel Prize and a great many lesser honors. He died in Messelhausen, Germany, on May 20, 1947.

Lenard left a mixed legacy to physics. He made substantial contributions to atomic physics but will also be remembered as one of the very few distinguished physicists to embrace Nazism.

See also MILLIKAN, ROBERT ANDREWS.

Further Reading

Keller, Peter. *The Cathode Ray Tube: Technology, History, and Applications.* New York: Palisades Books, 1992.

⊠   **Lenz, Heinrich Friedrich Emil**
     (1804–1865)
     Russian
     *Theoretical and Experimental Physicist
     (Classical Electromagnetism,
     Geophysics)*

Heinrich Lenz was a Russian physicist who contributed a significant chapter to the evolving story of electromagnetism that physicists throughout Europe were piecing together in the 1800s. The fundamental law that he discovered, which later became known as Lenz's law, revealed that the phenomenon of electromagnetic induction, as described earlier by MICHAEL FARADAY, obeyed the law of conservation of energy.

Lenz was born in Dorpat, now Tartu, Estonia, on February 12, 1804. After completing his secondary education at the age of 16, he entered the University of Dorpat, where he studied chemistry and physics. In 1923, while still a student, he managed to obtain the post of geophysical scientist on Otto von Kotzebue's third expedition around the world. For the next three years he took advantage of this unique opportunity to investigate geophysical phenomena. He studied climatic conditions such as barometric pressure, finding the areas of maximal and minimal pressure that exist in the Tropics and determining the overall climatic pattern. He also made extremely accurate measurements of the salinity, temperature, and specific gravity of seawater. He discovered areas of maximal salinity on both sides of the equator in the Atlantic and Pacific Oceans and established the differences in salinity between these oceans and the Indian Ocean. On a later expedition, to the Caucasus in southern Russia in 1829, he studied the area's natural resources, measured mountain heights, and measured the level of the Caspian Sea. No one would obtain observations of the ocean more accurate than Lenz's until the next century.

When he returned to Saint Petersburg, with his expedition findings to his credit, Lenz was admitted to the prestigious Saint Petersburg Academy of Science, on an apprentice level; by 1834, he would rise to the status of full academician. He began to study electromagnetism in 1831, after Faraday and JOSEPH HENRY independently discovered electromagnetic induction. His first major discovery, published in 1833, was the famous Lenz's law, which states that the direction of a current that is induced by an electromagnetic force always opposes the direction of the electromagnetic force that produces it. This occurs because a current induced by a moving magnet or coil flows in such a direction that it, in turn, induces a magnetic field that opposes the motion of the magnet or coil inducing the current.

Lenz described the dynamics of this process as follows: thrusting a pole of a permanent bar magnet through a coil of wire induces an electric current in the coil; the current in turn sets up a magnetic field around the coil, making it a magnet. Lenz's law indicates the direction of the induced current. Because like magnetic poles repel each other, when the north pole of the bar magnet approaches the coil, the induced current flows in such a way as to make the side of the coil nearer the pole of the bar magnet itself a north pole to oppose the approaching bar magnet. When the bar magnet is withdrawn from the coil, the induced current reverses itself, and the near side of the coil becomes a south pole to produce an attracting force on the receding bar magnet. A small amount of work, therefore, is done in pushing the magnet into the coil and in pulling it out against the magnetic effect of the induced current. The small amount of energy represented by this work manifests itself as a slight heating effect, the result of the induced current's encountering resistance in the material of the coil. Lenz's law thus upholds the general principle of the conservation of energy. If the current were induced in the opposite direction, its action would spontaneously draw the bar magnet into the coil in addition to producing the heating effect, thereby violating conservation of energy.

In 1833, Lenz investigated the way electrical resistance of metals changes with temperature and showed that an increase in temperature increased resistance. Around the same time, he studied the heat generated by current flowing in metals and discovered, independently of JAMES PRESCOTT JOULE, the law now known as Joule's law, which describes the proportional relationship between the production of heat and the square of the current. In the context of these experiments, he worked on establishing the unit for the measurement of resistance.

During the same period, he also worked on the application of certain theoretical physics principles to engineering design and on formulation of programs for geographical expeditions. In

1836, he became a professor at Saint Petersburg University and from 1840 to 1843 was dean of mathematics and physics. He authored a number of very successful books. Lenz died of a stroke just before his 61st birthday, while on vacation in Rome, on February 10, 1865.

Lenz's law helped HERMANN LUDWIG FERDINAND VON HELMHOLZ formulate the law of conservation of energy. It is applied today in electrical machines such as generators and electric motors.

Further Reading

Keithley, Joseph F. *The Story of Electrical and Magnetic Measurements: From Early Days to the Beginnings of the 20th Century.* New York: Wiley-IEEE Press, 1998.

⊠   **Lorentz, Hendrik Antoon**
(1853–1928)
Dutch
Theoretical Physicist (Classical Electromagnetism, Optics), Relativist

Hendrik Lorentz extended JAMES CLERK MAXWELL's theory of electromagnetism by proposing that since light is generated by oscillating electric currents, the presence of a strong magnetic field would have an effect on the charged particles that make up the oscillating currents. Specifically, it would result in a splitting of spectral lines by causing the wavelengths of the lines to vary. In 1896, his student PIETER ZEEMAN experimentally confirmed Lorentz's theory, and they shared the Nobel Prize in physics in 1902. Of equal importance, Lorentz developed a set of equations that mathematically predicted the increase of mass, shortening of length, and dilation of time for a physical object in motion with velocities ranging from zero to the speed of light. These equations, which later became known as the Lorentz transformation, played a major role in ALBERT EINSTEIN's development of the theory of special relativity.

Lorentz was born in Arnhem, the Netherlands, on July 18, 1853. His childhood was marked by the death of his mother when he was four and his businessman father's remarriage five years later. Hendrik was educated at local schools and entered the University of Leiden in 1870; in less than two years he had earned the B.Sc. degree in mathematics and physics. He left Leiden at the age of 19 to return to Arnhem as a night school teacher, while writing his Ph.D. thesis, which refined the electromagnetic theory of James Clerk Maxwell so that it more satisfactorily explained the reflection and refraction of light. He received his degree in 1875 at the age of 22.

When Lorentz was 24, in 1878, he assumed the chair of theoretical physics, newly created for him, at Leiden, where he would remain for the next 39 years. In 1881, he married Aletta Catharina Kaiser. The couple had a son and two daughters, one of whom became a physicist. During that time he published an essay on the relationship between the velocity of light in a medium and the density and composition of the medium. The resulting formula, developed almost simultaneously by the Danish physicist Ludwig Lorenz, has become known as the Lorenz–Lorentz formula.

In the process of attempting to develop a unified theory to explain the relationship of electricity, magnetism, and light, Lorentz proposed that if electromagnetic radiation is produced by the oscillation of electric charges, the source of light might be traced to the oscillation of charges inherent in atoms of matter. If this were the case, a strong magnetic field should have an effect on the oscillations and thus on the wavelength of the light generated. The theory was confirmed in 1896 by the discovery of the Zeeman effect, in which a magnetic field splits spectral lines. This made it possible to apply the molecular theory to the theory of electricity and to explain the behavior of light waves passing through moving transparent bodies.

Lorentz's electron theory was not successful in explaining the null result of the Michelson–Morley experiment that attempted to measure the "luminiferous ether," the hypothetical medium in which light waves were thought to propagate. The experimenters had hoped to measure the ether by observing changes in the speed of light on the basis of interference patterns created when a light beam was split in two and the separated beams were guided along perpendicular paths and then recombined. To their dismay, no changes in the speed of light were observed. In an attempt to salvage the concept of the ether, Lorentz and the Irish physicist FRANCIS GEORGE FITZGERALD, working independently, developed the idea, now known as the Lorentz–FitzGerald contraction hypothesis, that the length of an object moving through the ether contracts in the direction of the motion. When Lorentz heard of FitzGerald's work, he was glad to have an ally, commenting, "I have been rather laughed at for my idea over here." Both were scrupulous in acknowledging one another's work, each believing that the other had published first.

In 1899, Lorentz showed that the Lorentz–FitzGerald contraction resulted from a process delineated in a group of equations, which became known as Lorentz transformations, that mathematically predict the increase of mass, shortening of length, and dilation of time for an object traveling at near the speed of light. Taken together, they compose the revolutionary picture of space and time that Einstein would present in 1905 in his theory of special relativity. Einstein would base his theory on two postulates: (1) the laws of physics take the same form in all inertial frames; (2) in any inertial frame, the velocity of light c is the same whether the light is emitted by a body at rest or by a body in uniform motion. From these he deduced both the Lorentz transformations and the Lorentz–FitzGerald contraction.

However, even though Lorentz was recognized as a major contributor to the mathematics of special relativity, he himself never felt comfortable with Einstein's conclusions:

> [I find] a certain satisfaction in the older interpretation according to which the ether possesses some substantiality, space and time can be sharply separated, and simultaneity without further specification can be spoken of.

His attitude toward the second great physical revolution of the day, quantum mechanics, was similarly conservative. In 1911, he chaired the first Solvay Conference in Brussels, which met to consider the problem of having two different approaches, classical and quantum

Hendrik Lorentz developed a set of equations, later known as the Lorentz transformation, that played a major role in Albert Einstein's development of the theory of special relativity. *(AIP Emilio Segrè Visual Archives)*

physics. His private hope was that it would be possible somehow to reincorporate quantum theory into classical physics.

From 1912, when he accepted a double position as curator at the Teyler Institute in Haarlem and secretary of the Dutch Society of Sciences, onward, Lorentz continued at Leiden as extraordinary professor, delivering his famous Monday morning lectures for the rest of his life. In 1919, he assumed a leading role in the project to reclaim the Zuyderzee in the Netherlands, a great work of hydraulic engineering. His theoretical calculations played an important role in the project. Enjoying great prestige in Dutch governmental circles, he initiated steps leading to the creation of a society for applied scientific research. In 1923, he was elected to membership in the International Committee of Intellectual Cooperation of the League of Nations and became its chairman in 1925.

He died in Haarlem on February 5, 1928. On the day he was buried the state telegraph and telephone services in Holland were suspended for three hours as a tribute to "the greatest man Holland has produced in our time." The renowned British physicist ERNEST RUTHERFORD delivered an appreciative graveside oration.

See also MICHELSON, ALBERT ABRAHAM.

Further Reading

De Haas-Lorentz, G. L. H. A. *Lorentz: Impressions of His Life and Work*. Amsterdam: North-Holland, 1957.

M

⊠ Mach, Ernst
(1838–1916)
Austrian
Experimentalist (Optics, Acoustics, Classical Mechanics), Philosopher of Science

Ernst Mach was a brilliant experimentalist, who is best known for deducing the Mach number, the ratio of the velocity of an object to the velocity of sound, which plays a crucial role in aerodynamics. Mach was an important figure in turn-of-the-century controversies about the very nature of matter. He was the leading advocate for the energeticist point of view, which rejected the belief of the atomists that matter is composed of invisible atoms. His study of the subjective factors in scientific observations, which led him to question the Newtonian assumption that space and time are absolute, influenced ALBERT EINSTEIN's development of the general theory of relativity.

Ernst Mach was born in Chirlitz-Turas in the Austro-Hungarian Empire (now Moravia) on February 18, 1838. His parents educated him at home until he was 14, when he was sent to the local gymnasium (high school). He entered the University of Vienna at 17 and designed optical and acoustic experiments to study the then-controversial Doppler effect: the relationship between the perceived frequency of sound and the motion of the observer relative to that of the source. After receiving his Ph.D. in 1860, at age 22, he stayed on in Vienna teaching mechanics and physics until 1864.

While Mach was in Vienna his exposure to the work of Gustav Fechner on the physiological processes of perception launched him on a life-long quest to understand the physiological, psychological, and sensory dynamics that shape the acquisition of scientific knowledge. In 1863, he published a book on psychophysics. The following year he was appointed professor of mathematics at the University of Graz, where he would remain for three years. Lacking both equipment and funding for physics experiments, he continued to study vision, hearing, and the subjective sense of time.

Mach's most important work would be done, between 1867 and 1895, at Charles University in Prague, where, as professor of experimental physics, he would carry out a broad spectrum of studies, which resulted in 90 publications and six books. His experimental work focused mainly on acoustics and physical optics, including studies of interference, diffraction, polarization, and refraction of light in different media. During this period he developed optical and photographic techniques for the measurement of sound waves and wave propagation. In

1007, he published work on the photography of projectiles in flight, which showed the shock wave produced by the gas around the tip of the projectile. Mach determined that the angle between the direction of motion and the shock wave varies with the speed of the projectile: the flow of gas changes its character when the projectile reaches the speed of sound. He then formulated what came to be called the Mach number to describe the ratio of the velocity of an object to the speed of sound in the medium in which the object is moving. The Mach number would become a central concept of aerodynamics, particularly of supersonic flight. Thus, an aircraft flying at twice the speed of sound in air is said to be flying at Mach 2.

It was also in Prague, in 1883, that Mach would publish his most influential book, *The Development of Mechanics*, which contains his critique of Newtonian physics. Although others before him had challenged Newton's concept of absolute space and time, Mach's incisive treatise had a strong impact, appearing as it did just after JAMES CLERK MAXWELL'S work on the electromagnetic field had challenged the mechanistic conception of the world. Mach's central argument, for which he offered precise evidence, was that all measurements of spatial distances, time intervals, motion velocities, and masses of physical objects are basically measurements of relative quantities. Mach's contention that mass is not an absolute measure of matter but is conditioned by the surrounding environment is called the Mach principle. It states that an object cannot have inertia in a universe devoid of all other objects, since inertia depends on the reciprocal interaction of objects, however distant. Einstein, who wrote of the book's "profound influence" on him, searched for a mathematical expression of the Mach principle. But Mach disliked the theory of general relativity and was far from pleased to see his own work praised as its predecessor. He intended to write a book criticizing Einstein's work but never managed to do so.

At the same time, Mach found himself embroiled in another of the central controversies of the day: the atomist–energeticist debate. His conviction that phenomena under scientific investigation can be understood only in terms of experiences, or "sensations," present in the observation of phenomena led him to insist that no scientific hypothesis is admissible unless it is empirically viable. His scorn for hypothetical entities that could be neither seen nor tested experimentally made him vehemently oppose those physicists who were reaching for a deeper explanation of what they saw in the laboratory by postulating a dynamic submolecular world of colliding atoms. "I don't believe in atoms," he bluntly told LUDWIG BOLTZMANN, the leading atomist, in January 1897 at a meeting of the Imperial Academy of Sciences in Vienna. As it happened, Mach was wrong about the atom, whose existence would soon be demonstrated by ERNEST RUTHERFORD and whose properties would be explored by the new quantum mechanics. His view of scientific discovery was validated, however, since precise mathematical and statistical measurement would be required to prove the atom's existence.

In 1895, Mach returned to the University of Vienna as professor of inductive philosophy. Two years later, he suffered a stroke that paralyzed his right side. Despite this, he continued to do research until 1901, when he was appointed to the Austrian parliament, a post he held for 12 years. He continued to lecture and write in retirement, publishing *Knowledge and Error* in 1905 and an autobiography in 1910.

In 1913 he moved to his son's home near Munich, where he continued to write books. He died there a day after his 78th birthday, on February 19, 1916. He is remembered not only for his ingenious and exacting experimental studies, but for his contributions to the debate over the nature of scientific investigation, which continues to this day.

See also DOPPLER, CHRISTIAN.

Further Reading

Cohen, R. S., ed. *Boston Studies in the Philosophy of Science*. Vol. 6. *Ernst Mach: Physicist and Philosopher*, Boston: Kluwer Academic, 1975.

Lindley, David. *Boltzmann's Atom: The Great Debate That Launched a Revolution in Physics*. New York: Free Press, 2000.

⊠ Maxwell, James Clerk
(1831–1879)
Scottish
Theoretical Physicist (Classical Electromagnetism, Optics, Thermodynamics)

James Clerk Maxwell is considered the greatest theoretical physicist of the 19th century. He brilliantly synthesized the key findings of his predecessors in four equations that unified the description of electromagnetic processes. In so doing he discovered that light consists of electromagnetic waves and led the way to exploration of the whole spectrum of electromagnetic radiation. A multifaceted genius who made fundamental contributions to the study of every problem he addressed, he also formulated the statistical kinetic theory of gases, solved the mystery of Saturn's rings, and discovered the principles governing color vision.

Maxwell was born in Edinburgh, Scotland, on November 13, 1831, but spent his boyhood on his family's country estate in Glenlair. From his earliest years he showed a high degree of curiosity and a passion for finding out how things worked. After his mother died, plans for his home education were dropped and the boy was sent to Edinburgh Academy, where he studied from age 10 to age 16. A shy, solitary boy, absorbed in mathematical and mechanical pursuits, Maxwell quickly proved to be a precocious, prize-winning student. When he was only 14, he discovered an original method for drawing a perfect oval based on his generalization of the definition of an ellipse and submitted his work, in his first paper, to the Royal Society of Edinburgh in 1846.

Upon graduating from the academy in 1847, he entered the University of Edinburgh and, still an undergraduate, began doing independent research on the theory of color. In 1849, building on the work of THOMAS YOUNG and HERMANN LUDWIG FERDINAND VON HELMHOLTZ, he demonstrated that colors could be built up from mixtures of the three primary colors—red, green, and blue—by spinning disks containing sectors of these colors in various sizes. He would develop this work for many years, inventing a color box in which the primary colors could be selected from the Sun's spectrum and combined. This model explained how all colors are produced by adding and subtracting the primary colors. Turning to the problem of color vision, he confirmed Young's theory that the eye has three kinds of receptors sensitive to the primary colors and showed that color blindness is due to defects in the receptors. The culmination of his investigations would occur in 1861, when he produced the first color photograph to use a three-color process.

In 1850, Maxwell went on to study at Cambridge. He graduated with a degree in mathematics from Trinity College in 1854 and was awarded a fellowship allowing him to continue his work at Cambridge. Maxwell's genius for pursuing several research problems at once was already evident. In 1856, Maxwell moved to Scotland in order to be near his ailing father and became professor of natural philosophy at Marischal College, Aberdeen. He competed for the Adams Prize offered by Cambridge on 1857, awarded to whoever could offer a satisfactory explanation for the rings of Saturn that would result in a stable structure. He won the prize by showing that whereas a solid ring would collapse and a fluid ring would break up, stability could be achieved if the rings consisted of numerous small solid particles, an explanation now confirmed by the *Voyager* spacecraft.

In 1859, he married Katherine Mary Dewar, daughter of the head of Marischal College; he then moved to London in 1860 as professor of natural philosophy and astronomy at King's College. There he continued the work begun at Cambridge, in 1855–1856, when he had established the foundation for his most important work by providing a mathematical formulation for MICHAEL FARADAY's theory that electric and magnetic effects result from field lines of force that surround conductors and magnets. His seminal paper, "On Faraday's Lines of Force," had compared the behavior of the lines of force with the flow of an incompressible fluid. This model implied the existence of a medium in which the field lines were established.

Over the next 15 years, Maxwell would publish a series of papers in which, step by step, he developed the four field equations that describe the physics of electrodynamics. He continued to develop the incompressible fluid model for the medium, assuming that it contained rotating vortices corresponding to magnetic fields separated by cells corresponding to electric fields. By considering how the motion of the vortices and cells could produce magnetic and electric fields, he explained all previously known effects of electromagnetism. In Maxwell's formulation, Coulomb's law required two equations for the electric field, and Ampère's law and Faraday's law required two more for the magnetic field. Thus, Maxwell's equations showed that (1) unlike charges attract each other; like charges repel (Coulomb's law); (2) there are no single, isolated magnetic poles (if there is a north, there will be a corresponding south pole) (Ampère's law); (3) electrical currents can cause magnetic fields (Ampère's law); and (4) changing magnetic fields can cause changing electric fields (Faraday's law).

However, Maxwell noted that these equations were not symmetrical in the manner in which electric and magnetic fields entered into them. To correct this, he postulated that if a changing magnetic field could produce a chang-

ing electric field, then symmetry required that a changing electric field could produce a changing magnetic field. This implied a new phenomenon, namely, that an oscillating combination of transverse electric and magnetic fields could propagate through space at the speed of light. This led Maxwell to claim that light, in fact, *is* a form of electromagnetic radiation:

> We can scarcely avoid the conclusion that light consists in the transverse undulations of the same medium, which is the cause of electric and magnetic phenomena.

Maxwell's equations demonstrated that electricity and magnetism are aspects of a single electromagnetic field, and that light itself is a variety of this field. In this way he unified what had been the separate studies of electricity, magnetism, and optics. In his 1873 *Treatise on Electricity and Magnetism* he summarized all of his work on the subject, establishing that light has a radiation pressure and suggesting that a whole family of electromagnetic radiations must exist, of which light is only one. This prediction was confirmed in 1888 when HEINRICH RUDOLF HERTZ discovered the existence of radio waves, which move at the speed of light.

At the same time that he was developing his ideas on electromagnetic theory, Maxwell continued work begun at Aberdeen in 1860 on the kinetic theory of gases. He built on the work of RUDOLF JULIUS EMMANUEL CLAUSIUS, who in 1857–1858 had shown that a gas must consist of molecules in constant motion colliding with each other and the walls of the container. He arrived at a formula to express the distribution of velocity in gas molecules, relating it to temperature and thus demonstrating that heat resides in the motion of molecules.

His kinetic theory did not fully explain heat conduction, and it was modified by LUDWIG BOLTZMANN in 1868, resulting in what became known as the Maxwell–Boltzmann distribution law. Maxwell also accurately estimated the size of molecules and invented a method of separating gases in a centrifuge. In addition, the kinetic theory had an important impact on the question of the validity of the second law of thermodynamics, which states that heat cannot spontaneously flow from a cooler body to a hotter one. For example, when two connected containers of gases have the same temperature, it is statistically possible for the molecules to diffuse spontaneously so that the faster-moving molecules all concentrate in one container while the slower molecules gather in the other, making the first container hotter and the second colder. Maxwell conceived this hypothesis, known as Maxwell's demon. Even though this process is statistically possible it is highly unlikely. In this context, the second law is not absolute but only highly probable.

In 1865, Maxwell's father died, and he returned to his family estate in Glenlair, Scotland, and devoted himself to research. He made periodic trips to Cambridge. In 1871, he was persuaded to move to Cambridge, where he became the first professor of experimental physics and set up the Cavendish Laboratory, which opened in 1874. He continued his lectures there until 1879, when he contracted cancer. That summer, he returned to Glenlair to be with his wife, who was also ill. He died at age 48, on November 5, 1879, in Cambridge.

The four partial differential equations now known as Maxwell's equations are among the great achievements of 19th-century mathematics. The year before Maxwell died, he suggested an experiment for measuring the effect of the ether. This inspired ALBERT ABRAHAM MICHELSON and Edward Morley to carry out their famous experiment in the 1880s, the results of which disproved the existence of the ether, the medium in which light waves were thought to be propagated. The discovery that there was no ether did not discredit Maxwell's work. His equations and descriptions of electromagnetic waves remained valid even though the waves require no medium. They paved the way for the discovery of special relativity and of the spectrum of electromagnetic radiation, such as X rays and gamma rays, that is at the core of modern physics.

See also AMPÈRE, ANDRÉ-MARIE; COULOMB, CHARLES-AUGUSTIN DE; EINSTEIN, ALBERT.

Further Reading

Harman, Peter M. *The Natural Philosophy of James Clerk Maxwell.* Cambridge: Cambridge University Press, 1998.

Hendry, John. *James Clerk Maxwell and the Theory of the Electromagnetic Field.* Bristol: England: Institute of Physics, 1986.

Hunt, Bruce J. *The Maxwellians.* Ithaca, N.Y.: Cornell University Press, 1995.

⊠ Meitner, Lise
(1878–1968)
Austrian/Swedish
Experimentalist (Radioactivity), Nuclear Physicist

Lise Meitner is famous for research she performed in 1938, which correctly interpreted and described the splitting or fission of the uranium nucleus under neutron bombardment. This result was pivotal to the development of nuclear physics. Ironically she was deprived of the Nobel Prize for this achievement, which went to Otto Hahn, who had done the initial splitting experiment but had incorrectly interpreted the results.

She was born on November 7, 1878, the third of eight children born to Philipp Meitner, a wealthy lawyer, and Hedwig, the mother to whom she owed her love of music. She would later say of her childhood in this cultured Jewish family,

Even today I am filled with deep gratitude for the unusual goodness of my par-

Lise Meitner correctly interpreted and described the splitting or fission of the uranium nucleus under neutron bombardment, a pivotal discovery in the development of nuclear physics. *(AIP Emilio Segrè Visual Archives, Herzfeld Collection)*

ents, and the extraordinarily stimulating intellectual atmosphere in which my sisters and brothers and I grew up.

Lise showed an early interest in science and, choosing MARIE CURIE as her heroine, aspired to become a physicist and study radioactivity. Her parents, however, insisted she qualify as a French teacher, so she could support herself, before she studied physics. After passing the examination in French, she entered the University of Vienna in 1901.

In 1905, she completed a thesis on the physics of thermal conduction under the direc-

tion of LUDWIG BOLTZMANN and became the second woman to receive a Ph.D. in physics from the university.

In 1907, Meitner entered the University of Berlin to study under MAX ERNEST LUDWIG PLANCK, the founder of quantum mechanics. There she met Otto Hahn, a chemist, who was looking for someone to work with him on radioactivity at the Kaiser Wilhelm Institute of Chemistry in Dalhem. When Hahn's supervisor refused to allow a woman to work in his laboratory, she and Hahn set up a small lab in a carpenter's workroom. Following in the footsteps of ANTOINE-HENRI BECQUEREL and Marie Curie, they set out to analyze the physical properties of radioactive substances. Hahn's main interest was the discovery of new elements, whereas Meitner wanted to examine radiation emissions. Together they determined the beta emission spectra of numerous nuclear isotopes. By 1912, Meitner was a member of the Kaiser Wilhelm Institute and an assistant to Planck at the Berlin Institute of Theoretical Physics.

When World War I broke out in 1914, Hahn remained at the university doing war research while Meitner joined the Austrian army as an X-ray technician. During this period she continued sporadic work with Hahn, arranging her leave to coincide with his. In 1918, they announced their discovery of a new element, protactinium, the second heaviest element then known, which made them famous. Meitner later won the Leibniz Medal from the Berlin Academy of Science and the Leiben Prize from the Austrian Academy of Science and became head of a new department in radioactive physics at the Kaiser Wilhelm Institute. In 1922, she was made a lecturer at the University of Berlin, where four years later she became Germany's first woman to be appointed a full physics professor. Einstein called her "the German Marie Curie."

Working on her own during this period, she studied the relationship between nuclear beta and gamma radiation and the basis for the continuous beta spectrum and correctly determined that beta particles are electrons ejected from the nucleus. On the basis of conservation of energy and momentum, WOLFGANG PAULI, using these results, realized that there had to be a third neutral particle in the process of atomic decay; this insight led him to postulate the existence of the neutrino. Meitner was also the first to describe the atomic emission of Auger electrons, which occurs when an electron rather than a photon is emitted after one electron drops from a higher to a lower electron shell in the atom.

After 1933, when the Nazis rose to power, Meitner stayed in Berlin, despite the growing official anti-Semitism, because she was protected by her Austrian citizenship. Remaining intensely involved in her research, she used a Wilson cloud chamber to photograph positron production by gamma radiation. In 1935, she and Hahn resumed their collaboration and began to study the effects of neutron bombardment on uranium. ENRICO FERMI had shown that the nucleus of an atom, under the right conditions, could capture a neutron and emit a beta particle, thus becoming an atom of the next heaviest element. Meitner and Hahn wanted to confirm these results and use the neutron bombardment on uranium to produce transuranic elements, that is, elements with atomic numbers greater than that of uranium (92). They used a hydrogen sulfide precipitation method to remove elements with atomic numbers between 84 and 92 from their neutron-irradiated sample of uranium and succeeded in finding evidence for existence of elements with atomic numbers of 93, 94, 95, and 96.

In 1938, the Nazis annexed Austria and deprived Meitner of Austrian citizenship, forcing her to become a German Jew subject to German laws. With her life in danger, she escaped to Holland, with the help of Dutch scientists and a family heirloom diamond ring from Hahn, who "wanted her to be provided for in an emergency." Shortly afterward, she moved to Denmark, as the

guest of NIELS HENRIK DAVID BOHR. She then accepted a position Bohr had found for her, at the Nobel Physical Institute in Stockholm, where a cyclotron was being built. Not knowing the Swedish language and suffering from the loneliness of exile, however, she found it difficult to focus on her work. Meanwhile, she got word from Hahn that he and his new partner, a young German chemist named Fritz Strassmann, had found that the radioactive elements produced by neutron bombardment of uranium had properties like those of radium. But if Hahn was excited at the prospect of discovering new elements, Meitner found the energetic reactions necessary to produce such new elements unexpected and increasingly hard to explain. From Stockholm, she asked him for firm chemical evidence of the identities of the products.

Delving further, Hahn and Strassmann were surprised to learn that the neutron bombardment had produced not transuranic elements but three isotopes of barium, which has an atomic number of 56. Hahn asked Meitner whether she could explain this. Meitner and her nephew, Otto Frisch, who was working at Bohr's institute in Copenhagen, realized that Hahn and Strassmann's results indicated that the uranium nucleus had been split into smaller fragments. They found that Hahn's results could be explained if, rather than simply adding or subtracting a particle or two from the uranium nucleus, Hahn had split it almost in half. Doing so would produce barium and krypton, a gaseous element that would be harder to detect.

Their result was startling, since physicists had thought splitting atomic nuclei was not possible because of the very powerful forces that held the nucleus together. Most thought the nucleus was like a drop of water, held together by the equivalent of surface tension: the electric charge of a heavy nucleus generates a repulsive force that partially offsets the attractive nuclear force. Recalling that the repulsive force of the nuclear electric charge, that is, the atomic number, diminishes this surface tension, Meitner and Frisch calculated that at an atomic number of about 100 the surface tension of the nucleus disappears; therefore, uranium at 92 must be fairly close to that instability. For these reasons they concluded that the uranium nucleus might be "very wobbly, like a large, thin-walled balloon filled with water," and under these conditions a single neutron could cause it to split apart, a process they named nuclear fission.

They had also discovered the reason no elements beyond uranium exist naturally in the world: the electric and nuclear forces working against each other in the nucleus eventually cancel each other out. They also calculated, according to Einstein's relativity equation, $E = mc^2$, which shows how to convert mass into energy, that splitting a uranium nucleus should release 200 million electron volts, 20 million times more than an equivalent amount of TNT. (An electron volt is the energy necessary to accelerate an electron through a potential difference of one volt). This energy, which would be released in the form of radiation and heat, would not be obvious in samples of uranium as small as those Hahn had used, but it could be detected with the right instruments. Frisch duplicated the experiment and found evidence of this form of nuclear energy.

In January 1939, Frisch and Meitner published two papers describing their analysis, in which they expressed the idealistic hope that their work would herald a "promised land of atomic energy." Instead, it set in motion a series of discoveries leading to the development of the first nuclear reactor in 1942 and the development of the atomic bomb in 1945. Although Frisch did become a part of that effort, Meitner herself lived in semiretirement in neutral Sweden during World War II and refused to participate in atomic bomb research. She hoped that such a weapon would not be feasible and was devastated when it was successfully developed and dropped on Japan. After this, her remaining hope was that the real-

ity of nuclear devastation would make humanity realize there must be an end to war.

Ironically, in 1944, Hahn won the Nobel Prize in chemistry for his work on nuclear fission, and, although he never did anything to correct the injustice of Meitner's work's being overlooked, they remained friends. In 1947, the Swedish Atomic Energy Commission established a laboratory for Meitner at the Royal Institute of Technology. Later she moved to the Royal Swedish Academy of Engineering Science, to work on an experimental nuclear reactor. In 1949, she became a Swedish citizen and was awarded the Max Planck Medal. She continued to study the nature of fission products and contributed to the design of an experiment whereby fission products of uranium could be collected. Later she performed cyclotron research on the production of new radioactive species, which led to the development of the shell model of the nucleus.

In 1960, Meitner retired and settled in Cambridge, England, to be near Frisch, who taught there. Finally, in 1966, she became the first woman to receive the Fermi Award from the U.S. Atomic Energy Commission, which she shared with Hahn and Strassmann.

Lise Meitner was a brilliant scientist who made great contributions to nuclear physics, as well as a humanist who lived according to her convictions. In old age, she wrote:

> I believe all young people think about how they would like their life to develop. When I did so I always arrived at the conclusion that life need not be easy provided only that it was not empty. And this wish I have been granted.

She died in Cambridge on October 27, 1968, just before her 90th birthday. In 1982, when element 109 was created, it was named *meitnerium* in her honor.

Further Reading

Frisch, Otto. *What Little I Remember.* Cambridge: Cambridge University Press, 1991.

Rife, Patricia. *Lise Meitner and the Dawn of the Nuclear Age.* Cambridge, Mass.: Birkhauser, 1998.

Sime, Ruth Lewin. *Lise Meitner: A Life in Physics.* Berkeley: University of California Press, 1996.

⊠   Michelson, Albert Abraham
(1852–1931)
American
Experimental Physicist (Optics)

Albert Michelson was the great experimentalist who, together with Edward Morley, performed the classic Michelson–Morley experiment for light waves, which disproved the existence of the ether, the mysterious medium in which light waves were thought to propagate. With the instrument he designed for this experiment, the interferometer, Michelson pioneered a new field capable of making a broad range of highly exact measurements. He developed extremely sensitive optical equipment, which enabled him to make very precise measurements of the velocity of light. When, in 1907, he won the Nobel Prize in physics for these achievements, he became the first American to be honored in this way.

Michelson was born in Strelno, Prussia (now Poland), on December 19, 1852, the son of a Jewish merchant, who escaped persecution by emigrating with his family to the United States when Albert was two years old. The family started out in New York and eventually settled in Virginia City, Nevada, and San Francisco, where his father's business interests prospered and Albert attended school. When his father heard of an opportunity for his son to gain admission to the United States Naval Academy in Annapolis, Albert duly applied but was not selected. The young man pleaded his case before President Ulysses S. Grant, who ordered that he be admitted in 1869. But Michelson was not des-

tined for a naval career. In a class of 29, he placed first in optics, 25th in seamanship. When he graduated in 1873, he was ordered aboard the USS *Monongahela,* a sailing ship, for a two-year voyage through the Caribbean and down to Rio. After his return in 1875 he was appointed as an instructor in physics and chemistry at the Naval Academy. Two years later, he married Margaret McLean Heminway, with whom he would have three children. (Michelson would later divorce her, remarry, and have three more children with his second wife.)

From this time on, he would devote his life to science. Fascinated by the problem of measuring the speed of light, he would create ever more refined instruments for achieving his goal. He made his first attempt in 1878, at Annapolis, when he improved the rotating-mirror method of JEAN-BERNARD-LÉON FOUCAULT to determine the velocity of light. Michelson made his measurement by timing a flash of light traveling between mirrors. Using very high-quality lenses and mirrors to focus and reflect the beam of light, he found that light moved at a velocity of 186,355 miles per second, with a possible error of ±30 miles per second or so. This measurement was 20 times more accurate than Foucault's and remained the best to be obtained for a generation; when it was improved upon, in the 1920s, Michelson was the one who improved it, by making very precise measurements over a 23-mile path between two mountain peaks in California.

Although his work at Annapolis had given him renown, Michelson decided he needed a greater mastery of optics before attacking the next challenge he had set for himself: the measurement of the ether. Between 1880 and 1882, he pursued postgraduate studies in Berlin under HERMANN LUDWIG FERDINAND VAN HELMHOLTZ and later in Paris. It was in Berlin, in 1881, that Michelson built his first interferometer, in order to carry out an experiment suggested by JAMES CLERK MAXWELL in 1878. Maxwell, as did most physicists, believed that light waves, by analogy with sound waves propagating through the atmosphere, move through the ether, a material medium that surrounds and permeates all space, including outer space. To date, however, no one had figured out how to measure it.

Michelson's ingenious optical system was designed to do just that. It consisted of an L-shaped apparatus in which a beam of light was split in two. The separated beams were guided along perpendicular paths of identical length and then recombined in such a way that the beams "interfered" with each other: that is, the directions and distances of their light paths were so closely meshed that the beams could interact and create observable interference fringes. If the Earth was moving through a universal ether, the times needed by the two parts of the split light

Albert Abraham Michelson performed the classic Michelson–Morley experiment for light waves, which disproved the existence of the ether in which light waves were thought to be propagated. *(AIP Emilio Segrè Visual Archives, Michelson Museum)*

beam to transverse a sample of the ether current from perpendicular directions should differ slightly, and that difference should be detectable in the resulting pattern of interference fringes. Since the Earth is revolving about its axis as well as moving around the Sun, the flow of ether seen in the laboratory should be periodically changing and hence the speed of light should be changing as well. This change in the speed of light should cause the interference fringes in Michelson's interferometer to shift in a periodic manner, which should be observable.

To his consternation, Michelson could detect no change in the interference pattern, neither in this first experiment, nor in 1885 or 1887, when he repeated the experiment, using a more sensitive interferometer built with Morley. The results of the Michelson–Morley experiments were paradoxical in term of Newtonian physics and triggered passionate debate. Evidently, the speed of light plus any other added velocity was still equal only to the speed of light.

To explain this, fundamental and long-held physical ideas had to be reexamined. Some physicists continued to believe in the ether and claimed its effect could not be detected because it moved with the Earth. However, this was disproved by stellar aberration experiments performed by Oliver Lodge in 1893. A more promising theory was put forth by GEORGE FRANCIS FITZGERALD, who suggested that objects moving through the ether contract slightly in the direction of their motion and thereby hide the effect of the change in the velocity of light. A radically different interpretation of the FitzGerald contraction hypothesis would become part of the special theory of relativity proposed by ALBERT EINSTEIN in 1905. Michelson, however, believed that the physics of his day was founded on a rock and was skeptical of theories that promised to overthrow its basic tenets:

The more important fundamental laws and facts of physical science have all been discovered, and these are now so firmly established that the possibility of their ever being fully supplanted in consequence of new discoveries is exceedingly remote. . . . Our future discoveries must be looked for in the sixth place of decimals.

When he returned to the United States, disappointed by the results of his experiments, he left the navy and took up a position as professor of physics at the Case School of Applied Science in Cleveland, Ohio. In 1889, he joined the physics department at Clark University in Worcester, Massachusetts, where, in 1892, he and Morley redefined the length of the standard meter kept in Paris in terms of a certain number of wavelengths of monochromatic light, using a red line in the cadmium spectrum. This method of defining the standard unit of length was finally adopted in 1960, and a krypton line is now used.

Michelson made his final academic shift in 1892, accepting an appointment as professor of physics at the University of Chicago, where he would remain until 1929. During these years he developed his interferometer into a precise instrument for measuring the diameters of heavenly bodies and in 1920 announced the size of the giant star Betelgeuse, the first star to be measured.

He died on May 9, 1931, in Pasadena, California. Earlier that year Einstein had paid him tribute:

Dr. Michelson, it was you who led the physicists into new paths, and through your marvelous experimental work paved the way for the development of the theory of relativity.

The Michelson–Morley experiment was perhaps the most significant in the history of science. It marked a turning point in theoretical physics, demonstrating the counterintuitive fact that the velocity of light is constant, independent of the motion of the observer. Despite his desire to preserve the foundations of 19th-century physics, Michelson's gift for extremely precise experimental work ignited a revolutionary chain of advances in physics, which would destroy the theory of the luminescent ether and create the groundwork for relativity.

Further Reading

Jaffe, Bernard. *Michelson and the Speed of Light*. Garden City, N.Y.: Doubleday, 1960.

Livingston, Dorothy Michelson. *Master of Light: A Biography of Albert A. Michelson*. New York: Scribner, 1973.

⊠   **Millikan, Robert Andrews**
(1868–1953)
American
Experimentalist, Atomic Physicist

Robert Andrews Millikan performed the famous oil droplet experiment, which represented the first accurate determination of the charge on the electron. For this achievement, as well as for his measurement of Planck's constant, he was awarded the Nobel Prize in physics in 1923.

He was born on March 22, 1868, in Morrison, Illinois, the second son of the Reverend Silas Franklin Millikan, a minister, and Mary Jane Andrews Millikan, a former dean of women at Olivet College. His Scottish–Irish ancestors had pioneered in settling the Midwest. He had a rural childhood, helping out on the family farm. At 14, he worked 10-hour days during the summer at a local barrel head factory, earning a dollar a day. After graduating from Maquoketa High School in Iowa, he entered Oberlin College in 1886; there his Greek professor sparked his interest in physics. A 12-week physics course, which he would later call "a complete loss," was his only formal physics instruction at Oberlin. It was apparently enough, however, to win him an offer to teach elementary physics there; he needed the money and accepted. After earning a

B.A. from Oberlin in 1891 and an M.A. in 1893, he embarked on graduate study at Columbia University, which would award him a Ph.D. in physics in 1895.

Millikan then spent some time in Germany, studying with MAX ERNEST LUDWIG PLANCK in Berlin and WALTHER NERNST in Göttingen. He married Greta Erwin Blanchard, with whom he would have three sons. In 1896, Millikan accepted an appointment as assistant professor of physics at the University of Chicago, where, in 1906, he was made associate professor and, four years later, full professor.

In 1908, Millikan began constructing experiments to measure the charge on the electron. His approach was conceptually simple but hard to achieve in practice, and it took him five years of trying before he was able to determine the electron's charge. He first investigated the rate at which water droplets fall when exposed to an electric field. He began by filling a cloud chamber with ions generated by X rays. He then rapidly let air back into the cloud chamber, causing tiny water droplets to condense around the ionized particles. Next, by applying an electrostatic field and observing the behavior of the water droplets, he was able to measure the rate at which the water drops either fell, under the influence of gravity alone, or were balanced, under the influence of gravity plus an electric field. Since the ionized droplets would absorb multiple integers of the electronic charge, the value of which he was trying to determine, the strength of the electric field required to counteract the gravitational force on the droplets would also be an integral multiple of the electronic charge. By measuring a range of values of the electric field needed to balance a large number of different size droplets against gravity, and then calculating the least common integral multiple of these values, he could determine the value of the electronic charge.

By 1909, Millikan had an approximate value for the electronic charge, but the water droplets evaporated too quickly to make precise measurement possible, so he switched to oil droplets, which were much more stable. In his experiment he produced oil droplets with variable charge, which fell through fine holes in the upper of two charged metal plates. He measured the range of values of the electric field needed to balance a large number of different charged oil droplets against gravity. Then, calculating the least common integral multiple of these values, he found that the charges on the droplets were integral multiples of the charge on the electron. The fact that the oil droplets did not evaporate allowed him to determine the value of the electronic charge more accurately than in his previous work with water droplets. The calculations Millikan obtained from his work with oil droplets, which he completed in 1913, were used for many years to measure the charge of the electron.

During this period Millikan also studied the photoelectric effect, investigating ALBERT EINSTEIN's hypothesis that the kinetic energy of an electron emitted by incident radiation is proportional to the frequency of the radiation multiplied by Planck's constant, another fundamental constant in nature. He found a way to perform his experiments on photoelectricity when he observed that alkali metals are sensitive to a very wide range of electromagnetic frequencies. After working for three years to improve the sensitivity of his apparatus, in 1916, he was able to validate Einstein's 1905 equation, thereby obtaining an accurate value for Planck's constant. This established beyond doubt the validity of Einstein's linear relationship between energy and frequency, as well as the photon hypothesis underlying the nature of Einstein's photoelectric equation.

During World War I, Millikan was director of research for the National Research Council, which was involved in defense research. He helped develop submarine detection and destruction devices. In 1921, he moved from Chicago to Pasadena, where he became director of the Norman Bridge Laboratory at the California Institute of Technology. In the 1920s, he studied the ultraviolet spectra of many elements, extending the frequency range and identifying many new spectral lines. In 1925, he began a thorough study of cosmic rays, which he named after proving that they originated in outer space, a hypothesis first put forth by V. F. Hess, who discovered them in 1912. Millikan was skeptical of this hypothesis until he himself proved it experimentally. He did this by comparing the intensity of ionization in two lakes at different altitudes and in the process was able to demonstrate that the rays producing the ionization must have passed through the atmosphere from above and could not have originated on Earth.

However, the nature of cosmic rays remained a puzzle. At first Millikan argued that in the hydrogen clouds in interstellar space hydrogen atoms are being continually fused together to produce helium and heavier elements, releasing a large amount of energy in the form of photons. The hypothesis was widely accepted until 1929, when it was shown that the primary cosmic rays consist mostly of hydrogen and helium nuclei. Later Millikan asserted that cosmic rays are electromagnetic waves, but ARTHUR HOLLY COMPTON disproved this in 1934 when he demonstrated that they consist of charged particles.

Millikan remained the head of the Norman Bridge Laboratory until 1945, when he retired. A prolific author, he wrote, among other books, *Science and Life* (1924); *Evolution in Science and Religion* (1927); *Science and the New Civilization* (1930); *Time, Matter, and Values* (1932); *Electrons, Protons, Photons, Neutrons, Mesotrons, and Cosmic Rays* (1947); and *Autobiography* (1950). He was a religious philosopher and lectured on the reconciliation of science and religion. He died in Pasadena on December 19, 1953.

Millikan's experimental genius led him to discover accurate ways to determine two of the most fundamental constants of modern physics: the charge on the electron and the value of Planck's constant.

See also COULOMB, CHARLES-AUGUSTIN DE; WILSON, CHARLES THOMSON REES.

Further Reading
Kargon, Robert H. *The Rise of Robert Millikan: Portrait of a Life in American Science.* Ithaca, N.Y.: Cornell University Press, 1982.

Millikan, Robert Andrews, and I. Bernard Cohen. *Autobiography of Robert A. Millikan.* Manchester, N.H.: Ayer, 1980.

⊠　**Mössbauer, Rudolf Ludwig**
(1929–)
German
Experimentalist, Particle Physicist

Rudolf Ludwig Mössbauer won the 1961 Nobel Prize in physics for his discovery of what came to be known as the Mössbauer effect, the recoil-free resonance absorption of gamma radiation by atoms in a crystal. Mössbauer's discovery spawned the enormously productive field of Mössbauer spectroscopy, which, in addition to providing an accurate method for the experimental verification of ALBERT EINSTEIN's theory of relativity, continues to find applications as a research tool in a wide spectrum of scientific fields.

He was born on January 31, 1929, in Munich, to Ludwig Mössbauer and Erna Ernst. After completing his secondary studies in Munich in 1948, he took a year's break from his schooling, working in industrial laboratories. He then began his physics studies at the Munich Institute of Technology and received his undergraduate degree in 1952. Over the next two years Mössbauer completed his master's thesis at the Institute's Laboratory of Applied Physics. From 1955 to 1957, he worked on his doctoral thesis, carrying out a series of experiments in Heidelberg at the Max Planck Institute for Medical Research; he received a Ph.D. in 1958. At the suggestion of his thesis adviser, Professor Maier-Leibnitz, he investigated resonance absorption and had some unexpected results, which he investigated systematically. That year he announced the Mössbauer effect, which he discovered during his postdoctoral research.

Physicists had long known that incoming radio waves can only be received if the receiver is tuned to the same frequency as the sender. Under these conditions it is said that resonance absorption is taking place. In 1953, Mössbauer tried to observe the corresponding phenomenon for gamma rays impinging on nuclei. The method was to allow gamma radiation from some kind of nuclei to act on other nuclei of exactly the same kind. The absorption of a gamma ray by an atomic nucleus usually causes it to recoil, thereby affecting the wavelength of the reemitted ray. Mössbauer discovered that at low temperatures crystals absorb gamma rays of a specific wavelength and resonate, so that the crystal as a whole recoils while the nuclei do not. Mössbauer's experiment showed that after an atomic nucleus emits a gamma ray photon (during radioactive decay), the absorption of the momentum of the atom by the whole of its crystal lattice occurs because the atom is so firmly bound to the lattice. The same effect occurs with the absorption of gamma rays. Because of the very small width of the gamma ray absorption lines, the resonance absorption in the crystal is very sharp. More importantly, in terms of practical applications, the resonance absorption can be influenced by the Doppler effect associated with the moving of the source or the absorber of the gamma radiation at velocities that can be as slow as a few millimeters per hour. This essentially recoilless nuclear resonance absorption became known as the Mössbauer effect. A lengthening of the gamma ray wavelength in a gravitational field, verifying the predictions of the general theory of relativity, was observed experimentally in 1960.

Mössbauer traveled to the United States in 1960 and, the following year, became professor of physics at the California Institute of Tech-

nology, Pasadena, where he continued his studies of gamma absorption. He simultaneously held a professorship at Munich. He married Elisabeth Pritz, with whom he had two children, Peter and Regine.

Because the link between the Mössbauer effect and the electron structure of the sample can be exploited in the study of many types of materials, it is used primarily to study the electron structure of materials. Other important applications involve measuring the separation and displacement of nuclear energy levels that occur in solids as a result of the influence of the environment surrounding the nucleus. Mössbauer spectroscopy is extremely versatile and can also be used in solid state physics, surface physics, medicine, chemistry, biochemistry, and geology.

See also BETHE, HANS ALBRECHT; DOPPLER, CHRISTIAN; FERMI, ENRICO; RUTHERFORD, ERNEST.

Further Reading
Fraunfelder, H. *The Mössbauer Effect.* New York: W. A. Benjamin, 1963.
Maddock, Alfred G. *Mössbauer Spectroscopy: Principles and Applications of the Techniques.* West Sussex, England: Horwood, 1998.

⊠   **Mott, Sir Nevill Francis**
(1905–1996)
British
Theoretical Physicist, Solid State Physicist

Sir Nevill Francis Mott was a pioneer in solid state physics, who shared the 1977 Nobel Prize in physics with JOHN HOUSBROOK VAN VLECK and PHILIP WARREN ANDERSON for his work on the electromagnetic properties of noncrystalline, or amorphous, semiconductors.

He was born on September 30, 1905, in Leeds, England, to Charles Francis Mott and Lillian Mary Reynolds, who had met as students working under JOSEPH JOHN (J. J.) THOMSON at the Cavendish Laboratory at Cambridge University. The spirit of discovery ran in his family: His great-grandfather was Sir John Richardson, the arctic explorer. Mott attended Clifton College, Bristol, before going on to Cambridge's Saint John's College in 1923; there he majored in mathematics and physics. He found himself bored by laboratory experiments and was far more excited by the revolutionary developments in theoretical physics. His study of the new quantum theory was a formative moment in his development as a physicist; throughout his career, he would apply the intuitive understanding of its formalism he gained in those years. After earning a bachelor's degree in 1927, he spent a year in Copenhagen working with NIELS HENRIK DAVID BOHR, the father of quantum theory. He then went on to the University of Göttingen, the other major center of the new physics, and performed calculations there that predicted how quantum mechanics modified ERNEST RUTHERFORD's classical model of the scattering of alpha particles with helium atoms. These quantum predictions were later verified experimentally by the work of SIR JAMES CHADWICK.

In 1929, Mott spent a year as lecturer at Manchester University, working with Lawrence Bragg, a pioneer in X-ray diffraction; he then was invited back to Cambridge's Gonville and Caius College in 1930. That year he earned a master's degree and married Ruth Eleanor Horder, with whom he would have two daughters. He spent three years in Cambridge, working on problems in nuclear physics. Because of complications associated with the absence of his adviser, Ralph Fowler, Mott left without earning a Ph.D. However, Cambridge would make good this unfortunate omission in 1995 by awarding him an honorary doctorate.

In 1933, he moved to the University of Bristol to become Melvin Wills Professor of Theoretical Physics; at Bristol he switched from nuclear to solid state physics and built one of the leading groups in the new field. There he worked on metals and metal alloys, semiconductors and photographic emulsions. He devised the theoretical description of the effect of light on a photographic emulsion at the atomic level. He and R. W. Gurney proposed the first complete theory of the process that occurs when a photographic film is exposed to light, which was built on the hypothesis that light produces free electrons and holes that roam around the crystal. When they run into imperfections—dislocations or foreign atoms—the electrons become trapped. They then attract interstitial silver ions to form silver atoms and produce the latent image. In the presence of a developer, the entire grain may be catalyzed into free silver by the initially formed specks of silver.

Mott described his research in a series of books that are still widely used, including *The Theory of Atomic Collisions*, with Harrie Massie (1934); *The Theory of Properties of the Metals and Alloys*, with Harry Jones (1936); and *Electronic Processes in Ionic Crystals*, with R. W. Gurney (1940).

During World War II, Mott headed a theoretical physics group at Fort Halstead, working on problems related to munitions, such as deformation in metals due to projectiles. His social consciousness was aroused during this period, and in 1946 he helped to found the Atomic Scientists' Association, organized to inform the public about the realities of atomic energy. He also became an active member of the Pugwash Conference, the international group set up to explore ways of preventing nuclear war.

Mott remained at the University of Bristol until 1948, when he became director of the Henry Herbert Wills Physical Laboratories, also in Bristol. From 1954 to 1971, he was Cavendish Professor of Physics at Cambridge, where he nurtured the impressive growth of solid state physics and radio astronomy. He also became master of his Cambridge college, Gonville and Caius, from 1959 to 1966 and played a leading role in the reform of science education in Great Britain.

When he gave up the exhausting duties of his master's position, he was once more able to engage in research and became interested in the new field of noncrystalline semiconductors (materials whose resistive properties lie somewhere between those of a metal and those of an insulator). Mott's studies of electrical conduction in various metals led him to explore the conductivity potential of amorphous materials (i.e., materials with irregular atomic structures). When Mott attacked the problem, the electronic structure of crystalline solids had for some time been understood, in terms of the effects of diffraction from the translationally symmetric lattice. But solid state physicists did not know how to apply this description to amorphous materials such as glass, which have no such symmetry. Could such a disordered material be an electrical conductor? What determined whether a material could be a metal or an insulator? To answer these fundamental questions, Mott began collaborating with Philip Anderson, then at Bell Laboratories in New Jersey, enabling him to work several years at the Cavendish as a visiting professor. Together they devised a theoretical framework for the rapidly increasing pool of experimental data.

In their Nobel Prize–winning work Mott and his colleagues developed the quantum theory describing the transitions that glass and other amorphous substances can make between electrically conductive (metallic) states and insulating (nonmetallic) states, thereby functioning as semiconductors. In their groundbreaking research they discovered special electrical characteristics in glassy amorphous semiconductors and formulated fundamental laws of behavior for their materials. These glassy substances, which are relatively simple and cheap to produce, eventually replaced the more expensive crystalline semiconductors in many electronic switching and memory devices. This development ultimately led to the manufacture of more affordable personal computers, pocket calculators, copying machines, and other similar devices. Amorphous semiconduc-

tors also provided a source of cheap and reliable material that could be used to improve the performance of electronic circuits, increasing computer memory severalfold. The use of amorphous semiconductors has also led to a revolution in the transistor industry. More efficient photovoltaic cells, based on these materials, were produced with the capacity to convert solar energy into electricity; this advance opened the way for a wide range of new developments in electronics, including cheaper methods of solar heating.

In 1978, Mott became president of the London-based scientific publishing house Taylor & Francis, Ltd. He was Senior Research Fellow at Imperial College, London, from 1971 to 1973 and in 1971 published his important book *Metal Insulator Transitions*. He also wrote an autobiography, *A Life in Science*, and collaborated with several others on a science–religion interface, *Can Scientists Believe?*

In his later years, Mott's overriding research interest was the field of high-temperature superconductors, which were discovered in 1986. He became professor emeritus at Cambridge and retired to Milton Keynes, Buckingham, where he died at the age of 91 on August 8, 1996.

Nevill Mott shaped the course of solid state physics from the time of its emergence as an independent field in the 1920s, when the techniques of quantum mechanics first enabled physicists to study the behavior of matter in the solid state. His work on noncrystalline amorphous semiconductors in the 1970s helped make solid state physics into a force for rapid technological development.

Further Reading

Mott, Nevill. *A Life in Science*. London: Taylor & Francis, 1987.

Mott, N. F., and A. S. Alexandrov, eds. *Sir Nevill Mott—65 Years in Physics*. World Scientific Series in 20th Century Physics, Vol. 12. Singapore: World Scientific, 1995.

N

Ne'eman, Yuval
(1925–)
Israeli
Theoretician, Particle Physicist

Yuval Ne'eman is an eminent physicist, who in 1961 independently originated the theory of unitary symmetry, SU(3), for classifying the huge array of high-energy unstable particles discovered in the 1950s and 1960s. At the same time, MURRAY GELL-MANN independently made the same discovery and gave it the enduring name the *"eightfold way."* Later, Ne'eman and Gell-Mann collaborated on a book on the subject of SU(3) symmetry and elementary particles. Parallel to his life in science, Ne'eman has had an important and controversial career as an Israeli soldier, a pioneer in developing Israel's nuclear capability, and a politician and policymaker on the far Right of the political spectrum.

Ne'eman was born on May 14, 1925, in Tel Aviv, in what was then Palestine, to Gedalia and Zipora Ne'eman. He received a B.Sc. degree in engineering in 1945, and a diploma in mechanical engineering in 1946, from the Israeli Institute of Technology (the Technion) in Haifa. He worked as a hydrodynamical design engineer at a pump factory during these years. Ne'eman interrupted his studies to join Israel's struggle to become a sovereign state; as an activist working

against the British, he was a member of the Jewish underground known as the Hagana. He fought in the War of Independence in 1948 as an infantry commander. When the war ended with the formation of the Israeli state, he resumed his studies while continuing to serve with the Israeli Defense Forces (IDF).

On June 28, 1951, Ne'eman married Dvora Rubinstein, with whom he had three children. The couple traveled to Paris, where he received a diploma in military engineering, in 1952, from the École de Guerre (war college). Advancing through the ranks of the IDF, he became a colonel in 1955. As his military career progressed, he served as deputy director of the intelligence division of the IDF from 1955 to 1957 and as military attaché to London from 1958 to 1960. While in London, he studied at the Imperial College of Science and Technology, where his general adviser was ABDUS SALAM, one of the founders of the electroweak theory. In 1961, in his doctoral thesis, working independently of Gell-Mann, who made the same discovery that year, Ne'eman uncovered the basic symmetry of the subatomic particles, known as the unitary symmetry SU(3), which later became known as the eightfold way.

In his quest to understand how the new particle SU(3) symmetry worked, Ne'eman proposed a new physical attribute (which Gell-Mann in his independent, equivalent work called strangeness) of the strong interactions analogous to electric charge. Whereas in electromagnetic particle collisions electric charge is conserved—the total going in equals the total going out—Ne'eman assumed that strangeness is conserved in strong and electromagnetic interactions, but *not* in weak interactions. In this theory, strong interactions create a pair of particles with equal and opposite values of strangeness, which cancel out each other. The separated members of such a pair cannot spontaneously decay through a strong interaction, because that action would violate conservation of strangeness. Thus, the slower weak interaction, in which the violation of conservation of strangeness *is* allowed to occur, takes over and causes radioactive decay of the particles.

The idea of a theory of nuclear force based on a limited conservation of strangeness proved to be an organizing principle. The unitary symmetry SU(3) inherent in the concept of strangeness led Gell-Mann and Ne'eman directly to a method of sorting all known particles, according to certain general characteristics, into eight "families." The logic of this system was so tight that it revealed the obvious missing family members. In 1953, Gell-Mann published a series of papers predicting specific, as yet undiscovered new particles, as well as other particles he insisted could not be discovered. His timing could not have been better. Successive experiments confirmed each of his positive predictions and failed to contradict the negative ones.

Ne'eman and Gell-Mann continued to work along the same logical trajectory. As Ne'eman explains:

After Gell-Mann and I came out with our theory, the question arose of why the particles follow this particular order. In 1962, I wrote a paper together with Haim Goldberg-Ophir in which we proposed a structural explanation: the existence of three kinds of fundamental "building blocks" which make up protons and neutrons within the atomic nucleus (proton = aab, neutron = abb, omega minus = ccc, and so on). The following year, this model was improved upon by Gell-Mann, and independently by George Zweig, and Gell-Mann named the building blocks "quarks."

From 1961 to 1963, Ne'eman was science director of the Soreq Research Institute, part of Israel's Atomic Energy Commission. In 1963, he arrived for two years as a visiting professor at the California Institute of Technology, where he became a good friend of RICHARD PHILLIPS FEYNMAN, who, along with Gell-Mann, was Caltech's leading light. This gave him a ringside seat for what he called "the childish fight" into which the passionate collaboration of the eminent American physicists degenerated. Ne'eman published *The Eightfold Way* with Gell-Mann in 1964.

Ne'eman then returned to Israel, where, from 1965 to 1972, he was founder and head of Tel Aviv University's School of Physics and Astronomy. From 1971 to 1975, he served as president of the university. At the same time, Ne'eman became increasingly involved in national concerns, serving from 1974 to 1976 as Israel's chief defense scientist. He has been an influential voice in the debate on Israel's preferred security borders. In 1981, he became a member of the Knesset, Israel's parliament; he served as Israel's minister of science and development from 1982 to 1984, and again from 1990 to 1992, when he also served as minister of energy. He founded the Israeli Space Agency in 1983 and has served as its director. From 1979 to 1997, he was director of the Mortimer and Raymond Sackler Institute of Advanced Studies. He was chairman of the Mediterranean-Dead Sea Canal Committee, in which he directed the planning of 110-kilometer canal that supplies hydroelectric power to the region.

Ne'eman is an associate at the International Center for Theoretical Physics, Trieste, Italy; and a member of the Israeli and American Physical Societies and the Institute of Strategic Studies, London. A prolific author, he has written *The Past Decade in Particle Theory*, coedited with E. C. G. Sudarshan (1973); *Group Theoretical Methods in Physics*, coedited with L. Horwitz (1980); *To Fulfill a Vision* (1981); *Membranes and Other Extendons*, written with Elena Eizenberg (1992); and *The Particle Hunters*, written with Yoram Kirsh (1996), among other titles.

Dividing his time between Israel and the United States, from 1968 to 1990, he was director of the Center for Particle Theory at the University of Texas, Austin, where he continues to be a member of the faculty. As a member of the Institute for Advanced Studies at Tel Aviv University, he also participates extensively in the activities of the Inter-University Center for Terrorism Studies, established in 1997, in Washington, D.C., and Holon, Israel, to study the nature and impact of international and domestic terrorism.

Further Reading

Ne'eman, Yuval, and Murray Gell-Mann. *The Eightfold Way*. Cambridge, Mass.: Perseus, 2000.

Ne'eman, Yuval, and Yoram Kirsh. *The Particle Hunters*. Cambridge: Cambridge University Press, 1996.

⊠   **Nernst, Walther**
(1864–1941)
Prussian
Experimentalist (Thermodynamics),
Physical Chemist

Walther Nernst was a remarkable physicist and physical chemist, who discovered the third law of thermodynamics, also known as the Nernst heat theorem, which states that the entropy of a pure substance tends to zero as its thermodynamic temperature approaches zero. He received the Nobel Prize in chemistry for this work in 1920.

Nernst was born in Briesen, West Prussia, on June 25, 1864, the son of a district judge. He obtained his early schooling in Graudentz and then attended the Universities of Zurich, Berlin, and Graz, where he studied physics and mathematics. He then went on to study in Wurtzburg, where in 1887 he received a doctorate for a thesis on electromotive forces produced by magnetism in heated metal plates. His first position was at Leipzig University, where he joined a distinguished company of physical chemists. In 1892, he married Emma Lohmeyer. They had two sons, both of whom would be killed in World War I, and three daughters.

In 1894, Nernst accepted the chair in physical chemistry in Göttingen, where he founded the Institute for Physical Chemistry and Electrochemistry and became its director. Always looking for industrial applications of physical research, Nernst produced ingenious devices. In 1898, he invented a metallic-filament lamp, known as the Nernst lamp. A link between the carbon lamp and the incandescent lamp, it was superseded by the development of tatalum and tungsten filaments. He also invented an electrical piano, which replaced the sounding board with radio amplifiers.

Having rapidly developed a reputation for brilliance as well as egocentricity, Nernst accepted a position at the University of Berlin, as professor of chemistry and physics, in 1905. With his students he made many important physical and chemical measurements, including determinations of specific heats of solids at very low temperatures and of vapor densities at high temperatures. All these were considered from the point of view of the quantum hypothesis.

During this period, Nernst made his Nobel Prize–winning contribution, completing the foundations of thermodynamics, which had been laid in the mid-19th-century. In 1847, HERMANN LUDWIG FERDINAND VON HELMHOLTZ had derived a general equation that led to the first law of thermodynamics, which states that the total energy of a system and its surroundings remain constant even if it changes from one form of energy to another. Then, in 1850, RUDOLF JULIUS EMMANUEL CLAUSIUS had formulated the second law of thermodynamics, the law of entropy, which states that heat cannot spontaneously pass from a cooler to a hotter body.

Nernst's third law of thermodynamics, formulated in 1906, states that it is impossible to cool a body to absolute zero by any physical process. Although one can approach absolute zero, one cannot actually reach this limit. If one could reach absolute zero, all bodies would have the same entropy. Nernst and his collaborators tested his theorem by measuring the specific heats and other thermodynamic quantities of a number of substances at low temperatures. By early 1910, they found that specific heats not only decreased with lowering temperature, as Nernst had proposed, but appeared to be decreasing to zero. This outcome had been predicted by ALBERT EINSTEIN three years earlier on the basis of the quantum hypothesis.

Through his colleague MAX ERNEST LUDWIG PLANCK, Nernst made the acquaintance of Einstein. He and Planck would later be instrumental in luring Einstein to the University of Berlin in 1914. Nernst played an important role in organizing the first of the Solvay Conferences, in which the best minds of physics would grapple with the mysteries of the newly emerging quantum theory.

Nernst made fundamental contributions in electrochemistry, osmotic pressure, the theory of solutions, solid state chemistry, thermochemistry, electroacoustics, and photochemistry. He wrote numerous monographs on these topics as well as a standard textbook, *Introduction to the Mathematical Study of the Natural Sciences*. He became director of the new Physical Chemistry Institute in 1924, a position he held until his retirement in 1933. He died in Berlin on November 8, 1941.

As the founder of modern physical chemistry, Nernst was an important figure in the transition from classical to modern physical science.

Further Reading

Barkan, Diana. *Nernst: Architect of Physical Revolution*. Cambridge, Mass.: Cambridge University Press, 1999.

Mendelssohn, Kurt. *The World of Walther Nernst: The Rise and Fall of German Science, 1864–1941*. Pittsburgh: University of Pittsburgh, 1973.

⊠   **Newton, Sir Isaac**
(1642–1727)
British
Mathematical Physicist (Mechanics, Gravitation), Experimentalist (Optics), Astronomer

Sir Isaac Newton launched the modern age of scientific discovery and invention that continues to this day. His was a rare genius, capable of creating the mathematical tools—the calculus—he needed to pursue his physical investigations. His three laws of motion and his law of gravitation, clearly defining the nature of mass, weight, force, and acceleration, remain the foundation of our understanding of the mechanical dynamics of the macroscopic world. His mathematical description of gravitational force gave a physical basis to the Sun-centered universe proposed by Copernicus and defended by GALILEO GALILEI and erased the artificial boundary separating terrestrial and celestial events. His breakthroughs in optics, including the discovery that white light is composed of a spectrum of colors and the invention of the reflecting telescope, would suffice to secure the place of a lesser figure in the annals of physics; in Newton's case, they rank among the secondary achievements of an unparalleled career.

Isaac Newton was born in Woolsthorpe, in Lincolnshire, England, on December 25, 1642, by the old Julian calendar, the year Galileo died. His father, an illiterate but propertied farmer, had died three months earlier and Isaac himself, a premature, sickly baby, surprised everyone by surviving. His mother's remarriage to a wealthy, elderly clergyman in the next town, when Isaac was three, was a blow to the boy. Left in the care of grandparents, who treated him as an orphan, he found solace in making drawings and diagrams of mechanical things and in executing a few, such as water clocks, kites with fiery lanterns attached, and a model mill powered by a mouse. He was sent to school in nearby Grantham at age 12 but forced to withdraw four years later when his newly widowed mother demanded his services as manager of her estate. When Isaac proved wholly unsuited to the task, an uncle succeeded in reenrolling him at school, where he studied for his university entrance examinations.

In 1661, Newton entered Trinity College, Cambridge, where, despite his mother's wealth, he was obliged to work his way through the first three years by waiting tables and cleaning rooms for better subsidized students. He wrote, in a journal of his thoughts, "Plato is my friend, Aristotle is my friend, but my best friend is truth." He plunged avidly into his studies and in 1664 was elected a scholar, a status that included four years of financial support. But the bubonic plague would interrupt his university career. Spreading across Europe, it reached Cambridge in 1665, forcing the university to close its doors.

Newton would later call the next two years, spent at home in seclusion, "the prime of my age for invention." In light of what he accomplished, this seems an understatement. During this period he laid the foundations of differential and integral calculus, which he called the "method of fluxions." Newton was led to invent this new branch of mathematics by his desire to describe motion using analytic geometry. The problem was that plane geometry was capable of describing only linear motion. To solve this problem, he reasoned that, by dividing the straight lines of a

Isaac Newton launched the modern age of scientific discovery and invention. Newton's three laws of motion and his law of gravitation remain the foundation of our understanding of the mechanical dynamics of the macroscopic world. *(AIP Emilio Segrè Visual Archives, Physics Today Collection)*

polygon into infinitesimally smaller straight-line segments, a circle would result in the limit of an infinitely large number of line segments of vanishingly small length. From this insight flowed the calculus, the algebra of infinitesimal quantities, which enabled him to describe curvilinear motion mathematically.

The calculus was the mathematical language in which he would express his theory of universal dynamics. And, indeed, it was during this extraordinary two-year period that Newton experienced his major physical insights: he formulated the three laws of motion and the essentials of his gravitation theory, in addition to completing important work on optics. Amazingly, however, he would not share his findings with the world until several years later. True to his own complex personal dynamics, the withdrawn young man, fearing criticism and cherishing his "peace of mind," was roused to publish only with external encouragement, most frequently someone else's claiming credit for what Newton knew he had already discovered. At such times, he would defend his claim to primacy fiercely and even ruthlessly.

Thus, the publication in 1667 by Nicolas Mercator of a book with some methods for dealing with infinite series spurred Newton to write his own treatise, *De Analysi*, expounding his own wider-ranging mathematical results. By now he was back in Cambridge and had been elected a fellow of Trinity College. It was his mentor Isaac Barrow who disseminated Newton's work to the mathematics community, establishing the young man's reputation. When Barrow resigned as Lucasian Professor of Mathematics in 1669 to pursue divinity studies, Newton, age 26, was given this prestigious chair. Newton would remain at Cambridge for almost 30 years, living modestly and never marrying. An indifferent lecturer, whose classes were sparsely attended, Newton devoted his days and nights to his solitary studies, which, in addition to physics and mathematics, included chemistry, alchemy, theology, and mysticism. Newton would remain a passionate student of theology all his life, asserting the existence of a deity as the "first cause" of all natural phenomena.

Despite his quest for anonymity Newton soon became famous for his invention of the reflecting telescope. In need of a telescope for observing the motion of comets and planets, Newton was dissatisfied with the Galilean-style refractor telescope, then the only one in use, which had a large lens at the front end to gather light. Refractors tended to introduce spurious colors (i.e., chromatic aberration); to eliminate this, Newton used a mirror instead of a lens to collect light. The resulting efficient and inexpensive instrument became the most popular telescope in the world and remains the proto-

type of today's huge astronomical reflecting telescopes. When, in 1672, Newton presented one to the Royal Society, the most influential of numerous scientific societies that were formed in the 17th century, it elected him a fellow and urged him to write a paper on his work in optics. Newton obliged, at last publishing the results of his 1665–1666 experiments with light and color. He reported that sunlight, when passed through a prism, dispersed into a spectrum of colors; when passed through a second prism, the colors in the spectrum combined and once more formed white light. In this way he proved that colors are a property of light and not of the prism. He also investigated other optical phenomena including thin film interference effects such as Newton's rings. Although this paper was generally well received, criticism by the eminent ROBERT HOOKE made Newton once more recoil from the ordeal of publishing. Later Hooke would claim that Newton had stolen some of his optical results. Newton's response was to wait until Hooke was dead before publishing his *Opticks* in 1704.

Given his aversion to subjecting his work to public scrutiny, Newton might never have published his most important findings without the intervention of the eminent astronomer Edmond Halley, who urged him to write his magnum opus and then saw that it was published. Over a two- to three-year period, between 1684 and 1686, Newton wrote his *Mathematical Principles of Natural Philosophy*, known today as the *Principia*, which was published in 1687. Here, in what is considered the greatest scientific treatise ever written, Newton proposed his laws of motion, and, most centrally, the theory of gravity. In developing his system, he built upon a synthesis of Kepler's laws of planetary motion and Galileo's laws of motion and gravitation. He would later say

If I have seen further than other men, it is because I have stood on the shoulders of giants.

Newton "saw further" by divining the unifying physical principles underlying his predecessors' observations. He invented the concept of mass, a physical property of every object in the universe, and said that all objects with mass possess inertia, the tendency to resist any change in its state of motion. In his first law, the law of inertia, he states: "Every body remains at rest or in uniform motion in a straight line unless it is compelled to change that state." Here Newton affirms Galileo's contention, contradicting Aristotle, that no force is needed to sustain an object in motion. In Newton's universe, when an object is set in motion or changes its velocity or direction of motion, a force is responsible. Newton defined force in his second law, the law of acceleration, which states, "A force accelerates a body by an amount proportional to its mass." Since acceleration is the rate of change of velocity with time, the law can be stated as, Force equals mass times acceleration or $F = ma$. Newton's third law of motion states, "To every action there always exists an equal and opposite reaction." The force of action and the counterforce of reaction are always mutual; that formulation indicated how objects could be made to move and led to the law of conservation of momentum.

Newton then took these concepts and, by wedding them to his theory of the universality of the gravitational force, explained the dynamics of the entire solar system, validating Kepler's laws of planetary motion. According to legend, Newton saw an apple fall in his orchard around 1665–1666, thought of it in terms of an attractive gravitational force toward the Earth, and realized the same force might extend as far as the Moon. He was familiar with Galileo's work on projectiles and suggested that the Moon's motion in orbit could be understood as a natural extension of that theory. Since the law of inertia tells us that the Moon would move in a straight line unless acted upon by an outside force, Newton reasoned that a force—gravity—is acting on it. Calculating the force needed to hold the Moon in its orbit, as compared with the force pulling an object to the ground, he deduced his famous inverse-square law of gravitation: "The force of gravitational attraction between two bodies is proportional to their masses, and decreases with increasing distance between them, as the inverse of the square of that distance, so if the distance is doubled, the force is down by a factor of four." In order for the law to work, Newton had made the key assumption that gravity emanates from the center of the Earth. Using exact computation, he calculated the relative masses of heavenly bodies from their gravitational forces. Since comets were shown to obey the same laws, he conjectured that they might periodically return. Using his law of action–reaction, he was also able to describe the tides as resulting from the gravitational pull of both the Sun and the Moon: the action force holds the Moon in its orbit; the reaction force of gravity from the Moon moves the tides around the Earth. However, he never pretended to understand what actually caused gravitation, suggesting to those who found the idea of attraction across empty space objectionable that it might be caused by the impact of unseen particles.

The publication of the *Principia* had an electrifying effect throughout Europe and turned Newton into a figure of awe and reverence. After its appearance he seems to have grown bored with Cambridge. As a firm opponent of the attempt by King James II to make the universities into Catholic institutions, he was elected a member of Parliament for the University of Cambridge to the Convention Parliament of 1689 and sat again in 1701–1702. The excessive strain of his studies and the attendant disputes caused him to suffer severe depression in 1692, when he was described as having "lost his reason."

Four years later, in 1696, he moved to London as Warden of the Royal Mint. In 1699, he became Master of the Mint, an office he retained until his death. The Royal Society of London first elected him president in 1703 and annually reelected for the rest of his life. He was knighted by Queen Anne in 1705.

His major work, *Opticks,* in which he summed up his life's work on light, appeared in 1704. Although he held that light rays were corpuscular in nature, he integrated into his ideas the concept of periodicity, holding that "ether waves" were associated with light corpuscles. The corpuscle concept lent itself to an analysis by forces and established an analogy between the action of material bodies and that of light, reinforcing the universalizing tendency of the *Principia.* However, in the 1800s, the investigation of interference effects by THOMAS YOUNG led to the establishment of the wave theory of light.

As Newtonian science gained acceptance in Europe, he became the most highly esteemed natural philosopher in Europe. His last decades were spent revising his major works, polishing his studies of ancient history, and defending himself against critics, such as Leibnitz, with whom he engaged in bitter dispute over who had invented the calculus. He died at age 84 on March 20, 1727, and was buried with great pomp in Westminster Abbey.

Newton seemed to understand that his work heralded the beginning of an era in which the scientific method would continue to unlock the basic laws governing the universe. He wrote:

To myself I seem to have been only like a boy playing on the seashore and diverting myself in now and then finding a smoother pebble or a prettier shell than ordinary, whilst the great ocean of truth lay all unexplored before me.

Despite this apparent modesty, Newton would have been pleased, but not altogether surprised, to hear the reply of *Apollo 8* astronaut Bill Anders who was asked who was "driving" the spacecraft to the Moon by his eight-year-old son. "I think," said the astronaut, "Isaac Newton is doing most of the driving now."

Further Reading

Berlinski, David. *Newton's Gift: How Sir Isaac Newton Unlocked the System of the World.* New York: Free Press, 2000.

Fauvel, John, ed. *Let Newton Be!* Oxford: Oxford University Press, 1988.

Ferris, Timothy. "Newton's Reach," in *Coming of Age in the Milky Way.* New York: William Morrow, 1988, pp. 103–122.

Newton, Isaac. *The Principia: Mathematical Principles of Natural Philosophy.* Translated by Bernard Cohen and Anne Whitman. Berkeley: University of California Press, 1999.

Speyer, Edward. *Six Roads from Newton: Great Discoveries in Physics.* Wiley Popular Science. New York: John Wiley & Sons, 1996.

Westfall, Richard S. *Never At Rest: A Biography of Isaac Newton.* Cambridge: Cambridge University Press, 1983.

O

Ohm, Georg Simon
(1789–1854)
Bavarian
Theoretical and Experimental Physicist
(Classical Electromagnetism)

Georg Simon Ohm's formulation of the relationship between current, electromotive force (voltage), and resistance, known as Ohm's law, was a seminal contribution to the understanding of electricity and opened the door to an era of invention in which scientists could design electric circuits for specific functions. Although his name would later be synonymous with the unit of resistance (the ohm), he was underappreciated in his lifetime and frustrated in his academic ambitions.

Ohm was born in Erlangen, Bavaria (now Germany), on March 16, 1789. His Protestant parents had no formal education, but Ohm was fortunate in his father, a self-educated master locksmith who gave his children a solid education in mathematics, physics, chemistry, and philosophy. Young Georg gained far more from this home schooling than he did from the uninspiring, learning-by-rote methods used at the Erlangen Gymnasium, where he was sent at age 11 for his secondary education.

On entering the University of Erlangen in 1803, the future pioneer of the theory of electricity dissipated his opportunity for higher education in a nonstop round of social pleasures, which included dancing, ice-skating, and playing billiards. His irate father demanded that he drop out after three semesters and sent him to earn his living as a schoolteacher and private tutor in Switzerland. During his years in Switzerland, Ohm, guided by a former professor, carried out an exhaustive program of independent mathematical study, reading the works of Euler, Laplace, and Lacroix, among others. The fact that he was awarded his Ph.D. in October 1811, after returning to the University of Erlangen in April of that same year, gives some idea of how well he succeeded in mastering the material independently.

Although the university immediately engaged him as an unpaid lecturer, an unimpeded path to the scholarly career he desired was not to be. Ohm was forced to take on a paying job as a teacher in a mediocre school in Bamberg. In 1817, he found a somewhat better position in Cologne, teaching mathematics and physics at the Jesuit Gymnasium. Ohm doggedly pursued his self-education, reading the work of the French mathematicians and acquainting himself with current work on electricity. In 1820, when the Danish physicist HANS CHRISTIAN ØRSTED announced his groundbreaking discovery that an electric current can generate a magnetic force, Ohm began using the gymnasium's well-equipped physics laboratory to perform his own experiments.

His work would culminate in 1825, when, at age 36, Ohm published his first paper, examining the decrease in the magnetic force produced by a current flowing in a wire as the length of the wire increases. He used a voltaic pile (electric battery) to produce a current and connected different lengths of wire to it; he then measured the magnetic force that was generated by the current, using the magnetic needle of a galvanometer. In this way, that is, by measuring the magnetic force, he was, in fact, measuring the electric current flowing in the wire. The results confirmed his expectations: a longer wire produced a greater loss in the magnetic force. This implied that a longer wire had a smaller current flowing in it and, therefore, had greater *resistance*, the term coined by Ohm to designate the opposition of the material to the flow of charge.

In 1826, he repeated this experiment, this time generating the current by using a thermocouple (a pair of wires of different conductors, welded or soldered at one end, used to measure temperature). This technique had the advantage of producing a constant electric current, as distinct from the fluctuating current produced by the voltaic pile. He found that the magnetic force was equal to the electromotive force produced by the thermocouple divided by the length of the wire plus the resistance of the remainder of the circuit, including the thermocouple itself. He expressed his findings in terms known today as Ohm's law, which states that the current is equal to the electromotive force divided by the overall resistance of the circuit.

In the context of this experiment, Ohm pointed out that an electric current flows through a conductor of varying resistance to produce a potential difference, just as heat flows through a conductor of varying conductivity from one temperature to another to produce a temperature difference. In 1827, Ohm published his results and his complete theory of electricity in his great work, *Die Galvanische Kett, mahtematisch bearbeitet* (The Galvanic circuit, investigated mathematically).

Despite the immense importance of Ohm's work, recognition by the scientific community continued to elude him. Few of his peers were capable of understanding his mathematics, and those few German physicists who could appreciate the rigor of Ohm's formulations doubted the correctness of his approach. Ohm remained a schoolteacher, in Berlin, until 1833, when he moved to Nuremberg and became a professor of physics at the respectable but undistinguished Polytechnic Institute.

Recognition, when it was finally given him, began in England. In 1841, Ohm was awarded the Royal Society's Copley Medal. In 1849, he began to lecture at the University of Munich. He finally became a professor of physics at the university in 1852 but died only two years later in Munich on July 6, 1854.

Ohm's law, together with the laws of electrodynamics discovered by ANDRÉ-MARIE AMPÈRE at about the same time, pioneered the way to future theoretical investigation of electricity. His name is remembered in both the unit of resistance, the ohm, and its inverse, the unit of conductivity, the mho, *Ohm* spelled backward, also called the siemens.

Using Ohm's law, scientists could for the first time calculate the amounts of current, voltage, and resistance in electric currents, measuring changes in one of these variables through changes in the others.

Further Reading

Jungnickel, Christa, and Russell McCormmach. *Intellectual Mastery of Nature: Theoretical Physics from Ohm to Einstein: The Torch of Mathematics, 1800–1870.* Vol. 1. Chicago: University of Chicago Press, 1986.

Keithley, Joseph F. *The Story of Magnetic Measurements: From Early Days to the Beginning of the 20th Century.* New York: Wiley-IEEE Press, 1998.

Oppenheimer, J. Robert
(1904–1967)
American
Theoretical Physicist, Quantum Theorist, Atomic Physicist, Nuclear Physicist

J. Robert Oppenheimer was a brilliant theoretical physicist who made important contributions to quantum mechanics and nurtured the development of an outstanding theoretical physics community at the University of California, Berkeley. His name is forever associated with the development of the first atomic bomb in Los Alamos, New Mexico, where he was director of the Manhattan Project, and, in particular, with the moral and ethical controversies faced by physicists who participated in the creation of weapons of mass destruction.

He was born in New York City on April 22, 1904, the elder of Ella and Julius Oppenheimer's two sons. His father was a German Jew who immigrated to the United States at the age of 17 and made a fortune importing textiles. Robert was a frail, studious child, who would later describe himself as an "unctuous, repulsively good little boy," whose schooling failed to prepare him for life's cruelties. He attended the Ethical Culture School in New York, which was founded on the principle "Man must assume responsibility for the direction of his life and destiny." His interest in science was sparked by his grandfather's gift of a mineral collection. At age 12, the precocious child gave a lecture at the New York Mineralogical Club.

Oppenheimer entered Harvard University in 1922 and graduated summa cum laude in only three years. He would later describe his undergraduate years as the most exciting of his life, when he "intellectually looted the college," usually taking six courses and auditing four. He had a talent for languages, excelling in Latin, Greek, French, and German. He published poetry and studied Eastern philosophies—subjects he would pursue all his life. While majoring in chemistry, he was quickly drawn to understanding the physics that underlay the chemistry and began working in the laboratory of the future physics Nobel laureate PERCY WILLIAMS BRIDGMAN.

On the strength of a letter from Bridgman, who did not conceal the fact that Oppenheimer's strengths were analytical, not experimental, he was accepted by the Cavendish Laboratory at the University of Cambridge. He sailed to England in 1925, an exhilarating time for a young physicist to be in Europe. Postwar work on quantum theory was just getting under way, and Oppenheimer's work at the laboratory, which, under the

J. Robert Oppenheimer made important contributions to quantum mechanics. His name is forever associated with the development of the first atomic bomb in Los Alamos, New Mexico. *(AIP Emilio Segrè Visual Archives and Los Alamos Scientific Laboratory)*

direction of ERNEST RUTHERFORD, was internationally renowned for its pioneering studies on atomic physics, exposed him to the latest ideas. His acquaintance with NIELS HENRIK DAVID BOHR, the guiding spirit of quantum mechanics, was the turning point in his decision to commit himself to theoretical physics.

Oppenheimer had already published two papers on quantum mechanics when he accepted MAX BORN's invitation to study in Göttingen, Germany, in 1926. He earned his Ph.D. there in 1927 with the dissertation "On the Quantum Theory of Continuous Spectra," which was published in the prestigious German journal *Zeitschrift fur Physik*. The professor in charge of his oral examination is said to have expressed relief when the ordeal was over, fearing that the probing Oppenheimer was about to question *him*. Between 1926 and 1929, the young physicist published 16 papers. Together with Born he developed the quantum theory of molecules. When he returned to the United States, at age 25, he enjoyed an international reputation.

Awarded a National Research Fellowship, he headed back to Europe, honing his mathematical skills with Paul Ehrenfest in Leiden and his analytical edge with WOLFGANG PAULI in Zurich. In 1929, he was appointed assistant professor at both the California Institute of Technology (CalTech) and the University of California at Berkeley. For the next 13 years he would commute between these two campuses, inspiring a generation of physicists and transforming the face of American physics. Under his charismatic leadership, an outstanding group of researchers attacked the problems of theoretical physics with an intensity previously unknown at American universities. His most important research involved the study of the relativistic equations for the atom developed by PAUL ADRIEN MAURICE DIRAC. In 1930, Oppenheimer showed that the Dirac equation predicted that a positively charged particle with the mass of an electron could exist. This particle, detected in 1932 and called the positron, was the first example of antimatter. He also did research on the energy processes of subatomic particles, including electrons, positrons, neutrons, and mesons. In addition, his analyses anticipated many later finds in astrophysics, including neutron stars and black holes.

During these California years, in the early thirties, Oppenheimer's political consciousness was awakened by the rise of fascism in Europe. He became deeply concerned about the fate of Jews in Germany, where he had relatives, many of whom he would later help to escape. He formed friendships with Spanish students who were Communists and in 1936 he sided with the republic during the Spanish Civil War. He would stop short of joining the Communist Party, however, finding he could make "no sense" of its dogma.

In 1939 Oppenheimer met his future wife, Katherine Puening, known as Kitty, who would bear him a son and a daughter. Also in that year Bohr, who had escaped from Denmark to England, brought word that the Germans had split the atom. ALBERT EINSTEIN and Leo Szilard wrote their famous letter warning President Franklin Delano Roosevelt that the Nazis could be the first to make a nuclear bomb. When Roosevelt established the Manhattan Project in 1942, giving the United States Army responsibility for the joint efforts of British and U.S. physicists to develop an atomic bomb, Oppenheimer became its director. In 1943, he set up a new research station at Los Alamos, New Mexico, where he assembled a team of first-rate scientific minds, including HANS ALBRECHT BETHE, ENRICO FERMI, and EDWARD TELLER. The quality of Oppenheimer's supervision of the more than 3,000 people working at Los Alamos is generally believed to have been a crucial factor in the project's success. According to Teller,

> Oppenheimer was probably the best lab director I have ever seen, because of the

great mobility of his mind, because of his successful effort to know about practically everything important invented in the laboratory, and also because of his unusual psychological insight.

On July 16, 1945, Oppenheimer was present at "Trinity," the first test of an atomic bomb, in the New Mexico desert. He described his initial reaction with masterful understatement: "We knew the world would not be the same." With three other scientists who had been consulted on how to deploy the bomb, Oppenheimer recommended that a populated "military target" be selected. Within a month, two bombs were dropped on Nagasaki and Hiroshima, leading to the Japanese surrender on August 10, 1945. Oppenheimer would later write:

> I have no remorse about the making of the bomb and Trinity. That was done right. As for how we used it . . . I do not have the feeling that it was done right. . . . Our government should have acted with more foresight and clarity in telling the world and Japan what the bomb meant.

After the war, as chairman of the United States Atomic Energy Commission (AEC), Oppenheimer continued to experience the limits of his influence over the technology he had played so central a role in developing. Along with most members of the commission, he opposed the creation of a hydrogen bomb. But when the Soviets exploded a nuclear device in the summer of 1949, President Truman gave the H-bomb project a green light. Although Oppenheimer's offer to resign was not accepted, his opposition to the hydrogen bomb was to have dramatic consequences. In 1953, in the heat of the McCarthy era, when anti-Communist hysteria reigned, Oppenheimer was accused of having Communist sympathies. The Federation of American Scien-

tists rushed to his defense, to no avail. President Eisenhower ordered he be stripped of his security clearance, and Oppenheimer's influence on national science policy came to an end.

In the eyes of the world, Oppenheimer has come to epitomize the plight of the scientist, attempting to assume moral responsibility for the consequences of his discoveries, who becomes the target of a witch-hunt. Yet there is a darker side to Oppenheimer's struggle; four years earlier, in 1949, perhaps wishing to shore up his own position, he had joined forces with the witchhunters, denouncing several of his colleagues as Communist sympathizers before the House Un-American Activities Committee.

In his last years, as director of the Institute of Advanced Study at Princeton, Oppenheimer devoted much of his time to writing about the problems of intellectual ethics and morality. In 1963, President Johnson awarded him the Enrico Fermi Prize, the highest award the AEC confers, as an attempt to make amends for his unjust treatment. In 1966, he retired from Princeton, where he died on February 18 of the following year, of throat cancer. His final printed words were "Science is not everything, but science is very beautiful."

See also FEYNMAN, RICHARD PHILLIPS; RABI, ISIDOR ISAAC.

Further Reading

Herken, Gregg. *Brotherhood of the Bomb: The Tangled Lives and Loyalties of Robert Oppenheimer, Ernest Lawrence and Edward Teller.* New York: Henry Holt, 2002.

Rhodes, Richard. *Dark Sun: The Making of the Hydrogen Bomb.* New York: Simon & Schuster, 1995.

———. *The Making of the Atomic Bomb.* New York: Simon & Schuster, 1986.

Schweber, Silvan S. *In the Shadow of the Bomb.* Princeton, N.J.: Princeton University Press, 2000.

Smith, Alice K., and Charles Weiner, eds. *Robert Oppenheimer: Letters and Recollections.* Cambridge, Mass.: Harvard University Press, 1980.

⊠   **Ørsted, Hans Christian**
(1777–1851)
Danish
*Experimental Physicist (Classical
Electromagnetism), Physical Chemist*

Hans Christian Ørsted exhilarated the European scientific community in July 1820 with the announcement of his discovery that an electric current produces a magnetic field. Physicists, among them ANDRÉ-MARIE AMPÈRE, lost no time following up on Ørsted's groundbreaking work, developing the fields of electromagnetism and electrodynamics.

The man who would initiate the main thrust of 19th-century physics was educated as a pharmacist. At the age of 17, he moved to Copenhagen from Rudkøbing, Langeland, Denmark, where he had been born on August 14, 1777. His studies at the University of Copenhagen represented his first experience of formal education. He entered the university in 1794, earned a doctorate in pharmacology in three years, and settled down to practice his profession. But two years later he abandoned pharmaceutical work for an extensive tour of Europe. When he returned, he began to exercise his talent as a teacher, successfully offering a series of public lectures. Ørsted lived in an age in which specialization was far less developed than in our own, and when it was common for scientists to teach and do research in more than one field. In 1806, presumably on the strength of his teaching abilities, he became a professor of physics at the University of Copenhagen, a position he would retain until 1829.

In the late 18th century, interest in magnetism was spurred, in part, by the need for a compass that would be accurate near the polar regions and could thus be used in the search for the Northwest Passage (a sea passage through the Arctic regions of North America that would connect the Atlantic and Pacific Oceans). Ørsted began seeking the connection between electricity and magnetism around 1813, philosophically convinced that all forces in nature have a unified essence and can therefore be converted into one another. As early as 1600, William Gilbert, a pioneer of magnetism, had proposed such a connection. In Ørsted's time, many physicists looked for a magnetic force acting in the direction of the electric current but failed to find it. Ørsted succeeded because he hypothesized that the magnetic force would be acting in a direction perpendicular to that of the current.

Ørsted performed his history-making experiment before a lecture hall filled with his students, in April 1820. The demonstration was a simple one: he placed a compass needle beneath a wire connected to a battery. The needle moved faintly toward the wire. Ørsted was convinced of his success but, given the small size of the effect, proceeded cautiously. Not until July, after performing further experiments showing that a circular magnetic force, which aligns itself perpendicular to the current, is produced around the wire, did he report his findings to Europe's leading scientific journals. Here at last was definite experimental evidence of the relationship between electricity and magnetism. Within weeks of learning of Ørsted's results, Ampère in France would develop a mathematical description of the magnetic force between two electric currents, founding the new science of electrodynamics. The race to develop a unified theory of electrodynamics was on, although Ørsted himself would play no further role in it.

In 1829, he left his university post to become director of the Polytechnic Institute in Copenhagen, where he remained until his death on March 2, 1851. As a teacher and writer, he played a dynamic role in elevating the level of scientific education and research in Denmark. In 1824, he founded a society devoted to the spread of scientific knowledge among the general public, the Danish Society for the Promotion of Natural Science. His name was given to the oersted (symbol Oe), the unit of magnetic-field strength in the centimeter–gram–second system of physical units.

Further Reading

Jelved, Karen, ed. *Selected Scientific Works of Hans Christian Ørsted*. Princeton, N.J.: Princeton University Press, 1998.

Keithley, Joseph F. *The Story of Magnetic Measurements: From Early Days to the Beginning of the 20th Century*. New York: Wiley-IEEE Press, 1998.

P

⊠ **Pauli, Wolfgang**
(1900–1958)
Austrian-born Swiss
*Theoretical Physicist, Quantum
Theorist, Particle Physicist*

Wolfgang Pauli was one of the pioneering generation of 20th-century physicists who discovered and refined the basic principles of atomic behavior as revealed by the laws of quantum mechanics. His most far-reaching contributions flowed from his insight that no two electrons in an atom can occupy the same energy state. Enunciated in 1925, the Pauli exclusion principle, as it came to be called, explained numerous phenomena on both the atomic and nuclear levels and garnered Pauli the 1945 Nobel Prize in physics. He is also famous for his 1930 prediction of the existence of the neutrino, a particle without charge or mass, which was discovered almost 30 years later.

Wolfgang Pauli was born in Vienna on April 25, 1900. His father, a physician and professor of chemistry at the University of Vienna, was of Jewish origin but had his son baptized a Catholic. His godfather was the famous Austrian physicist ERNST MACH. A precocious high school student in Vienna, Wolfgang entered the University of Munich in 1918, having already mastered both special and general relativity. Encouraged by his professor, the eminent

ARNOLD JOHANNES WILHELM SOMMERFELD, Pauli wrote the first comprehensive monograph on the subject, a 200-page work, which prompted ALBERT EINSTEIN himself to observe, "Whoever studies this mature and grandly conceived work might not believe that its author is a twenty-one year old man." The Munich years were formative for Pauli. It was then that he was exposed to ideas far stranger than those he had found in Einstein's work. As he would later recall, "I was not spared the shock, which every physicist accustomed to the classical way of thinking experienced when he came to know Niels Bohr's basic postulate of quantum theory for the first time."

He formed a lifelong friendship with WERNER HEISENBERG, whose uncertainty principle would add the notion of nondeterminacy to the startling picture of nature emerging from quantum mechanics in the 1920s. Their long and copious correspondence would later be published, providing a wealth of insights into one of the most exciting periods in the history of physics. Pauli received a doctorate in 1922 for a thesis, supervised by Sommerfeld, on the quantum theory of ionized molecular hydrogen.

That same year he went to Göttingen as an assistant to MAX BORN but soon moved to Copenhagen to study with NIELS HENRIK DAVID BOHR; there he began, in his own words, "a new phase of my scientific life." When he returned to Germany in 1923, accepting a position at the University of Hamburg as a lecturer, his exposure to Bohr's work on the structure of the atom and the problems that work was encountering had become the focus of his research.

Bohr's model of the atom, modified by Sommerfeld, proposed that the electrons of an atom are arranged in groups, which have different mean distances from the nucleus and are each characterized by three quantum numbers, which describe the rotation of the electron around the nucleus. Problems with the model arose when these numbers failed to account for magnetic anomalies in matter. Pauli solved the problem by adding a fourth quantum number to the three already in use (n, l, and m). This number, s, would represent the electron's rotation around an axis through its own center of gravity, what came to be called spin, and would have two possible values: $+1/2$ or $-1/2$. Building on this hypothesis, in 1925, he enunciated his Pauli exclusion principle, which stated that no two electrons in the same atom could have the same values for their four quantum numbers. One of these quantum numbers describes one of the two possible directions for the electron's intrinsic spin. Thus, two electrons that are in the same energy level as described by the other three quantum numbers are differentiated from each other because they have opposite spins. The exclusion principle proved essential not only in explaining atomic structure: by uncovering the significance of ordering elements by their atomic number, it provided an explanation for the periodic table of the elements.

Pauli's important work led to his appointment, in 1928, as professor of experimental physics at the Federal Institute of Technology, Zurich, a position he would hold for the rest of his life. During the years before World War II, Pauli developed the institute into a vibrant international center for physical research. At the beginning of his Zurich years he made

Wolfgang Pauli discovered the fact that no two electrons in an atom can occupy the same energy state. Known as the Pauli exclusion principle, it explained numerous phenomena on both the atomic and the nuclear level. *(AIP Emilio Segrè Visual Archives)*

another major contribution in his prediction of a new elementary particle: the neutrino. Pauli's hypothetical particle was offered as a solution to the dilemma posed by the observation that electrons appeared to be emitted in a continuous stream whereas theory called for a discontinuous spectrum. Pauli accounted for this discrepancy by proposing that the emission of an electron in beta decay is accompanied by the production of an unknown particle. Because Pauli's particle had neither charge nor mass, the model explained why it had never been detected. It was ENRICO FERMI who in 1934 confirmed Pauli's view and dubbed the new particle the neutrino. Two years before Pauli's premature death, on December 15, 1958, in Zurich, his prediction was experimentally validated by the American physicists FREDERICK REINES and Clyde L. Cowan, who recognized neutrinos by their impact with subnuclear particles in mineral water.

Pauli left Europe when hostilities broke out and spent the war years in the United States, at the Institute for Advanced Study, Princeton. He acquired both Swiss and U.S. citizenship and spent his later years in Princeton and in Zurich.

An intuitively gifted scientist, universally respected by colleagues such as Bohr and Heisenberg, who relied on his advice, Pauli was also a lifelong student of the great Swiss psychoanalyst Carl Gustav Jung. When his brief first marriage failed in 1929, Pauli underwent analysis with Jung and later continued corresponding with him. In 1952, Jung and Pauli coauthored *The Explanation of Nature and the Psyche*, which discusses the influence of Jungian archetypes on the work of the great astronomer Johannes Kepler.

Further Reading

Hermann, A. *Wolfgang Pauli: Scientific Correspondence with Bohr, Einstein, Heisenberg.* Vol. 191. New York: Springer-Verlag, 1979.

Laurikainen, Kalervo Vihtori. *Beyond the Atom: Philosophical Thought of Wolfgang Pauli.* New York: Springer-Verlag, 1988.

Pauli, Wolfgang, et al., eds. *Writings on Physics and Philosophy.* New York: Springer-Verlag, 1994.

⊠ Penzias, Arno Allan
(1933–   )
German/American
Experimentalist, Radio Astronomer, Astrophysicist

Arno Allen Penzias shared the 1978 Nobel Prize in physics with Robert W. Wilson for their discovery of cosmic microwave background radiation, which provided strong support for the big bang theory of the origin of the universe.

He was born into a Jewish family on April 26, 1933, in Munich, Germany, the year the Nazis rose to power. Until the age of six, he was, in his own words, "an adored child in a closely-knit middle class family." The mirage of safety evaporated when the family was rounded up, along with other Jews of Polish origin, for deportation to Poland. The Poles refused to accept more immigrants and the family was returned to Munich, but from that point on they thought only of emigrating to America. When he was six, Arno and his little brother were put on a train for England, where they were soon joined by their parents. Within six months, the family sailed for America, arriving in January 1940 in New York, where his father found carpentry work and the boys attended public schools in the Bronx. Arno went on to attend Brooklyn Technical High School, a specialized public school for highly promising students. He then studied physics at the City College of New York, which had long been a gateway to the middle class for struggling immigrants, and graduated in 1954 in the top 10 percent of his class. He married his lifelong partner, Anne Pearl Barras, with whom he would have three children, and spent the next two years as a radar officer in the United States Signal Corps. In 1956, his army experience qualified him for a research assistantship in the Columbia Radiation Laboratory, where exciting research in microwave physics was under way. Working under CHARLES HARD TOWNES, the discoverer of the maser and the laser, he built a maser amplifier in a radio astronomy experiment and completed his Ph.D. thesis in 1961.

Penzias then took a job with Bell Laboratories in Holmdel, New Jersey, anticipating that it would be a temporary position. Instead, Bell became his professional home for the next 37 years. He started out doing research in radio communications and participated in the pioneering *Echo* and *Telstar* communication satellite experiments. In 1963, he was joined in his research in radio astronomy by Robert Wilson, a young physicist from the California Institute of Technology, with whom he formed his historic partnership. Taking over the state-of-the-art horn-shaped radio antenna that had been used for radio communication with *Echo* and *Telstar*, the two men soon detected a faint microwave signal with unique properties. The signal corresponded to the radiation emitted from a blackbody with a temperature of three degrees above absolute zero, pervaded all of space, and never wavered over time. They looked into the possibility that the noise was originating in the antenna itself—which had been clogged by pigeon dung. But, even when the offending matter had been removed, the signal persisted. When other possible sources—such as the Milky Way and the Sun—had been systematically eliminated, they became aware, in 1965, through discussions with the Princeton physicist Robert H. Dicke, of an electrifying explanation for their findings: the big bang theory of cosmic creation.

At the time, there were two competing cosmological theories. The steady-state theory, espoused by Hermann Bondi, Thomas Gold, and Fred Hoyle, held that the universe was as it had always been: homogeneous in space and time. The big bang theory arose from Edwin Hubble's 1929 finding that the galaxies are rapidly moving away from each other. This led physicists such as GEORGE GAMOW to argue that galaxies must have been closer to one another in the past and that in the farthest reaches of the past—the cataclysmic moment in which the universe was created some 15 billion years ago—they were merged in a single, infinitely dense, hot mass. This theory postulated that a background radiation at three Kelvin had survived from the infancy of the universe and pervaded all of space. Penzias's and Wilson's mysterious signal fit the big bang's prediction in every respect. Their findings offered hard experimental evidence for the big bang theory, a result that has held up until the present day.

From the mid-1960s, Penzias served as a thesis advisor to Princeton graduate students in astrophysics, while continuing to do research at Bell. During this period he focused on research in interstellar chemistry and discovered the presence of key chemicals in interstellar space. Using the techniques he had pioneered, he was able to observe millimeter-wave radio spectra emanating from carbon monoxide and several other simple molecules in the dusty clouds in interstellar space. In 1973, working with Keith Jefferts, Penzias made the amazing discovery of the existence of deuterium (heavy hydrogen) in outer space, which earned the nickname "Arno's white whale"—a reference to Moby Dick, the elusive whale in Herman Melville's great novel.

Between 1972 and 1982, Penzias continued the academic side of his career as a visiting member of the astrophysics department at Princeton University. But in the early 1970s, he became increasingly involved in managerial responsibilities at Bell Labs. He became head of the Radio Physics Research Department in 1972 and, four years later, director of the Radio Research Lab, devoted to understanding radio and its communication applications. Shortly after receiving his Nobel Prize in 1978, he was promoted once more, to executive director of the Communications Sciences Research Division.

The year 1981 was a watershed: Penzias's scientific research career came to an abrupt end when the Justice Department and AT&T decided to settle their antitrust case by breaking up the Bell Laboratory system. Penzias agreed to serve as vice president of research, a position he held for 14 years, a crucial period when Bell Labs Research was attempting to transform itself into a competitive market player while trying to maintain its scientific excellence. He left that job in 1995, to become vice president of AT&T Bell Laboratories.

At this time he turned from scientific research to exploration of the impact of technology on society. A prolific writer, in addition to over 100 scientific papers, he published two books, two science fiction stories, and numerous technical and business articles. Penzias describes his first book, *Ideas and Information*, which examines the impact of information technology on business and society, as a depiction of computers "as a wonderful tool for human beings, but a dreadful role model." *Digital Harmony*, his second book, envisions machines that will work in harmony with each other, their human users, and the natural environment. It explores how emerging technologies will change the way people work and live. As vice-chairman of the Committee of Concerned Scientists, he devotes much time to defending the rights of scientists living under repressive regimes.

The discovery of the cosmic microwave background by Penzias and Wilson turned the study of cosmology into a respected, empirical science. On November 18, 1989, the National Aeronautics and Space Administration (NASA) sent the *Cosmic Microwave Background Explorer* (COBE) satellite into orbit to carry out more advanced studies of this phenomenon and its clues to the nature and history of the universe.

Further Reading

Penzias, Arno. *Digital Harmony: Business, Technology and Life After Paperwork.* New York: Harper Business, 1995.

Penzias, Arno Allan. *Ideas and Information: Managing in a High-Tech World.* New York: W. W. Norton, 1989.

Silk, Joseph. *The Big Bang.* New York: Henry Holt, 2000.

⊠   **Perrin, Jean-Baptiste**
(1870–1942)
French
Experimentalist, Physical Chemist, Atomic Physicist

Jean Perrin was a great experimentalist, who produced the first empirical evidence for the existence of atoms by observation of Brownian motion, the random, continuous movement of particles in a liquid. His groundbreaking work ended the passionate dispute that had long divided physicists over whether matter is fundamentally continuous or is made up of particle-like atoms and won him the 1926 Nobel Prize in physics.

Perrin was born in Lille, France, on September 30, 1870. His father was a member of a family of peasants in the Lorraine region and served as an infantry officer. He received his early education in Lyons and Paris; in 1891, he entered the École Normale Superieure, where he earned his doctorate six years later. During these years, he began his research into cathode rays: the phenomenon that produced a fluorescent glow when most of the air was pumped out of a glass tube with wires embedded at each end and a high voltage was sent across it.

At the time physicists were questioning whether cathode rays were charged particles or some undefined wavelike phenomenon in the ether. To investigate this issue Perrin designed experiments in which cathode rays that were generated in discharge tubes were allowed to penetrate thin sheets of glass or aluminum and collected in a hollow cylinder produced a negative charge on a fluorescent screen onto which the cathode rays were focused. As the negative charge was increased, the intensity of fluorescence fell. The finding that they could be slowed by an electric field demonstrated that the cathode rays were particle-like and carried a negative electric charge. In these experiments Perrin was also able to establish crude values for the charge-to-mass ratio of an electron, e/m. Later on, when the measurement of the e/m ratio of cathode rays was improved upon by the classic experiments of JOSEPH JOHN (J. J.) THOMSON, the identification of cathode rays with the existence of the first elementary particle, the electron, was firmly established.

In 1897, he began a lectureship in physical chemistry at the Sorbonne, where, 12 years later, he published his most important contribution, work on atomic theory. These were the years when what is now known as the atomist–energeticist debate was at its height. Because there was no technology available to verify their existence at the turn of the century, atoms had been relegated by most physicists to the realm of speculation. ERNST MACH was the leading advocate for this energeticist school of thought. For Mach the purpose of science was to measure and demonstrate only that which it can observe. Mach and his colleagues were content to measure the expansion of gases and empirically deduce a simple law relating temperature, pressure, and volume. They were not disturbed by their inability to explain *why* these properties were related in this particular way. LUDWIG BOLTZMANN, the leading atomist, on the other hand, believed that by hypothesizing a dynamic submolecular world of colliding atoms he could explain why gas expands and by how much. The starting point for Perrin's research was the work of Robert Brown, who in 1827 had observed through a microscope the random motion of small grains in a fluid, a phenomenon that became known as Brownian motion. Brown noted that the motion of the particles increased when the temperature increased but decreased when larger particles were used. To energeticists who believed in continuous matter, Brownian motion, with its discrete bumps, was particularly disturbing. In one of his landmark 1905 papers, ALBERT EINSTEIN increased the consternation of the energeticists when he explained the phenomenon of Brownian motion as the effect of large numbers of molecules bombarding the particles in a manner that allowed him to predict the movement and size of the particles.

Four years later, Perrin was able to verify Einstein's explanation experimentally. Perrin reasoned that since both the grains and the water molecules were very small, both would behave as gas molecules and the principles of the kinetic theory of gases could be used to explain their collision process. In his classic experiment

he used gamboge particles obtained from vegetable sap, and the distribution of these suspended particles in a container was analyzed by depth. He found that their number decreased exponentially with height. Using principles proposed in Einstein's paper on Brownian motion, he was able to deduce a definite value for the Avogadro number (the number of particles in one mole of a substance) that agreed substantially with experimental values obtained in other ways. The fact that the Avogadro number could be accurately determined in this experiment showed that Brownian motion implied the existence of atoms. Although atoms were not visible, Perrin's results were accepted as definitive proof of their existence.

In 1910, Perrin was made a full professor of physical chemistry at the Sorbonne, but his academic life was soon interrupted by World War I. Between 1914 and 1918, he served as an officer in the engineer corps. Afterward, he returned to the Sorbonne, where he continued to pursue his research. He had a deep interest in astrophysics and, in 1920, published results demonstrating that only the fusion of hydrogen atoms into helium atoms can account for the origin of solar energy. In addition, during the 1930s, he was instrumental in creating several vital scientific institutions: the National Scientific Research Center, which offered talented French scientists an alternative to hard-to-obtain university positions; the Palace of Discovery; the Institute of Astrophysics in Paris; the Institute of Physical and Chemical Biology; and the large Observatory in Haute Provence.

In 1926 he received the Nobel Prize in physics and became a Commander of the Legion of Honor, Commander of the British Empire, and of the Belgian Order of Leopold.

A prolific author of books and scientific papers, he was especially known for his 1913 work *Les Atomes*, explicating the body of experimental evidence, of which his own work was so vital a part, that established the existence of atoms. In that work, Perrin eloquently characterized the nature of his own achievement:

> To divine . . . the existence and properties of objects that still lie outside our ken, *to explain the complications of the visible in terms of invisible simplicity,* is the function of the intuitive intelligence which, thanks to men like Dalton and Boltzmann, has given us the doctrine of atoms.

Perrin remained at the Sorbonne until 1940, but his outspoken opposition to the occupying Nazi regime forced him to flee to the United States in December 1941. Soon afterward he became ill and died in New York City on April 17, 1942. After the war, in 1948, France paid homage to him by carrying home his remains on the battleship *Jeanne d'Arc* and burying them in the Panthéon.

Further Reading

Perrin, Jean. *Atoms.* Reprint ed. Woodbridge, Conn.: Ox Bow Press, 1990.

⊠   **Planck, Max Ernest Ludwig**
(1858–1947)
German
*Theoretical Physicist
(Electromagnetism), Quantum Theorist*

The work of Max Planck laid the foundation for the paradigm shift that ushered in the era of modern physics: the discovery of a world of discrete, discontinuous "quanta" of energy beneath the apparent continuity of classical Newtonian mechanics. The inescapable logic that led him, in 1900, to postulate the universal constant in nature that came to be known as Planck's constant led directly to the formulation of quantum mechanics 20 years later. Planck himself, whose intellectual roots were firmly planted in 19th-century notions of the absolute nature of physical laws, only reluctantly accepted the implications of the discovery that earned him the 1918 Nobel Prize in physics.

Max Ernest Ludwig Planck was born on April 23, 1858, in Kiel, Germany, into a distinguished scholarly family with a long history of devotion to church and state. When his father, a renowned law professor, accepted a position at the University of Munich, young Max entered the city's famous Maximilian Gymnasium, where he excelled in all subjects. Graduating at 17, he was torn between a career in music, which remained a passion throughout his life, and one in physics. He chose physics because, he explained, "The outside world is something independent from man, something absolute, and the quest for the laws which apply to this absolute appeared to me as the most sublime scientific pursuit in life." Inspired by the discovery that "pure reasoning can enable man to gain an insight into the mechanism of the world," Planck resolved to devote his life to theoretical physics when it was not yet a formal discipline.

As a student at the University of Munich, he encountered the attitude, not uncommon among physicists of the time, that the study of physics was essentially a dead end, since everything of importance had already been discovered. Following his own interests, Planck found this idea to be anything but accurate. His fascination with absolutes in nature led him to focus on the laws of thermodynamics, which, in turn, led him to the problem of blackbody radiation. Classical physics thought of radiation (light) as a continuous wave in a field. Physicists knew that whereas the color of a cool object is due to its surface light, that is, the reflected light emitted by an external source, a heated object glows with the light inside it. In this context, in 1859–1960, GUSTAV ROBERT KIRCHHOFF introduced the idealized concept of a blackbody: an object that does not reflect any surface light and is, therefore, a perfect emitter and absorber of radiation at all frequencies. Blackbody radiation is the full spectrum of light frequencies inside a heated body that has no surface reflection. By the 1890s, classical physicists had made several experimental and theoretical approaches to determining spectral energy distribution: a curve showing how much light energy is emitted at different frequencies for a given temperature of the blackbody. A formula derived by Planck's colleague WILHELM (CARL WERNER OTTO FRITZ FRANZ) WIEN proved valid only at high frequencies, and a similar equation found by LORD RAYLEIGH (JOHN WILLIAM STRUTT) was accurate only at low frequencies. In 1900, at the age of 42, Planck succeeded in combining these two equations, producing a formula that related the energy of the radiation to its frequency.

But, while the physics community hailed Planck's radiation law as indisputably correct, Planck himself was uneasy with the fact that it was based on nothing more solid than a "lucky intuition." Before it could be taken seriously, he believed, it had to be derived from first principles. No other episode in Planck's career better illustrates his uncompromising scientific integrity. In order to ground his derivation in physical principles, he had to relinquish his cherished belief in the absolute nature of the second law of thermodynamics (entropy) in favor of LUDWIG BOLTZMANN's statistical interpretation of that law. His second unwelcome realization was that a sound derivation could only be based on the postulate that the energy of radiation is emitted and absorbed not continuously, but in discrete packets, which he called *quanta*. He assumed that the energy of quanta of light was proportional to the frequency of the light: that is, the higher the frequency, the higher the energy. The proportionality constant in this relationship between the quantum energy of radiation and the frequency of light emitted turned out to be a universal constant, which kept showing up for all

kinds of blackbodies and all temperatures. This number, 6.626, 196 × 10⁻³⁴ J/s (joules per second), came to be called Planck's constant; it is designated by the symbol h.

The fact that Planck's constant was not zero, but had a small finite value, gradually led physicists to understand that the world of the very small, the microphysical world, could not be described by classical Newtonian mechanics. No one greeted this revelation with less enthusiasm than Planck himself did. Over the years he devoted his efforts to undercutting the new quantum mechanics but succeeded only in confirming its necessity.

Planck's law of radiation set the stage for later applications of the paradigm shift that light

Max Ernest Ludwig Planck created the foundation for the idea of the quantum hypothesis: the discovery of a world of discrete, discontinuous quanta of energy beneath the apparent continuity of classical Newtonian mechanics. *(AIP Emilio Segrè Visual Archives)*

energy is quantized. In 1905, ALBERT EINSTEIN used it in his explanation of the photoelectric effect. Light, he said, was composed not only of waves, but also of particles, which he named *photons*. In 1913, NIELS HENRIK DAVID BOHR applied the quantum theory to the structure of the hydrogen atom. By the 1920s, quantum mechanics had evolved into a full system.

But Planck would play no further role in the revolution he had inadvertently ignited. As professor of physics at the University of Berlin, from 1989 until his retirement in 1926, he made contributions to optics, thermodynamics, statistical mechanics, and physical chemistry. In his later years, he turned to philosophical writings, in which he rejected the quantum mechanical view that observer and observed are inextricably linked and continued to see nature as existing in total independence of human beings. As did Einstein and ERWIN SCHRÖDINGER, he found the quantum mechanical view of matter as statistical and non-deterministic alien to his most deeply held convictions about the nature of reality.

If the physics of the new century was uncongenial to Planck's worldview, its history was nothing less than personally devastating to him. In 1909, he lost his much-loved first wife, Marie Merck, who had borne him two sons and twin daughters. The next year he married Marga von Hoesslin, with whom he had a son. But he was to witness the tragic deaths of four of his children. His twin daughters, Margerete and Emma, died in childbirth in 1917 and 1919, respectively. Remaining in Germany throughout two world wars, he saw his eldest son, Karl, killed in action in 1916. When the Nazis gained power, Planck did what he could to preserve German physics, pleading with Hitler directly to reverse his racist policies. Throughout these trials, Planck had been upheld by his stoicism and religious conviction. But when, in 1945, his youngest son, Erwin, who had been implicated in an attempt to assassinate Hitler, was brutally executed by the Gestapo, Planck lost the will to live. He died in 1947, at the age of 89, in Göttingen.

Further Reading

Heilbron, J. L. *The Dilemmas of an Upright Man: Max Planck and the Fortunes of Germany.* Cambridge, Mass.: Harvard University Press, 2000.

Klein, M. J. "Max Planck and the Beginnings of Quantum Theory." *Archive for History of Exact Sciences* 1: 459–479, 1962.

Planck, Max. *A Survey of Physical Theory.* New York: Dover, 1994.

⊠ Poynting, John Henry
(1852–1914)
British
*Theoretician and Experimentalist
(Classical Electromagnetism,
Gravitation)*

John Henry Poynting is famous for his discovery of the Poynting vector, a key element in the electromagnetic conservation of energy equation. In this formulation, Poynting determined the rate of outward flow of electromagnetic energy from a volume containing charge and currents. He also produced a measurement of SIR ISAAC NEWTON's gravitational constant, which remains accurate today.

He was born in Monton, Lancashire, near Manchester, on September 9, 1852, and attended the secondary school where his father taught. In 1867, he entered Owens College in Manchester (later Manchester University), where he earned a bachelor of science degree in 1872. He did his graduate work at Trinity College, Cambridge University, from 1872 to 1876 and, on completing his studies, served as a physics demonstrator at Owens College.

Then, in 1878, Poynting had the opportunity to work as a researcher at the Cavendish Laboratories in Cambridge under the great

JAMES CLERK MAXWELL, who had produced a comprehensive theory of electromagnetism. He worked with Maxwell until 1880, when he became professor of physics at Mason College, Birmingham (later Birmingham University). At Birmingham, in 1884, he made his own significant contribution to electromagnetism, when he published his paper "On the Transfer of Energy in the Electromagnetic Field." It contained an equation (which he worked out by using Maxwell's electromagnetic field theory), by which the magnitude and direction of the flow of electromagnetic energy—later called the Poynting vector—can be determined. Mathematically, the Poynting vector is given by

$$S = (1/\mu)EB \sin \theta$$

where S is the Poynting vector, μ is the permeability of the medium, E is the strength of the electric field vector, B is the strength of the magnetic field vector, and θ is the angle between the electric and magnetic field vectors. Interestingly, Poynting's work did not assume the existence of an ether, the all-pervasive cosmic "wind" in which electromagnetic waves were thought to propagate by most physicists of the time; instead he used the concept of flux of electricity between lines of force, suggested by MICHAEL FARADAY.

Poynting made his second important contribution to physics in 1891, when he published the results of an extensive series of experiments, begun during his Cambridge days, designed to measure Newton's gravitational constant (on the basis of which the Earth's mean density can be calculated). Poynting's results, which he obtained by using an ordinary beam balance, are almost identical with the numbers accepted today. In 1895, Charles Boys improved upon Poynting's experimental method by using a quartz fiber torsion balance. Poynting used Boys's apparatus for several later studies of his own, such as the measurement of radiation pressure.

In 1903, he did important work on radiation in which he hypothesized that the Sun's radia-tion causes small particles orbiting the Sun to move increasingly closer to the Sun, until they eventually fall into it. This discovery, which is now known as the Poynting–Robertson effect, was later developed and related to the theory of relativity by the American physicist Howard Robertson. Poynting also devised a method for measuring the radiation pressure from a body, which can be used to determine the absolute temperature of celestial objects.

Poynting remained at the University of Birmingham until his death on March 30, 1914, in Birmingham, of diabetes-related causes.

Poynting's work, by determining the manner in which conservation of electromagnetic energy occurred in Maxwell's equations, was a vital stepping stone in the confirmation and verification of Maxwell's electromagnetic theory.

Further Reading

Schultz Slusher, Harold, and Stephen J. Robertson. *The Age of the Solar System: A Study of the Poynting–Robertson Effect and Extinction of Interplanetary Dust.* El Cajon, Calif.: Institute for Creative Research, 1982.

⊠ **Purcell, Edward Mills**
(1912–1997)
American
Nuclear Physicist, Solid State Physicist, Astrophysicist, Biophysicist

Edward Mills Purcell's 1945 discovery, with his colleagues Henry C. Torrey and Robert V. Pound, of nuclear magnetic resonance (NMR) led to his sharing of the 1952 Nobel Prize in physics with FELIX BLOCH, who had made the same discovery by using a different approach. His other famous achievement was detection of the emission of 21-cm-wavelength radiation from atomic hydrogen in interstellar space, which became a fundamental measuring tool in astrophysics.

He was born on August 30, 1912 in Taylorville, Illinois, to Mary Elizabeth Mills, a high school Latin teacher, and Edward A. Purcell, the manager of a local telephone company. When he was 14, the family moved to nearby Matoon, Illinois, where his father became general manager of the Illinois Southeastern Telephone Company. His father's involvement in the telephone industry played an important role in his life, providing him with discarded equipment to investigate and exposing him to the scientific articles in the *Bell System Technical Journal*. He would later describe these as "a glimpse into some kind of wonderful world where electricity and mathematics and engineering and nice diagrams all came together."

Purcell entered Purdue University in 1929, intending to become an electrical engineer. But when Karl Lark-Horovitz became head of the physics department, the study of physics at Purdue flourished and Purcell signed up for a course of independent study. His professor set him to work rebuilding a spectrometer and an electrometer to measure nuclear half-lives, an experience that heightened his growing fascination with physics. It solidified when Lark-Horovitz let him take part in experimental research on electron diffraction. After his 1933 graduation with a bachelor's degree in electrical engineering, Purcell won an exchange fellowship to the Technische Hochschule in Karlesruhe, Germany. On the ocean voyage to Germany, he met Beth C. Busser, an exchange student from Bryn Mawr College on her way to Munich to study German literature. They would marry in 1937 and have two sons, Dennis and Frank.

In 1934, Purcell returned to the United States and entered Harvard University, which became his lifelong academic home, for his graduate studies. It was there that JOHN HOUSBROOK VAN VLECK's course on electric and magnetic susceptibilities influenced the future course of his career. He received his Ph.D. in 1938 for an experimental thesis under Kenneth T. Bainbridge on the focusing properties of the electric field in the space between two concentric metal spheres forming a spherical condenser (what is now called a capacitor).

Purcell was a physics instructor at Harvard for two years, before going to work at the Radiation Laboratory at the Massachusetts Institute of Technology (MIT), which had just been organized for the purpose of radar development. During World War II, between 1941 and 1946, he was the leader of the advanced developments group, responsible for moving the radar systems to shorter wavelengths. He was among the few scientists who stayed on after the MIT facility closed in order to publish a series of books preserving the technology developed over the five years of its existence. He gained invaluable experience at MIT, where he knew scientists such as ISIDOR ISAAC RABI who were interested in the study of molecular and nuclear properties by radio methods.

By then Rabi had theoretically determined that magnetic resonance, associated with groups of atoms shot through a region of strong magnetic fields as a beam, was a measurable phenomenon. But instruments that were capable of measuring magnetic resonance in liquids or solids had not yet been designed. Purcell, armed with the latest information on energy states in nuclear particles and on microwave energy, surmised that with a strong magnetic field he could move the spinning nuclear particles of a specimen into alignment, then use microwaves to find their resonant frequency and magnetism. A three-scientist team, consisting of Purcell; Torrey, who had worked in Rabi's lab; and Pound, a colleague from the MIT lab, improvised an experiment using inactive Radiation Lab equipment, which they moved to the Research Laboratory of Physics at Harvard. Working mostly evenings and weekends, they succeeded in detecting the absorption of radio frequency by protons in paraffin wax by magnetic resonance on December 15, 1945. The great advantage of

Purcell's new experimental approach was that it could be applied to solid, liquid, and gaseous substances. Its extraordinary sensitivity also made it ideal as a micromethod in many scientific and technical fields. Seven years later, when he was awarded the Nobel Prize for this work, Purcell commented that the reward of the discoverer is "to see the world as something rich and strange."

In 1946, he returned to Harvard as a tenured associate professor and began developing the new field of magnetic resonance of nuclei. One of his first graduate students was NICOLAAS BLOEMBERGEN, who helped him in developing more sensitive instrumentation for future nuclear magnetic resonance (NMR) studies. Bloembergen, Purcell, and Pound used the new instruments for a series of important experiments, which explained the thermal relaxation of the nuclear spins and the collision narrowing of NMR lines in liquids, gases, and certain solids. Purcell became full professor at Harvard in 1949.

In the late 1940s, Purcell encouraged a graduate student, Harold I. Ewen, to look for an astronomical spectral line based on the hyperfine splitting of the ground state of the interstellar atomic hydrogen in the galaxy. Then, in 1951, Dutch astronomers predicted that in the near-vacuum of space, lonely hydrogen atoms, after being moved to a slightly higher energy level by radiation or collision, would emit a radio wave of 21-cm wavelength as they dropped back to the lowest energy level. Ewen and Purcell, working at night and with borrowed equipment, help from a university carpenter, and a $500 grant, built a horn antenna on the roof of Harvard's Lyman Laboratory. On March 24, 1951, they detected the elusive spectral line radio emissions from clouds of hydrogen in space for the first time. Their observations matched predictions that the emissions would have a wavelength of 21 cm. The same observation was then made by the Dutch scientist Jan Oort, who, by observing cloud motions, was able to map the otherwise invisible far side of the rotating Milky Way

galaxy. Since hydrogen is the most abundant substance in the universe, its 21-cm wavelength became a fundamental tool in astrophysics. As a prominent landmark in radio astronomy, it has been the focus of efforts to detect signals from extraterrestrial life.

Purcell continued working in the field of nuclear magnetism, especially relaxation phenomena, molecular structure, measurement of atomic constants, and nuclear magnetic behavior at low temperatures. During the 1960s, as a result of his contribution to radio astronomy, he became involved in astrophysics. The problem of understanding the mechanisms of the interactions of interstellar dust and light propagating through the galaxy was a central research focus in his later years. He also did important work on biophysics; with Howard Berg he developed a machine based on concentric rotating cylinders for separating molecules in liquid states according to their masses. He and Berg later shared the Biological Physics Prize of the American Physical Society in 1984 for their work on the hydrodynamics of *Escherichia coli* bacteria.

A devoted educator, who wrote a textbook on electricity and magnetism in the early 1960s, Purcell wrote a pedagogical column for the *American Journal of Physics* between 1983 and 1988. Entitled "Back of the Envelope," it presented thought problems, which he could solve quantitatively in a few lines.

He served on the President's Science Advisory Committee to Presidents Eisenhower, Kennedy, and Johnson. Under Eisenhower, he chaired the subcommittee on space, which was influential in the organization of the National Aeronautics and Space Administration (NASA) and the development of the space exploration program. He resigned from Johnson's committee in 1965 because of his stance against the Vietnam War and two years later served as spokesperson for a group of antiwar Harvard professors who met with Johnson to explain their position. He retired from Harvard in 1977. He died of respiratory failure on March 7, 1997, in Cambridge, Massachusetts.

Both of Purcell's major discoveries have had far-reaching consequences. NMR has become a major research tool in material sciences, chemistry, and medicine, in which magnetic resonance imaging (MRI) is a fundamental diagnostic tool. Radio spectroscopy of atoms and molecules in space, whose development resulted from the detection of the hyperfine transition in hydrogen, the first element studied, plays an increasingly greater role in radio astronomy.

Further Reading

Matson, James. *The Pioneers of NMR and Magnetic Resonance in Medicine: The Story of MRI*. Jericho, N.Y.: Dean, 1996.

Pound, Robert V. "Edward Mills Purcell," in *Biographical Memoirs*. Vol. 78. Washington, D.C.: National Academy Press, 2000, pp. 182–209.

R

Rabi, Isidor Isaac
(1898–1988)
American
Theoretician, Experimentalist, Solid State Physicist

Isidor Isaac Rabi was a major figure in 20th-century physics, who is best known for his invention of the technique of atomic and molecular beam magnetic resonance (ABMR). This technique is also the basis for nuclear magnetic resonance, a method for determining the magnetic moments of atoms and nuclei. He won the 1944 Nobel Prize in physics for this work. Intimately involved with the Manhattan Project, which developed the first atomic bomb, he later became a strong advocate for arms control.

He was born on July 29, 1898, in Galicia, Poland, the son of David Rabi and Janet Teig. The following year his family immigrated to the United States, where Isidor grew up on the Lower East Side of Manhattan. His father worked in the sweatshops of Manhattan, making women's blouses, until he had saved enough to move his family to Brooklyn, where he bought a grocery store. The Rabi family, which had fled the poverty and anti-Semitism of Eastern Europe, was Yiddish-speaking and practiced a fundamentalist form of Orthodox Judaism. As a child, Isidor did not know that the Earth revolved around the Sun until he read it in a library book. His secularizing exposure to modern scientific knowledge in the public schools of Manhattan and Brooklyn quickly transformed his sense of the world and led him to a life in science.

In 1919, he received his bachelor's degree in chemistry from Cornell University in Ithaca, New York. He worked for three years in jobs unrelated to science before beginning his graduate studies at Cornell in 1921, and later at Columbia University. In 1926, he married Helen Newmark, with whom he would have two daughters. The following year he was awarded his Ph.D. for a dissertation on the magnetic properties of crystals. He then spent two years in Europe, supported by fellowships, which gave him the opportunity to work with such giants of early-20th-century physics as ARNOLD JOHANNES WILHELM SOMMERFELD; NIELS HENRIK DAVID BOHR; OTTO STERN; WOLFGANG PAULI; and WERNER HEISENBERG. When he returned to the United States, Rabi became a lecturer in theoretical physics at Columbia, where he rose steadily through the ranks, becoming a full professor in 1937.

Rabi began his work on magnetic phenomena in atoms at the outset of his career. In 1930, he began investigating the magnetic properties of atomic nuclei, developing Stern's molecular beam method to a new level of precision as a

Isidor Rabi is best known for his invention of the atomic and molecular beam magnetic resonance method for determining the magnetic moments of atoms and nuclei. *(AIP Emilio Segrè Visual Archives)*

tool for measuring these properties. His apparatus was based on the production of ordinary electromagnetic oscillations of the same frequency as that of the physical precession, that is, rotation, of the magnetic moments of atomic systems in an external magnetic field. By an ingenious application of an electromagnetic resonance principle, he succeeded in detecting and measuring single states of rotation of atoms and molecules and in determining the magnetic moments of the nuclei. He had theoreti-

cally determined that magnetic resonance was a measurable phenomenon. But instruments capable of measuring magnetic resonance in liquids or solids had not yet been designed. When FELIX BLOCH and EDWARD MILLS PURCELL succeeded in doing so in the late 1940s, the field of nuclear magnetic resonance (NMR), which would give rise to magnetic resonance imaging (MRI), was launched.

In 1940, Rabi took a leave from Columbia to become associate director of the newly opened

Radiation Laboratory at the Massachusetts Institute of Technology, created to develop radar technology. When he was invited by J. ROBERT OPPENHEIMER, director of the Manhattan Project charged with development of the atomic bomb, to become associate director of the work at Los Alamos, Rabi refused. For one thing, he believed that radar was more immediately important to the progress of the war. More fundamentally, however, as he told Oppenheimer, he was unwilling to devote himself to working full-time "to make the culmination of three centuries of physics" a weapon of mass destruction. Eventually, however, he did participate in the project, as a visiting consultant. He was present at Trinity, the first test of the atomic bomb in the Nevada desert.

When, at the war's conclusion, he returned to Columbia as head of the physics department, he became involved with the Brookhaven National Laboratory for Atomic Research on Long Island, New York, which was dedicated to research into the peaceful uses of atomic energy.

In 1945, Rabi was the first to suggest the idea of using the hyperfine states of an atom as an atomic clock. The hyperfine state of an atom is associated with the quantum mechanical interaction of electrons and nuclei in atoms. The atomic clock uses oscillations occurring in the electromagnetic field generated by the changes that occur between the magnetic moments of two different hyperfine quantum mechanical states of atoms.

During this period, he made fundamental investigations of the hyperfine structure of the electron in an external magnetic field. The electron had long been known to have the property of acting as small magnet whose strength, as measured by its magnetic moment, was determined by PAUL ADRIEN MAURICE DIRAC's theory of the electron to be equal to a quantity known as the Bohr magneton. However, at the beginning of 1947, Rabi and his collaborators found that the hyperfine structure of the electron in external magnetic fields did not entirely conform with the predictions of Dirac's theory. The physicist Gregory Breit hypothesized that this might be due to the fact that the value of the magnetic moment of the electron was larger than the value of one Bohr magneton as predicted by the Dirac equation. POLYKARP KUSCH, working in Rabi's laboratory, then did the experiment that confirmed this hypothesis. This result spurred the reformulation of quantum electrodynamics (QED), the theory of the interaction of electrons and electromagnetic radiation.

Rabi worked toward control of nuclear weapons as a member of the General Advisory Committee to the Arms Control and Disarmament Agency, and of the United States National Commission for the United Nations Educational, Scientific, and Cultural Organization (UNESCO). In 1955, he served as the U.S. delegate and vice president of the International Conference on Peaceful Uses of Atomic Energy, in Geneva. He was also a member of the Science Advisory Committee of the International Atomic Energy Agency.

A man of principle, Rabi defended Oppenheimer, when he was persecuted as a Communist and deprived of his security clearance, in strong terms as

> a man who had done greatly for his country. A wonderful representative. . . . He was a man of peace and they destroyed him. He was a man of science and they destroyed this man. A small, mean group.

In 1970, reflecting on the development of nuclear weapons, Rabi wrote:

> The lesson we should learn from all this and the frightening thing that we did learn in the course of the war, was . . . how easy it is to kill people when you turn your mind to it. When you turn the resource of modern science to the problem of killing people, you realize how vulnerable they really are.

By the time of Rabi's death in 1988, it was clear that his experimental technique for determining the magnetic moments of atoms and nuclei, as well as his atomic clock, had lain the foundations of precision atomic measurements in modern physics.

Rabi had a deeply emotional commitment to science and thought that physics was "infinite." He expressed disappointment that young physicists, more interested in technique, seemed oblivious of "the mystery of it: how very different it is from what you can see, and how profound nature is."

See also RAMSEY, NORMAN F.

Further Reading

Bernstein, Jeremy. "Physicist." *The New Yorker* Part I: Oct 15, 1975; Part II: Oct 20, 1975.

Rabi, Isidor Isaac. *My Life and Times as a Physicist.* Claremont, Calif.: Clermont College, 1960.

———. *Science: The Center of Culture.* New York: World Publishing, 1970.

———. *Oppenheimer.* New York: Scribner's & Sons, 1969.

Rigden, John S. *Rabi: Scientist and Citizen.* New York: Basic Books, 1987.

⊠   **Raman, Sir Chandrasekhara Venkata**
(1888–1970)
Indian
*Experimentalist (Optics, Acoustics),
Spectroscopist*

Sir Chandrasekhara Venkata Raman, the father of Indian science, was a brilliant experimentalist, who made fundamental contributions to the understanding of the nature of light. He won the 1930 Nobel Prize in physics for his discovery of what came to be called the Raman effect: the scattering of monochromatic photons of light as they pass through a transparent medium in which the interaction of the photons with the molecules of the medium causes the scattered photons to have wavelengths different from that of the incident photons. Raman scattering spectra can be used to determine information on the structure of molecules.

Raman was born on November 7, 1888, in Trichinopoly, Madras, in southern India, into a highly educated family. His father was a professor of physics, and his nephew, SUBRAMANYAN CHANDRASEKHAR, was destined to become a Nobel Prize winner in astrophysics. An exceptional student from the start, Raman passed his matriculation exam at the age of twelve. In 1902, he entered Presidency College, Madras, where he earned his B.A. two years later, along with the gold medal in physics. By 1907, at the age of 19, he had completed the requirements for the M.A., with the highest distinction.

Continuing his scientific studies in those years would have entailed travel to England, a step he could not take because of poor health. Since India offered no opportunity for a scientific career, Raman took a job with the financial division of the civil service. For 10 years, he worked as an accountant in Calcutta; he managed to continue to pursue his research during this time, publishing an astonishing 30 papers. Using the laboratories of the Indian Association for the Cultivation of Science in Calcutta, he investigated diffraction, vibrations in sound, and the theory of musical instruments, which remained a lifelong interest. The attention he gained through this work generated the offer of a professorship in physics at the University of Calcutta.

In 1917, Raman began a new life in academia, remaining at Calcutta for the next 16 years, the period when he would do his most important work. In 1921, when he was returning by ship from a conference in England, the intense blue color of the sea inspired his work on diffraction. He was led to question the theory of LORD RAYLEIGH (JOHN WILLIAM STRUTT) that the sea's color was due to the scattering of light by particles suspended in the water. Back in Calcutta, he showed that it was the scattering of light by water molecules that caused the sea to be blue. In 1922, he published his paper "Molecular Diffraction of Light," in which he described his experiments on the diffusion of sunlight in its passage through water, transparent blocks of ice, and other materials.

In 1923, ARTHUR HOLLY COMPTON discovered the Compton effect, which X rays are scattered when they pass through matter and emerge with a longer wavelength. He explained this phenomenon by proposing that X-ray particles or photons had collided with electrons and lost some energy. In 1925, WERNER HEISENBERG predicted that the Compton effect should also be observed with visible light—a conclusion Raman had already reached two years earlier on the basis of actual observation. Raman used monochromatic light from a mercury arc and the spectroscope to study the nature of diffused radiation emerging from the material under examination. In 1928, after refining his experiments for the scattering of monochromatic light in dust-free air and pure liquids, Raman announced the existence of a Compton-like scattering phenomenon for visible light photons, which came to be known as Raman scattering. He explained the scattering effect as being due to the internal motion of the molecules encountered, which can either transfer energy to the light photons or absorb energy from them in the resulting collisions. The Raman scattering effect was then used in what is now known as Raman spectroscopy, which gives precise information on the motion and shape of molecules.

A shaping force in the development of physics in his native country, he established the *Indian Journal of Physics* in 1926. In 1928, he sponsored the founding of the Indian Academy of Sciences and became its first president. He also initiated the *Proceedings* of that academy, in which most of his own work was published. The British government knighted him in 1929, a year before he became the first Indian to win a Nobel Prize in physics in 1930. Although he never earned a doctorate, he was awarded several honorary doctorates and memberships in scientific societies.

In 1934, Raman became head of the physics department at the Indian Institute of Science. He remained there until 1948, serving a term as president from 1933 to 1937. He left in order to become the first director of the Raman Research Institute, built for him by the Indian government in Bangalore. One of his colleagues at the institute was HOMI JEHANGIR BHABHA, a leading particle physicist and influential figure in India's scientific development. During this vital period of his career, Raman made important contribu-

Chandrasekhara Venkata Raman discovered what came to be called the Raman effect, in which the interaction of the photons with a transparent medium causes the scattered photons to have wavelengths different from those of the incident photons. *(AIP Emilio Segrè Visual Archives)*

tions to the understanding of many different kinds of physical phenomena, including the effects of sound waves on the scattering of light; the vibration of atoms in crystals; the optics of gemstones, particularly diamonds, and of minerals; and the physiological mechanisms of human color vision. Raman remained head of the institute until his death in Bangalore, on November 21, 1970.

The leading Indian scientist of his generation, Raman was also a devoted educator who trained large numbers of his compatriots, thus significantly enhancing the standards and stature of Indian physics. His discovery of the Raman scattering effect afforded a method for the analysis of molecular structure, as well as demonstrating conclusively that visible light photons behave as particles, thereby offering confirmation of the quantum theory.

Further Reading

Venkataraman, G. *Journey into Light: Life and Science of C. V. Raman*. Bangalore, India: Indian Academy of Sciences, 1989.

⊠ **Ramsey, Norman F.**
(1915–   )
American
Experimentalist, Particle Physicist, Quantum Theorist

Norman F. Ramsey did pioneering work in atomic and molecular physics by extending and developing the atomic beam magnetic resonance method developed by ISIDOR ISAAC RABI. He won the 1989 Nobel Prize in physics for the invention of the separated oscillatory fields method and its use in the hydrogen maser, a method of storing and studying atoms, and in the cesium atomic clock, now used as the modern time standard.

Ramsey was born on August 27, 1915, in Washington, D.C., the descendant of German and Scottish immigrants. His mother had been a university mathematics instructor; his father was a West Point graduate and officer in the Army Ordnance Corps. As a military family, the Ramseys traveled a great deal, both in the United States and abroad. In the midst of these frequent moves, Norman managed to skip grades and graduated from high school at the age of 15. When the Ramseys moved once more, this time to New York City, Norman entered Columbia University in 1931, in the midst of the Great Depression, starting out as an engineering major. Soon, however, finding that he craved a deeper understanding of nature, he switched to mathematics, in which he excelled.

After graduation in 1935, he received a Kellett Fellowship from Columbia, which took him to Cambridge University, England, where he enrolled as a physics undergraduate. He thrived at Cambridge's Cavendish Laboratory, then enjoying a golden age under the directorship of ERNEST RUTHERFORD. Having become interested in molecular and atomic beams, after receiving a bachelor's degree—his second—from Cambridge, he returned to Columbia to work with Rabi, who was then developing his famous atomic beam magnetic resonance (ABMR) method, the basis of nuclear magnetic resonance (NMR). Thus the novice experimentalist, as Rabi's first graduate student, was able to participate in the new field of magnetic resonance and to share in the discovery of the electric quadrupole moment charge distribution on the deuteron (a nuclear particle made up of the bound state of a proton and a neutron). He was awarded a Ph.D. for this work and went on the Carnegie Institution in Washington, D.C., to study neutron–proton and proton–helium scattering.

Ramsey married Elinor Jameson of Brooklyn, New York, on June 3, 1940; the couple would have four daughters. They set out to begin life together at the University of Illinois. Shortly afterward, however, World War II intervened, and Ramsey became involved in the war effort, working at the Massachusetts Institute of Technology (MIT) Radiation Laboratory in Cambridge, Massachusetts, as head of a radar development group. After a period in Washington, D.C., where he served as radar consultant to the secretary of war, in 1943 the Ramseys moved to Los Alamos, where the Manhattan Project to build an atomic bomb was under way. Over the next two years, he did important work on bomb delivery systems.

With the war's conclusion Ramsey rejoined Rabi at Columbia and helped him in reviving the molecular beam laboratory. He and Rabi also collaborated in establishing the Brookhaven National Laboratory on Long Island, New York, where in 1946 Ramsey became the first head of the physics department. He remained in this position for only a year, however, before joining the physics department at Harvard University, where he would teach for the next 40 years, directing the research dissertations of no fewer than 86 graduate students. He would leave Harvard only briefly, for visiting professorships at Middlebury College, Oxford University, Mount Holyoke College, and the University of Virginia.

Soon after arriving at Harvard, Ramsey began working on improvements to Rabi's ABMR technique, a method of obtaining data on the quantum energy levels of the atom through measuring its optical spectra. In ABMR, when a beam of atoms passes through a homogeneous magnetic field with a superimposed oscillating electromagnetic field, the latter can induce the atoms to undergo specific quantum transitions if the frequency of the oscillating field is in resonance with the frequency of light that is emitted in this transition. The time the atoms spend in the oscillating field determines the width of the resonance line—the longer the time, the narrower the line—provided that the magnetic

Norman F. Ramsey did pioneering work in atomic and molecular physics by extending and developing the atomic beam magnetic resonance method developed by Isidor I. Rabi. *(AIP Emilio Segrè Visual Archives. Physics Today Collection)*

field is sufficiently homogeneous to allow this measurement to be made.

When he set up his molecular beam lab at Harvard, Ramsey had the goal of producing more accurate measurements by increasing the homogeneity of the magnetic fields. Inspired by his initial failures, he invented a new technique: the separated oscillatory field. Ramsey's modified ABMR method worked by introducing two separated magnets, each generating an oscillatory magnetic field. This caused a spatial interference pattern in the magnetic fields to appear between the two magnets; its sharpness depended only on the distance between the magnets, independent of the degree of homogeneity of the individual magnetic field generated by each magnet. The interference pattern of the magnetic fields thus created an artificial homogeneity, which greatly increased the accuracy of measurement. Ramsey and his students then used this method to measure in many molecules a number of molecular and nuclear properties, including nuclear spins, nuclear magnetic dipole and electric quadrupole moments, rotational magnetic moments of molecules, spin-rotational interactions, spin–spin interactions, and electron distribution in molecules.

During this period, Ramsey also worked with groups engaged in the application of the separated oscillatory field method to atomic clocks. This work produced the cesium clock, in which transitions between two very closely spaced hyperfine levels in the cesium atom are observed. The accuracy of such a clock, which is the present modern time standard, is today about one part in 10,000 billion. Since 1967, one second has been defined as the time during which the cesium atom makes exactly 9,192,631,770 oscillations.

Working with his student Daniel Kleppner, Ramsey also used his methods to invent the hydrogen maser, a method of storing and studying atoms. Atoms of hydrogen in an excited state are fed into a cavity, which can be put into a state of self-oscillation if properly tuned. The line width is determined by the average time the atoms spend in the cavity—about one second. The walls of the cavity are covered with Teflon to reduce the effect of wall collisions. The hydrogen maser was first used to study the hyperfine structure of hydrogen with extreme precision. Although it is inherently more stable than the cesium clock, its absolute accuracy is inferior. It is thus used mainly as a secondary standard and in measurement of frequency shifts when extreme precision is needed. A dramatic example of this was the ver-

ification of the gravitational red shift, the effect of gravitation on electromagnetic radiation predicted by ALBERT EINSTEIN's general theory of relativity. By comparison of the frequencies of a hydrogen maser on a rocket to those of one on the Earth's surface, the predictions of the general theory of relativity have been verified to one part in 10,000.

When Ramsey was at the Institut Laue-Langevin in Grenoble, France, he and his students accurately measured the magnetic moment of the neutron and set a low limit to the dipole moment of the neutron (a test of time-reversal symmetry), as well as discovering and measuring the parity nonconserving rotations of the spins of neutrons passing through various materials. During this same period, he was director of the Harvard Cyclotron and participated in proton–proton scattering experiments there. He was later chair of the joint Harvard–MIT committee managing construction of the six-giga-electron-volt Cambridge electron accelerator, which he used for various particle physics experiments. For what he has called "sixteen exciting years," he was on half-time leave from Harvard as president of Universities Research Association. Working with the laboratory directors, LEON M. LEDERMAN and Robert W. Wilson, he played a major role in the construction and operation of the 200-GeV accelerator at the Fermi Laboratory, Batavia, Illinois.

After the death of his wife, Elinor, in 1983, he married Ellie Welch of Brookline, Massachusetts, with whom he has enjoyed a shared family of seven children and six grandchildren. Although retired from Harvard since 1986, Ramsey has remained an engaged physicist, teaching at Williams College and at the Universities of Chicago and Michigan and conducting research at the Joint Institute for Laboratory Astrophysics at the University of Colorado, in Grenoble, and in his Harvard office. He has published over 300 research papers; his books include *Experimental Nuclear Physics*, with Emilio Segrè (1953); *Nuclear Moments* (1953); *Molecular Beams* (1956 and 1985); and *Quick Calculus* (1965 and 1985).

A brilliant experimentalist, Ramsey has also published fundamental theoretical physics papers on parity and time-reversal symmetry, NMR chemical shifts, nuclear interactions in molecules, and thermodynamics and statistical mechanics at negative absolute temperatures. His experimental methods have contributed to the knowledge of magnetic moments, the structural shape of nuclear particles, the nature of nuclear forces, and the thermodynamics of energized populations of atoms and molecules (e.g., those in masers and lasers). Applications of Ramsey's methods have been used in such fundamental measurements as testing the principles of quantum electrodynamics and the general theory of relativity.

See also BOHR, NIELS HENRIK DAVID.

Further Reading

Itano, Wayne M., and Norman F. Ramsey. "Accurate Measurement of Time," *Scientific American* 269: 56–65, 1993.

Ramsey, Norman F. *Spectroscopy with Coherent Radiation: Selected Papers of Norman F. Ramsey.* Singapore: World Scientific, 1997.

⊠ **Rayleigh, Lord** (John William Strutt)
(1842–1919)
British
*Theorist and Experimentalist
(Electrodynamics, Thermodynamics,
Statistical Mechanics)*

Lord Rayleigh was one of the last great classical physicists, who made important contributions in several fields, including electromagnetic theory, thermodynamics, and statistical mechanics. Although his work was influential in the development of both quantum theory and relativity, he remained steadfastly opposed to both theories. He is most renowned for discovering, together with William Ramsay, the inert gas argon, for which he was awarded the Nobel Prize in physics in 1904. Not least among his discoveries is the phenomenon of Rayleigh scattering of sunlight in the atmosphere, on the basis of which he explained why the sky is blue and the sunset red.

He was born John William Strutt on November 12, 1842, in Langford Grove, Essex, the first of seven children born to John James Strutt, the second Baron Rayleigh, and Clara Elizabeth La Touche. The pursuit of science was neither traditional nor highly valued in this family of hereditary aristocrats and landowners. A frail child, whose schooling at Eton and Harrow was interrupted by bouts of illness, John William showed no early signs of scientific aptitude. He was privately tutored until 1861 and seemed to have only average abilities. Not until he entered Trinity College, Cambridge University, to study mathematics did he excel. He studied under Edward Routh, an applied mathematician, who provided him with the mathematical skills he would later use to good purpose in his scientific career, particularly the ability to identify the appropriate methods for solving individual physical problems. Although he performed no experiments himself, watching his physics professor, GEORGE GABRIEL STOKES, do so opened up the world of the laboratory to him. His intellectual powers blossomed in this atmosphere, and he emerged from his fourth year exams as "Senior Wrangler," the top student. When he graduated in 1865, he was elected a fellow of Trinity College.

Whereas it was standard practice for young people of means to make a "grand tour" of Europe before settling down, Rayleigh, after graduating, headed west, to the United States, instead. When he returned to England in 1868, he set up his own laboratory at the family home at Terling Place, Witham, in Essex. In this way, he announced his decision to commit his life to science, despite the disapproval of his family and social class. Rayleigh was an economical man, and, since these were hard times for agriculture in England, his reduced income forced him to work with cheap, unsophisticated equipment in his home laboratory. Nonetheless, he was able to do his first important work on properties of waves in optics and acoustics, extending the research of HERMANN LUDWIG FERDINAND VON HELMHOLTZ on resonance and vibration. While investigating the nature of light and color in 1871, he discovered the physical explanation of why the sky is blue and the sunset red, in terms of the effect of the scattering of light by dust particles. Rayleigh's reasoning was based on the fact that small particles in the atmosphere interact with and scatter the sunlight incident on them. By analyzing this process using JAMES CLERK MAXWELL's equations for the electromagnetic field, he found that the intensity of scattered radiation varies inversely with the fourth power of the wavelength. This meant that the shorter blue wavelengths of sunlight would be scattered more efficiently into the atmosphere than the longer red wavelengths of sunlight: thus, the sky appears blue; the sunset, red.

During this period, he also showed that the resolving power of diffraction gratings is determined by the total number of lines in the grating multiplied by the order of the spectrum, and not by the closeness of the lines. This discovery led to improvements in the spectroscope, which in the 1870s was rapidly becoming an essential instrument for investigating solar and chemical spectra.

While he was at Cambridge, Raleigh's fellow student, the earl of Balfour, who would later become prime minister, had introduced him to his sister, Evelyn. In 1871, they married. Their long union would produce three sons, the eldest of whom would become a physicist. Shortly after the marriage, Rayleigh had rheumatic fever and, at his doctor's urging, went to Egypt with his new wife to recuperate. The trip nurtured both

his health and his scientific career: While sailing down the Nile, he began writing his classic book, *The Theory of Sound*, which he would complete five years later.

When his father died in 1873, he became the third baron of Rayleigh. The new title did nothing to dampen his passion for scientific research. That year he was elected a fellow of the Royal Society. In 1876, despite his progressive management of his estate, he handed those duties over to his younger brother, an arrangement that allowed him to devote all his time to science. In need of additional income, in 1879, he agreed to succeed Maxwell as Cavendish Professor of Experimental Physics at Cambridge, a position he held until 1884. There he improved the teaching of experimental physics, causing an explosion of interest in the subject, which led to the flowering of British physics. In 1884, continuing the work of Maxwell, he established the accurate standardization of the three basic electrical units, the ohm, the ampere, and the volt.

His income having improved, in 1884, he returned to his Terling laboratory, to the research life he loved most. His professorship at the Royal Institution in London, from 1887 to 1905, did not require him to leave home often. From then on he conducted most of his experimental research in his private laboratory. He wrote a paper in 1879 on traveling waves that in the modern era was developed into the theory of soliton waves (waves that can travel through a nonlinear medium without changing shape). In 1885, he published his famous paper on Rayleigh wave theory, elucidating his discovery that elastic waves can be guided by a surface. Whereas he thought it would have potential value only in seismology; it later led to the explosive growth of the field of electronic signal processing.

Rayleigh's most famous work involved a careful study of the measurement of the density of different gases. According to the accepted theory of his time, known as Prout's hypothesis, all elements have atomic weights that are multiples of hydrogen and have atomic weights in integer multiples of the weight of nitrogen. However, Rayleigh's results did not support this theory. He observed that when he measured the density of nitrogen in air, it was 0.5% greater than the density of nitrogen from any other source. After ruling out all possible explanations for the source of the impurity causing the increase, he found himself devoid of new ideas for solving the mystery. He published a short note in *Nature* in 1892, inviting suggestions. He and Ramsay studied the problem for the next three years and jointly announced the discovery of a new element, *argon*, from the Greek word for "inactive." Chemists had previously failed to detect it, because argon is chemically inert, that is, virtually devoid of the ability to make chemical combinations with other elements. In 1904, Rayleigh received the Nobel Prize in physics for this work, while Ramsay was awarded the Nobel Prize in chemistry.

In 1900, he published the Rayleigh–Jeans equation, which described the distribution of wavelengths in blackbody radiation, accounting for only the longer wavelengths. Later, WILHELM (CARL WERNER OTTO FRITZ FRANZ) WIEN produced an equation to describe the shorter-wavelength radiation. It was MAX ERNEST LUDWIG PLANCK who would account for the distribution of all wavelengths by postulating the existence of indivisible energy packets called quanta, thereby sparking the quantum revolution. Rayleigh resisted the quantum theory, because it overturned classical radiation theory; for similar reasons he could not accept NIELS HENRIK DAVID BOHR's explanation of the hydrogen spectrum, which furthered quantum theory. He also opposed ALBERT EINSTEIN's special theory of relativity, because it rejected the classical idea of the ether as the medium in which light waves moved. Ironically, Rayleigh's own unsuccessful attempt in 1901 to detect the ether inadvertently lent support to relativity.

Lord Rayleigh published 446 papers from 1869 to 1920, each a model of clarity and elegance, on a vast range of topics in physics and applied mathematics. His amazing scope is reflected in the sheer number of terms used in modern physics that bear his name: Rayleigh distribution (statistics), Rayleigh–Jeans law (blackbody radiation), Rayleigh scattering (coloration of the sky), Rayleigh damping (damped vibrational behavior), Rayleigh quotient (elastodynamics), Rayleigh–Ritz process (elastomechanics), Rayleigh waves (surface waves), Rayleigh fading and Rayleigh distance (propagation of electromagnetic waves), Rayleigh criterion (resolving power of telescopes), and Rayleigh number (natural convection).

Although he worked at home, he was far from isolated and had many connections with the scientific community. From 1885 to 1896, he served as secretary of the Royal Society; he was president from 1905 to 1908. Through his wife, he was connected with the political scene and held many advisory roles on national committees, including one on aeronautics. In 1908, he was appointed chancellor of Cambridge University, a post he retained for the rest of his life. Despite his renown, he maintained humility, as demonstrated by his comments in 1902, at the coronation of King Edward VII, when he received the Order of Merit:

> The only merit of which I personally am conscious was that of having pleased myself by my studies, and any results that may be due to my researches were owing to the fact that it has been a pleasure for me to become a physicist.

A generous man, he donated the cash award for the 1904 Nobel Prize to Cambridge University, to build an extension to the Cavendish Laboratories. He died at home on June 30, 1919, of a heart attack.

Opposed to relativity and quantum mechanics, Lord Rayleigh made great contributions to the consolidation and advancement of the branches of classical mechanics that remained valid. Further, his discovery of argon led to the uncovering of a whole family of elements (the noble gases) that are of great importance in themselves and to an understanding of chemical bonding. Among the applications of argon in modern industry are its uses in welding, as a shield of the molten metal; in nuclear reactors, as an inert cooling agent; in electric light bulbs; and in lasers, in the manufacture of semiconductor crystals, in which an inert atmosphere plays an important part.

See also FRAUNHOFER, JOSEPH VON; KIRCH-OFF, GUSTAV ROBERT; MICHELSON, ALBERT ABRAHAM.

Further Reading

Humphrey, A. T. "Lord Rayleigh—the Last of the Great Victorian Polymaths." *GEC Review* 7: 167–179, 1992.

Lindsay, R. B. *Lord Rayleigh, the Man and His Works*. London: Oxford University Press, 1970.

⊠   **Reines, Frederick**
(1918–1998)
American
Experimentalist, Particle Physicist

Frederick Reines shared the Nobel Prize in physics in 1995 for his historic 1956 experiment, with Clyde Cowan, in which he discovered the elusive neutrino, a particle first predicted by WOLFGANG PAULI in the mid-1930s. Throughout his career, he continued to explore the neutrino's properties and interactions, in the context of both elementary particle physics and astrophysical processes.

Reines was born on March 16, 1918, the youngest of the four children of Israel Reines and Gussie Cohen, who had immigrated to the United States (where they met) from the same small town in Russia. Reines's childhood centered around the general store his father ran in

the town of Hillburn, New York, where he developed his mechanical and musical interests and first became interested in science when he became aware of the phenomenon of light diffraction. Because older siblings were studying medicine and law, the household was filled with books. Although initially attracted to literature, he declared in the yearbook in his senior year in high school that his main ambition was "to be a physicist extraordinaire."

Reines attended the Stevens Institute of Technology in Hoboken, New Jersey, where he studied engineering, while finding time to develop his interests in drama, dance, and, most enduringly, choral music. For a while he considered a career as a professional singer, and, although he abandoned this idea, he would continue singing during his physics career, performing in Gilbert and Sullivan operettas while working at Los Alamos. While living in Cleveland, he performed in the chorus of the Cleveland Symphony Orchestra under George Szell—an experience that stood out for him as "the peak of my musical endeavors." In between receiving a bachelor's degree in engineering in 1939 and a master's degree in mathematical physics in 1941, both from the Stevens Institute, Reines married Sylvia Samuels. Theirs would be a lasting union, producing two children and six grandchildren.

Reines continued his graduate work at New York University, first concentrating on experimental cosmic ray physics. However, the thesis he submitted in 1944 for his Ph.D. was a theoretical study, directed by R. D. Present, "The Liquid Drop Model for Nuclear Fission." By this time, he was already working for RICHARD PHILLIPS FEYNMAN on the Manhattan Project to develop an atomic bomb in Los Alamos, where he would remain for the next 15 years. He soon became a group leader in the theoretical division and later was made director of Operation Greenhouse, a series of Atomic Energy Commission experiments at Eniwetok atoll. He analyzed the results of bomb tests at Eniwetok, Bikini, and the Nevada testing ground and coauthored a study with John von Neumann on the effects of nuclear blasts.

In the mid-1930s, physicists studying the nuclear radioactive beta decay process (e.g., the radioactive decay of a neutron into a proton) became aware that the nucleus appears to emit only an electron when it disintegrates by radioactive beta decay, thereby apparently violating conservation of energy and momentum. This problem led Pauli to postulate the existence of a new elementary particle. In what he termed his "desperate solution," he hypothesized that another subatomic particle, which lacked electric charge and hence reacted very weakly with its environment, was also emitted in the nuclear beta decay process, taking part of the energy and momentum with it and then disappearing again into nothingness. The name neutrino ("little neutral one") was bestowed upon Pauli's new particle by ENRICO FERMI, thereby distinguishing it from the neutron. Fermi developed a successful theory of weak nuclear processes, in which the neutrino played a central role. But the neutrino's extremely weak coupling to matter posed a formidable challenge to detecting it. Most physicists preferred not to try.

Reines, however, had been considering this challenge for several years, and, during a sabbatical from his Los Alamos work in 1951, he made the decision to take it on. A Los Alamos colleague, Clyde Cowan, joined him in this work. They considered different sources of neutrinos—a nuclear bomb test, the nuclear reactor at Hanford, Washington—but finally settled on the Savannah River reactor facility in South Carolina, which was completed in 1955. The following year they succeeded in detecting the electron neutrino through inverse beta decay in which a neutrino hitting a proton makes it into a neutron. The neutrino appeared to have the properties of an uncharged electron; however, at that time the question of whether the neutrino had a nonzero mass was unknown. (Current experiments now seem to imply that the electron neu-

trino has a very small but finite mass.) When Cowan left Los Alamos, their collaboration ended, but Reines carried forward their groundbreaking experimental work. After Cowan died, Reines received the Nobel Prize for this work, which was described as "a feat considered to border on the impossible. They had raised the neutrino from its status as a figure of the imagination to an existence as a free particle."

After this, Reines worked on gamma ray astronomy, as well as beginning the first of a series of experiments at Savannah River to study the properties of the neutrino. This work ushered in the era of "neutrino physics" when neutrinos, whether accelerator-produced or reactor and cosmic ray neutrinos, could be used to investigate the weak interactions, the structure of

Frederick Reines discovered the elusive neutrino, a neutral spinning particle with zero mass first predicted by Wolfgang Pauli in the mid-1930s. *(AIP Emilio Segrè Visual Archives and Los Alamos Scientific Laboratory)*

protons and neutrons, and the properties of their internal constituents, quarks.

Reines ended his long period at Los Alamos by serving, in 1958, as a delegate to the Atoms for Peace conference in Geneva. The following year he joined the Case Institute of Technology, Cleveland, where he was involved in neutrino experiments at reactors and in pioneering studies deep underground to search for atmospheric and cosmic particles. His group worked in reactor neutrino physics, double beta decay, electron lifetime studies, and searches for nucleon decay. They also conducted bold experiments in a gold mine in South Africa that made the first observation of the neutrinos produced in the atmosphere by cosmic rays. The experiments were designed to explore the properties of the neutrino and to probe the limits of fundamental symmetry principles and conservation laws, such as the conservation of charge, baryon number, and lepton number. In carrying out these experiments, the group necessarily became expert in the operation of deep underground laboratories, where, according to Reines, "the projects also drew us into developing innovative detector techniques, including the use of large liquid scintillator and water Cherenkov detectors."

In 1966, he took his group with him to the University of California, Irvine, where he became the founding dean of the School of Physical Sciences, a position he held until 1974, when he returned to full-time teaching and research. He became distinguished professor of physics in 1987 and professor emeritus in 1988. The "neutrino group" still plays a leading role in major neutrino experiments, including the famous "IMB" (Irvine/Michigan/Brookhaven) underground detector and its proton decay experiment. The IMB detector was the first to observe double beta decay in the laboratory. It was used to study neutrino physics, primarily employing neutrinos produced by interactions of cosmic rays in the atmosphere. The detector's impressive size and neutrino detection capability led to the historic detection of the burst from the supernova SN1987A—providing conclusive proof that neutrinos play a role in stellar collapse. The IMB group shared the Bruno Rossi Prize of the American Astronomical Society (1989) with the Kamiokande experiment in Japan for their joint observation of this phenomenon, which led to the birth of neutrino astronomy.

This work led Reines to explore many other intriguing experimental ideas and research areas, including the search for relic neutrinos from the big bang; the "neutrino Mössbauer effect," in which a photon is replaced by a neutrino; a neutrino technique to improve the accuracy of the measurement of the gravitational constant G, the most poorly measured fundamental constant, by several orders of magnitude; a spherical neutrino lens space telescope; an attempt to set more stringent limits on violation of the Pauli exclusion principle; exploration of the brain using ultrasound; and a variety of new detector ideas.

When Reines died in August 1998, the neutrino was far less elusive than it had been at the start of his career. He left a rich legacy of contributions, including the first detection of neutrinos produced in the atmosphere; the first study of muons induced by neutrino interactions underground; the first observation of the scattering of electron antineutrinos with electrons; the detection of the weak neutral current interactions of electron antineutrinos with deuterons; investigations looking for neutrino oscillations (the possibility of neutrino transformation from one type to another); and the first detection of a neutrino from a supernova. Current studies have shown that neutrino oscillations actually do occur; that means that the neutrino must have nonzero mass. This discovery will have profound implications for cosmological models in terms of the estimated missing mass of the universe.

See also CHERENKOV, PAVEL ALEKSEYEVICH.

Further Reading

Franklin, Allan. *Are There Really Neutrinos? An Evidential History.* Cambridge, Mass.: Perseus, 2000.

Kropp, W., J. Schultz, M. Moe, L. Price, and H. Sobel, eds. *Neutrinos and Other Matters: Selected Works of Frederick Reines.* Singapore: World Scientific, 1996.

Solomey, Nickolas. *The Elusive Neutrino: A Subatomic Detective Story.* New York: W. H. Freeman, 1997.

Sutton, Christine. *Spaceship Neutrino.* Cambridge: Cambridge University Press, 1992.

⊠ Richter, Burton
(1931–)
American
Experimentalist, Particle Physicist

Burton Richter is an outstanding experimental particle physicist, whose career has centered on high-energy electrons and electron–positron colliding beams. He shared the 1976 Nobel Prize in physics with SAMUEL CHAO CHUNG TING for his discovery of what is now called the J/ψ particle, an excited state of nuclear matter exhibiting the physical attributes associated with the existence of charmed quarks, which confirmed the standard quark–lepton model of electroweak and nuclear forces.

He was born on March 22, 1931, in Brooklyn, New York City, the elder child of Abraham and Fanny Richter. He was a member of that extraordinary generation of physicists, children of European Jewish immigrants, who grew up in New York in the years of the Great Depression and World War II, who included MURRAY GELL-MANN, SHELDON LEE GLASHOW, LEON M. LEDERMAN, and STEVEN WEINBERG. After making an excellent record for himself in the New York public schools, he was accepted in 1948 at the highly competitive Massachusetts Institute of Technology (MIT) in Cambridge. Undecided at first whether to major in physics or chemistry, he chose physics as a result of the strong influence

of Francis Friedman, one of his professors, who revealed to him the inherent beauty of the subject. He was introduced to the electron–positron system, which would be important in his future career, during the summer following his junior year, when he worked with Francis Bitter in MIT's magnet laboratory. There he assisted Martin Deutsch in his classical positronium experiments, using a large magnet. Later, Bitter agreed to direct Richter's senior thesis on the quadratic Zeeman effect in hydrogen.

For his graduate studies, begun in 1952, Richter remained at MIT and continued working with Bitter and his group. Initially, he worked with the group on a measurement of the isotope shift and hyperfine structure of mercury isotopes. Soon, however, he became more interested in the nuclear and particle physics problems he had studied as an undergraduate. After six months at the Brookhaven National Laboratory on Long Island, New York, working at the three-giga-electron-volt proton accelerator, he knew that particle physics research was what he wanted to do. When he returned to MIT, it was to the synchrotron laboratory. He would later write, "This small machine was a magnificent training ground for students, for not only did we have to design and build the apparatus required for our experiments, but we also had to help maintain and operate the accelerator." Working under L. S. Osborne, he completed a doctoral thesis on the photoproduction of pi mesons from hydrogen in 1956.

From 1956 to 1960, he was a research associate at the High-Energy Physics Laboratory (HEPL) at Stanford University. He was drawn to HEPL's 700-MeV electron linear accelerator, which would allow him to pursue his interest in experimentally testing quantum electrodynamics (QED, the quantum field theory describing the interaction of electrons and electromagnetic radiation) and, specifically, to investigate the short-distance behavior of the electromagnetic interaction. His very first experimental study at

HEPL, which looked at electron–positron pairs by using gamma rays, established that QED was correct to distances as small as about 10^{-13} cm.

In 1960, Richter became an assistant professor in the physics department at Stanford and married Laurose Becker, with whom he would have two children, Elizabeth and Matthew. For the next few years, Richter worked with G. K. O'Neill of Princeton, W. C. Barber, and B. Gittelman on the construction of the first colliding beam device, which became the prototype for all future colliding beam storage rings. Using the HEPL linear accelerator as an injector, their device allowed them to study electron–electron scattering at a center-of-mass energy 10 times larger than that used in Richter's earlier pair experiment. In 1965, they conducted an experiment resulting in extension of the validity of QED down to less than 10^{-14} cm.

Richter's next goal was to create a high-energy electron–positron colliding beam machine that would allow him to study the structure of the hadronic (strongly interacting) particles associated with the excited states of protons, neutrons, and mesons. At the Stanford Linear Accelerating Center (SLAC), which he had joined in 1963, he and his group designed the machine and embarked on the long struggle for funding. When it finally came through in 1970, they built the Stanford Positron Electron Accelerator Ring (SPEAR), including the storage ring and a large magnetic detector. Experiments began in 1973.

In November 1974, Richter and his team detected a new kind of massive hadronic particle state that had the physical properties of a meson. They gave this excited hadronic state the name ψ. The particle was more than twice as heavy as any comparable hadronic particle and yet a thousand times more narrow in its energy spectrum (which, according to the uncertainty principle, meant that it was a long-lived particle). They published their results in a 35-author paper (characteristic for high-energy experimental teams) in *Physical Review Letters*.

Meanwhile, in August 1974, at the Brookhaven Laboratory, Samuel Ting and his team had made a startling discovery: the first of a totally unpredicted new group of extremely heavy long-lived mesons. In November, after rechecking his results, Ting announced his discovery, which he called the J particle (based on the symbol for the electromagnetic current). Just before publishing his findings, Ting attended a conference at Stanford University with scientists working at SLAC, where he learned of Richter's almost simultaneous discovery. It was at this conference that the new particle state was named the J/ψ particle.

The significance of Richter and Ting's discovery was considerable. In 1961, Murray Gell-Mann and YUVAL NE'EMAN had devised the eightfold way, a system for classifying the myriad newly discovered elementary particles into eight families of elementary building blocks called quarks, of which, at that time, there were three types. The unique feature of the newly discovered J/ψ particle was that it did not belong to any of the families, as they were known before 1974. Its detection confirmed Sheldon Glashow's earlier prediction of a fourth quark (i.e., a "charmed quark"), which was needed in order to make the Gell-Mann SU(3) quark theory of the strong interactions consistent with this observation. Ting and Richter's discovery of the J/ψ particle changed the landscape of particle physics. Two years later, they shared the Nobel Prize for their work.

Later in the 1970s, working with physicists at CERN; Novosibirsk, Russia; and Cornell University, Richter developed the ideas that led to the creation of the SLAC Linear Collider, which he describes as "a kind of hybrid machine, with both electrons and positrons accelerated in the same linear accelerator, and with an array of magnets at the end to separate the two beams and then bring them back into head-on collisions." Completed in 1987, it began to be used for experiments in 1990. Richter predicted,

Probably the most lasting contribution that this facility makes to particle physics will be the work of accelerator physics and beam dynamics that has been done with the machine and which forms the basis of very active R&D programs aimed at the TeV [trillion electron volt] scale linear colliders for the future.

While bringing this major project to fruition, Richter became a scientific administrator, serving as technical director of SLAC from 1982 to 1984, and then as director from 1984 until his retirement in 1999.

Richter, who has over 300 publications in high-energy physics, accelerators, and colliding beam systems, is a fellow of the American Physical Society and served as its president in 1994. In reviewing his life in physics, Richter has written:

> [I] realize what a long love affair I have had with the electron. Like most love affairs, it has had its ups and its downs, but for me the joys have far outweighed the frustrations.

By experimentally verifying the need for a charmed quark, Burton Richter's groundbreaking experiment opened the door to the further elaboration of the fully developed Standard Model of electromagnetic, weak, and nuclear forces, which currently contains six quarks: up, down, strange, charm, top, and bottom.

See also SALAM, ABDUS; SCHWINGER, JULIAN SEYMOUR; FEYNMAN, RICHARD PHILLIPS; DYSON, FREEMAN.

Further Reading

Kane, Gordon. *The Particle Garden: Our Universe as Understood by Particle Physicists*. Cambridge, Mass.: Perseus, 1995.

Ne'eman, Yuval, and Murray Gell-Mann. *The Eightfold Way*. Cambridge, Mass.: Perseus, 2000.

Ne'eman, Yuval, and Yoram Kirsh. *The Particle Hunters*. Cambridge: Cambridge University Press, 1996.

⊗ **Röntgen, Wilhelm Conrad**
(1845–1923)
German
Experimental Physicist
(Electrodynamics, Optics)

Wilhelm Conrad Röntgen discovered X rays in 1895, initiating a new era in physics and medicine. For this momentous achievement, he became the first recipient of the Nobel Prize in physics in 1901.

Röntgen was born in Lennep, in the Lower Rhine region of Germany, on March 27, 1845, the only child of a cloth manufacturer. Charlotte Constanze Frowein, his mother, was from Amsterdam, and, when Wilhelm was three, the family moved to Apeldoorn, the Netherlands, where he received his early education at a boarding school. As a boy, he loved to explore the countryside and showed an early aptitude for making mechanical devices. His boyhood was marred by the injustice of being expelled from the Utrecht Technical School for drawing a caricature of one of the teachers; the prank, in fact, had been committed by another student. He overcame this setback, however, and gained admission to the Federal Polytechnic Institute in Zurich, Switzerland, in 1866, to study mechanical engineering; there he received his diploma in 1868.

In 1869, Röntgen received a Ph.D. from the University of Zurich. While in Zurich, he met Anna Bertha Ludwig, the daughter of a café owner, and married her in 1872. They would not have children of their own; in 1887, they adopted the six-year-old daughter of Anna Röntgen's only brother. In his immediate postdoctoral years, while serving as assistant to the German experimental physicist August Kundt, Röntgen decided to pursue pure science, specializing in physics. In order to be near Kundt, he held a series of positions, in Würzburg and Strasbourg; in 1875, he became professor of physics and mathematics at the Agricultural Academy of Hohenheim. He returned to Strasbourg the following year to teach physics. From 1879 to 1888, he was professor of physics at Giessen. Then, in 1888, he became professor of physics at Würzburg, where his colleagues included HERMANN LUDWIG FERDINAND VON HELMHOLTZ and HENDRIK ANTOON LORENTZ.

During this period, Röntgen carried out significant research on an impressive number of questions. He published his first paper, dealing with the specific heats of gases, in 1870. Between 1876 and 1895, he made important studies of the characteristics of gases and crystalline substances. These included thermal conductivity of gases, electrical and other characteristics of quartz, the influence of pressure on the refractive indices of various fluids, modification of the planes of polarized light by electromagnetic influences, variations in the behavior of the temperature and compressibility of water and other fluids, and the phenomena accompanying the spreading of oil drops on water.

However, this large body of work would be overshadowed by the great discovery Röntgen made in November 1895. He was studying the properties of cathode rays emitted by a partially evacuated glass Crookes tube (a glass vacuum tube with electrodes at each end). While experimenting with a barium compound, he was surprised to observe that it glowed even when the tube was encased in black cardboard. He took the chemical into another room; it continued to glow whenever the tube was activated. Röntgen concluded that these rays, with such high penetrating power, were entirely different from cathode rays. Later, when asked what his thoughts had been at the moment of his discovery, he said: "I didn't think. I investigated."

His investigations revealed that this new kind of cathode ray passed unchanged through cardboard and thin plates of metal, traveled in straight lines, and was not deflected by electric or magnetic fields. He called them X rays, since he could not explain their nature. Later MAX THEODOR FELIX VON LAUE and his students would show that X rays possess the same electromagnetic nature as light but have a higher frequency of vibration. Using X rays, Röntgen took his first "X-ray photograph," of his wife's hand; it showed the shadows thrown by the bones of her hand and of the ring she was wearing, surrounded by the penumbra of the flesh, which threw a fainter shadow since it was more permeable to the rays.

In January 1896, after two months of thorough investigation the properties of X rays, he revealed his discovery to the general public. Almost immediately Röntgen's announcement inspired further breakthroughs in both pure and applied science. Before long, development of X-ray equipment and photography for use in medical work was under way in Europe and the United States. Röntgen became the first recipient of the Nobel Prize in physics in 1901. He was showered with prizes, medals, honorary doctorates, and honorary and corresponding memberships to learned societies in Germany and abroad; several cities and streets were named after him.

In 1900, Röntgen became professor of physics and director of the Physical Institute at Munich, where he remained until he retired in 1920. He was a man of great integrity, uninterested in amassing personal wealth or honors. He refused the title *von Röntgen*, which would have made him a member of the German nobility, and donated the money for his Nobel Prize to the University of Würtzburg. He did, however, accept the honorary degree of doctor of medicine offered to him at Würzburg. Ironically, he never took out a patent on X rays and refused to benefit financially from the fruits of his discovery, since he believed that the products of scientific research should be available to everyone. He was almost bankrupt at the end of his life, during a period of runaway inflation in Germany that followed World War I. He died in Munich on February 10, 1923, of intestinal cancer.

Röntgen's discovery of X rays had an enormous impact on the development of 20th-century physics. It led ANTOINE-HENRI BECQUEREL to discover radioactivity that same year, that discovery in turn caused a revolution in ideas about the atom. It would later lead to the discovery of X-ray diffraction, which yielded methods of studying atomic and molecular structure. In contemporary physics, X rays continue to find new uses in a growing number of fields, including condensed matter physics, molecular biophysics, astrophysics, nuclear physics, relativity, plasma physics, and cosmology.

See also RUTHERFORD, ERNEST; THOMSON, JOSEPH JOHN (J. J.).

Further Reading

Glasser, Otto. Wilhelm Conrad Roentgen and the Early History of the Roentgen Rays. Novato, Calif.: Jeremy Norman, 1992.

Physics Today, November 1995 issue commemorating the centennial of Röntgen's discovery of X rays.

⊠   **Rubbia, Carlo**
(1934–   )
Italian
Experimentalist, Particle Physicist

Carlo Rubbia designed and headed the synchrotron experiments at the Center for European Nuclear Research (CERN) that led to the discovery of the intermediate vector boson particles W and Z. At the deepest level of matter these extremely heavy particles act as carriers of the weak interaction between quarks and leptons, thereby generating radioactive decay. Their existence was theoretically predicted by the electroweak quantum field theory developed by SHELDON LEE GLASHOW, ABDUS SALAM, and STEVEN WEINBERG, which unified the electromagnetic force associated with quan-

tum electrodynamics (QED) and the weak force associated with radioactive decay into a single theory. Rubbia shared the 1984 Nobel Prize in physics with his collaborator Simon Van der Meer for this work.

Rubbia was born on March 31, 1934, in the small town of Gorizia, near Trieste, in Italy. His father worked as an electrical engineer for the local telephone company, and his mother was an elementary school teacher. When, at the end of World War II, the province of Gorizia was taken over by Yugoslavia, the Rubbia family fled, first to Venice and then to Udine. As a boy, Rubbia early evinced the qualities that would lead him to experimental physics: he read everything he could find on the electrical and mechanical ideas that so fascinated him and was drawn to "the hardware and construction aspects" rather than to theoretical concerns.

His formal education had been badly disrupted by the war, however, and he failed his entrance examinations to the Scuola Normale in Pisa, where he had hoped to study physics. He had resigned himself to studying engineering in Milan when the withdrawal of one of the winning contestants allowed him to enter the Scuola Normale after all. Overjoyed by this happy accident, he moved to Pisa, where he struggled to make up for the shortcomings in his preparation. Under his thesis adviser Marcello Conversi, he participated in the construction of new instruments such as the first pulsed gas particle detector and earned his undergraduate degree with a thesis on cosmic ray experiments.

Eager to learn about giant particle accelerators, Rubbia, in 1958, went to the United States; there he worked at the Nevis Cyclotron Laboratory at Columbia University. With W. Baker, he performed the first of a series of experiments on weak interactions, which would be his primary focus. The two men measured the angular symmetry associated with the capture of polarized muons, thereby showing that parity violation occurs in this basic process.

Rubbia was attracted back to Europe two years later by the establishment of CERN in Geneva, Switzerland. Using a cyclotron superior to that at the Nevis Laboratory, Rubbia and his colleagues carried out a number of important experiments on the structure of weak interactions, notably including the discovery of the radioactive beta decay process of the positive pion and the first observation of the muon capture by free hydrogen.

In the early 1960s, Rubbia began working on the newly constructed proton synchrotron at CERN, where he determined the parity violation in the beta decay of the lambda hyperon (an excited state of the nucleus). After VAL LOGSDON FITCH announced the discovery of charge-parity symmetry violation (known as CP or T violation) in some particle processes, Rubbia began a long series of CP symmetry—related observations associated with the K^0 particle decay and on the K_L–K_S particle mass difference.

A few years later, in 1973, together with David Cline and Alfred Mann, Rubbia proposed a major neutrino experiment at the Fermi Laboratory in Batavia, Illinois. After more than a year of hard work they were able to observe cleanly the presence of the all-muon events in neutrino interactions needed to confirm experimentally new theoretical predictions made at CERN about the existence of a "charmed quark" and the ψ particle resonance.

By 1976, it was already clear that the unified electroweak theory of the symmetry type SU(2)×(1) had a good chance of predicting the existence and masses of the triplet of extremely massive intermediate vector bosons W and the neutral Z particle singlet that were required to carry the electroweak interaction. The problem was finding a practical way to discover them. To achieve high enough energies to create these bosons (roughly 100 times heavier than protons), experimentalists needed a radically new approach.

At CERN, under the leadership of Victor Weisskopf, a new type of colliding beam machine had been built with intersecting storage rings in which counterrotating beams of protons collide with one another. Rubbia and his collaborators transformed this high-energy accelerator into a colliding beam device in which a beam of protons and antiprotons counterrotate and collide head-on. For this purpose, they had to develop techniques for creating antiprotons, confining them in a concentrated beam, and colliding them with an intense proton beam. Rubbia did this with Van der Meer, Guido Petrucci, and Jacques Gareyte at CERN, where the first collisions were observed in 1981.

Hundreds of scientists were involved in these experiments in teams throughout the world. In 1983, Rubbia and collaborating teams of scientists announced the discovery of the W and Z particles on the basis of signals from detectors specially designed for this purpose. When the 1984 Nobel Prize was awarded for this work, Rubbia was cited for developing the idea; Van der Meer, the equipment.

Their work was the culmination of 50 years of research into the weak interaction, begun in 1934 with ENRICO FERMI's discovery that beta decay radiates into a final state involving an electron and a neutrino pair. Fermi assumed that the electron and the neutrino pair were created directly, when neutrons were transformed into protons. Rubbia's work showed that pair creation is a two-step process, in which the particles W and Z are emitted in the first intermediate step and then convert into an electron and a neutrino pair in the second and final step. (This process has an analogy in QED, in which two electron bodies interact with each other by exchanging photons.)

For many years Rubbia divided his time between Cambridge, Massachusetts, where he taught for one semester a year at Harvard University, where he had been appointed professor in

1960, and at Geneva, where he conducted experiments such as the UA-1 collaboration at the proton–antiproton collider. He served as director general of CERN from 1989 to 1993. Married to Marisa Rubbia, a high school physics teacher, he is the father of two and a grandfather.

In November 1993, he proposed his Energy Amplifier project, a search for new sources of nuclear energy, exploiting knowledge and skills from high-energy physics that suggested the development of a power plant in which energy would be produced in a subcritical reactor. In 1994 he began to explore a route to energy production through controlled nuclear fission.

Rubbia believes that future experiments will lead to the discovery of an ever-finer definition of nature's fundamental building blocks:

> I think that an elementary particle is something much more complex than a mathematical point. . . . For a long time we thought nuclei were elementary particles. Even the word "atom" means "that which cannot be divided." Later, we thought that protons were elementary particles. Now we know that the proton is made of quarks. In the history of physics, we have often over-simplified the structure of nature.

The fundamental impact of Carlo Rubbia's Nobel Prize–winning work, in verifying Glashow, Weinberg, and Salam's unified electroweak SU(2) × U(1) theory, was to open the door for physicists to begin searching for a global unification of the electroweak theory with the quantum chromodynamic SU(3) strong interaction theory of MURRAY GELL-MANN. The search for a "theory of everything" to complete this unification is one of the leading directions in particle physics today.

See also ANDERSON, CARL DAVID; LEE, TSUNG DAO; WU, CHIEN-SHIUNG; YANG, CHEN NING; YUKAWA, HIDEKI.

Further Reading

Forward, Robert L., and Joel Davis. *Mirror Matter: Pioneering Antimatter Physics*. Lincoln, Nebr.: iUniverse.com, 2001.

Taubes, Gary. *Nobel Dreams: Power, Deceit, and the Ultimate Experiment*. New York: Random House, 1987.

⊠   Rutherford, Ernest   (Lord Rutherford of Nelson)
(1871–1937)
New Zealander/British
Theorist and Experimentalist (Radioactivity), Atomic Physicist, Nuclear Physicist

Ernest Rutherford is famous for discovering the basic structure of the atom, showing that it consists of a central nucleus surrounded by orbiting electrons. He invented the language to describe the theoretical concepts of the atom and the phenomenon of radioactivity. He won the 1908 Nobel Prize in chemistry for his discovery that radioactivity is produced by the disintegration of atoms and that alpha particles consist of helium nuclei.

Rutherford was born near Nelson, on August 30, 1871, in a remote province of New Zealand, the fourth of 12 children. Both his father, a wheelwright and flax miller, and his mother, an English schoolteacher, had emigrated from Britain as children. Despite the family's limited means, he managed to get a good education, first in state schools and then at Nelson College, which he entered at age 16. In 1889, he won a scholarship to Canterbury College in Christchurch, part of the University of New Zealand, and went on to earn a string of degrees: a B.A. in 1892, an M.A. in 1893, and a B.Sc. in 1894. At Canterbury he became a pioneer in designing original experiments with high-frequency alternating currents. In the same year that Guglielmo Marconi began his radio experiments and six years after HEINRICH RUDOLF HERTZ discovered radio waves, Rutherford studied the magnetic properties of iron exposed to high-frequency electric discharges and constructed a very sensitive detector of radio waves.

Rutherford left New Zealand for England in 1895, on a scholarship enabling him to study at the Cavendish Laboratory, Cambridge, under JOSEPH JOHN (J. J.) THOMSON, the discoverer of the electron. Under Thomson's influence, he embarked on what would be his life's work: atomic and nuclear physics. He began by studying the effect of X rays on the discharge of electricity in gases, found that X rays form positive and negative ions in gases, and measured their mobility. In 1897, he made a similar study of the effects of ultraviolet light in gases, as well as the radioactivity produced by uranium minerals. Intrigued by radioactivity, he began to investigate its nature. In 1898, he discovered two kinds of radioactivity with different penetrating power. He called the less penetrating alpha rays and the more penetrating beta rays. Later these "rays" were found to consist of streams of particles and became known as alpha and beta particles.

In 1898, Thomson helped Rutherford obtain his first academic appointment, at McGill University in Montreal, Canada. Working at the university's superbly equipped MacDonald Laboratory, Rutherford and his colleagues would make McGill a focal point for early work in subatomic physics. In 1900, he married Mary Newton, with whom he would have a daughter, Eileen. That same year he discovered a third type of radioactivity with great penetrating power, which he called gamma rays. These were later found to be quanta of very high frequency electromagnetic fields; today they are still called gamma rays or gamma radiation.

Rutherford began to use a radioactive element called thorium as a source of radioactivity instead of uranium; he found that it produced an intensely radioactive gas, a process he called emanation. He studied the emanation of tho-

rium and discovered a new noble gas, an isotope of radon, later known as thoron. To identify these radioactive products he enlisted the aid of Frederick Soddy, who arrived at McGill in 1900 from Oxford. In 1903, experimental evidence led Rutherford and Soddy to propose the disintegration theory of radioactivity, which views radioactive phenomena as nuclear processes, that is, processes that occur in the nucleus of an atom. Together they discovered a number of new radioactive substances and fixed their positions in the series of radioactive transformations of the elements.

Rutherford found that the intensity of the radioactivity produced decreases at a rate governed by the element's half-life, a term he invented to designate the time it takes for half the nuclei of a sample of radioactive substance to decay. His cogent explanation of the revolutionary idea that atoms could change their identity led to its immediate acceptance. Next Rutherford wanted to identify the alpha rays, which he was sure consisted of positively charged particles—either hydrogen or helium ions. In 1903, his experiments on deflection of alpha rays in electric and magnetic fields proved that they were positive particles, but his apparatus was not sensitive enough to allow him to determine the amount of charge. In 1904, he worked out the series of transformations that radioactive elements undergo and showed that they end with the element lead. In 1907, he estimated the rates of change involved and calculated the ages of mineral samples, arriving at figures of more than a thousand million years. This was the first accurate calculation of the age of rocks, derived through the method of radioactive dating.

In 1907, Rutherford was offered a position at the University of Manchester and returned to Britain. There he would build a renowned laboratory and make his momentous discoveries of the nuclear atom and artificial transformation. He continued to explore alpha particles and, with HANS WILHELM GEIGER, developed ionization chambers and scintillation screens to count the particles produced by a source of radioactivity. They devised a method of detecting a single alpha particle and counting the number of particles emitted from radium. That same year Rutherford proved that alpha particles are helium ions: when he performed experiments in which he trapped alpha particles in a glass tube and sparked the gas produced, its spectrum showed that it was helium.

In 1908, Rutherford won the Nobel Prize in chemistry for "his investigations into the disintegration of the elements, and the chemistry of radioactive substances." Some said this was a mistake; he should have been given the prize for physics. But, as a presenter would say, the new science of nuclear physics was "at the same time, both physics and chemistry."

Rutherford's next major discovery emerged in 1909 from experiments that involved bombarding a thin gold foil with alpha particles. He used a scintillation counter that could be moved around the foil, which was struck by a beam of alpha particles from a radon source. He observed that although all the particles passed through the gold foil, one in 8,000 "bounced," backward, deflected through angles of more than 90 degrees by the foil. In Rutherford's words, it was "as if you fired a 15-inch naval shell at a piece of tissue paper and the shell came right back and hit you." From this simple observation, he concluded that the atom's mass must be concentrated in a small positively charged nucleus and the electrons must inhabit the farthest reaches of the atom. Rutherford was convinced that the explanation lay in the nature of the gold atoms in the foil, believing that each contained a positively charged nucleus surrounded by orbiting electrons. Only such nuclei could repulse the positively charged alpha particles that happened to strike them to produce such enormous deflections. From this he calculated that the nucleus must have a diameter of about 10^{-13} cm—100,000 times smaller than that of the atom—

Ernest Rutherford is famous for discovering the basic structure of the atom, showing that it consists of a central nucleus surrounded by orbiting electrons. *(AIP Emilio Segrè Visual Archives, Otto Hahn, and Lawrence Badash)*

and calculated the number of particles that would be scattered at different angles.

These predictions about the nuclear structure of the atom were confirmed experimentally in 1911. In announcing these results, Rutherford said that practically the whole mass of the atom and all of its positive charge were concentrated in the nucleus—a minute space at the center; the much lighter electrons revolved around it, as planets revolve around the Sun. Most physicists were skeptical that the atom could be almost entirely empty space. Rutherford's supporters, however, included the future father of quantum mechanics, NIELS

HENRIK DAVID BOHR. Bohr went to Manchester in 1912 to work with Rutherford and the next year produced his quantum model of the atom. In Bohr's model a central positive nucleus is surrounded by electrons orbiting at discrete energy levels. Other experiments, performed in 1913, verified the existence of the atomic number, which identifies elements, and showed that this number was equal to the number of positive charges on the nucleus, and, since atoms are electrically neutral, also equal to the number of electrons around it. Rutherford's view of the nuclear atom was thereby vindicated and universally accepted.

Rutherford would go on to make more important discoveries at Manchester. In 1914, he performed experiments that proved gamma rays are electromagnetic waves that can be diffracted with a crystal. The wavelengths of gamma rays were found to lie beyond X ray wavelengths in the electromagnetic spectrum. ANTOINE-HENRI BECQUEREL, in 1900, had identified beta rays with cathode rays, which were shown to be electrons, thus the nature of radioactivity was now fully revealed.

In 1919, the year he moved to Cambridge and became director of the Cavendish Laboratory, Rutherford made another great discovery. He found that the nuclei of certain light elements, such as nitrogen, could be "disintegrated" by the direct hit of energetic alpha particles from a radioactive source (a process he described as "playing with marbles"). He noted that during this process fast protons were emitted. (In 1920, at Rutherford's suggestion, the name *proton* was given to the hydrogen nucleus.) Six years later, PATRICK MAYNARD STUART, LORD BLACKETT, who developed the Wilson cloud chamber into an invaluable tool for studying nuclear reactions, would prove that the nitrogen in this process was actually transformed into an oxygen isotope. What Rutherford had done was to perform the first artificial transmutation of one element into another. Blackett also used a cloud chamber to record the tracks of disintegrated nuclei; he showed that the bombarding alpha particles combine with the nucleus before disintegration and do not break the nucleus apart as a bullet would.

Bombardment with alpha particles had its limits, since large nuclei repelled them without disintegrating. Rutherford directed the construction of an accelerator to produce particles of the required energy. The first one was built by JOHN DOUGLAS COCKCROFT and Ernest Walton and went into operation at the Cavendish in 1932. In the same year, another colleague, SIR JAMES CHADWICK discovered the neutron, whose existence Rutherford had predicted in 1920.

In 1934, Rutherford made his final important discovery when, using an isotope of water called deuterium, which had recently been discovered in the United States, he and Marcus Oliphant and Paul Harteck produced the first nuclear fusion reaction.

Rutherford would remain director of the Cavendish for the rest of his life. In addition to 150 original papers, he published a number of books, including *Radioactivity* (1904), *Radioactive Transformations* (106), *The Electrical Structure of Matter* (1926), *The Artificial Transmutation of the Elements* (1933), and *The Newer Alchemy* (1937). Not least among his achievements was his work during World War II in opening English academic life to Jewish scientists fleeing nazism.

Rutherford died suddenly in Cambridge of a strangulated hernia on October 18, 1937, at age 66. His ashes are buried in Westminster Abbey, not far from the remains of SIR ISAAC NEWTON and LORD KELVIN (WILLIAM THOMSON).

One of the great figures in early 20th-century physics, Ernest Rutherford was the father of nuclear physics.

See also BECQUEREL, ANTOINE-HENRI; RÖNTGEN, WILHELM CONRAD; WILSON, CHARLES THOMSON REES.

Further Reading

Asimov, Isaac. *Atom: Journey Across the Subatomic Cosmos.* New York: Dutton/Plume, 1992.

Oliphant, Mark. *Rutherford: Recollections of the Cambridge Days.* London: Elsevier, 1972.

Von Baeyer, Hans Christian. *Taming the Atom: The Emergence of the Visible Microworld.* New York: Dover, 2000.

⊠   **Rydberg, Johannes Robert**
(1854–1919)
Swedish
Mathematical Physicist, Atomic Spectroscopist

Johannes Rydberg discovered a mathematical formula, containing a constant now known as the Rydberg constant, that gives the frequencies of spectral lines for elements.

He was born in Halmstad, a port town in southwestern Sweden, on November 8, 1954, to Sven Rydberg and the former Maria Anderson. His father was a local merchant and small ship owner, who died when Johannes was only four. The boy received his schooling at the local gymnasium (high school). In 1873 he entered the University of Lund, where he studied mathematics and received his bachelor's degree two years later. He went on to earn his doctorate in mathematics, for a dissertation on conic sections, in 1879.

The University of Lund was to become his permanent academic home, although his scholarly involvement there would undergo a major transformation. While serving as a lecturer in mathematics from 1880 to 1882, he began working on problems in friction electricity and soon became a lecturer in physics. In 1886, he married Lydia Eleonora Matilda Carlsson, with whom he would have two daughters and a son.

Rydberg's major work, which was in spectroscopy, was motivated by the desire to understand the periodic table of the elements. By organizing existing data on spectral lines, he classified the lines into three categories: principal (strong, persistent lines), sharp (weaker, but well-defined lines), and diffuse (broader lines). At the time it was known that each spectrum of an element consists of several series of these lines superimposed on one another. Rydberg wanted to find a mathematical expression to define the relationship of frequencies in any one series of lines. In 1890, he succeeded with a formula introducing the concept of the wave number N, which is the reciprocal of the wavelength (i.e., 1 divided by the wavelength) and is a measure of the number of waves per centimeter. Rydberg's formula expressed the wave number in terms of a constant common to all series of lines, the so-called Rydberg constant; two additional constants that characterize the particular series; and an integer. The structure of the lines series emerged as a result of the changing value of the integer.

At this point in his investigations, Rydberg learned that a formula for a series of lines of the hydrogen spectrum had already been worked out by the Swiss physicist Johann Balmer; Balmer's formula proved to be a special case of Rydberg's more general one. His next goal was to discover a formula capable of expressing the frequency of *every* line in *every* series of an element. Eventually, he came up with

$$N = R[1/(n + a)^2 - 1/(m + b)^2]$$

where N is the wave number; R is the Rydberg constant; n and m are integers, m greater than n; and a and b are constants for a particular series. This formula succeeded in expressing five more series of hydrogen lines. Rydberg had no theoretical explanation for his result, and it would remain for NIELS HENRIK DAVID BOHR to show that Rydberg's equation described the quantum energy states required for the theory of atomic structure that he unveiled in 1913.

In 1879, Rydberg received a temporary appointment as full professor of physics at

Lund. That year he presented the idea of atomic number and its importance in revising the periodic table. His views were validated by Henry Moseley's research into X-ray spectra in 1913.

In 1902, Rydberg's appointment as full professor at Lund was made permanent. Although his health began to deteriorate in 1914, he remained at his post until a few weeks before his death in Lund on December 28, 1919, of a brain hemorrhage. Earlier that year he was elected as a member of the Royal Society of London. Although he never attained his goal of determining the structure of the atom, his work formed the basis for others, such as Bohr, to do so.

Further Reading
Siegbahn, Manne. "Johannes Robert Rydberg." In *Swedish Men of Science*. Stockholm: Almquist & Wiksell International, 1952, pp. 214–218.

S

⊠ Sakharov, Andrei Dmitriyevich
(1921–1989)
Russian
Theoretical Physicist, Nuclear Physicist

Andrei Sakharov embodied the 20th-century physicist's recognition of moral responsibility for the social consequences of his or her work. The man known as "the father of the Soviet hydrogen bomb" became a passionate advocate of nuclear disarmament and defender of human rights, who received the Nobel Prize in peace in 1975. As his political awareness evolved, he traded the privileged life of a Soviet weapons scientist for the hardships and internal exile of a dissident against the Communist system. Released by Premier Mikhail Gorbachev in the mid-1980s, he became the spiritual leader of the burgeoning democratic movement, only to die prematurely, at 68, at a crucial phase of that still ongoing struggle.

Andrey Dmitriyevich Sakharov was born on May 21, 1921, into a family of Moscow intellectuals, whose forebears included military nobility and Russian Orthodox priests. As a member of the first postrevolutionary generation, he believed in his country as the harbinger of international justice and social equality. Schooled at home until the seventh grade, he first learned physics from his father, Dmitri, a college profes-

sor and author of textbooks and popular science books. He enrolled as a student of physics at Moscow State University in 1938, at the height of the Stalinist purges, when the best of the physics faculty had been picked off. When the Germans invaded in 1941, young Sakharov, found unfit for military service, was evacuated to Central Asia, along with other remaining students and professors. After graduating with honors in 1942, he declined graduate school in favor of laboratory work at a munitions factory on the Volga. There he met Klavdia Vikhireva, a technician, and married her in 1943.

As the war was ending in 1945, the couple returned to Moscow, where Sakharov enrolled as a graduate student at FIAN, the Russian acronym for the renowned P. N. Lebedev Institute of Physics of the USSR Academy of Sciences. Under the Soviet system, the nation's scientific elite carried out research and mentored the next generation at academy institutes, rather than colleges and universities. The physicist who was to shape Sakharov's career in crucial ways was Igor Tamm, head of the theoretical department, who would become a Nobel laureate in physics in 1958. Eager to continue working with Tamm on fundamental science, Sakharov twice refused an invitation to work on the Soviet atomic bomb project. But after receiving his doctorate for work on particle

Andrei Sakharov, "the father of the Soviet hydrogen bomb," became a passionate advocate of nuclear disarmament and defender of human rights. He was awarded the Nobel Peace Prize in 1975. *(AIP Emilio Segrè Visual Archives and VNIIEF Museum and Archive)*

physics in 1947, he joined Tamm in a special group at FIAN under the leadership of Yakov Zeldovich, which was studying the feasibility of a thermonuclear, or hydrogen, bomb. The Zeldovich team was exploring the *Truba* (Russian for "tube") design, based on data obtained by spies from the Manhattan Project, the Los Alamos–based team engaged in developing atomic and hydrogen bombs. Convinced of the inadequacy of Truba and spurred by a sense of national emergency, Sakharov worked feverishly and within two months came up with an alternative design. Nicknamed *Sloyka* (Russian for "layer cake"), it was based on alternating layers of light elements (deuterium, tritium, and their chemical compounds) and heavy elements (uranium 238), which initiated a chain of energy-releasing events. fission–fusion fission. This structure would be basic to all future design variants. Another student of Tamm, Vitaly Ginzburg, proposed that lithium deuteride, instead of deuterium and tritium, be used in the fusion layer, thereby maximizing the efficacy of the thermonuclear explosive material in Sakharov's design. These conceptual breakthroughs made possible the first Soviet H-bomb, which was successfully tested on August 12, 1953. Although it yielded four hundred kilotons, 15% to 20% from fusion, it did not put the Soviets ahead of the American bomb makers, who already had a bigger atomic bomb, as well as megaton hydrogen bombs.

By then Sakharov had been transferred to Arzamas-16, a secret city, whose real name was Sarov, in the central Volga region. Sometimes nicknamed Los Arzamas after its American prototype, Los Alamos, the city had a special design bureau that was the nerve center of Soviet nuclear weapons research. While there, Sakharov would make major contributions to the so-called Czar-Bomb of 1961, the most powerful device ever exploded. His work there, however, was not all military-related. In 1950 Sakharov and Tamm invented the TOKOMAK (an acronym of the Russian term for the toroidal chamber with magnetic coil), a controlled thermonuclear fusion reactor. Soviet research on controlled fusion was eventually declassified, and TOKOMAK came to be regarded as a promising direction in the quest for creating an unlimited source of cheap energy.

In 1964, Sakharov returned to fundamental science, after a hiatus of 19 years. He published a paper on cosmology, which attempted to explain the asymmetry between the amount of matter and amount of antimatter in the universe by proposing that protons, generally assumed to be stable particles, can spontaneously decay. In a second paper, he took on the highly complex and still unresolved problem of constructing a unified field theory: one that unifies gravitation with other physical forces.

During his years at Arzamas-16, Sakharov was a highly rewarded member of a scientific community in which weapons-related research garnered the lion's share of prestige and resources. At 32, he became the youngest scientist ever to receive full membership in the prestigious Soviet Academy of Sciences. He was awarded several Hero of Socialist Labor Medals, a Stalin Prize, and a dacha (cottage) in the elite Moscow suburb of Zhukovka.

Sakharov's political conscience had expressed itself in his younger years through his aid to victims of repression and opposition to the faction within the academy supporting Lysenko's crusade against modern genetics. By the mid-1960s, political activism was becoming the dominant force in his life. While still pursuing research on new weapons, he became concerned about the dangers of nuclear testing; that concern led him to urge the Soviet leadership to accept American proposals for a moratorium on antiballistic missile defense. When his attempts to work within the system failed, he stepped outside it, publishing in *samizdat*, the underground dissident network, his famous 1968 essay "Reflections on Progress, Peaceful Co-Existence and Intellectual Freedom." In it, he proposed international cooperation in the interest of reducing nuclear arms and establishment of civil liberties in the Soviet Union. When the essay found its way into the Western press, Sakharov was banished from military-related research.

A year later, his wife, Klavdia, died of cancer. Left to care for their three children, then aged 24, 19, and 11, Sakharov accepted an offer to return to FIAN in Moscow, to work on academic topics. There his political activism accelerated. In 1970, he founded the human rights movement with other Soviet dissidents and met another activist, Yelena Bonner, the woman who would become his devoted companion and comrade in arms. They married in 1972. When Sakharov was refused permission to travel abroad to accept the Nobel Prize in peace in 1975, Yelena Bonner accepted it in his stead.

As he became the country's foremost defender of human rights, Sakharov was under increasing pressure from the regime. In 1980, after his strenuous and highly publicized protests against the war in Afghanistan, he was exiled to the city of Gorky (now renamed Nizhniy Novgorod), where, with Bonner, he would remain for close to seven years. During this ordeal of constant surveillance and isolation, he was buoyed by support from American physicists who found ways of getting reprints of their papers into his hands and campaigned for his release in the media. His three hunger strikes, protesting his and his family's persecution, were widely publicized in the West.

But Sakharov would not regain his freedom until the ascendance of Mikhail Gorbachev, who, in December 1986, in the spirit of *perestroyka* (rebuilding or reform), recalled him to Moscow and allowed him to resume his public role. He swiftly became the commanding figure in the struggle for democracy and, in April 1989, was elected to the Soviet Union's new parliament. As coleader of the democratic faction, Sakharov championed the adoption of a new constitution that would abolish the hegemony of the Communist Party and allow Russia to evolve peacefully toward a multiparty system. After a day of acrimonious debate, on December 14, 1989, Sakharov died of a sudden heart attack. His funeral, attended by 50,000 people, took place on a day of national mourning for the man who was revered as the "saint" of Russia's democratic rebirth.

Further Reading

Bonner, Yelena. *Alone Together*. New York: A. A. Knopf, 1986.

Holloway, David. *Stalin and the Bomb: The Soviet Union and Atomic Energy, 1939–1956*. New Haven, Conn.: Yale University Press, 1994.

Kapitsa, S. P., and S. Drell. *Sakharov Remembered*. New York: American Institute of Physics, 1991.

Rhodes, Richard. *Dark Sun: The Making of the Hydrogen Bomb*. New York: Simon & Schuster, 1995.

Sakharov, A. D. *Memoirs*. New York: A. A. Knopf, 1990.

———. *Progress, Coexistence, and Intellectual Freedom*. New York: Norton, 1968.

———. *Sakharov Speaks*. New York: A. A. Knopf, 1974.

⊠   **Salam, Abdus**
(1926–1996)
Pakistani
Theoretician, Particle Physicist,
Quantum Field Theorist

Abdus Salam shared the 1979 Nobel Prize in physics with STEVEN WEINBERG and SHELDON LEE GLASHOW for his groundbreaking work in unifying the electromagnetic and weak forces. The first Pakistani and the first Muslim scientist to win this most prestigious of awards, Salam was a champion in the cause of developing science in the Third World.

He was born on January 29, 1926, in Jhang, a small town in what is now Pakistan, into a Muslim family with a long tradition of piety and learning. His father, who worked in an impoverished farming district as an official in the Department of Education, had prayed to Allah for a son of outstanding intellect. Abdus did not disappoint him. At the age of 14, he became a legend in his hometown, when he received the highest marks ever recorded on the matriculation examination given at the University of Panjab. When he bicycled home after the news got out, the town's inhabitants turned out in force to welcome him.

Awarded a scholarship to Government College at the University of Panjab, Lahore, he earned his master's degree in 1946. That year, he left the country on a scholarship to Saint John's College at Cambridge University, England, where the great PAUL ADRIEN MAURICE DIRAC became his mentor and hero. Three years later, he received his B.A. with honors, with a double first place in mathematics and physics. In 1950, he won Cambridge's Smith Prize, awarded for the most outstanding predoctoral contribution to physics. He remained at Cambridge, at the Cavendish Laboratory, for his graduate studies, intending to become an experimentalist, but soon judged himself lacking in what he called "the sublime quality of patience—patience in accumulating data, patience with recalcitrant equipment." He began working under Herbert Kemmer and earned his doctorate in theoretical physics in 1952 for a thesis on the question of whether quantum electrodynamics (QED) was renormalizable, that is, capable of being altered by a mathematical procedure that cancels the unwanted infinities in a quantum field theory by introducing the appropriate renormalization constants.

Abdus Salam shared the 1979 Nobel Prize in physics with Steven Weinberg and Sheldon Glashow for his groundbreaking theoretical work unifying the electromagnetic and weak forces. *(AIP Emilio Segrè Visual Archives, Marshak Collection)*

Around this time, he came up with the startling idea that "all neutrinos are left-handed," which suggested the possibility that a violation of parity could occur in the weak interactions, before that revolutionary hypothesis was put forth by TSUNG-DAO LEE and CHEN NING YANG and experimentally confirmed by CHIEN-SHIUNG WU. When Salam described the idea to WOLFGANG PAULI, he summarily dismissed it, convincing Salam not to publish it. After this chastening experience, Salam determined never again to listen to "grand old men" and adopted a "publish or perish" policy, which resulted in the publication of over 300 papers.

As a newly minted Ph.D., who already had an international reputation, Salam returned to Pakistan, with the intention of being a physicist in his own land. He was made head of the mathematics department of the University of Panjab, where he wanted to establish a school of research. But prevailing conditions persuaded him there was no way to have the kind of research atmosphere and collegial stimulation he needed. In 1954, he returned to England as a lecturer at Cambridge and, at the invitation of LORD PATRICK MAYNARD STUART BLACKETT, in 1957, moved to Imperial College, London, where he founded the Theoretical Physics Group. Always receptive to strange new ideas, he encouraged the young Israeli physicist YUVAL NE'EMAN to develop the ideas that resulted in his independent discovery of the unitary symmetry SU (3), which MURRAY GELL-MANN, making the same discovery, would dub the "eightfold way" of classifying the growing number of elementary particles.

At the same time, Salam was determined to improve the situation for scientists in developing countries. In 1964, he founded the International Center for Theoretical Physics (ICTP) in Trieste, Italy. There he instituted a program of research associateships, which allowed deserving young physicists from Third World countries to spend three months a year at the center, interacting with leaders in their fields. He would remain director of the ICTP until his death.

At Imperial College, in the 1960s, he did his Nobel Prize–winning work. Salam was influenced by the research he did in 1961 with Steven Weinberg on spontaneous symmetry breaking, that is, asymmetric relations that have spontaneously arisen from the functioning of symmetrical laws. (An example of this is the asymmetric crystal structure of ice, which freezes out from the symmetric liquid structure of water when the temperature becomes low enough.) He also worked extensively with his colleague John Ward on unsuccessful attempts to unify the electromagnetic and weak forces.

In the early 1960s, Salam, like Weinberg and Glashow, was aware of the groundbreaking work of Yang and Robert Mills in creating the so-called non-Abelian Yang–Mills gauge theories with internal symmetries. In 1950, they had shown that the QED formalism developed earlier by JULIAN SEYMOUR SCHWINGER, RICHARD PHILLIPS FEYNMAN, and SIN-ITIRO TOMONAGA could be generalized to include internal dynamic symmetries that were more general than the standard spacetime C (charge), P (parity), and T (time-reversal) symmetries. Salam, Weinberg, and Glashow also knew that Peter Higgs had shown that coupling a scalar field to a Yang–Mills gauge theory could cause the spontaneous breaking of these internal symmetries to occur, thus creating new kinds of force-carrying particles, some of them massive.

Using these ideas and working independently with different theoretical approaches, the three physicists reached the same conclusion: if the virtual particles that carry the electromagnetic and weak forces (known collectively as the intermediate vector boson W and Z particles) were related by a broken internal symmetry in a Yang–Mills gauge theory, these new theoretical ideas might make it possible to estimate their masses in terms of the unified, more symmetrical force from which the two forces were thought to have arisen.

However, over the next four years this unified electroweak theory attracted scant attention since, unlike quantum electrodynamics, the electroweak theory had not yet been shown to be renormalizable. But in 1971, the Dutch physicist GERARD 'T HOOFT used computer algebra techniques to prove that the electroweak theory was indeed renormalizable.

Although experimental verification of the electroweak theory would not occur until 1983, 't Hooft's proof was the final step that led the Nobel Committee to the unusually insightful decision to award Salam, Glashow, and Weinberg the 1979 prize in physics. Salam upstaged his corecipients by appearing at the ceremonies in Stockholm in traditional clothing (including bejeweled turban, baggy pants, scimitar, and curly shoes), with his two wives (permitted him by Muslim law).

While pursuing his own theoretical work, Salam remained vitally involved in Pakistan's scientific development, serving as a member of the Pakistan Atomic Energy Commission, a member of the Scientific Commission of Pakistan, and chief scientific adviser to the president from 1961 to 1974.

Salam was a prominent and dedicated member of the international science community, a tireless worker who was known to sacrifice rest and recreation for his many causes. He served on a number of United Nations committees on the advancement of science and technology in developing countries. His contributions to peace and international scientific collaboration were recognized by the Atoms for Peace Award (1968) and many others. He used his money from both the Atoms for Peace prize and the Nobel Prize to assist physicists in developing countries.

Salam's books include *Supergravities in Diverse Dimensions*, with Ergin Sezgin (1990); *Science and the Third World* (1991); and *The Renaissance of Sciences in Islamic Countries* (1994).

When Abdus Salam died at age 70, of Parkinson's disease, on November 21, 1996, in Oxford, England, he was mourned by the physics community, Pakistan, and the world. He was survived by a wife, two sons, and four daughters. He was revered as both a brilliant physicist and a tireless advocate for the development of science, not only in Pakistan, but throughout the Third World. A devout Muslim, he made no distinction between his religion and his scientific pursuits:

> The Holy Quran enjoins us to reflect on the verities of Allah's created laws of nature; however, that our generation has been privileged to glimpse a part of His design is a bounty and a grace for which I render thanks with a humble heart.

See also RUBBIA, CARLO; VELTMAN, MARTINUS J. G.

Further Reading
Ghani, Abdul. *Abdus Salam*. Karachi: Ma'aref, 1982.
Hassan, Z., and C. H. Lai, (eds.) *Ideals and Realities: Selected Essays of Abdus Salam*. Singapore: World Scientific, 1983.
Salam, Abdus. *Science and the Third World*. Edinburgh: Edinburgh University Press, 1991.

⊠ **Schawlow, Arthur Leonard**
(1921–1999)
American
Experimentalist, Laser Spectroscopist

Arthur Leonard Schawlow was an experimentalist who, in collaboration with CHARLES HARD TOWNES, in 1958 invented the *laser*, an acronym for light amplification by stimulated emission of radiation. Their landmark discovery has led to advances in virtually every field of modern technology.

Schawlow was born in Mount Vernon, New York, on May 5, 1921, five years after his father had immigrated to the United States from Riga, Latvia. When Arthur was three, at the urging of his mother, who was Canadian, the family moved to Toronto, where he attended public schools. His passionate interest in science, particularly electronics, mechanical subjects, and astronomy, was born in those early years. His high school years coincided with the Great Depression of the 1930s and Schawlow's father, an insurance agent, was in no position to send him or his sister to college. Fortunately, both won scholarships to the Faculty of Arts at the University of Toronto. Arthur's was in physics, which seemed to him "pretty close" to his original choice of radio engineering. He would later consider this the correct path, since he did not have "the patience with design details that an engineer must have" and enjoyed the "chance to concentrate on concepts and methods" that physics provided.

When Canada entered World War II in 1941, Schawlow spent three years teaching classes to armed service personnel at the University of Toronto, before joining a group working on microwave antenna development at a radar factory. Returning to his graduate studies after the war, to an understaffed and underequipped physics department, he was attracted to an area in which Toronto had a solid tradition: optical spectroscopy. Running an atomic beam spectroscopy experiment in the basement of a campus laboratory, Schawlow would serenade his atomic beam on his jazz clarinet (he would pursue his love of jazz throughout his career). After a rewarding experience writing his dissertation under Malcolm F. Crawford, he was awarded the Ph.D. in 1949.

He then took a postdoctoral fellowship under Charles Townes in the physics department at Columbia University—an exhilarating place to be in those years, under the leadership of ISIDOR ISAAC RABI and with no fewer than eight future Nobel laureates in the department. His association with Townes, the leader of research on microwave spectroscopy, opened up Schawlow's life both professionally and personally; in 1951, he married Townes's youngest sister, Aurelia, a

musician, mezzo soprano, and choral conductor. The marriage would produce a son and two daughters; it had the immediate, unfortunate effect, however, of forcing Schawlow to leave Columbia, since the university's antinepotism rules prevented Townes from keeping him on the staff. He became a physicist at Bell Laboratories in New Jersey in 1951, working mainly on superconductivity, and thus was excluded from the exciting developments surrounding Townes's 1951 invention of the *maser*, an acronym for "microwave amplification by stimulated emission of radiation."

The idea of stimulated emission of radiation had originated with ALBERT EINSTEIN. While studying the theory of blackbody radiation in 1917, Einstein had found that the process of light absorption must be accompanied by a complementary process in which the absorbed radiation stimulates the atoms to emit the same kind of radiation. However, in order for amplification of the radiation by stimulated emission to occur in a physical medium, the stimulated emission must be larger than the absorbed radiation. Hence, the physical medium must have more atoms in a high-energy state than in a lower-energy state (i.e., an inverted population of excited atoms). Since energy spontaneously flows from a higher to a lower state, such an inverted population of excited atoms is inherently unstable. Moreover, to generate beams of coherent light from this inverted population of excited atoms, that is, intense light waves consisting of essentially one frequency and moving in the same direction, it would be necessary to find the specific atomic systems with the correct internal storage mechanisms.

Townes selected the simple case of ammonia molecules, because they can occupy only two energy levels. The ammonia molecules had the property of emitting radiation with a 1.25-cm wavelength, which lies within the microwave range of the electromagnetic spectrum. He reasoned that when an ammonia molecule in the high-energy state absorbed a photon of this frequency, the molecule would fall to the lower energy level, emitting two coherent photons of the same frequency, thus producing a coherent beam of single-frequency radiation with a 1.25-cm wavelength. By studying the behavior of ammonia molecules in a resonant cavity containing electric fields, Townes developed a method that separated the relatively few high-energy molecules from the more numerous low-energy ones and thus created the population inversion required for maser action. By 1953, he had constructed the first working model of the maser. Masers soon found a range of applications: in the most accurate timepieces available to this day, atomic clocks; in shortwave radios, where they serve as extremely sensitive receivers; in radio astronomy; and in space research, for recording the radio signals from satellites.

Although Schawlow, exiled in New Jersey, was not part of this work, he and Townes had not lost touch. On weekends in New York they continued working on their book *Microwave Spectroscopy* (1955), and their collaboration soon took a dramatic turn. Schawlow began to worked with Townes

> to see what would be needed to extend the principles of the maser to much shorter wavelengths, to make an optical maser or . . . laser. Thereupon, I began work on optical properties and spectra of solids, which might be relevant to laser materials and then to lasers.

In 1957, he and Townes began working on the principle of a device, the laser, light amplification by stimulated emission of radiation, that could operate at shorter wavelengths than the maser. Because the change from microwaves and visible light required a 100,000-fold increase in frequency, a fundamentally different operating structure was required. Schawlow had

the idea of building a chamber, or cavity, consisting of a synthetic ruby that would act as a kind of echo chamber for light. The atoms or molecules of a ruby or garnet crystal, or of a gas, liquid, or other substance, were excited by light in the cavity in which they were contained, so that more of them were at higher energy levels than were at lower ones. In order to achieve the necessary high radiation density, two mirrors at each end of the cavity forced the light to bounce back and forth until coherent light escaped from the cavity. With Schawlow focusing on the device and Townes on the theory, they coauthored a paper in the December 1958 *Physical Review*, "Infrared and Optical Masers," in which they showed theoretically that an optical maser, or laser, could be produced by a coherent beam of visible light, instead of a microwave beam. The publication set off an international competition to build the first working laser, which was won in 1960 by Theodore Maiman, working at the Hughes Research Laboratories in Malibu, California.

In 1961, Schawlow left Bell to become professor and then chair in 1966 of the department of physics at Stanford University, where his research focused on molecular spectroscopy. Both fatherly and inspiring, he gathered about him a following of devoted students and renowned visitors. He was famous for such aphorisms as "To do successful research, you don't need to know everything; you just need to know of one thing that isn't known." and "Anything worth doing is worth doing twice, the first time quick and dirty, and the second time, the best way you can." A serious scientist with an irrepressible sense of humor, he earned the nickname "Laser Man," because of his popular demonstrations of the tool he had helped to invent. In one, designed to show the laser's selectivity, he used a "ray gun" laser to shoot through a transparent balloon to pop a dark Mickey Mouse balloon inside without damaging the outer balloon.

It was Townes who, in 1964, shared the Nobel Prize in physics with the Russian physicists Nikolay Basov and Aleksandr Prokhorov, for invention of the laser. Schawlow did not become a Nobel laureate until 1981, when he shared the prize in physics with NICOLAAS BLOEMBERGEN for their "contributions to laser spectroscopy."

In addition to his physics career, Schawlow's life in California was focused on the Peninsula Children's Center for handicapped children, attended by his autistic son, Artie. The Schawlows became activists on behalf of autistic people and played a role in establishing California Vocations, a nonprofit organization dedicated to creating group homes for those afflicted with the disease. In 1991, the year he retired from teaching and became professor emeritus, his wife died in an automobile accident while on her way to visit their son. Schawlow died on April 28, 1999, at the age of 77, of congestive heart failure resulting from leukemia, in a hospital in Palo Alto, California, where he lived.

Schawlow and Townes had not anticipated the applications of the laser; nor did either of them, as Bell employees, profit from any of them. In 1998, a year before he died, Schawlow, with a characteristic lack of self-importance, commented on the avalanche of innovations that continues to flow from their great discovery:

> The thing that turns me on now, particularly, is that lasers now permit people to study single atoms and single photons, and we're really learning a lot more about just how strange the world of quantum mechanics is and finding new ways to test it. . . . And it's nice that there are medical uses. . . . One of the first applications of lasers was for surgery of the retina in the eye to prevent blindness. Neither Charlie [Townes] nor I had ever heard of surgery for detached reti-

nas . . . and if we had we probably wouldn't have been fooling around with stimulated emissions from atoms. . . . It's been a great 40 years.

See also CHU, STEVEN; KUSCH, POLYKARP.

Further Reading
Schawlow, Arthur L. "Advances in Optical Masers." *Scientific American*, July 1963, pp. 28, 34–45.
———. "Optical Masers." *Scientific American* June 1961, pp. 41, 52–61.

⊠ **Schrieffer, John Robert**
(1931–)
American
Theoretician, Solid State Physicist

John Robert Schrieffer is a solid state theoretician, who, as a 26-year-old graduate student, working with JOHN BARDEEN and Leon Cooper, discovered the Bardeen–Cooper–Schrieffer (BCS) theory, a comprehensive explanation of superconductivity, the phenomenon in which electricity flows through a substance without any loss due to electrical resistance.

He was born in Oak Park, Illinois, on May 31, 1931, to John H. Schrieffer and Louise Anderson Schrieffer. The family moved in 1940 to Manhasset, New York, and again, when John was 16, to Eustis, Florida, when his father quit his job as a pharmaceutical salesman and began a successful new career in the citrus industry. John remembers spending much of his time in Florida playing with gadgets: first home-made rockets, then ham radio, a hobby that sparked his interest in an electrical engineering career. He graduated from Eustis High School in 1949 and entered the Massachusetts Institute of Technology in Cambridge, where he started out as an electrical engineering major. In his junior year, he switched his field of concentration to physics. Working under John C. Slater, he wrote a bachelor's thesis on the energy level multiplet structure of heavy atoms. By now he knew that he was interested in pursuing a career in solid state physics. He entered the University of Illinois for graduate studies and began working with Bardeen, who had already discovered the transistor in 1946. After joining Bardeen's illustrious group of scientists working on semiconductor research, Schrieffer began working on a problem related to electrical conduction on semiconductor surfaces; he spent a year in the laboratory applying theoretical analysis to several surface problems. In the third year of graduate studies, working with Bardeen and Cooper, he developed as his doctoral dissertation the theory of superconductivity, which would later be known as the Bardeen–Cooper–Schrieffer (BCS) theory.

While riding the New York subway in January 1957, Schrieffer had an insight that would solve the mystery set in motion in 1911, when HEIKE KAMERLINGH ONNES first observed superconductivity: zero electrical resistance in some metals below a critical temperature. Since then physicists had been looking for a microscopic interpretation of this phenomenon. The methods that were successful in explaining the electric properties of normal metals were unable to predict the effect. At very low temperatures, metals were still expected to have a finite resistance, which was due to scattering of mobile electrons by the ions in the crystal lattice. The BCS solution to this problem was to show that in a crystalline material electrons have the potential to pair up quantum mechanically through an attractive interaction mediated by the crystal lattice, and that zero resistivity occurs when the thermal energy available is insufficient to break apart the pair.

Thus, for electrons embedded in a crystal, the normal Coulomb repulsion can be compensated for by this pairing effect when the temperature is below the critical value. The ion cores in the crystal lattice respond to the presence of a nearby electron, and the motion may result in another electron's being attracted to the ion. The net effect is an attraction between two electrons through the mediating response of the ions in the solid. The BCS theory was based on the idea that the interaction between the electrons and the lattice leads to the formation of bound pairs of electrons, which came to be called Cooper pairs. The different pairs are strongly coupled to each other, an arrangement that leads to a complex collective pattern in which a considerable fraction of the total number of conduction electrons are coupled to form a superconducting state. Because of this characteristic coupling of all the electrons, one cannot break up a single pair of electrons without also perturbing all the others, and this breaking requires an amount of energy that must exceed a critical value. This is the reason for the existence of a critical temperature below which superconductivity occurs. Many of the remarkable qualities of superconductors can be understood qualitatively from the structure of this correlated many-electron state.

The comprehensive BCS theory has the ability to explain all known properties associated with superconductivity. Although applications of superconductivity to magnets and motors were possible without the BCS theory, the theory is important for examining strategies to increase the critical temperature as high as possible, since, if the temperature could be raised above liquid nitrogen temperature, the economics of superconductivity would be transformed. In addition, the theory was an essential prerequisite for the prediction of Josephson junction quantum tunneling with its important applications in magnetometers, computers, and determination of the fundamental constants of physics. The BCS theory has had profound effects on nearly every field of physics from elementary particle to nuclear physics and helium from liquids to neutron stars. Schrieffer and his two colleagues shared the Nobel Prize in physics for their theory in 1972.

Schrieffer continued his research on superconductivity as a National Science Foundation Fellow at the University of Birmingham and the Niels Bohr Institute in Copenhagen during the 1957–1958 academic year. He spent the next year as an assistant professor at the University of Chicago, before joining the faculty of the University of Illinois in 1959. During a summer visit to the Bohr Institute in 1960, he became engaged to Anne Grete Thomsen, whom he married at Christmas of that year. They would have three children.

In 1962, Schrieffer took a position at the University of Pennsylvania in Philadelphia, where, two years later, he became Mary Amanda Wood Professor of Physics. He published *Theory of Superconductivity*, a book that would become a classic in its field, in 1964. Five years later, he was awarded a six-year term as Andrew D. White Professor-at-Large.

In 1980, Schrieffer joined the faculty of the University of California, Santa Barbara, where he was appointed Chancellor Professor in 1984. He served as director of the Institute of Theoretical Physics in Santa Barbara from 1984 to 1989. In 1992, he was appointed University Professor at Florida State University and Chief Scientist of the National High Magnetic Field Laboratory.

Despite its mathematical complexity, the BCS theory enabled physicists to develop new materials that conduct at temperatures that do not need to be supercold, resulting in the production of whole new classes of superconducting metals. By 1960, physicists had used the BCS theory to solve problems in nuclear physics concerning the unusual behavior of neutrons and protons in the atomic nucleus. When pulsars—small rotating stars that emit regular bursts of radio waves—were discovered in 1963, astrophysicists used BCS theory to interpret their behavior.

In 1987, however, the theory was challenged by the surprising discovery of a class of so-called high-temperature cuprate superconductors, which have the property of becoming superconducting at temperatures above 40 K. This violated the BCS theory, which asserted that superconductivity was theoretically impossible at these temperatures.

In his most recent work, Schrieffer has joined the group of physicists attempting to understand high-temperature superconductivity; he believes that rather than being negated by the new phenomenon, the BCS theory may provide its theoretical groundwork. Currently Schrieffer is continuing this research at the National High Magnetic Field Laboratory in Florida, directing a team of investigators whose research focuses on developing and extending the BCS formalism into a theory of high-temperature superconductivity, as well as on the dynamics of electrons in strong magnetic fields.

See also JOSEPHSON, BRIAN DAVID.

Further Reading
Schrieffer, J. Robert. *Theory of Superconductivity.* Cambridge, Mass.: Perseus, 1999.

⊠ Schrödinger, Erwin
(1887–1961)
Austrian
Theoretical Physicist (Quantum Mechanics)

Erwin Schrödinger was one of the great architects of quantum mechanics. The famous equation that bears his name, which describes the behavior of electrons in atoms, created wave mechanics, thereby setting quantum mechanics on a firm mathematical foundation. Schrödinger's work garnered him the 1933 Nobel Prize in physics, which he shared with PAUL ADRIEN MAURICE DIRAC. However, as were ALBERT EINSTEIN and MAX ERNEST LUDWIG PLANCK, Schrödinger was philosophically incapable of accepting the probabilistic view of nature to which his ideas led. He made his second most famous contribution to

20th-century physics, a thought experiment known as the cat paradox, in order to refute it.

Erwin Schrödinger was born in Vienna on August 12, 1887. His father, who ran a small linoleum factory inherited from his family, was a cultured man who studied chemistry, painted, and wrote papers on botany. His mother, the daughter of a chemistry professor, was half-English and half-Austrian. Young Erwin, an only child, who grew up speaking both English and German, was tutored privately until the age of 10, when he entered the gymnasium (secondary school) in Vienna. He loved mathematics, physics, German poetry, and ancient languages but resisted rote memorization, a sign, perhaps of his latent originality as a thinker.

Erwin Schrödinger, who discovered the famous equation that describes the behavior of electrons in atoms, created wave mechanics and set quantum mechanics on a firm mathematical foundation. (*AIP Emilio Segré Visual Archives*)

Entering the University of Vienna in 1906, Schödinger specialized in physics. The ideas he was exposed to as a student by Fritz Hasenohr, the successor of LUDWIG BOLTZMANN, who considered phenomena in terms of probability theory, made a strong impression on him and became the subject of his first published paper. He received a doctorate in 1910 for his theoretical dissertation "On the Conduction of Electricity on the Surface of Insulators in Moist Air." The following year, he accepted an assistantship in experimental physics with Max Wien at the University of Vienna's Second Physics Institute. He would later say that this early experimental work was invaluable to his future theoretical work.

World War I would decelerate his physics career without wholly derailing it. While serving as an artillery officer, first on the Italian border and then in Hungary, Schrödinger managed to write and submit theoretical papers. Sent back to the Italian front, he saw combat as a battery commander and received a citation for outstanding service. As the war was ending, Schrödinger returned to an academic post in Vienna, where he would remain until 1920, publishing his first results on quantum mechanics and doing important work in color theory. Because of the hardships of life in postwar Vienna, he moved to Jena and then Stuttgart, Germany, in 1920. That same year he married Annemarie Bertel, who was to remain with him until his death, despite his series of scandalous affairs.

The young couple would move again, in 1921, when Schrödinger was offered a professorship at the University of Zurich. There he would do his best work and enjoy the friendship of such illustrious colleagues as Peter Debye and Hermann Weyl, who was instrumental in developing his mathematical prowess. He wrote papers on a variety of subjects, including the theory of color vision, specific heats of solids, electrodynamics, and atomic spectra.

Schrödinger found himself dissatisfied with the ad hoc nature of NIELS HENRIK DAVID BOHR's model of the atom, in which electrons rotate around the nucleus in a certain number of discrete stable orbits, emitting or absorbing quanta of energy only when they jump from one orbit to another. There was no physical reason for Bohr's quantized orbits, but they "worked." They successfully explained the spectral lines of the hydrogen atom. For Schrödinger, who was familiar with LOUIS-VICTOR-PIERRE, PRINCE DE BROGLIE's, discovery that an electron or any other particle has an associated wave, the next breakthrough in understanding atomic structure would require an equation capable of combining within a unified framework Bohr's quantized electron leaps and de Broglie's wave–particle duality. He began searching for a type of partial differential equation—a wave equation, like those used to describe sound, for example—capable of describing the behavior of electrons and other particles. De Broglie himself was involved in a similar quest: in 1926, he and Schrödinger came up with the same equation in the attempt to find this synthesis, but it failed to predict the observed spectra. Schrödinger tried a different approach later that year, formulating his Schrödinger equation in terms of the energies of the electron and the field in which it was located. In the new model of atomic structure revealed by Schrödinger's wave function, the electron, although most likely to be found where Bohr expected it to be, did not follow an orbit. Instead, it existed within what he called an *orbital*, an electron cloud or spatial region, which was an extension of Broglie's model of electrons existing in standing waves around the nucleus. Schrödinger got rid of Bohr's disturbing concepts of quantum jumps and discontinuities by explaining "so-called electron orbits" as the vibration frequencies of electron "matter waves" around the nucleus of the atom.

Schrödinger's wave mechanics came on the scene when another mathematical formulation of atomic structure, the matrix mechanics of WERNER HEISENBERG and MAX BORN, was generating much excitement. In May 1926, Schrödinger published an article proving that wave and matrix mechanics were equivalent mathematically, while claiming the superiority of his own approach. The physics community at large agreed. Matrix mechanics, which was based on abstract calculation, without an accompanying picture of the atom, was far less accessible than Schrödinger's familiar visualization. Planck hailed wave mechanics as "epoch-making;" and Einstein dubbed it a "decisive advance."

Schrödinger's subsequent renown generated the offer of a professorship at the University of Wisconsin, Madison, where he delivered a brilliant series of lectures in 1927. He accepted Planck's former position as professor of theoretical physics in Berlin, then a hub of scientific debate and discovery. Schrödinger plunged wholeheartedly into discussions with an array of exceptional colleagues who included Einstein himself. He would remain there until the rise of the Nazis in 1933, when he decided that living in a country that persecuted Jews as a matter of national policy was not an option for him.

It was in Oxford, England, where he became a fellow at Magdalen College in the spring of 1934, that he learned he had won the 1933 Nobel Prize in physics for discovering wave mechanics. Refusing an offer from Princeton, he remained at Oxford for the next two years. In 1935, he published a three-part essay, "The Present Situation in Quantum Mechanics," in which he presented the cat paradox, his famous attempt to discredit the Copenhagen interpretation of his work, which asserted that nature contains a fundamental randomness, an idea he could never accept. Schrödinger believed that his wave function represented an actual electron, which was smeared out in the electron cloud. The Copenhagen interpretation, as formulated by Max Born, responded, "No, the wave function describes the *probability* of finding the electron in the cloud." According to this view, a particle exists in an indefinite state until it is observed, at which point it becomes a definite particle. To illustrate the absurdities to which this interpretation leads, Schrödinger devised a thought experiment in which a cat in a closed box either lived or died according to whether a quantum event occurred. The paradox arose from applying the conventional quantum mechanical view about the microscopic world to a macroscopic object (a cat), which is, after all, made of atoms and should, therefore, follow the same laws. Until the cat is observed and found to be either dead or alive, it must be viewed as existing in an indeterminate state—neither dead nor alive. This discomfiting paradox has never been satisfactorily resolved, although attempts to do so have led to different interpretations of quantum mechanics.

While wrestling with these complexities, Schrödinger came face to face with his personal paradox: his love for a homeland that had become both repulsive and inimical to him. His tenure at the University of Graz (soon to be renamed Adolf Hitler University), in Austria, lasted only two years. He fled with his wife in 1939, via Rome, to the Institute for Advanced Studies in Dublin, where he remained until 1956. He devoted himself to the problem of unifying gravitation and electromagnetism and in 1947 triumphantly presented his solution. When Einstein, with whom he had been corresponding on the issue, rejected his ideas, he was devastated and broke off correspondence.

Schrödinger returned to his beloved Vienna in 1956. But he had occupied his chair at the University of Vienna for only a year when he became seriously ill. He died in Vienna on January 4, 1961. In 1993, the International Erwin Schrödinger Institute for Mathematical Physics (ESI) was established in the building where he spent his last years.

In his later years, he turned his intellect to the task of applying philosophy to physics and the atom. In *Nature and the Greeks* (1954), he presented his vision of Greek science and philos-

ophy. His final book, *My View of the World*, published posthumously in 1961, expressed his personal metaphysical outlook.

Schrödinger's own words best describe his original genius: "The task is, not so much to see what no one has yet seen; but to think what nobody has yet thought, about that which everybody sees."

Further Reading

Gribben, John. *In Search of Schrödinger's Cat: Quantum Physics and Reality.* New York: Bantam Doubleday Dell, 1984.

———. *Schrodinger's Kittens and the Search for Reality: Solving the Quantum Mysteries.* Boston: Little, Brown, 1996.

Marshall, Ian N., Danah Zohar, and F. David Peat. *Who's Afraid of Schrödinger's Cat?: An A-To-Z Guide to All the New Science Ideas You Need to Keep Up With the New Thinking.* New York: William Morrow, 1998.

Milburn, Gerard J. *Schrödinger's Machines: The Quantum Technology Reshaping Everyday Life.* New York: W. H. Freeman, 1997.

Moore, Walter J. *A Life of Erwin Schrödinger.* London: Cambridge University Press, 1994.

Schrödinger, Erwin. *My View of the World.* Woodbridge, Conn.: Ox Bow Press, 1994.

———. *Space-Time Structure.* Cambridge, Eng.: Press Syndicate of the University of Cambridge, 1990.

⊠ **Schwinger, Julian Seymour**
(1918–1994)
American
Theoretician, Quantum Field Theorist

Julian Schwinger, one of the great theoretical physicists of his time, won the 1965 Nobel Prize in physics for his role as a prime architect of quantum electrodynamics (QED), the study of the interaction of electrons and electromagnetic radiation. Working concurrently with RICHARD PHILLIPS FEYNMAN and SIN-ITIRO TOMONAGA, Schwinger laid the foundations of relativistic QED and set the stage for FREEMAN DYSON's work, which reconciled his mathematical formalism with Feynman's diagrammatic formulation of QED.

Schwinger was born on February 12, 1918, in New York City, into a middle-class Jewish family, the younger of two sons. His father, Benjamin, who had immigrated to the United States in 1880, was a successful designer of women's clothing. His mother, Belle, had immigrated as a child from Lodz, then in eastern Poland. The family lived in Jewish Harlem, a well-to-do neighborhood, and later in the prosperous environs of Riverside Drive. A prodigy who discovered his obsession with physics at an early age, Schwinger attended Townsend Harris High School, a renowned institution affiliated with the City College of New York (CCNY), which he entered in 1934.

At the age of 16, he wrote his first paper (never published), "On the Interaction of Several Electrons," an insightful analysis of the electron field within the context of the quantum mechanics of the electromagnetic field. He was from the beginning an autodidact, who sat in the college's library teaching himself modern physics from original papers. He read all the papers of PAUL ADRIEN MAURICE DIRAC, who he later said was "by far the overwhelming influence" in his thinking. However, his failure to attend classes led to a mediocre record at CCNY.

Then, through the intervention of ISIDOR ISAAC RABI, whom he instantly impressed with his theoretical insights, Schwinger was admitted to Columbia University, where, despite an F in a chemistry course, he was elected to Phi Beta Kappa. At the age of 19, he published his first paper, "The Magnetic Scattering of Neutrons," which contained the core of his soon-to-be-completed dissertation, in *Physical Review* in 1937. He hid not receive a Ph.D. until two years later, however, since, refusing to attend classes, he had trouble fulfilling the formal requirements for the degree. For the next two years he was at the University of California at Berkeley, first as a National Research Council Fellow and then as an assistant to J. ROBERT OPPENHEIMER.

When World War II began, he was teaching elementary physics to engineering students at Purdue University, where, in 1942, he was made an assistant professor. The following year he was given a leave of absence by Purdue and went to work at the Radiation Laboratory at the Massachusetts Institute of Technology (MIT), Cambridge. He was soon sent to the University of Chicago's Metallurgical Laboratory, which was involved in the atom bomb project. During that summer he worked on improving the design of the Hanford nuclear reactor. His colleagues at Chicago found working with him to be a challenge, since Schwinger had a two-handed blackboard technique and when excited would be simultaneously solving two equations. Dissatisfied with this type of work, Schwinger drove back to Boston and was reinstated at the Radiation Lab; he worked in George Uhlenbeck's group, in which he became the prime force in the development of radar waveguides.

Later, he would say of his work at the Radiation Lab, "I first approached radar problems as a nuclear physicist, but soon I began to think of nuclear physics in the language of electrical engineering that would eventually emerge as the effective range formulation of nuclear scattering."

Having become aware of the large magnitude of microwave powers available, he began to think about electron accelerators, which led to the question of radiation by electrons in magnetic fields. He was especially impressed by the fact that at the classical level, the reaction of the electron's field alters the properties of the particle, including its mass. This property would be significant in the future development of QED. During this period he also developed variational techniques that produced major advances in several fields of mathematical physics.

In 1945, when the war ended, Schwinger resigned from his position at Purdue to become

Julian Seymour Schwinger was a prime theoretical architect of relativistic quantum electrodynamics (QED), the study of the interaction of electrons, positrons, and electromagnetic radiation. *(AIP Emilio Segrè Visual Archives. Physics Today Collection)*

associate professor at Harvard. Two years later he was promoted to full professor and married Clarice Carrol of Boston. This was the beginning of a legendary period of physics in Cambridge. Schwinger's friendship with the eminent theorist Victor Weisskopf, who was at MIT, forged the theoretical physicists at Harvard and MIT into a close-knit community. Schwinger's brilliant lectures and mentoring of outstanding graduate students helped Harvard rapidly become one of the most important training grounds for theoretical physicists. The notes from one of Schwinger's courses were compiled by John Blatt as "Advanced Theoretical Nuclear

Physics" and had a tremendous influence on graduate students in the late 1940s and 1950s.

While Schwinger was thriving at Harvard, in 1947, WILLIS EUGENE LAMB, working in Rabi's laboratory at Columbia University, began experimental investigations that would have a profound impact on the formulation of QED. Applying the art of spectroscopy with unprecedented precision, he shone a beam of microwaves onto a hot wisp of hydrogen gas blowing from an oven. He found that two fine structure levels in the next lowest group, which should have coincided with the Dirac theory, were in reality shifted relative to each other by a certain quantity (the Lamb shift). He measured it with great accuracy and later made similar measurements on heavy hydrogen. Lamb's announcement of his news to the participants of the 1947 Shelter Island (New York) Conference had an electrifying effect. Schwinger, who was present, commented:

> Everybody was highly euphoric. . . . The facts were incredible—to be told that the sacred Dirac theory was breaking down all over the place.

Another flaw in the Dirac theory was found that same year in the same Columbia lab, by POLYKARP KUSCH, who discovered a tiny discrepancy from what the theory predicted when he made highly accurate measurements of the magnetic moment of the electron.

Schwinger and other quantum theorists such as HANS ALBRECHT BETHE, RICHARD PHILLIPS FEYNMAN, and SIN-ITIRO TOMONAGA began to realize that what was missing from Dirac's theory was a proper interpretation of the unwieldy concept of the self-interaction of the electron, which by its very nature contains infinities, thus preventing a straightforward physical interpretation. When the electromagnetic field is quantized, according to the rules of quantum mechanics, particles of light called photons are generated. At the heart of the quantum electrodynamic process is the quantum exchange force by which different electrons interact by exchanging photons; in this context an electron can also exchange a photon with itself.

How were physicists to deal with this self-interaction? QED, as it was formulated in the mid-1940s, was not considered to be a relativistically covariant formalism (that is, it was not formally compatible with the rules of special relativity). This lack of relativistic covariance prevented a unique mathematical interpretation of the physical effects of self-interaction. Schwinger changed all this when he discovered a relativistically covariant form for QED, by introducing the concept of renormalization, which allowed a consistent mathematical interpretation of the self-energy infinities.

On the physical level, renormalization implied that physical particles are surrounded by a cloud of *virtual particles*, that is, ghostly particles that exist within the context of the uncertainty principle, whose energy, momentum, and charge modify the physical appearance of the bare original particle. In applying the method of renormalization, Schwinger found that the self-energy infinities could be subtracted out. This led to a fully consistent relativistic theory of QED that explained the Lamb shift as due to the virtual particle modification of the Coulomb force between the electron and the proton in the hydrogen atom. Using his new relativistically covariant QED formalism with renormalization, Schwinger was also able to calculate the anomalous magnetic moment of the electron.

After his groundbreaking work on QED, Schwinger concentrated on general theoretical questions, rather than specific topics of immediate experimental interest. Early in 1957, he anticipated the existence of two different neutrinos, associated, respectively, with the electron and the muon. This was later confirmed experimentally by LEON M. LEDERMAN. A related and somewhat earlier speculation, that all weak interactions are transmitted by heavy, charged, unit-spin particles, was also later confirmed by decisive experimental tests. Schwinger's habit of finding theoretical value in experimentally unknown particles led him to a revived concern with the possible existence of magnetically charged particles called magnetic monopoles, which later were found to be involved in the understanding of strong interactions.

In his later years, Schwinger backed away from his earlier work on quantum field theory and worked on a phenomenological theory of particles, which he called "source theory," which deals uniformly with strong interacting particles, photons, and gravitons, thus providing a general approach to all physical phenomena. He described this work in his two-volume *Particles, Sources, and Fields*. The theory's modest scientific goal was to move from solid knowledge of phenomena at accessible energies to that at higher energies. He perceived it as a sound and simple mathematical description of laboratory practice, without the difficulties that occur in the standard quantum field operator formalism. It incorporated no infinities and thus needed no renormalization; no new constants would appear, because all parameters were fixed when the class of phenomena under examination was fixed.

However, the physics community, which was evolving into the realm of the unified gauge theories of elementary particles, was less than enthusiastic about Schwinger's new development. In 1965, *Physical Review Letters* returned his submissions with scathing comments. In protest, he resigned from the American Physical Society. Schwinger grew increasingly isolated as he pursued his new theory. Nonetheless, his views influenced what is now known as the effective field theory (EFT) approach to particle processes. The rationalizing of EFT has produced a resurgence of interest in Schwinger's legacy, even if its long-term effects are currently unpredictable.

In 1972, Schwinger left Harvard and from then until his death in 1994, he taught at the

University of California, Los Angeles. An enormously respected and highly gifted lecturer, he supervised numerous gifted graduate students, including three future Nobel Prize winners. His books include *Discontinuities in Waveguides*, with D. Saxon (1968); *Quantum Kinematics and Dynamics* (1970); *Einstein's Legacy: The Unity of Space and Time*, (1987) and *Particles, Sources, and Fields* (1989). He died on July 16, 1994, in Los Angeles at the age of 76.

Schwinger's career spanned an unusual arc—from early recognition for his work on QED to post–Nobel Prize ostracism for his work on the source theory of elementary particles. Nonetheless, he was a legendary figure in mid-20th-century physics, who will be remembered for his reformulation of QED. His concept of renormalization made possible the first relativistically self-consistent framework for the quantization of fields from which physical consequences could be extracted and experimentally verified.

See also BOHR, NIELS HENRIK DAVID; HEISENBERG, WERNER.

Further Reading

Mehra, Jagdish, and Kimball A. Milton. *Climbing the Mountain: The Scientific Biography of Julian Schwinger.* Oxford: Oxford University Press, 2000.

Schweber, Silvan S. *QED and the Men Who Made It: Dyson, Feynman, Schwinger, and Tomonaga.* Princeton, N.J.: Princeton University Press, 1994.

⊠   Segrè, Emilio Gino
(1905–1989)
Italian/American
Experimentalist, Nuclear Physicist, Particle Physicist

Emilio Gino Segrè opened the door to the modern exploration of the world of antimatter with his discovery of the antiproton in 1955. He and his colleague Owen Chamberlain were honored for this work with the 1959 Nobel Prize in physics. During World War II, Segrè made a key contribution to the development of the atomic bomb by his confirmation that plutonium is fissionable.

He was born on February 1, 1905, in Tivoli, Rome, the son of Guiseppe Segrè, an industrialist, and Amelia Treves. The Segrès were a wealthy family of intellectuals, professionals, and businesspeople. After completing his secondary education in Rome, Emilio entered the University of Rome in 1922 to study engineering. Fascinated with the new developments in physics, he changed fields five years later. Segrè received a doctoral degree in 1928 as the first doctoral student of the great ENRICO FERMI, who remained his friend and collaborator for over three decades. After Fermi's death, Segrè became his biographer.

After being discharged from the Italian army, in which he served between 1928 and 1929, Segrè became a research assistant at the University of Rome. A Rockefeller Foundation Fellowship in 1930 gave him the opportunity to work with OTTO STERN in Hamburg, Germany, and PIETER ZEEMAN in Amsterdam. In 1932, he returned to Italy, where he was appointed associate professor at the University of Rome and began a period of intensive collaboration with Fermi and his group. Until 1934, except for a short period spent in investigating molecular beams, all his research focused on atomic spectroscopy. Between 1934 and 1935, he played a pioneering role in the discovery of slow neutrons, which became an important component in the discovery of nuclear reactors.

Segrè left Rome in 1936 to become director of the Physics Laboratory at the University of Palermo but was forced to leave Italy only two years later to escape Mussolini's anti-Semitic decrees. He decided to immigrate to the United States and eventually became a U.S. citizen. His American career began where it would end many years later: at the University of California, Berkeley. Starting out as a research assistant at the Radiation Laboratory, he went on to become a lecturer in the physics department. His prewar period at Berkeley was one of his most productive times in nuclear physics, when he worked with Glenn Seaborg on methods of separating nuclear isomers, that is, pairs of isotopes of the same proton and neutron numbers, but in different quantum states.

When World War II erupted, as were many scientists of the time, Segrè was concerned about Germany's possible military application of the new discoveries in nuclear fission. He undertook a study to prove that a bomb based not on separated isotopes of uranium, but on plutonium, discovered in 1943, was feasible. The results of this work put him in the center of the uranium project, which later became the Manhattan Project to build the first atomic bomb. Moving his family to the Los Alamos Laboratory, from 1943 to 1946, he directed a group whose task was to study spontaneous fission of uranium and plutonium isotopes. Present at the first test of an atomic bomb at Alamogordo in July 1945, he described it as "an awesome sight, comparable to great natural phenomena, [which] had a sobering impact on the beholders."

After the war, Segrè had several offers from universities; drawn by the prospect of new accelerators and the exciting experimental possibilities they offered, he chose to return to Berkeley as a full professor of physics. There, in 1955 he and his colleagues, Owen Chamberlain, Clyde Weigand, and Thomas Ypsilantis, discovered the antiproton. The notion of antiparticles originated with PAUL ADRIEN MAURICE DIRAC, who in 1931 had predicted the existence of a positron, or positive electron. Dirac's prediction was motivated by his discovery that the mathematics describing the electron contained twice as many states as were expected. He proposed that the positive energy states described the electron, and that the negative energy states could be physically interpreted as describing a particle with a mass equal to that of an electron but with an opposite (positive) charge of equal strength, that is, an antiparticle of the electron. Independent experiments by CARL DAVID ANDERSON and LORD PATRICK MAYNARD STUART BLACKETT, in 1932 and 1933, confirmed that a positron could be produced by a photon. Dirac's discovery of the positron led to the prediction of other antiparticles, such as the antiproton.

In 1954, the testing of this prediction could only be done by means of a high-energy proton accelerator, such as the Bevatron at Berkeley, which had reached its planned capacity of 6 billion electron volts (Bevs)—the energy required for the pair formation of protons–antiprotons. This, however, was only a starting point. Segrè would later recall:

> The chief difficulty of the experiment was the extraction of a reliable antiproton signal out of a huge noise produced by the many reactions occurring in the target. In fact, only about one in 50,000 of the negatively charged particles emerging from the target was an antiproton. Two lines of attack were possible: a determination of the e/m ratio for the particles produced, or an observation of the terminal event that was sufficiently detailed to identify the annihilation process.

Segrè and Chamberlain's ingenious methods of detection and analysis used both approaches to establish the existence of the antiproton. The colleagues shared the 1959 Nobel Prize in physics for this work.

Shortly afterward, another group at Berkeley discovered the antineutron, the antimatter counterpart of the neutron. From this point on, Segrè's major focus of research became antinucleons, the antimatter counterpart of the nuclear structure of matter. However, he would make no further basic discoveries. His group dispersed and Segrè himself, who preferred "to keep apparatus as simple and inexpensive as possible [and] to have a solid theoretical foundation," was never at home with the increasingly more powerful accelerators or with the larger and larger collaborations that became necessary in order to work with them.

Segrè was married, first to Elfriede Spiro, with whom he had a son and two daughters, and later to Rosa M. Segrè. During the 1960s and 1970s, he was editor of *The Annual Review of Nuclear Science*. After his retirement from Berkeley in 1972, he devoted much of his time to studying the history of physics and published *From X-Rays to Quarks: Modern Physicists and Their Discoveries* and *Falling Bodies to Radio Waves: Classical Physicists and Their Discoveries*. He died on April 29, 1989, of a heart attack.

Speaking of his 1964 text, *Nuclei and Particles*, which reflects the experience of his generation of physicists, Segrè offered an important historical perspective: "Nuclei and particles were not yet separated, and indeed, a good fraction of the particle physicists came from the ranks of nuclear physicists or cosmic-ray physicists." Segrè's own career both illustrated and furthered this progress. As a young physicist he probed the physics of the nucleus, then thought to consist only of neutrons and protons. In his mature work, through his experimental detection of the antiproton, he glimpsed the far greater complexity of the exotic world of elementary particles and became one of the first particle physicists.

See also LEON M. LEDERMAN; J. ROBERT OPPENHEIMER.

Further Reading

Segrè, Claudio G. *Atoms, Bombs and Eskimo Kisses: A Memoir of Father and Son.* New York: Viking Penguin, 1995.

Segrè, Emilio. *A Mind Always in Motion: The Autobiography of Emilio Segrè.* Berkeley: University of California Press, 1993.

⊠   **Shockley, William Bradford**
(1910–1989)
American
Theoretical Physicist, Solid State Physicist

William Bradford Shockley shared the 1956 Nobel Prize in physics with JOHN BARDEEN and Walter Battrain for his part in inventing the transistor, the basis for the most important technological advances of the 20th century.

He was born on February 13, 1910, in London, England, to American parents, William Hillman Shockley, a mining engineer, and May Bradford Shockley, a federal deputy surveyor of mineral lands. The family remained in London on business until William Jr. was three, when they returned to the United States and settled in Palo Alto, California. Shockley's parents schooled him at home until he was eight. In addition to his parents' encouragement of his scientific interests, he was stimulated by his relationship with a neighbor who taught physics at Stanford University. He spent two of his high school years at the Palo Alto Military Academy before enrolling briefly in the Los Angeles Coaching School to study physics. After rounding out his idiosyncratic secondary education at Hollywood High School, he entered the University of California at Los Angeles in 1927. After a year he transferred to the California Institute of Technology in Pasadena, where he studied with a number of distinguished physicists, including Linus Pauling. He earned a B.S. in 1932 and went on to the Massachusetts Institute of Technology (MIT) in Cambridge for his graduate work, on a teaching fellowship. In 1933, he married Jean Alberta Bailey, with whom he would three children. MIT gave him his grounding in solid state physics and awarded him a Ph.D. in 1936 for a dissertation on the energy band structure of sodium chloride.

Shockley then took a job with Bell Laboratories in Murray Hill, New Jersey, where he first proved himself by his vacuum tube research, which advanced the company's goal of using electronic switches for telephone exchanges instead of the mechanical switches that were currently in use.

When World War II broke out, Shockley turned to work on military projects at Bell, particularly the refinement of radar systems. He then became research director of the Antisubmarine Warfare Operations Research Group, set up by the United States Navy, at Columbia University. After the war, he continued to serve national interests as a member of the Scientific Advisory Panel of the United States Army, the Air Force Scientific Advisory Board, and the President's Scientific Advisory Board.

Returning to Bell Labs in 1945, Shockley was made head of a research group focusing on the basic physical nature of semiconductors that included John Bardeen and Walter Brittain. Through the application of quantum theory to solid state physics in the late 1930s physicists first understood the electrical properties of semiconductors: they became aware of the role of low concentrations of impurities in controlling the number of mobile charge carriers in materials. Current rectification (i.e., the conversion of oscillating current into direct current) at metal–semiconductor junctions had long been known, but the next step required was to produce amplification analogous to that achieved by vacuum tube technology. Shockley's group began a program to control the number of charge carriers at semiconductor surfaces by varying the electric field.

In the spring of 1947, Shockley asked them to investigate the reason for the failure of an amplifier he had designed, which was based on a crystal of silicon, later replaced by germanium. By observing Brittain's experiments, Bardeen realized that the assumption they had been making—that electrical current traveled through all parts of the germanium in the same way—was incorrect. On the contrary, electrons behave differently at the surface of the metal. If they could control what was happening at the surface, the amplifier should work. On the basis of this insight they were able to demonstrate the effects of amplification by using two metal contacts 0.05 mm apart on a germanium surface. Large variations of the power output through one contact were observed in response to tiny changes in the current through the other.

Using this technique Bardeen and Brittain succeeded in building the first point-contact transistor on December 23, 1947; it was the forerunner of the many complex devices now available through silicon chip technology. They called this first successful amplifying device the transistor because it combined the notions of charge transfer and electrical resistance. Apparently stung by the discovery, in which he had not directly participated, Shockley hastened to make improvements to the original transistor in 1950, developing the junction transistor, which made mass production of transistors possible.

The year 1955 was a time of transition. Shockley divorced his first wife and married Emmy Laning. He also resigned as director of the Transistor Physics Department at Bell Labs. In 1956, he shared the Nobel Prize in physics for the invention of the transistor with Bardeen and Battrain. He then served as visiting professor and consultant at various universities and corporations. For a while he directed his own lab for developing transistors and related devices, however, it folded in 1968. In 1963, he was appointed professor of engineering at Stanford University, where he taught until 1975.

He was the author of numerous scientific articles, as well as two books on semiconductors, and the holder of more than 90 U.S. patents. His 1950 publication, *Electrons and Holes in Semiconductors*, became an essential guide for researchers.

During the latter part of his career he became increasingly immersed in a system of racially noxious ideas on the relationship between intelligence and genetics, which strongly resembled the program of the eugenics movement in the 1910s and 1920s. His argument that the higher birth rate among those with lower IQs posed a threat to the future of the population was couched in racial terms: he claimed that blacks were genetically inferior to whites. Further, he advocated that eugenic measures be taken to prevent this outcome. Shockley's stance made him increasingly controversial and an object of widespread hatred. In 1980, he sued the *Atlanta Constitution* for accusing him of ideas similar to those of the Nazi genetics experiments and won a $1 million settlement. Running on the 1982 California Republican ticket for the Senate on the single issue of "dysgenics," he finished eighth.

Shockley died in Stanford, California, on August 11, 1989. His scientific insights eventually led to the mass marketing of silicon chip electronic devices that enabled computers and other electronic equipment to be produced more cheaply, more rapidly, and more reliably and ultimately led to the birth of Silicon Valley.

Further Reading

Bernstein, Jeremy. *Three Degrees Above Zero: Bell Labs in the Information Age*. New York: Scribner & Sons, 1984.

Riordan, Michael, and Lillian Hoddeson. *Crystal Fire: The Invention of the Transistor and the Birth of the Information Age*. New York: W. W. Norton, 1998.

⊠   **Snell, Willibrord**
(1580–1626)
Dutch
Experimentalist (Optics)

Willebrord Snell is best known for discovering the law of refraction, also known as Snell's law, which describes the way light rays are deviated when they pass from one material to another.

He was born, in 1580, in Leiden, the Netherlands, where his father taught mathematics at the university. While pursuing his interest in mathematics, he studied law at the University of Leiden. From 1600 to 1604, he traveled throughout Europe, collaborating with such scientific luminaries as the astronomers Tycho Brahe and Johannes Keppler. He received his degree from Leiden in 1607 and, six years later, assumed his father's position there.

In 1617, Snell published his method of determining distances by triangulation, using his house and the spires of nearby churches as reference points. He used a large quadrant more than two meters long to determine angles and, by building up a network of triangles, obtained a value for the distance between two towns on the same meridian. In this way he was able to determine accurately the radius of the Earth, thus founding the science of geodesics.

Snell made his great contribution to optics in 1621, when, after exhaustive experimental work, he discovered his law of refraction, which relates the angle of incidence of an incoming light ray as it enters a material medium to the angle of refraction of an outgoing light ray as it leaves a material medium. The physical process of refraction causes a light ray to change its direction as it passes from one material medium to another. Snell defined the index of refraction (n) of a transparent medium as the speed of light in a vacuum (c) divided by the speed of light in a transparent material (v). Then $n = c/v$. Since the speed of light in a transparent material is slower than the speed of light in a vacuum, Snell found that this quantity is greater than or equal to 1. Snell did experiments with two different transparent media (i.e., medium 1 and medium 2 with indices of refraction n_1 and n_2, respectively) separated by a flat surface. In these experiments he defined the angle of incidence as the angle between the perpendicular to the flat surface and the light ray leaving medium 1 and the angle of refraction of the angle between the perpendicular to the flat surface and the light ray entering medium 2. On the basis of these experiments Snell formulated the following physical law:

$$n_1 \sin(\text{angle of incidence}) = n_2 \sin(\text{angle of refraction})$$

where the symbol *sin* represents the trigonometric sine of the angle.

Unfortunately, Snell did not publish this result at the time of its discovery and it remained unpublished for many years after his death in Leiden on October 30, 1626, at the age of 46. Through the work of René Descartes and CHRISTIAAN HUYGENS, Snell's law of refraction became an essential element of the modern science of optics. Its ability to predict how light is bent as it passes from air to glass, for example, makes it an invaluable tool in designing lenses.

Further Reading

Joyce, W. B., and A. Joyce. "Descartes, Newton, and Snell's Law." *Journal of the Optical Society of America* 66, no. 1: 1–8, 1976.

⊠   ## Sommerfeld, Arnold Johannes Wilhelm
(1868–1951)
Prussian
Theoretical Physicist (Statistical Mechanics, Quantum Mechanics), Atomic Physicist

Arnold Sommerfeld was an influential and revered member of the community of European-born physicists who produced the quantum revolution in the first third of the 20th century. As a researcher, he made the major contribution of refining NIELS HENRIK DAVID BOHR's model of the atom, thereby setting that model on firmer ground and solidifying the claims of quantum theory. As a gifted teacher, he introduced the new physics to generations of students and nurtured the careers of several of its major architects.

Arnold Johannes Wilhelm Sommerfeld was born on December 5, 1868, in Königsberg, Prussia (now Kaliningrad, Russia). After completing his studies at the gymnasium in his native city, he entered Königsberg University, where he chose to specialize in mathematics, despite Königsberg's preeminence as a school for theoretical physics. After receiving his Ph.D. in 1891, he became an assistant at the Mineralogical Institute at Göttingen University. Four years later, he won the post of lecturer at Göttingen, where his research moved in several directions: the mathematical theory of diffraction, propagation of electromagnetic waves in wires, and the study of the field produced by a moving electron. His next academic position took him, in 1897, to the Mining Academy in Clausthal, where he taught mathematics. While there he began working with Felix Klein on the 13-year study of gyroscopes that would result in their four-volume work on the subject. He then moved on to the Technical Institute in Aachen, in 1900; there he served as professor of technical mechanics.

Sommerfeld's years of academic wandering came to an end in 1906, when the University of Munich appointed him director of the Institute of Theoretical Physics, which had been created there for him. His ability to attract brilliant students and faculty, including the future Nobel Prize winners WOLFGANG PAULI, WERNER HEISENBERG, and HANS ALBRECHT BETHE, made Munich an internationally respected center. It was Sommerfeld who directed MAX THEODOR FELIX VON LAUE and his colleagues when they made their famous discovery of X-ray diffraction in 1912.

His own research began to focus on quantum theory, which he promoted as a fundamental law of nature. He is credited with inspiring Bohr to apply the theory to the structure of the atom. In 1913, Bohr proposed that electrons are confined to a certain number of stable orbits, in which they neither emit nor absorb energy. Only when it jumps from one discrete orbit to a lower one does the electron lose energy: it sends off an individual photon (particle of light). Since an electron in the innermost orbit has no orbit with less energy to jump to, the atom remains stable. Bohr's theory explained many of the spectral lines for hydrogen. This model began to break down, however, when it failed to account for new observations that revealed the fine structure of the spectral lines of hydrogen. In 1916, Sommerfeld, who had been studying atomic spectra, suggested a modification of the Bohr model, which would permit electrons to move in elliptical orbits as well as circular ones. The new model required the introduction of a second quantum number, the orbital quantum number, l, which describes the electron's angular momentum, in addition to the principal Bohr quantum number, n. Sommerfeld later introduced the magnetic quantum number, m, which determines the orientation of the electron orbital in a magnetic field. When spectroscopic studies confirmed Sommerfeld's predictions, acceptance of Bohr's quantum theory of the atom accelerated.

The golden years of Sommerfeld's institute ended in the early 1930s, when the Nazis gained power. The man who had boldly defended ALBERT EINSTEIN and other Jewish scientists would see his school closed in 1940, when he was 71. He would return to his post after World War II and remain in Munich until he died on April 26, 1951, after being struck by a car.

In the later part of his career, Sommerfeld continued to demonstrate his versatility as a researcher. He successfully replaced HENDRIK ANTOON LORENTZ's 1905 explanation of the electronic properties of metals, which was based on classical physics, with an approach based on statistical mechanics. Among the many influential books in which he presented state-of-the-art knowledge in several fields, the most famous are *Atomic Structure and Spectral Lines*, published in 1919 and reprinted in a number of editions in the 1920s, and *Wave Mechanics*, 1929.

See also BOLTZMANN, LUDWIG.

Further Reading

Sommerfeld, Arnold. *Electrodynamics*. New York: Academic Press, 1952.
———. *Mechanics*. New York: Academic Press, 1964.
Wheaton, Bruce R. *The Tiger and the Shark*. Cambridge: Cambridge University Press, 1990.

⊠ Stark, Johannes
(1874–1957)
Bavarian
Theoretician and Experimentalist, Atomic Physicist

Johannes Stark is famous for his discovery of what is now known as the Stark effect, the division of spectral lines in an electric field, work that was recognized by the 1919 Nobel Prize in physics.

Stark was born in Schikenhof, Bavaria, on April 15, 1874. His father, a landowner, sent the boy to secondary schools in Bayreuth and Regensburg. Stark then went on to the University of Munich, where he studied physics, mathematics, chemistry, and crystallography. He wrote his dissertation on the phenomenon of Newton's rings in a dim medium, and received his Ph.D. in 1897. For the next three years, he worked as a research assistant at the Physics Institute at the University of Munich.

In 1900, Stark moved to the University of Göttingen to work as an unsalaried lecturer. However, two years later, his fortunes changed when he predicted that the high-velocity canal rays (positively charged ions produced in a cathode ray tube) should exhibit the Doppler effect (an apparent change in the frequency of a wave motion, caused by relative motion between the source and the observer). In 1905, he confirmed this prediction by demonstrating the frequency shift in hydrogen canal rays. On the basis of this work, he was made a professor at the Technische Hochschule in Hanover, in 1906 and, three years later, obtained a similar position at the Technische Hochschule in Aachen, where he remained until 1917.

More than a decade earlier, PIETER ZEEMAN had demonstrated that a division of the spectral lines emitted by atoms can be caused by the influence of a magnetic field, the so-called Zeeman effect. Stark built on this work by demonstrating that the same division of spectral lines can be produced by the influence of an electric field. His technique was to photograph the spectrum emitted by canal rays, consisting of hydrogen and helium atoms, as they passed through a strong electric field. The electric field caused the electrons in the radiating atoms to change position relative to the nucleus and distorted the electron orbital motion. Since light is emitted when an excited electron moves from a higher- to a lower-energy orbit, distortion of the orbits also distorts the emitted light, which manifests itself as division of the spectral lines. Stark announced his discovery of this spectral phenomenon, now known as the Stark effect, in 1913.

In that same year, Stark made yet another important contribution when he generalized the photoelectric law proposed by ALBERT EINSTEIN in 1906. This principle, now called the Stark–Einstein law, states that each molecule involved in a photochemical reaction absorbs only one quantum of the radiation that causes the reaction.

Stark went on to become professor of physics at the University of Greifswald. His discovery of the Stark effect garnered him the 1919 Nobel Prize in physics, and he used his prize money to set up his own laboratory. Surprisingly, at this high point in his career, he made a detour into the world of commerce, attempting to set up a porcelain factory in northern Germany. The German economy was depressed, however, and Stark was forced to abandon his venture and return to the world of science.

Stark married Luise Uepler, with whom he had five children. The later part of his scientific career was marred by his involvement in the rabid anti-Semitism sweeping Germany, which led him to join the Nazi Party in 1930. He became an important figure in German physics, serving as president of both the Reich Physical–Technical Institute and the German Research Association, from 1933 to 1939, when a conflict with government authorities caused him to resign. After World War II, he was sentenced, for his Nazi

activities, by a German denazification court, to four years in a labor camp.

A prolific writer, Stark published more than 300 scientific papers, as well as books on electricity, elementary radiation, and electrical spectroscopic analysis of chemical atoms. He founded the *Jahrbuch der Radioaktivitat und Electronik* (Yearbook of radioactivity and electronics), which he edited from 1904 to 1913. In his final years, he worked in his private laboratory on his country estate near Traunstein, West Germany, studying the effect of light deflection in an inhomogeneous electric field. He died on June 21, 1957, in Traunstein.

The discovery of the Stark effect, together with the Zeeman effect, provided support for the quantum mechanical model of the atom and acted as an experimental template against which the development of quantum mechanics in general, and the evolution of the Bohr model of the atom, could be refined and advanced.

See also BARKLA, CHARLES GLOVER; BOHR, NIELS HENRIK DAVID; LAUE, MAX THEODOR FELIX VON; LENARD, PHILIPP VON; SCHRÖDINGER, ERWIN.

Further Reading
Born, Max. *Atomic Physics*. New York: Dover, 1989.

⊠   **Stefan, Josef**
(1835–1893)
Austrian
Experimentalist (Electromagnetism, Thermodynamics)

Josef Stefan made a lasting contribution to physics in 1879 when he empirically determined the relationship between the amount of energy a blackbody radiates and its temperature. Five years later, in 1884, Stefan's brilliant student LUDWIG BOLTZMANN would formulate the theoretical basis for the relationship, which became known as the Stefan–Boltzmann law. This work on blackbody radiation proved crucial in the evolution of ideas leading up to the formulation of the quantum theory.

Stefan was born on March 24, 1835, in Saint Peter, Austria, and studied at the secondary school there. At the age of 19 he entered the University of Vienna, which would remain his academic home for the rest of his life: in 1858, after completing his studies, he became a lecturer; in 1863, he was promoted to the rank of professor of higher mathematics and physics. He also became director of Vienna's Institute for Experimental Physics in 1866.

Stefan's inquiries into the laws of radiation arose from a troubling inconsistency in the laws of cooling derived by SIR ISAAC NEWTON. According to Newton, the rate of cooling of a hot body is proportional to the difference in temperature between the body and its surroundings; however, researchers had found that Newton's formulation greatly underestimated the amount of heat bodies give out at very high temperatures. By measuring the heat energy radiated by a blackbody, that is, a body that absorbs all the radiant energy that falls on it, Stefan determined that the power emitted per unit area was proportional to the fourth power of the absolute temperature in degrees Kelvin. Boltzmann derived this same law from thermodynamic principles and the kinetic theory developed by JAMES CLERK MAXWELL. It became known as the Stefan–Boltzmann law, and the universal proportionality constant in the equation was called the Stefan–Boltzmann constant. By treating the Sun as an approximate blackbody, Stefan was able to use his equation to determine its surface temperature to be about 6000°C.

Later Stefan would help to confirm Maxwell's kinetic theory by making accurate measurements of the conductivity of gases. He also served as vice president of the Imperial Academy of Sciences from 1885 until his death, in Vienna, on January 7, 1893.

See also PLANCK, MAX; ERNEST LUDWIG.

Further Reading
Kuhn, Thomas S. *Black-Body Theory and the Quantum Discontinuity.* Chicago: University of Chicago Press, 1987.

⊠ Stern, Otto
(1888–1969)
German/American
*Theoretician and Experimentalist,
Atomic Physicist, Physical Chemist*

Otto Stern was a distinguished experimentalist whose pioneering work with the molecular beam method, a powerful tool for studying the properties of molecules, atoms, and atomic nuclei, provided the first evidence that beams of atoms and molecules have wave properties, and determined the magnetic moment of the proton. He won the 1943 Nobel Prize in physics for these contributions.

He was born in Sohrau, Upper Silesia, Germany (now Zory, Poland), on February 17, 1888, the son of a mill owner. When he was four, the family moved to Breslau, where Stern entered the Johannes Gymnasium in 1897 and obtained a diploma in 1906. He studied physical chemistry in universities: in Frieburg im Breisgau, Munich, and Breslau; in Breslau he had an apprenticeship with the eminent physicist ARNOLD JOHANNES WILHELM SOMMERFELD and earned a Ph.D. in 1912 for a dissertation on physical chemistry. He became intrigued with theoretical physics and managed to become a postdoctoral associate to ALBERT EINSTEIN in Prague. He followed Einstein to Zurich in 1913, taking a job as an unpaid lecturer at the technical high school.

Then, in 1914, Stern moved on to the University of Frankfurt am Main as an unpaid lecturer of theoretical physics. When World War I began, Stern joined the army and was assigned to meteorological work on the Russian front. He continued his theoretical work in his spare time;

toward the end of the war he was transferred to a laboratory in Berlin. After the war he returned to Frankfurt, where he remained until 1921.

Between 1912 and 1918, Stern published theoretical work on statistical thermodynamics and quantum theory. His most significant contributions, however, were in experimental physics, through his development of the molecular beam method, discovered in 1911 by Louis Dunoyer. In Stern's molecular beam method a tiny opening is made in a heated container held inside a region of high vacuum. The vaporized molecules inside the heated container flow out, without colliding with one another, to form a straight beam of moving particles. The molecular beam could be narrowed still further by the use of slits, and a system of rotating slits could be used to select only those particles traveling at particular velocities. In 1919, Stern used this molecular beam method to verify Maxwell's law of velocity distribution in gases.

Stern then used his technique to resolve a dilemma that had arisen in physicists' understanding of the influence of magnetic fields on the electric forces within the atom. ANDRÉ-MARIE AMPÈRE had shown that the intrinsic magnetic fields associated with atoms can generally be ascribed to the presence of electric currents in the atoms. However, the classical theory and the quantum theory generated very different predictions about the behavior of these atomic magnets in an external magnetic field. The classical theory predicted that the atomic magnets assume all possible directions with respect to the direction of the external magnetic field. On the other hand, the quantum theory predicted that they can take only two directions: parallel and antiparallel to the external magnetic field. In particular, using NIELS HENRIK DAVID BOHR's quantum model of the atom, Sommerfeld had derived a formula for the magnetic moment of the silver atom and predicted that silver atoms in a magnetic field could orient in only two directions with respect to that field. In 1920, in

collaboration with Walter Gerlach, Stern designed an experiment to show which of these theories was correct.

In the historic Stern–Gerlach experiment, a narrow beam of silver atoms was passed through a strong magnetic field. According to classical theory, this field would cause the beam to broaden. However, the spatial quantization, that is, the splitting of the beam of atoms in an external magnetic field, predicted by the quantum theory implied that the beam would split into two distinct beams. The experimental result, showing a split beam, was the first clear evidence for space quantization. This experiment also made it possible to obtain a measurement of the silver atom's magnetic moment, the strength of the magnetic field generated by a current-carrying loop. The magnetic moment proved to be in close accord with the universal quantized unit of atomic magnetic moment, the so-called Bohr's magneton.

During 1921 and 1922, Stern was associate professor of theoretical physics at the University of Rostock, and, in 1923, he became professor of physical chemistry at the University of Hamburg and director of the Institute of Physical Chemistry, where he remained until 1933. The 10 years between 1923 and 1933 were the peak of his career, during which he made important contributions to quantum theory and built up a thriving research group. During this period Stern improved his molecular beam technique and was able to detect the wave nature of particles in the beams. This was an important confirmation of the wave–particle duality proposed by LOUIS-VICTOR-PIERRE, PRINCE DE BROGLIE, in 1924 for the electron.

In 1933, Stern measured the magnetic moment of the proton and the deuteron. The magnetic moment of the proton had been predicted by PAUL ADRIEN MAURICE DIRAC to be one nuclear magneton. It was known that the atomic nucleus, as does the electron, possesses an intrinsic rotation of its own, known as spin. However, the intrinsic spin of a charged proton creates current loops that can generate an intrinsic magnetic field, which gives the proton a magnetic moment. Only indirect, approximate measurements of the tiny nuclear magnet—estimated to be a couple of thousand times smaller than that of the electron—could be made. Because the proton, together with the neutron, formed the basic constituents of matter, physicists ascribed great importance to determining the magnetic properties of the hydrogen nucleus, or proton. If the proton and neutron were to be regarded as true elementary particles, then the proton's magnetic factor would be as many times smaller than the electron's as its mass is greater than the electron's. Stern determined that the proton factor was about 2.3 times greater than anticipated.

With the rise of the Nazis, Stern, who was Jewish, found himself in increasingly precarious circumstances. At first he found a measure of security in his personal wealth and international reputation and devoted himself to finding positions abroad for his Jewish coworkers. But conditions worsened, and Stern knew it was time to emigrate when he was ordered to remove a portrait of Einstein from his laboratory. In 1933, he left Germany for the United States.

Stern was appointed research professor of physics at the Carnegie Institute of Physics, Pittsburgh, where he set up a new department for the study of molecular beams. In 1939, he became a U.S. citizen and worked as a consultant to the United States War Department during World War II. In 1945, he became professor emeritus at Carnegie and moved to Berkeley, California, where he died of a heart attack on August 17, 1969.

Stern's own words, contained in his Nobel Prize acceptance speech, best summarize the nature of his contribution to modern physics:

> The most distinctive characteristic property of the molecular ray method is its simplicity and directness. It enables us to make measurements on isolated neutral atoms or molecules with macroscopic tools. For this reason, it is especially valuable for testing and demonstrating directly fundamental assumptions of the [quantum] theory.

See also MAXWELL, JAMES CLERK.

Further Reading

Campargue, Roger, ed. *Atomic and Molecular Beams: The State of the Art 2000*. New York: Springer-Verlag, 2001.

Estermann, Immanuel, ed. *Recent Research on Molecular Beams: A Collection of Papers Dedicated to Otto Stern on the Occasion of his 70th Birthday*. New York: Academic Press, 1959.

⊠   **Stokes, George Gabriel**
(1819–1903)
Irish
Mathematical Physicist
(Hydrodynamics, Optics)

George Gabriel Stokes left his mark on modern physics through his discovery of what is called Stokes's law, which relates the force acting on a body moving a body through a liquid to the velocity and size of the body and the viscosity of the fluid. In the field of optics, he coined the term *fluorescence* and made vital contributions to the understanding of that phenomenon.

He was born on August 13, 1819, in Skreen, Sligo, in Ireland, the youngest of six children in a religious Protestant family. His father, Gabriel, was rector of the Church of Ireland, and his mother was the daughter of a minister. George's four older brothers served the church. The family valued education, and George was taught at home, by his father and the clerk of Skreen Parish, who noted that the boy "worked out for himself new ways of doing sums, better than the book." He entered school in 1832 in Dublin, where he lived with his uncle. Despite the death of his father, money was found to continue his education. Three years later, he moved to England and attended Bristol College for two years, preparing for his entrance into Cambridge University. After winning prizes in mathematics, in 1837, he entered Pembroke College, the third oldest at Cambridge. There he was influenced by William Hopkins, a renowned academic coach, who urged him to use his mathematical prowess for understanding the material universe. When he graduated four years later with a degree in mathematics, he was "Senior Wrangler," the top math student in the whole university, and was immediately awarded a fellowship by Pembroke College.

Stokes began to work in the field of hydrodynamics, publishing papers on the motion of incompressible fluids in 1842 and 1843, and on the internal friction of fluids in motion in 1845. A report to the British Association for the Advancement of Science in 1846 on recent research in hydrodynamics established his reputation as a mathematician. He became known for his contribution to the Navier–Stokes equation, which is the primary equation of computational fluid dynamics, relating pressure and external forces acting on a fluid to the response of the fluid flow.

Over the next few years, while engaged in work on viscous fluids, he would make his greatest discovery. In a seminal 1851 paper on hydrodynamics, he derived the equation that determines the movement of a small sphere through viscous fluids of varying density. The equation, which became later known as Stokes's law, was given by

$$F = 6\pi \, r \, \eta v$$

where F is the force acting on the sphere, r is the radius of the sphere, v is the velocity of the sphere, and η is the coefficient of viscosity of the fluid. Stokes's law enables physicists to assess the resistance of fluids to motion and to determine

the terminal velocity of a body. It proved to be of great importance in ROBERT ANDREWS MILLIKAN's oil-drop experiments, performed between 1909 and 1913, which determined the charge on the electron.

In 1848, Stokes was led by his studies of fluid dynamics to look into the question of the ether, the hypothetical fluid permeating the universe through which light waves were thought to propagate. Stokes showed that the laws of optics held if the Earth dragged the ether with it in its motion through space; this caused him to propose that the ether was an elastic substance that flowed with the Earth. Later, in the 1880s, the Michelson–Morley experiments would disprove the existence of the ether altogether.

In 1849, Stokes was appointed to the prestigious position of Lucasian Professor of Mathematics at Cambridge, previously held by SIR ISAAC NEWTON, which he would retain for the rest of his life. He began giving lecture courses on hydrostatics that he would continue for the next 53 years. A famous anecdote about him claimed that he would always answer questions with a plain yes or no, when something more elaborate was expected. This habit was said to date from the time he transferred from Irish to English schools, when his brothers warned him that if he gave long Irish answers, he would be laughed at by his fellow students. To supplement the professorship's meager income, Stokes took an additional position as professor of physics at the Government School of Mines in London. That same year, he made an important contribution to the field of geodesy when he published a study of the variation of gravity at the surface of the Earth.

In 1852, Stokes offered the first explanation of a newly discovered phenomenon in matter, which he called fluorescence. In this research he had found that when illuminated by ultraviolet light, a solution of quinine sulfate that is normally colorless emits fluorescent blue light in certain circumstances. He was able to use fluorescence as a method to study ultraviolet spectra. He based his explanation on the idea that fluorescence results from absorption of ultraviolet light and emission of blue light by the molecules of matter. Ironically, his explanation was based on 19th-century concepts that incorrectly assumed the existence of an elastic ether that vibrates as a result of the illuminated molecules.

In 1854, he realized that the Sun's spectrum is made up of spectra of the elements it contains and concluded that the dark Fraunhofer lines are the spectral lines of elements absorbing light in the Sun's outer layers. This important idea would be developed by Robert Bunsen and GUSTAV ROBERT KIRCHHOFF, who proposed the method of spectrum analysis in 1860.

In 1857, Stokes moved from his highly active theoretical research period to administrative and experimental work. This change was connected to his marriage that year to Mary Susanna Robinson, without whom, he wrote to his bride-to-be, "I should go to my grave a thinking machine unenlivened and uncheered and unwarmed by the happiness of domestic affection." He turned away from his life of intense mathematical research at this point. He had to give up his fellowship at Pembroke College, whose rules required that fellowship holders be unmarried; when the rules changed, he resumed his fellowship in 1862. The marriage was said to be a happy one, but it was marred by the deaths of three of their four children, two in infancy.

A gifted mathematician, Stokes helped develop Fourier series. His pure mathematical results arose from the needs of the physical problems he studied independently and with others. To a great extent, however, Stokes's activities as a mathematician were driven by the needs of industrial applications in his own time. In addition to his connection to the School of Mines, he served as consultant to a renowned optical works and adviser on lighthouse illuminants and on wind pressure on railway structures.

Stokes was a deeply religious man, who had always been interested in the relationship between religion and science. From 1886 to 1903, he was president of the Victoria Institute, whose aim was to examine this broad issue.

He served as master of Pembroke College from 1902 until his death in Cambridge, on February 1, 1903.

Stokes was a brilliant mathematical physicist whose broad range of studies in the physics of fluids exerted an important formative influence on the next Cambridge generation. Most prominent among them was JAMES CLERK MAXWELL, who used Stokes's mathematical discoveries in formulating his theory of electromagnetism, now known as Maxwell's equations. Stokes was known for his generosity in putting aside his own research to help others and serving as a sounding board for many famous scientists, such as LORD KELVIN (WILLIAM THOMSON), with whom he carried on an extensive correspondence. The Navier–Stokes equation is currently used in computations for aircraft and ship design, weather prediction, and climate modeling.

See also BECQUEREL, ANTOINE-HENRI; FRAUNHOFER, JOSEPH VON; MICHELSON, ALBERT ABRAHAM.

Further Reading

Wilson, D. B. *Kelvin and Stokes: A Comparative Study in Victorian Physics.* Bristol, England: Adam Hilger, 1987.

Wood, Alistair. "George Gabriel Stokes 1819–1903: An Irish Mathematical Physicist." *Irish Mathematical Society Bulletin* 35: 49–58, 1995.

T

⊠ Taylor, Joseph H., Jr.
(1941–)
American
Astrophysicist, Relativist, Radio Astronomer

Joseph H. Taylor Jr. shared the 1993 Nobel Prize in physics with his graduate student, Russell A. Hulse, for their 1974 discovery of the first binary pulsar, a rapidly spinning neutron star orbiting around a companion star. Their radio pulse measurements of a binary pulsar system provided the first experimental confirmation of the existence of the gravitational waves, the traveling wave ripples in the geometric curvature of space and time, predicted by ALBERT EINSTEIN in 1916 in his general theory of relativity.

Taylor was born on March 29, 1941, in Philadelphia, Pennsylvania, the second son of Joseph Hooton and Sylvia Evans Taylor. Eventually, the family would expand to six children and move back to the family farm in Cinnaminson Township, New Jersey. Taylor describes a wholesome, carefree childhood, steeped in his parents' Quaker values of frugal simplicity, tolerance, and concern for others. When he was not collecting stone arrowheads, he and his brother, Hal, took pleasure in erecting large rotating ham radio antennas and working with ham radio transmitters and receivers. He attended mostly Quaker

schools, including Moorestown Friends School, and Haverford College, where he fell in love with mathematics, was an avid and versatile athlete, and, in spite of the poor quality of his physics and chemistry courses, began to understand "the delights of what science is really all about." For his senior honors project Taylor built a working radio telescope, an activity that allowed him to combine "a working knowledge of radio-frequency electronics with an awakening appreciation of scientific inquiry." It also led him to the field of physics, which he would pursue in his graduate studies.

After receiving his B.A. from Haverford in 1963, Taylor won admission to Harvard University, where, in the departments of astronomy, physics, and applied mathematics, he maintained a grueling work schedule, especially during his first year. His doctoral thesis, supervised by Alan Maxwell, taught him the signal processing techniques that later played an important role in his discovery of pulsars and earned him a Ph. D. in astronomy in 1968. The following year Taylor joined the physics faculty at the University of Massachusetts, Amherst, where, in 1974, he and his graduate student Russell Hulse would plan the work that would take them to Arecibo, Puerto Rico, to use the 300-meter radio telescope to search systematically for pulsars.

Pulsars were discovered in 1967 by the British physicist Antony Hewish, who observed that certain radio sources in space emitted radio signals (pulses) that were repeated with great regularity at intervals of about a second. This discovery made it possible to establish the existence of neutron stars, which physicists had been speculating about since the 1930s. A neutron star is the extremely dense remnant of a supernova explosion, which occurs when a massive star runs out of fuel at the end of its life. The combination of large mass and small size results in an extremely large gravitational field, estimated to be about 100 billion times that on Earth. When a supernova explodes, a newly formed neutron star can be sent rapidly spinning. Moreover, such a star can have a magnetic field many orders of magnitude larger than those found on Earth. If a neutron star is also surrounded by a plasma environment, such an object is called a pulsar. If the spin axis of the neutron star is properly aligned, it is possible for the spinning highly magnetized neutron star to generate enough electric potential to accelerate charges in the plasma surrounding the star, resulting in a beam of nonthermal radio emission that rotates with the star. To an observer, this radio beam appears as a series of radio pulses as the beam sweeps across the line of sight, similarly to the rotating beacon of a lighthouse.

Using the immense radio telescope at Arecibo, Taylor and Hulse picked up an unusual series of radio pulses that varied in a regular pattern. The time between pulses, 59 milliseconds, was not constant: it changed, periodically decreasing and increasing over an eight-hour period. This signal pattern led them to surmise that they were observing a pulsar, which must be alternately moving toward and away from the Earth. In accordance with the Doppler effect, when the pulsar moved toward the Earth, the antenna picked up a higher-frequency signal; when it moved away from the Earth, a lower-frequency signal was received. This meant that the pulsar must be moving around a companion star as part of a binary system where the companion must also be a neutron star.

Taylor and Hulse called this binary pulsar system by the code name PSR 1913 + 16, to indicate its celestial coordinates. They found that the pulsar whirled around its companion at a speed of up to 300 kilometers per second—10 times faster than the speed at which the Earth orbits the Sun. More importantly, they realized that they had stumbled across an unparalleled opportunity—a cosmic laboratory—to test the prediction of the general theory of relativity that objects traveling at extremely high speeds emit gravitational radiation, or gravitational waves: ripples in the curvature of spacetime. With its enormous interacting gravitational fields, the binary pulsar should emit gravitational waves, and the resulting energy drain should diminish the orbital distance between the two stars, as indicated by the change in the electromagnetic signals that the orbiting pulsar emits. This is precisely what Taylor and Hulse observed. In 1978, they announced that the orbiting period of the pulsar around its companion diminishes over time by a minuscule amount, in agreement with Einstein's prediction to an accuracy of 0.5%. This finding provided the first experimental confirmation of Einstein's gravitation theory. Taylor and his group went on to repeat the measurements of PSR 1913+16 carefully and to discover several other binary pulsars. More than 30 years after the discovery of the first pulsar, about 750 pulsars have been detected and cataloged.

Taylor joined the faculty of Princeton University in 1981 and was named James S. McDonnell Distinguished University Professor in 1986.

Taylor, who is married to Marietta Bisson Taylor, continues to work with the pulsar group at Princeton, which has pioneered the application of pulsar studies in a wide range of topics in gravitational physics, stellar evolution, cosmology, fundamental astrometry, and time-keeping metrology.

Further Reading

Hewish, Antony, et al, eds. *Pulsars as Physics Laboratories*. Oxford: Oxford University Press, 1994.

Lyne, Andrew G., and Francis Graham-Smith. *Pulsar Astronomy.* Cambridge: Cambridge University Press, 1989.

Manchester, Richard N., and Joseph H. Taylor. *Pulsars.* San Francisco: W. H. Freeman, 1977.

Piran, Tsvi. "Binary Neutron Stars." *Scientific American* 272 52, 1995.

⊠ **Teller, Edward**
(1908–)
Hungarian/American
Theoretical Physicist (Quantum Mechanics), Nuclear Physicist

Edward Teller's vociferous support for the development of an American hydrogen bomb and his crucial role in designing one earned him the title of "father of the hydrogen bomb." Hungarian-born and German-educated in the tumultuous first third of the 20th century, Teller fled both communism in his native land and Nazi fascism for U.S. shores. In America he turned his prodigious scientific inventiveness toward the creation of weapons of mass destruction, making significant contributions to the development of the atomic bomb. A polarizing figure, who later lobbied for the development of a far deadlier nuclear fusion bomb when the majority of his colleagues opposed such a weapon, Teller has consistently championed the use of advanced technology for purposes of national defense.

Edward Teller was born on January 15, 1908, in Budapest, when Hungary was a part of the Austro-Hungarian Empire. His family belonged to the community of affluent Hungarian Jews that produced an unusual number of outstanding physicists, including EUGENE PAUL WIGNER and John Von Neumann. Teller relates his childhood attraction to mathematics to the linguistic confusion of the Teller household, where he was taught both German and Hungarian at the same time and "did not catch on" at first. He found the world of numbers more congenial. When he

was 10, a retired mathematics professor, Leopold Klug, whom he would call "the man who had the greatest influence on my life," gave him a copy of Leonhard Euler's *Algebra*, which became his favorite book. When he graduated from secondary school he knew he wanted to be a mathematician. He compromised with his father, however, who insisted that his son have a "practical" career, and agreed to study chemical engineering at the Institute of Technology in Karlsruhe, Germany.

While in Karlsruhe Teller changed direction after he was exposed to the new ideas of quantum theory. In 1928, he transferred to the University of Munich, where he studied briefly with "the most famous teacher of quantum mechanics," ARNOLD JOHANNES WILHELM SOMMERFELD. In Munich he suffered a personal tragedy, losing his left foot in a streetcar accident. When he had recovered and learned to walk using a prosthesis, he moved to the University of Leipzig, to study with WERNER HEISENBERG. After receiving a Ph.D. in physics in 1930, he became a research consultant at the University of Göttingen and published his first paper on the hydrogen molecular ion, which describes the orbiting of one electron around two nuclei. It proved to be an early statement of what remains the most widely held view of this molecule.

When the Nazis came to power, Teller was obliged to truncate his promising career in Germany. He would later say, "No one could have had a greater influence on me than Hitler . . . who made it entirely clear to me that one could not ignore politics, and very particularly one could not ignore the worst evils in politics."

In 1934, Teller left for London and, after a brief stay, emigrated to Denmark. He joined NIELS HENRIK DAVID BOHR's Institute for Theoretical Physics in Copenhagen, where he participated in the quest to unlock the secrets of the atom, using the radical insights of quantum theory. Teller would later say of this period, "Of all the strange and important things that I have witnessed in my life, this was the most strange and the most important."

It was in Copenhagen, too, that Teller met his future wife, Augusta Harkany, and formed a close friendship with another political refugee, the Russian physicist GEORGE GAMOW. In 1935, Gamow, who had gone on to George Washington University in Washington, D.C., invited Teller to join him. When Teller and his wife eagerly accepted, a period of collaboration between the two men, which resulted in the Gamow–Teller rules for classifying subatomic particle behavior in radioactive decay, began.

In 1941, Teller became a U.S. citizen and joined the illustrious group of physicists working on the Manhattan Project, which, under the auspices of the United States Army, was coordinating British and American efforts to develop an atomic bomb. Teller worked in Chicago with ENRICO FERMI and Leo Szilard, before moving to J. ROBERT OPPENHEIMER's laboratory in Los Alamos, New Mexico. Among Teller's important contributions to atomic bomb research was a calculation that reassured physicists who feared that an uncontrolled nuclear reaction might continue indefinitely and devastate the entire planet. Teller's numbers showed that only a limited area would be destroyed. The "limited area," as history has recorded, proved to be the Japanese cities of Hiroshima and Nagasaki. Their destruction led to the Japanese surrender and the end of World War II.

By this time Teller had a new obsession. Since Fermi had suggested the idea to him in 1941, he had passionately championed the development of a thermonuclear explosive, a "superbomb." At Los Alamos, in 1944, he had done a theoretical study of the possibility of using an atomic bomb to ignite a mass of deuterium, a weapon he called the "Super," which might explode with a force equivalent not to thousands of tons of trinitrotolvene (TNT) but to millions. Such a bomb would use the intense heat generated by nuclear fission, the instanta-

neous disintegration of heavy atomic nuclei into lighter ones, that underlies the atomic bomb in order to provide a catalyst for the generation of nuclear fusion, the instantaneous merging of light atomic nuclei into heavier ones.

After the war, Teller urged his superiors at Los Alamos to go forward with work on the hydrogen bomb. Oppenheimer, who had been made chairman of the Atomic Energy Commission, and most commission members opposed the development of such a fusion weapon. But in 1949, after it was learned that the Soviets had exploded their first atomic bomb, President Harry Truman ordered the team at Los Alamos to begin developing the next generation of nuclear weapons.

A breakthrough occurred in 1951, when Stanislaw Ulam came up with the idea of using compression as a way to improve the reaction rates and staging as a way to achieve this. Teller then proposed using the photon radiation emitted by the primary fission device, rather than the neutron emission, to compress the secondary thermonuclear device. Use of a radiation shock wave instead of a material shock wave to implode the secondary thermonuclear device would cause a faster and longer-sustained compression of the fusion fuel to greater density. This Ulam–Teller invention tipped the balance in the argument about building the H-bomb; physicists now knew they could do it. Teller, who was loath to share credit for the breakthrough, worked to make it his own. Later he added a crucial additional stage: a second fission component positioned within the thermonuclear second stage to increase the efficiency of thermonuclear burning.

But, in 1951, Teller left Los Alamos when he was passed over for the job of heading the thermonuclear program, in favor of Marshall Holloway. Because of what had been called his "wild Hungarian temperament," Teller was not considered administrator material. Thus, when in 1952 Los Alamos scientists held the first successful test of a staged thermonuclear device on Eniwotok Atoll in the Pacific Ocean, Teller was not there to watch his former colleagues explode it. He did claim paternity, however, in a telegram announcing, "It's a boy."

Meanwhile, claiming that Los Alamos scientists were unenthusiastic about their new mission, Teller had lobbied for a second weapons laboratory. In 1952, the Lawrence Livermore Laboratory in northern California was established. Teller would be associated with Livermore from its founding until his retirement in 1975, acting as director between 1958 and 1960. During the heated debates of those times over whether to weaponize the thermonuclear device (as was later accomplished at Livermore), Teller's was a leading voice urging hydrogen bomb testing; Robert Oppenheimer's was the leading voice against it.

When, at the height of anti-Communist hysteria, in 1954, Oppenheimer was tried before McCarthy's House Un-American Activities Committee for disloyalty, Teller was asked to testify. Stopping short of calling his former boss a traitor, he admitted his mind would rest easier knowing that Oppenheimer's security clearance had been revoked. Oppenheimer did lose his clearance and influence on science policy. Rather than any personal antagonism, the source of Teller's uneasiness appears to have been Oppenheimer's stance against developing a "usable" hydrogen bomb. His early experiences with Hungarian communism gave his fear of the Soviet regime a special intensity. Whatever its motivation, there were many in the scientific community—and elsewhere—who never forgave Teller for his damning testimony.

Over the years Teller has been a leading advocate of a strong national defense; in the 1980s, he was an outspoken proponent of President Ronald Reagan's strategic missile defense system (otherwise known as SDI or "Star Wars"). He was also a leading advocate of nuclear energy as a power source for peaceful ends. He has authored more than a dozen books on nuclear energy and defense issues. In 1960, he accepted a joint appointment as professor of physics at the University of California and as associate director of the Lawrence Livermore Laboratory, a post he held until his retirement in 1975. He then became a senior fellow at the conservative Hoover Institute for the Study of War, Revolution and Peace at Stanford University. Now in his 90s, he lives in Palo Alto, California.

Looking back, Teller has defended, in straightforward terms, his advocacy of the most destructive weapons known to human beings:

> What would have happened if Stalin got the hydrogen bomb and we did not? . . . If I claim credit for anything, I think I should not claim credit for knowledge, but for courage. It was not easy to contradict the great majority of the scientists who were my only friends in a new country.

Further Reading

Herken, George. *Brotherhood of The Bomb: The Tangled Lives and Loyalties of Robert Oppenheimer, Ernest Lawrence and Edward Teller.* New York: Henry Holt, 2002.

Rhodes, Richard. *Dark Sun: The Making of the Hydrogen Bomb.* New York: Simon & Schuster, 1995.

Teller, Edward, with Judith L. Shoolery. *Memoirs: A Twentieth-Century Journey in Science and Politics.* New York: Perseus Books, 2001.

⊠   Tesla, Nikola
(1856–1943)
Serbian/American
Experimental Physicist (Classical Electromagnetism), Electrical Engineer

Nikola Tesla was a great experimentalist and inventor whose work ushered in the age of electrical power. One of the pioneers of the use of alternating currents, he invented the alternating current induction motor and the high-frequency coil that bears his name.

He was born at midnight on July 9, 1856, in the village of Smiljan, Croatia, then part of the Austro-Hungarian Empire, to Serbian parents. Nikola was supposed to follow in the footsteps of his father, a Serbian Orthodox priest, but his natural inclinations led him along other paths. The combination of poetic imagination and scientific discipline that would characterize his amazing career manifested itself early in life. A clever child, he liked poetry (he would later write his own poetry and translate Serbian poetry into English) and scientific experimentation. At the age of five he built a small waterwheel with an unusual design that he would recall years later when designing his bladeless turbine. In secondary school he developed an interest in science and, on graduating, entered the Technical University in Graz, Austria, where he studied engineering. In 1878, after seeing a demonstration of a direct current electric dynamo and motor, he conceived the idea of improving the machine by removing the commutator and sparking brushes, which were sources of wear.

Tesla went on to the University of Prague in 1880 but was forced to leave without graduating when his father died the following year. Tesla then took a job in Budapest as an engineer for a telephone company. That year, while strolling in the park with a friend, he visualized the principle of the rotating magnetic field, which is produced by two currents running out of step with one another, or alternating. This was the principle that lay at the heart of his idea of the induction motor: an iron rotor spinning between stationary coils that were electrified by two out-of-phase alternating currents producing a rotating magnetic field. Before a working motor had been built, Tesla had conceptualized an array of practical applications for it. Years later, he would describe the manner in which an invention would come to life in his imagination:

Before I put a sketch on paper, the whole idea is worked out mentally. In my mind I change the construction, make improvements, and even operate the device. Without ever having drawn a sketch I can give the measurements of all parts to workmen, and when completed all these parts will fit, just as certainly as though I had made the actual drawings.

Tesla would not build an actual working induction motor until 1883, when he had moved to Paris and was working for the Continental Edison Company. Finding little interest in developing it in Europe, however, in 1884, he sailed to the United States. He arrived in New York with a few coins in his pocket and a letter of recommendation from a European business associate to the eminent inventor Thomas Edison at his Menlo Park, New Jersey, research laboratory. The letter said, "I know two great men and you are one of them; the other is this young man." Edison was impressed with this and gave the young man a job, but he was a staunch supporter of direct current, which moves in one direction only, rather than alternating current, which regularly reverses its direction. After a year Tesla and Edison parted ways.

Tesla lost little time in obtaining patents for the generation, transmission, and use of alternating current electricity, which he sold, in May 1885, to George Westinghouse, head of the Westinghouse Electric Company in Pittsburgh. He received $60,000 for his patents and a contract promising him $2.50 per horsepower of electricity sold. But when Westinghouse's company found itself in financial trouble, Tesla agreed to forgo any royalties, as a gesture of friendship and support, "so that you can develop my inventions." His selfless act would later prevent him from realizing the financial benefits of many of his other inventions.

The Westinghouse purchase precipitated an intense power struggle between Edison's direct current systems and the Tesla–Westinghouse alternating current approach. Because of its higher efficiency, alternating current, which can be transmitted over much longer distances than direct current, eventually won out. Westinghouse demonstrated the superiority of alternating current by using his system for lighting at the 1893 World Columbian Exposition in Chicago. In 1896, Westinghouse won the contract to install the first power machinery to generate electricity, at Niagara Falls, using Tesla's system to supply and deliver alternating current to Buffalo, 22 miles away.

Tesla himself had meanwhile become involved in a number of projects in the personal laboratory and workshop he had set up in 1887. He experimented with shadowgraphs, similar to those used by WILHELM CONRAD RÖNTGEN, in 1895, when he discovered X rays, as well as working on a carbon button lamp, the power of electrical resonance, and various types of lighting. To show that there was nothing to fear from alternating currents, he gave exhibitions in his laboratory in which he lighted lamps without wires by allowing electricity to flow through his body. As his fame grew, he was in demand for lectures in the U.S. and abroad.

At the time he became an American citizen, in 1891, Tesla directed his inventive powers to developing alternating currents at very high frequencies, believing they would be useful for lighting and communication. He experimented with high-frequency alternators before designing what came to be called the Tesla coil, which produces high-voltage oscillations from a low-voltage direct current source. The Tesla coil is widely used in radio and television sets and other electronic equipment. At the peak of his creative powers, during this period he also invented fluorescent lighting, the bladeless turbine, and radio.

One of his most fervid interests was radio communication. In 1898, he demonstrated remote control of two model boats before large crowds in Madison Square Garden in New York. He later extended this to remote controlled weapons, in particular, torpedoes.

During an 1899–1900 stay in Colorado Springs, Colorado, Tesla made what he viewed as his most significant discovery: terrestrial stationary waves, with which he proved that the Earth could be used as a conductor and would be as responsive as a tuning fork to electrical vibrations of a certain frequency. Using high-frequency alternating currents, he was also able to light more than 200 lamps over a distance of 25 miles without using intervening wires. Because gas-filled tubes are readily energized by these high-frequency currents, this kind of light could easily be operated within the field of a large Tesla coil. Later, he used the Tesla coil to produce artificial lightning: an electric flash 135 feet long. However, sometimes his imaginative leaps seemed excessive to his contemporaries: when he announced that he had received signals from another planet in his Colorado laboratory, he was ridiculed in scientific journals.

In 1900, with support from the financier J. Pierpont Morgan, he began building a wireless world broadcast station on Long Island, New York. His ambition was to provide worldwide communication and to furnish facilities for sending pictures, messages, weather warnings, and stock reports. When his backers found the project too expensive, he reluctantly abandoned this idea, considering it to be his greatest defeat.

Tesla created several other inventions, including electrical clocks and turbines, many of which remained on the drawing board for lack of financial backers to produce them. One of his most ambitious ideas was to transmit alternating current electricity to anywhere in the world without wires, using the Earth itself as an enormous oscillator. His scheme for detecting ships at sea was later developed into what is now known as radar.

In 1917, he received the Edison Medal, the highest honor awarded by the American Institute of Electrical Engineers. The standard international unit of magnetic flux density was named the *tesla* in his honor.

Eccentric and withdrawn, he lived his last years as a recluse. He was known for his loyalty to his few close friends, one of whom was Mark Twain. When he died in New York City on January 7, 1943, he owned more than 700 patents. His funeral was attended by a large crowd, and a flood of telegrams paid tribute to his genius and contributions.

It would be difficult to overstate the impact of Tesla's inventive genius. His discovery that alternating current can be transmitted over much greater distances than direct current underlies our capability to power the machinery of the modern world.

Further Reading

Cheney, Margaret. *Tesla: Man out of Time*. New York: Dell, 1981.

Hunt, Inez, and Wanetta W. Draper. *Lightning in His Hand: The Life Story of Nikola Tesla*. Thousand Oaks, Calif.: Sage Books, 1964.

Tesla, Nikola. *My Inventions: The Autobiography of Nikola Tesla*. Edited and with an introduction by Ben Johnston. Williston, Vt.: Hart Bros., 1944.

⊠   **Thomson, Joseph John (J. J.)**
(1856–1940)
British
Experimentalist (Classical Electromagnetism), Atomic Physicist

J. J. Thomson was a brilliant experimentalist whose exploration of the nature of cathode rays led him to the discovery of the electron. He was awarded the 1906 Nobel Prize in physics for his research on the conduction of electricity through gases.

Born in Manchester, England, on December 18, 1856, Thomson became a physicist by default. Intending to become an engineer, he enrolled at Owens College (later to become Manchester University) at the age of 14. But when his father, a seller of antique books, died two years later, he found himself without the means to finance an engineering apprenticeship. Instead, he managed to win a scholarship to study mathematics, physics, and chemistry and, in 1876, entered Trinity College, Cambridge.

After graduating in mathematics in 1880, he worked at the Cavendish Laboratory under LORD RAYLEIGH (JOHN WILLIAM STRUTT), whom he succeeded as Cavendish Professor of Experimental Physics in 1884. Over the next 35 years he would develop the Cavendish Laboratory into the world's leading center for subatomic physics. It was at the Cavendish that Thomson, applying his genius for designing apparatus turning to the mysteries of atomic structure, made his breakthrough discoveries.

His *Treatise on the Motion of Vortex Rings*, which won him the Adams Prize in 1884, approached the subject from the point of view, espoused by many of Thomson's peers, that atoms exist in vortex rings within a hypothetical "ether," which was thought to permeate all of space. From there he turned his attention to the current debate over the nature of cathode rays, that is, the phenomenon that produced a fluorescent glow when most of the air was pumped out of a glass tube with wires embedded at each end and a high voltage was sent across it. Physicists asked whether cathode rays were charged particles or some undefined wavelike process in the ether. When Thomson performed his first experiment, prior evidence seemed to favor waves. The German physicist HEINRICH RUDOLF HERTZ had apparently shown that cathode rays were not deflected by an electric field; this indicated that they did not possess electric charge and, therefore, did not have a particle nature. In 1897, Thomson invalidated Hertz's results, showing that they were caused by his use of an insufficiently evacuated cathode ray tube. When Thomson used a new technique to try to bend cathode rays with an electric field, extracting nearly all the gas from a tube, he succeeded where Hertz had failed.

He then sought to determine the basic properties of the cathode ray particles. Although he lacked the means to measure the mass or electric charge of such particles, he could measure how much the rays were bent by a magnetic field and how much energy they carried. From these data he could calculate the ratio of the electric charge of a particle to its mass. He collected data by using a variety of tubes as well as different gases. Whatever gas he used, he found that the ratio of

the charge divided by the mass of the cathode ray particles was constant and had a value nearly 1000 times larger than that of a charged hydrogen atom. These results led Thomson to a logical fork in the road: either the cathode rays carried an enormous charge, as compared with a charged atom, or else they were extraordinarily light relative to their charge. The question was settled by PHILIPP VON LENARD in experiments on how cathode rays penetrate gases. He showed that if cathode rays were particles they had to have a mass much smaller than the mass of any atom. Subsequent experiments by ROBERT ANDREWS

Joseph John (J. J.) Thomson's exploration of the nature of cathode rays led him to the discovery of the electron. *(AIP Emilio Segrè Visual Archives)*

MILLIKAN measured the charge directly and confirmed Lenard's conclusions.

At a historic meeting of the Royal Institution in 1897 Thomson announced his results. Having shown that cathode rays were fundamental, negatively charged particles with a mass much less than that of the lightest atom known, he went on to suggest that these material particles were the building blocks of the atom, the basic unit of all matter in the universe that physicists had long been seeking. For Thomson, however, this discovery only led to another, deeper mystery:

> I can see no escape from the conclusion that [cathode rays] are charges of negative electricity carried by particles of matter . . . [but] what are these particles? Are they atoms, or molecules, or matter in a still finer state of subdivision?

What Thomson called "corpuscles" or "particles" would later be given the name *electrons*, and physicists would devise numerous theories to explain how they combined to form atoms. Thomson himself suggested a model for the atom called the "plum pudding" or "raisin cake model," in which thousands of tiny, negatively charged corpuscles move inside a cloud of positive charge. Later on, using alpha particles (a different kind of particle beam), Thomson's former student ERNEST RUTHERFORD disproved Thomson's model, replacing it with his solar system model of the atom: a massive, positively charged center circled by only a few electrons.

After investigating the nature and properties of electrons for several more years, Thomson began researching "canal rays," streams of positively charged ions (i.e., atoms that have lost one of their negatively charged electrons), which he named *positive rays*. In the process he learned how to use positive rays for separating different kinds of atoms and molecules. His technique led to the development of the mass spectroscope, an instrument that measures the charge-to-mass ratio. Using magnetic and electric fields to deflect these rays, Thomson found, in 1912, that ions of neon gas were deflected by different amounts, indicating that they consist of a mixture of ions with different charge-to-mass ratios. He confirmed the existence of isotopes, earlier proposed by the British chemist Frederick Soddy, when in that same year he identified the isotope neon 22. Later many more isotopes would be discovered.

In 1919, Thomson resigned his post at the Cavendish Laboratory, after being elected Master of Trinity College. He would remain at Trinity until his death on August 30, 1940, at the age of 84. As an educator he devoted much attention to the problems of teaching science at secondary and university levels. Many universities used his prolific writings on electricity, magnetism, and other topics. His profound personal impact on his students is reflected in the fact that seven of them—including his own son, George—went on to win the Nobel Prize in physics.

Most importantly, however, through his discovery of the first known elementary particle, the electron, Thomson initiated a period of exploration that would lead 20th-century physicists to uncover a new universe of the infinitesimally small.

Further Reading

Buchwald, Jed Z. *Histories of the Electron (Dibner Institute Studies in the History of Science and Technology)*. Cambridge, Mass.: MIT Press, 2001.

Dahl, Per F. *Flash of the Cathode Rays: A History of J. J. Thomson's Electron*. Philadelphia: Institute of Physics, 1997.

⊠   **Ting, Samuel Chao Chung**
(1936–   )
Chinese/American
Experimentalist, Particle Physicist

Samuel C. C. Ting is a leading experimentalist, best known for his pioneering work in particle physics. His discovery of the J/ψ particle, an excited state of nuclear matter, confirmed the standard quark lepton model of electroweak and nuclear forces. For this work, he shared the 1976 Nobel Prize in physics with BURTON RICHTER, who had independently made the same discovery.

He was born on January 27, 1936, in Ann Arbor, Michigan, the first of three children of Kuan Hai Ting and Tsun-Ying Wong, who were both foreign graduate students at the University of Michigan. The accident of his premature birth made Samuel an American citizen. Two months later, his family returned to Beijing, China, where his father became a professor of engineering, and his mother a professor of psychology. World War II prevented him from attending school until he was nine, although his home was always filled with his parents' colleagues, who imbued him with an early desire to be part of academic life. With both parents working, he was raised by his maternal grandmother, a widow who had struggled to educate herself and her daughter during turbulent years in China's history. Ting writes of his grandmother and mother, "Both of them were daring, original, and determined people, and they have left an indelible impression on me." After the war, the family moved to Taiwan, where Ting continued his schooling, excelling in mathematics, science, and history.

In quest of a better education, in 1956, Ting returned to Ann Arbor as an undergraduate at his parents' alma mater. He had heard that American students earned their own way through college and told his parents he would do likewise. Knowing little English, he arrived with $100 in his pocket and an invitation to stay with his father's former engineering professor. But he managed to obtain scholarships and worked very hard to retain them. Entering as an engineering student, he soon switched his major to physics. Only three years later, in 1959, he received a B.S. in physics and mathematics. He stayed at

Michigan for his graduate studies and received a Ph.D. in physics in 1962 for an experimental thesis supervised by Lawrence Jones and the future Nobel laureate Martin Perl.

In 1963, Ting was awarded a Ford Foundation Fellowship to work at the European Center for Nuclear Research (CERN) in Geneva, where he collaborated with Giuseppi Cocconi on experiments using the 28-billion-volt proton synchrotron. Cocconi taught him a great deal about physics and deeply impressed him with his "simple way of viewing a complex problem" and the care with which he conducted experiments.

Ting returned to the United States in the spring of 1965, to spend an exciting year as an instructor at Columbia University, where he had the chance to observe such eminent physicists as LEON M. LEDERMAN, TSUNG-DAO LEE, ISIDOR ISAAC RABI, and CHIEN-SHIUNG WU. The following year, however, he took leave from Columbia in order to lead an experimental group at the Deutsches Elektronen Synchrotron (DESY) in Hamburg, Germany. His goal was to reproduce an experiment performed that year at the Cambridge Electron Accelerator at Harvard University on electron–positron pair production by photon collision with a nuclear target. (Positrons are the antiparticles of electrons.) The experiment seemed to indicate findings inconsistent with what Ting knew of quantum electrodynamics (QED), which describes the properties of the interaction of matter and electromagnetic radiation. With the new detector they built, Ting's group confirmed the results of the Harvard group.

This was a turning point in Ting's career: from then on he would devote himself to the physics of electron or muon pairs, studying the quantum electrodynamic production and decay of new photonlike particles, which decay to electron or muon pairs. He explains, "These types of experiments are characterized by the need for a high-intensity incident flux, for high rejection against a large number of unwanted background events, and at the same time the need for a detector with good mass resolution." To achieve these conditions, he took his group back to the United States in 1971, to the Brookhaven National Laboratory in Stonybrook, New York.

In the spring of 1972, Ting's group modified the detector they had designed in Germany, increasing its sensitivity to the specific energy signal of the electron–positron pairs in order to discern its specific signal amid the noise of millions of other particle collisions. The Nobel Prize committee would later liken this feat to being able to "hear a cricket close to a jumbo jet taking off." With their newly improved detector, Ting's team bombarded a beryllium target with a proton beam, took measurements, and looked for the signature of the electron–positron pairs. In August 1974, they made a startling discovery: the first of a totally unpredicted new group of extremely heavy long-lived mesons. After rechecking his results, in November of that year, Ting announced his discovery, which he called the J particle (a name based on the symbol for the electromagnetic current).

Just before publishing his findings, Ting attended a conference at Stanford University with scientists working at the Stanford Linear Accelerator Center and made another astounding discovery: The physicist Burton Richter, using a wholly different experimental approach, had found the same particle, which he called the psi particle. Two years later, both men shared the Nobel Prize for their work in detecting what came to be called the J/ψ particle, determined to be more than twice as heavy as any comparable particle and yet a thousand times more narrow in its energy spectrum (and that, according to the uncertainty principle, meant that it was a long-lived particle).

In 1961, MURRAY GELL-MANN and YUVAL NE'EMAN had devised the eightfold way, a system for classifying the myriad newly discovered elementary particles into eight families of elemen-

tary building blocks called quarks, of which there were three types. The unique feature of the newly discovered J/ψ particle was that it did not belong to any of the families as they were known prior to 1974. Its detection confirmed SHELDON LEE GLASHOW's prediction of a fourth quark (i.e., the charmed quark, which was needed in order to make the Gell-Mann SU(3) quark theory of the strong interactions consistent with this observation).

In 1969, Ting was appointed professor of physics at the Massachusetts Institute of Technology (MIT) in Cambridge, where, in 1977, he was selected as the first recipient of the Thomas Dudley Cabot Institute Professorship. In 1985, he married Dr. Susan Marks, with whom he had a son, Christopher, the following year. He also has two daughters, Jeanne and Amy, from a previous marriage.

Ting continues to spend most of his time at CERN, where he heads the L3 Experiment, launched in 1982, involving more than 500 physicists from about 33 universities and institutions throughout the world. Another major current project in which he is involved is the construction of a three-ton detector, called the Alpha Magnetic Spectrometer (AMS), designed to search for the existence of antimatter nuclei among cosmic rays, which will operate in the *International Space Station*. Since antimatter is expected to be extremely difficult to detect, experimentation in empty space, where "background noise" is considerably less, is essential. In June 1998, a prototype AMS was tested on the Space Shuttle *Discovery*.

The AMS may also play a role in investigating the mysterious dark matter, which, though estimated to make up the greater part of the universe, has so far eluded detection. The AMS may be sensitive to certain properties of weakly interacting particles that some physicists believe to be the essence of dark matter.

Ting and Richter's discovery of the J/ψ particle changed the landscape of particle physics.

By verifying the need for a charmed quark, it opened the door to the further elaboration of the full Standard Model, which contains six quarks: up, down, strange, charm, top, and bottom.

See also ANDERSON, CARL DAVID; DIRAC, PAUL ADRIEN MAURICE; SALAM, ABDUS; WEINBERG, STEVEN.

Further Reading

Kane, Gordon. *The Particle Garden: Our Universe as Understood by Particle Physicists.* Cambridge, Mass.: Perseus, 1995.

Ne'eman, Yuval, and Murray Gell-Mann. *The Eightfold Way.* Cambridge, Mass.: Perseus, 2000.

Ne'eman, Yuval, and Yoram Kirsh. *The Particle Hunters.* Cambridge: Cambridge University Press, 1996.

⊠   **Tomonaga, Sin-Itiro**
(1906–1979)
Japanese
Quantum Field Theorist, Particle Physicist

Sin-Itiro Tomonaga was a great Japanese theoretician, who laid the foundations of a relativistic quantum electrodynamics (QED), the study of the quantum mechanical interaction of electrons, positrons, and photons. He shared the 1965 Nobel Prize in physics with RICHARD PHILLIPS FEYNMAN and JULIAN SEYMOUR SCHWINGER, who, working concurrently, achieved the same result, using different approaches.

He was born in Tokyo, Japan, on March 31, 1906, the eldest son of Sanjuro and Hide Tomonaga. When he was seven, the family moved to Kyoto, where his father was appointed a professor of philosophy at Kyoto Imperial University. He was a sickly boy, who did poorly in athletics and was teased by his peers as a "crybaby." He discovered his passion for science early and spent his free time with a friend building simple electric circuits with parts they had

found in junkyards. After graduating from the renowned Third Higher School, he entered Kyoto Imperial University in 1923. After earning their *rigakushi* (bachelor's degrees) in physics in 1929, Tomonaga and his good friend, another future Nobel Prize winner, HIDEKI YUKAWA, traveled to Tokyo to listen to a series of lectures by WERNER HEISENBERG and PAUL ADRIEN MAURICE DIRAC, which made a strong impression on them.

He remained at Kyoto for three years of graduate studies, after which he was taken on as a research associate by Yoshio Nishina, one of Japan's leading physicists at the Institute for Physical and Chemical Research (known as Riken) in Tokyo. Nishina, who introduced research on quantum mechanics to Japan, was a role model and father figure for the young Tomonaga, who later reminisced:

The Nishina laboratory in those days was full of freshness. All the members were young; even our great chief Nishina was still in his early forties. We all got together for lunch every day, an eager group of people discussing various matters, not only physics, but also such things as plans for beer parties, excursions and so on.

In this stimulating atmosphere, Tomonaga made his first contribution to QED in a 1933 paper, coauthored with Nishina, on photoelectric pair creation. With the other members of the theoretical group, he studied Dirac's work, translating his textbook on quantum mechanics into Japanese.

In the mid-1930s, Tomonaga's interests shifted to nuclear physics. From 1937 to 1939, he was in Leipzig, Germany, studying nuclear physics and quantum field theory under Heisenberg. With Heisenberg's encouragement, he attacked the problem of the description in the Bohr liquid-drop model of the heating of a

Sin-Itiro Tomonaga shared the 1965 Nobel Prize in physics with Richard Phillips Feynman and Julian S. Schwinger for his work in quantum electrodynamics (QED), the study of the quantum mechanical interaction of electrons, positrons, and photons. *(AIP Meggars Gallery of Nobel Laureates)*

nucleus when it absorbs a neutron. He had hoped to continue working with Heisenberg, but when war broke out on September 1, 1939, he returned to Japan. The results of his Leipzig research formed a large part of his doctoral thesis, which he submitted in December 1939 at Tokyo Imperial University. The following year, he married Ryoko Sekiguchi, with whom he would have two sons and a daughter.

With his country at war, Tomonaga, who, in 1941, became professor of physics at Tokyo University of Education (Bunrika), managed to press on with his fundamental research on QED. He addressed himself to the problem that, whereas every physical theory must obey the rules of special relativity, QED, in its early Dirac formulation, was not fully relativistic. Some physicists believed that the infinities (also called divergences) that QED yielded, which were physically nonsensical, could be dealt with more realistically if the theory could be made relativistically covariant, that is, compatible with the rules of special relativity. In work done between 1941 and 1943, Tomonaga proposed a relativistic formulation of quantum field theory, in which the concepts of the quantum state vector (the quantum field theoretic equivalent of the Schrödinger wave function) and its equation of motion, concepts having relativistic spacetime meaning, were generalized so as to be relativistically covariant.

In 1943, Tomonaga published this work, developed in isolation from his Western counterparts, who would not learn of it until four years later, just as he was forced to turn away from basic research to problems related to military needs. Until the end of the war, he worked on the properties of magnetrons and the behavior of microwaves in waveguides and cavity resonators, developing a unified theory of the systems consisting of waveguides and cavity resonators.

In the midst of Japan's postwar devastation, Tomonaga and his family found themselves homeless and hungry. Nonetheless, there was a sense of communal responsibility for national reconstruction among the group of young scientists who gathered around Tomonaga in Tokyo. In 1946, he returned to the problems of quantum field theory, with the goal of summarizing and applying the covariant field theory to actual physical systems. He was sure that the divergence problems in QED could be overcome by finding a way to handle the infinite mass and charge due to field reaction. Elimination of the divergences was to be accomplished by a renormalization procedure that first identified the divergent terms according to their relativistic and gauge transformation properties, then showed that these divergent contributions could be absorbed in a redefinition of the mass and charge parameters entering the original formulation of the theory. If this could be done, then the renormalized formalism could make predictions about observable phenomena.

In 1947, Tomonaga learned about the Lamb shift and read HANS ALBRECHT BETHE's article in *Physical Review*, presenting his nonrelativistic Lamb shift calculation. He recognized its importance and went on to work out his mass renormalization method, substituting the experimental mass for the fictive mechanical mass that appeared in the equations of QED. He performed a similar renormalization of the electric charge. He also deduced a correct formula for the Lamb shift, which agreed with measurements. In 1948, he wrote about his work to J. ROBERT OPPENHEIMER, who urged him to submit a summary to *Physical Review* for publication.

In a scientific community only slowly overcoming the barriers imposed by global war, Tomonaga's discovery did not influence QED research in the United States. Schwinger independently made the same advances and Feynman was unaware of his Japanese colleague's work. However, Tomonaga did resume his collaboration with Western physicists when, in 1949, he was invited to the Institute for Advanced Study, Princeton; there he turned his attention to nuclear physics, investigating a one-dimensional fermion system. In so doing he clarified the nature of collective oscillations of a quantum mechanical many-body system and opened a new frontier of theoretical physics, the modern many-body problem. In 1955, he published an elementary theory of quantum mechanical collective motions.

Tomonaga was a leader in establishing the Institute for Nuclear Physics, Tokyo, in 1955. From 1956 to 1962, he was president of the Tokyo University of Education. In 1963, he became president of the science council of Japan and director of the Institute for Optical Research at the Tokyo University of Education. He held many key positions on government committees dealing with scientific research and policy making.

In 1965, Tomonaga shared the Nobel Prize in physics with Schwinger and Feynman but was prevented by an accident from being present in Stockholm.

He published widely in scientific journals on QED, the meson theory, nuclear physics, cosmic rays, and the many-body problem. His book *Quantum Mechanics* was published in Japan in 1949; an English-language edition appeared in 1962. His memoir, *Development of Quantum Electrodynamics: Personal Recollections*, became available in English translation in 1966.

Tomonaga, who died on July 8, 1979, in Tokyo, was a physicist of great intellectual powers, an extraordinary teacher in the Socratic tradition whose dictum was that "if you formulate the problem correctly, that is, if you ask the right question, the answer emerges spontaneously." He possessed keen aesthetic sensibilities and a love of communing with nature. He will be remembered as a strong advocate of nonproliferation of nuclear weapons and the peaceful uses of atomic energy.

See also EINSTEIN, ALBERT; KUSCH, POLYKARP; LAMB, WILLIS EUGENE.

Further Reading
Schweber, Silvan S. *QED and the Men Who Made It: Dyson, Feynman, Schwinger, and Tomonaga.* Princeton, N.J.: Princeton University Press, 1994.
Tomonaga, Sin-Itiro. *The Story of Spin.* Chicago: University of Chicago Press, 1998.

⊠ Townes, Charles Hard
(1915–)
American
Theoretician and Experimentalist (Optics), Laser Spectroscopist

Charles Hard Townes is a giant of 20th-century physics, who, together with ARTHUR LEONARD SCHAWLOW, invented the laser, the revolutionary device that creates and amplifies a narrow, intense beam of coherent light. Townes independently invented the laser's forerunner, the maser. He shared the 1964 Nobel Prize in physics for these accomplishments with the Russian physicists Nikolay Basov and Aleksandr Prokhorov, who independently made the same discoveries.

He was born on July 28, 1915, in Greenville, South Carolina, to Henry Keith Townes, an attorney, and Ellen Hard Townes. A child who had to know how things work, he once instructed his older sister to "buy out a hardware store" for his Christmas present. After attending

Charles Hard Townes invented the laser, the revolutionary device that creates and amplifies a narrow, intense beam of coherent light. *(AIP Emilio Segrè Visual Archives, and AT&T Bell Labs)*

the Greenville public schools, he gained early admission at age 16 to Furman University in Greenville, where he earned both a B.Sc. in physics and a B.A. in modern languages summa cum laude in 1935. In addition to his broad academic interests, he was an "all-around" student, a member of the swimming team, the college newspaper, and the football band. From early childhood, when he took long walks in the woods with his father and carried home a collection of pet frogs, caterpillars, lizards, and snakes, he had a strong love of nature; at Furman, he served as curator of the natural history museum and was collector for the summer biology camp. But it was physics, with its "beautifully logical structure," that most engaged him. Townes decided to pursue graduate work at Duke University and, in 1937, received his master's degree in physics. From there he went on to the California Institute of Technology, where he wrote a dissertation on isotope separation and nuclear spin, earning a Ph.D. in 1939.

Townes then began work for Bell Telephone Laboratories in New York City and, in 1941, married Francis H. Brown, with whom he would have four daughters. During World War II, he did important work using microwave techniques to design radar bombing systems, using the shorter wavelength of microwave radiation, which permitted radar beams to reveal the shape of a target more accurately. When the war ended, sensing in this technology a potentially revolutionary method for studying atomic and molecular structure, as well as for controlling electromagnetic waves, he set about finding peaceful applications in microwave spectroscopy for his discoveries. When he left Bell Laboratories in 1947 to join the physics faculty at Columbia University, he continued his work on microwaves; by 1950, he had become a professor of physics and director of the Columbia Radiation Laboratory.

At the time, physicists around the world were trying to find a way to produce extremely short waves for measuring the properties of matter; the vacuum technology of the time was not capable of doing this. In 1951, while sitting on a park bench in Washington, D.C., where he was attending a conference devoted to this problem, Townes suddenly saw the solution and scribbled it on an envelope he found in his shirt pocket. He had conceived the idea of the maser, an acronym for *microwave amplification by stimulated emission of radiation*, and the laser, a similar verbal construction, with the word *light* substituted for *microwave*.

The concept of stimulated emission originated with ALBERT EINSTEIN, who, while studying the theory of blackbody radiation in 1917, had found that the process of light absorption must be accompanied by a complementary process in which the absorbed radiation stimulates the atoms to emit the same kind of radiation. However, in order for amplification of the radiation by stimulated emission to occur in a physical medium, the stimulated emission must be larger than the absorbed radiation. For this to happen, the physical medium must have more atoms in a high-energy state than in a lower one. Since the law of conservation of energy implies that energy spontaneously flows from a higher to a lower state, this physical situation, which involves an inverted population of excited atoms, is inherently unstable. Moreover, in order to use this method to generate beams of coherent light (i.e., intense light waves consisting of essentially one frequency and moving in the same direction), it would be necessary to find the specific atomic systems with the right internal storage mechanisms.

In order to do this, Townes selected the simple case of ammonia molecules, because they can occupy only two energy levels. The ammonia molecules had the property of emitting radiation with a 1.25-cm wavelength, which lies within the microwave range of the electromagnetic spectrum. He reasoned that when an ammonia molecule in the high-energy state absorbed a photon of this frequency, the molecule would fall to the lower energy level, emitting two coherent photons of the same frequency, thus producing a coherent beam of single-frequency radiation with a 1.25-cm wavelength.

By studying the behavior of ammonia molecules in a resonant cavity containing electric fields, Townes developed a method that separated the relatively few high-energy molecules from the more numerous low-energy ones and thus created the required population inversion needed for maser action to occur. By 1953, he had constructed the first working model of the maser. Masers soon found a range of applications: in the most accurate timepieces available to this day—atomic clocks; in short wave radios, where they serve as extremely sensitive receivers; in radio astronomy; and in space research, for recording the radio signals from satellites.

Townes continued his research at Columbia, where he was made chairman of the physics department in 1952. In December 1958, he and his former research student (now his brother-in-law) Arthur Schawlow published a landmark paper, "Infrared and Optical Masers," in *Physical Review*, in which they showed theoretically that an optical maser, or laser, could be produced by a coherent beam of visible light, instead of a microwave beam. Because the change from microwaves and visible light required a 100,000-fold increase in frequency, a fundamentally different operating structure was required. The atoms or molecules of a ruby or garnet crystal or of a gas, liquid, or other substance are excited by light in the cavity in which they are contained, so that more of them are at higher energy levels than are at lower ones. In order to achieve the necessary high-radiation density, two mirrors, one at each end of the cavity, force the light to bounce back and forth until coherent light escapes from the cavity. The first working laser was built in 1960 by Theodore Maiman at the Hughes Research Laboratories.

At this high point of his career, Townes took a leave of absence from Columbia University and served as vice-president and director of research at the Institute for Defense Analysis, a nonprofit organization in Washington, D.C., from 1959 to 1961. He was then appointed provost and professor of physics at the Massachusetts Institute of Technology (MIT), where, in addition to shaping MIT's overall research program, he engaged in research in quantum electronics and astronomy. In 1964, he shared the Nobel Prize in physics with Basov and Prokhorov for invention of the laser. Schawlow was inexplicably excluded from this select group and would not have his contribution recognized by the Nobel Committee until 1981.

In 1967, Townes joined the faculty of the University of California, Berkeley, where his research focused on the study of radiation from space, using principles of radio and infrared astronomy. He and his colleagues discovered water and ammonia in interstellar space, the first complex molecules found outside our solar system, and showed that water was producing intense natural maser activity there. This discovery prompted him to comment:

> For billions of years [astronomical masers] have been sending out intense radiation in all directions . . . but we have only recently gotten their message. . . . The whole field of masers and lasers may well have originated from discovery there, and the development . . . might have had a very different history.

Townes served as chairman of the NASA Science Advisory Committee for the Lunar Landing.

With a team of researchers, he designed instruments used to analyze infrared radiation from space. He became professor emeritus in 1986. He and his wife continue to reside in Berkeley.

The evolution of the early lasers into the multitude of laser devices in the modern world led Townes to comment:

> People said, "It's a nice idea, but what can it do?" My view then was that it would touch many applications because it combined electricity and light. But it is amazing how it is used today. . . . I've had people come up to me and say "lasers saved my eyesight." That's a very emotional experience for me. It's so personal, this connection.

Lasers are so pervasive in daily life—in the bar-code scanners at supermarket checkout counters, in laser printers, scanners, and compact disc technology—that the term *Laser Age* has been coined to describe the present era. Although it is probably true, as Townes believes, that lasers are still in their "adolescence," they are already powerful tools for research in many fields, widely used in industry for cutting and boring metals and other materials; in medicine for surgery; in communication, scientific research, and holography.

Further Reading

Chiao, Raymond Y. *Amazing Light: A Volume Dedicated to Charles Hard Townes on His 80th Birthday.* New York: Springer-Verlag, 1996.

Townes, Charles Hard. *How the Laser Happened: Adventures of a Scientist.* Oxford: Oxford University Press, 1999.

V

Van Allen, James
(1914–   )
American
Astrophysicist

James Van Allen is a pioneering astrophysicist who played a vital role in the early development of the U.S. space program. His work in magnetospheric physics led to the discovery of a zone of high-level radiation around the Earth caused by the presence of trapped charged particles, known as the magnetosphere or the Van Allen belt.

He was born in Mount Pleasant, Iowa, on September 7, 1914, the second of four sons of Alfred Van Allen and Alma Olney Van Allen. Although his father was an attorney in a family of lawyers, he encouraged the boy to follow his own path. Van Allen writes:

> My boyhood activities were all centered around our closely knit family, which had a strong resemblance to that of other pioneer families. The virtues of frugality, hard work, and devotion to education were enforced rigorously and on a daily basis by my father. My mother exemplified the pioneer qualities of affection and nurture for her husband and their children and of comprehensive self-reliance.

A studious boy, with a consuming interest in science and mechanical and electrical devices, he was an avid fan of the magazines *Popular Mechanics* and *Popular Science*. When he graduated from Mount Pleasant High School in 1931, he was class valedictorian.

From there he went on to Iowa Wesleyan College, where he was drawn to both physics and chemistry and received a B.S. in 1935. The opportunity to work in a high-pressure research lab assisting his professor helped him to decide in favor of physics. He then went on to the University of Iowa, where he earned an M.S. in 1936 and a Ph.D. in 1939, writing a dissertation on nuclear disintegration. From 1939 to 1942, Van Allen worked as a research fellow at the Carnegie Institute in Washington, D.C., in the Department of Terrestrial Magnetism, where he studied the photodisintegration process. It was there that he "crossed the culture gap from nuclear physics to the department's traditional research in geomagnetism, cosmic rays, auroral and ionospheric physics," which he resolved to make his future fields of research.

During World War II, he "plunged into the war effort with the patriotic fervor of those days." In April 1942, he moved to the applied physics department at Johns Hopkins University, where he worked on the creation of a rugged vacuum tube. He helped to develop the proximity fuze, a

James Van Allen discovered a zone of high-level radiation around the Earth caused by the presence of trapped charged particles, known as the magnetosphere or the Van Allen belt. *(AIP Emilio Segrè Visual Archives)*

device attached to a missile such as an antiaircraft shell that detonated when it neared the target, when the radio waves it emitted were reflected back to it with sufficient intensity from the target. The proximity fuse allowed the detonation to occur even if the missile had not been aimed accurately enough for a direct hit and greatly increased the effectiveness of antiaircraft weapons. By the fall of 1942, Van Allen had been commissioned as an officer in the United States Navy and was sent to the Pacific to field test and complete operational requirements for the proximity fuze.

In 1946, the war now ended, Van Allen married his lifetime companion, Abigail Fithian Halsey, with whom he would have five children. For the next four years he was involved in high-altitude research, as supervisor of the high-altitude research group and proximity fuse unit at

Johns Hopkins, working on utilization of the unused German stock of V-2 rockets and of the Aerobee rockets for research purposes. The group developed devices for measuring the levels of cosmic radiation that were sent to the outer reaches of the atmosphere in these rockets, which radioed the data back to Earth. Van Allen's experience in miniaturization of electronic equipment enabled him to include a maximum of instrumentation in the limited payload of these rockets, an achievement that was to be crucial in the early stages of the U.S. space program.

After a year at the Brookhaven National Laboratory on a Guggenheim Fellowship, in 1951, he became professor of physics and head of the physics department at the University of Iowa. Working with his students, he invented the rocket–balloon (rockoon), which went into use in 1952. It consisted of a small rocket that was lifted by means of a balloon into the upper regions of the atmosphere and then fired off, thus reaching heights otherwise attainable only by a much larger rocket.

From 1949 to 1957, Van Allen also organized and led scientific expeditions to Peru (1949), the Gulf of Alaska (1950), Greenland (1952 and 1957), and Antarctica (1957) to study cosmic radiation. During an Arctic expedition in 1953, rockoons fired by Van Allen and his students were the first to detect a hint of the radiation belts surrounding the Earth.

He became part of the organizing panel of International Geophysical Year (IGY), July 1957–December 1958, whose members actively promoted the adoption of scientific satellites of the Earth as an element of the IGY program and laid the foundations for the scientific program of the National Aeronautics and Space Administration, created in 1958. In the mid-1950s, the U.S. government had first seriously considered the possibility of sending a rocket into orbit around the Earth, thereby creating an artificial satellite. In 1955, President Eisenhower announced the Vanguard Program, designed to put an artificial satellite into orbit within two years, to coincide with the scheduled IGY, which was itself timed to coincide with a peak in solar activity. Van Allen was given responsibility for the instrumentation of the project.

On October 4, 1957, the day that the Soviet Union launched *Sputnik 1*, the world's first artificial satellite, Van Allen was on a scientific expedition to the Antarctic. On hearing the news, the American team lost no time in launching *Explorer 1*, on January 31, 1958. It was carrying a Geiger counter Van Allen had intended to use to measure the levels of cosmic radiation, but at a height of about 500 miles the counters registered a radiation level of zero. This nonsensical reading led Van Allen to suspect instrument failure. When the same result was recorded on *Explorer 3*, launched on March 26, 1958, he realized that the zero reading could have resulted from the counters' being swamped with very high levels of radiation. On July 29, 1958, *Explorer 4* was sent up with a counter shielded with lead in order to allow less radiation to penetrate, and this method showed clearly that parts of space contained much higher levels of radiation than previously suspected. Van Allen studied the size and distribution of these high-radiation zones and found that they consisted of two toroidal (i.e., donut-shaped) radiation belts around the Earth, which arise by the trapping of charged particles in the Earth's magnetic field. The inner belt was made up of high-energy protons, the outer belt of high-energy electrons and other particles. These radiation belts were found to be part of the tear-shaped magnetic region around the Earth called the magnetosphere. The radiation belts started at an altitude of several hundred miles from the Earth and extended for several thousand miles into space. In 1993, other scientists discovered a third belt, enclosed by the inner belt, containing ions of oxygen, nitrogen, and neon.

Van Allen was elected to the National Academy of Sciences in 1952 and became president of the American Geographical Union in 1982. In 1959, he was made professor of astronomy at Iowa and, from 1972 to 1985, was Carver Professor of Physics; after retirement, he eventually became professor emeritus there. At age 85, he was still mentoring graduate students, doing research, and monitoring transmissions from the system he invented on *Pioneer 10*, launched in 1972, as the first artificial object to carry sounds and images from Earth into outer space.

As an astrophysicist, James Van Allen played a vital role in the early development of the U.S. space program. In one capacity or another, he was involved in the first four *Explorer* probes, the first *Pioneers*, several *Mariner* efforts, and the orbiting geophysical observatory. From 1966 to 1970, as a member of the Space Science Board and the Lunar and Planetary Missions Board, he became a leading advocate for missions to the outer planets, especially Jupiter. The first fruits of these efforts were the National Aeronautics and Space Administration's (NASA's) two missions to Jupiter, *Pioneer 10* and *Pioneer 11*, launched in 1972 and 1973.

Further Reading

Gillmor, C. Stewart, and John R. Spreiter. *Discovery of the Magnetosphere*. Washington, D.C.: American Geographical Union, 1997.

Newton, David E. "James A. Van Allen," in Emily J. McMurray, ed., *Notable Twentieth-Century Scientists*. New York: Gale Research, 1995, pp. 2070–2072.

Van Allen, James. "Autobiography," *Annual Review of Earth and Planet Sciences*, 18: pp. 1–26, 1990.

———. *Origins of Magnetospheric Physics*. Washington, D.C.: Smithsonian Institution Press, 1983.

⊠ **Van de Graaff, Robert Jemison**
(1901–1967)
American
Experimentalist, Particle Physicist

Robert Jemison Van de Graaff made a major contribution to advances in high-energy physics through his invention of what came to be called the Van de Graaff generator, an electrostatic high-voltage accelerator capable of accelerating electrons to millions of volts of kinetic energy.

He was born on December 20, 1901, in Tuscaloosa, Alabama, to Minnie Cherokee Hargrove and Adrian Sebastian Van de Graaff. His early education was in the public schools of Tuscaloosa. Intending to become a mechanical engineer, he studied at the University of Alabama, where he received his B.S. degree in 1922 and his M.S. the following year. He then took a job as a research assistant with the Alabama Power Company but left after only a year to travel to the Sorbonne in Paris to continue his education. The trip was to change the direction of his life. In Paris he attended the lectures of MARIE CURIE, one of the great pioneers of the study of radioactivity, who exposed him to the ongoing revolution in atomic physics.

He continued his European education as a Rhodes Scholar at Oxford University, under John Sealy Edward Townsend, a major figure in developing the study of the kinetics of electrons and ions in gases. Van de Graaff worked primarily on these topics at Oxford, where he was awarded a B.Sc. in 1926 and a Ph.D. in 1928. During this period, as he became aware of the work of ERNEST RUTHERFORD and others trying to fathom the mysteries of atomic structure, the pressing need for a high-energy subatomic particle accelerator, capable of disintegrating atomic nuclei, became increasingly clear to him. He realized that a high potential could be built up by storing electrostatic charge within a hollow sphere and that this could be achieved by depositing charges on a moving belt that carries the charges into the sphere, where a collector transfers them to the outer surface of the sphere. Van de Graaff's idea led him to develop a device, now known as a Van de Graaff generator, that consisted of a large smooth metal sphere on top of a hollow insulating cylinder within which an endless insulating belt ran between pulleys at

each end of the cylinder. Electric charge was sprayed from metal points connected to a high-voltage source onto the bottom of the belt; it was then carried up to the top of the belt, where it was collected by other metal points and accumulated on the outside of the sphere, causing it to become highly charged at a high electrostatic potential difference. Charged particles falling through this large electrostatic potential difference would then be accelerated to very high values of kinetic energy.

While at Oxford, Van de Graaff began working on designs for his generator. When he returned to the United States in 1929 he built his first working model, operating at 80,000 volts, at the Palmer Physics Laboratory at Princeton University, where he was a National Research Fellow from 1929 to 1931. He demonstrated an improved model, which operated at 1 million volts, at the inaugural dinner of the American Institute of Physics in November 1931.

That year he moved to Cambridge, Massachusetts, as a research associate at the Massachusetts Institute of Technology (MIT), where, in 1934, he would become an associate professor of physics. In an aircraft hangar in South Dartmouth, Massachusetts, he constructed his first large generator, able to produce 7 million volts, which was demonstrated on November 28, 1933. He developed the generator so that it could be used to accelerate subatomic particles to very high velocities. The previous year JOHN DOUGLAS COCKCROFT and Ernest Walton had built the first particle accelerator, which produced the first nuclear transformations by means of artificially accelerated particles, at the Cavendish Laboratory at Cambridge. Van de Graaff's machine was more compact, simpler in design, and capable of producing higher voltages (and, therefore, higher particle accelerations) than the Cockcroft–Walton accelerator.

In 1936, he married Catherine Boyden, with whom he would have two sons, John and

William. Having obtained the patent for his generator, he began working with John G. Trump, a professor of electrical engineering at MIT, in designing a modified generator capable of producing X rays for treating cancerous tumors. It had its first clinical trial at Harvard Medical School in 1937.

During World War II, the U.S. Navy commissioned the production of five generators with a 2-million-volt capacity for use in the X-ray examination of munitions. With the experience of this work behind him, together with Trump, he was able to set up the High Voltage Engineering Company (HVEC), in 1946, for the commercial production of generators. The company developed the Van de Graaff generator for the wide variety of scientific, medical, and industrial research purposes it has today. Van de Graaff was honored with the Duddel Medal of the Physical Society of Great Britain in 1947.

Numerous improvements of the generator were to follow. The tandem principle of particle acceleration, developed by LUIS W. ALVAREZ in 1951, was incorporated into a tandem version of the Van de Graaff machine, in which the high-voltage terminal is able to accelerate the ion twice. In the late 1950s, Van de Graaff invented a new insulating core transformer that generated high currents by using magnetic flux rather than electrostatic charging. He also developed techniques of controlling beams during and after acceleration, allowing physicists to adapt them for specific research needs. The improved accelerators yielded a mass of data on nuclear disintegrations and reactions, which led directly to advances in the theory of the atomic nucleus.

In 1960, he resigned from MIT in order to devote all his time to HVEC. He was awarded the American Physical Society's Tom W. Bonner Award in 1966 for his generator.

Van de Graaff, who died in Boston on the morning of January 16, 1967, at the age of 65, left a substantial legacy to modern science and technology. Since its invention in the early 1930s, the Van de Graaff generator has played a vital role in many fields of physics, in astrophysics, and in medicine and industry.

Further Reading

Burrill, E. Alfred. "Van de Graaff, the Man and His Accelerators," *Physics Today* 2, no. 20 (February 1967): 49–52.

Livingston, M. Stanley. *Particle Accelerators: A Brief History*. Cambridge, Mass.: Harvard University Press, 1967.

⊠ Van Vleck, John Housbrook
(1899–1980)
American
Theoretician, Solid State Physicist

John Housbrook Van Vleck, who developed the first complete quantum mechanical theory of magnetism in matter in the early 1930s, is known as "the father of modern magnetism." He shared the 1977 Nobel Prize in physics with PHILIP WARREN ANDERSON and SIR NEVILL FRANCIS MOTT for his fundamental theoretical investigations of the electronic structure of magnetism in the solid state of matter.

He was born on March 13, 1899, in Middletown, Connecticut, into a family of scientists. As a child, John reacted to having a father who was a professor of mathematics and a grandfather who was a professor of astronomy by rebelling against the academic tradition and vowing not to continue it. When his father accepted a position at the University of Wisconsin, the family moved to Madison, where John attended public schools.

He did his undergraduate work at the University of Wisconsin and his graduate studies at Harvard University, where he took courses with PERCY WILLIAMS BRIDGMAN and Edwin C. Kemble. He wrote his dissertation, on the binding energy of the helium atom, under Kemble, whom he described as "the one person in America at that time qualified to direct purely theoretical research in quantum atomic physics," and earned his doctorate in 1922. He had long before outgrown his "childish prejudices," as he called them, and realized that he was best qualified for the life of a physicist in an academic environment.

For the next year Van Vleck served as an instructor at Harvard; he then took a position as assistant professor at the University of Minnesota, where he had the opportunity, rarely given to a new faculty member, of teaching only graduate courses. In 1927, he was promoted to full professor and married Abigail Pearson; their union would last more than 50 years. The young couple then set out on a bit of academic wandering: in 1928, they moved to Wisconsin, where Van Vleck was a professor of physics at his alma mater and remained until 1934. During this period, Van Vleck was a Guggenheim Foundation Fellow. In 1930, they moved to Harvard, where he did his groundbreaking work on magnetism and published his classic work *The Theory of Electric and Magnetic Susceptibilities* in 1932.

The focus of his long, illustrious career was magnetism and its relationship to the quantum theory of atomic structure. As quantum wave mechanics was being developed in the early 1930s, Van Vleck set out to understand its implications for magnetism. As a result, he constructed a theory, using ERWIN SCHRÖDINGER's wave mechanics, that offered a precise explanation of the magnetic properties of individual atoms in a series of chemical elements. He went on to propose the notion of temperature-independent susceptibility in paramagnetic materials, that is, materials with small susceptibility to the external magnetic field. This phenomenon, which came to be called Van Vleck paramagnetism, drew attention to the importance of electron correlation (the interaction between the wave mechanical motion of electrons) in the appearance of localized magnetic moments (tiny quantum mechanical magnets in metals). His former student Philip Anderson would further develop these ideas to explain how local magnetic moments can occur in metals such as copper and silver, which in pure form are not magnetic at all.

Van Vleck's most important work was his formulation of the quantum mechanics of chemical bonding in crystals, used for interpreting the patterns of chemical bonds in complex compounds. This theory was able to explain the magnetic, electrical, and optical properties of many elements and compounds by considering the influences exerted on the electrons in particular atoms by other atoms nearby. It gave the following description of how a perturbing (foreign) ion or atom behaves in a crystal: at first the electrons of a perturbing ion feel the influence of the electric field that is generated by the atomic nuclei and the electrons of the host crystal; then, through the action of its electrons, the perturbing ion enters into chemical bonding with its environment. In this context he also found that a perturbing atom in a crystal could sometimes replace a host atom without essential changes in the surrounding lattice. However, under certain circumstances the electronic structure of the perturbing atom is so incompatible with the symmetry of the environment that it leads to a local distortion of the lattice. This phenomenon, which later became known as the Jahn–Teller effect, played an important role in the development of the physics of the solid state. At age 78, Van de Graaff was awarded the Nobel Prize in physics for this work.

During World War II, he worked on radar, discovering that water molecules in the atmosphere absorb radar waves with a wavelength of about 1.25 cm and that oxygen molecules have a similar effect on 0.5-cm radar waves. This finding played an important role in the design of effective radar systems and later in microwave communication and radio astronomy.

In 1951, Van Vleck became Hollis Professor of Mathematics and Natural Philosophy at Harvard. He also served as Dean of Engineering and Applied Physics from 1951 to 1957. He took sabbaticals from Harvard to serve as Lorentz Professor at Leiden University, Holland, in 1960, and as Eastman Professor at Oxford University, in 1962–1963. In 1969, when he retired, he became professor emeritus. He died in his sleep at his home in Cambridge, Massachusetts, on October 27, 1980, at the age of 81.

Van Vleck's groundbreaking research on the electronic structure of magnetism in the solid state of matter led to the development of the laser, new industrial uses of glass, and copper spirals used in birth control devices. The quantum chemical methods have now become routine tools, particularly in inorganic chemistry, with important extensions to molecular biology, medicine, and biology. His research into molecular spectra and the theoretical problems associated with their fine structure contributed to advances in radio astronomical investigations of molecular gases in space.

Further Reading

Mailis, Norberto. *The Quantum Theory of Magnetism.* Singapore: World Scientific, 2001.

⊠ **Veltman, Martinus J. G.**
(1931–)
Dutch
Quantum Field Theorist, Particle Physicist

Martinus J. G. Veltman is a pioneer in the development of the quantum field theoretic gauge theories of particle physics. He shared the 1999 Nobel Prize in physics with his former student, GERARD 'T HOOFT, for work they did in 1971, which proved that the quantum field theoretic structure of gauge theories, undergoing spontaneous symmetry breaking interactions with a Higgs scalar field, was renormalizable. Veltman's development of his "Schoonschip" symbolic manipulation computer program, which alge-

braically simplified the complicated Feynman diagrammatic equations found in quantum field theories, was the foundation of this work.

Veltman was born on June 27, 1931, in the small town of Waalwijjk, in the south of the Netherlands, the fourth of what would become a six-child family. Veltman considers himself the inheritor of both the practicality of his mother's family of tradespeople and the studiousness of his father's family of pedagogues. Veltman's father was head of the local primary school, a comfortable position that exempted his family from the hardships of the Depression years. Veltman saw the German army marching in, as they invaded his country in 1940, and remembers how they requisitioned his father's school for the billeting of soldiers. He was fond of playing with the ammunition left carelessly about by Allied troops but somehow survived this pastime, as well as the bombing of his hometown; the memory of that carnage would remain with him. Liberated in the fall of 1944, the southern part of the Netherlands escaped the brutal winter of 1944–1945, in which many Dutch people died of hunger.

In the midst of the war, in 1943, Veltman entered high school, where conditions, as he put it, "were marked by irregularities," such as the substitution of a horse stable for a more conventional classroom. During these years he learned about electronics from the local plumber and became the town radio repairman. But his academic performance declined as a result of his lack of aptitude for foreign languages. In 1948, at the age of 17, he narrowly passed his final examinations.

Through the benevolent intervention of his high school physics teacher, Veltman's parents were persuaded to send him to the University of Utrecht, a 90-minute commute from his home. He made the round trip regularly for three years, although the teaching of physics there, in the difficult postwar years, when many physicists had been killed or had left the country, was uninspiring. He then moved to Utrecht, supporting himself meagerly by typing lectures notes but generally happy in his life of "mainly bumming around." He took five years, rather than the usual three, to pass his candidate's exam.

At this point, a revelation occurred to him in the form of a popular book on the theory of relativity—a subject that had not been touched upon in his physics courses. He obtained a copy of ALBERT EINSTEIN's *The Meaning of Relativity*, and from then on he was "hooked." While supporting himself as a part-time teacher at a lower technical school, teaching physics to plumbers, he embarked on graduate studies in physics.

Veltman began studying experimental physics but soon realized that this was not his "real destiny." He switched to theoretical work with Leon Van Hove in 1955 but was soon interrupted by two years of military service. When he returned to the university in February 1959, Van Hove took him on as his Ph.D. student, despite his "relatively advanced age" of 27. Since Van Hove did statistical mechanics and Veltman wanted to do particle physics, he supplemented his education by attending summer schools in Naples and Edinburgh.

In 1961, he followed Van Hove to the European Center for Nuclear Research (CERN), in Geneva. He had married his wife, Anneke, in 1960, and she remained in the Netherlands for the birth of their daughter, Helene (who would one day do a thesis on particle physics at Berkeley), before joining him. In Geneva, Veltman completed a dissertation on both unstable particles and Coulomb corrections to the production of vector bosons produced by neutrinos, which earned him the Ph.D. from Utrecht in 1963. While at CERN, Veltman became deeply involved with the neutrino experiments being conducted there and was spokesperson for the group at the Brookhaven Conference in 1963. The experience left him with a lasting fascination for experiments.

Veltman's development of his "Schoonschip" symbolic computer program, while at the Stanford Linear Accelerator (SLAC) in Stanford in 1963, was a direct response to the frustration he and colleagues had felt at the work involved in doing error-free algebraic calculations for vector boson production. He returned to CERN in 1964 and remained there until just after the birth of his son, Hugo, in 1966, after which he spent a short period at the Brookhaven National Laboratory.

Veltman then returned to the University of Utrecht, where he succeeded Van Hove as professor of theoretical physics and began building up the particle physics group, which thrived under his leadership. Hoping to alleviate his relative isolation, he took over as editor of *Physics Letters* but quit after two years, oppressed by the large amount of "junk" he received and felt obliged to reject.

In the scholarly quiet of a one-month visit in April 1968 to Rockefeller University, Veltman began the work he would successfully complete back at Utrecht, in 1971, with his doctoral student, Gerard 't Hooft. When "Tini," as everyone called Veltman, took 't Hooft on as his student, the first material he gave him to study was a 1950 paper by CHEN NING YANG and Robert Mills, telling him, "This stuff you must know." Yang and Mills had shown that the quantum electrodynamic (QED) formalism developed earlier by JULIAN SEYMOUR SCHWINGER, RICHARD PHILLIPS FEYNMAN, and SIN-ITIRO TOMONAGA could be generalized to include internal dynamic symmetries that were more general than the standard spacetime C (charge), P (parity), and T (time-reversal) symmetries.

In the early 1960s, SHELDON LEE GLASHOW, ABDUS SALAM, and STEVEN WEINBERG used this new generalization of QED to unify the electroweak forces into one quantum field theoretic formalism. At the heart of their quest was the fascinating phenomenon of so-called broken symmetries—asymmetric relations that have spontaneously arisen from the functioning of symmetrical laws (e.g., the asymmetric crystal structure of ice that freezes out from the symmetric liquid structure of water when the temperature becomes low enough)—that seem to permeate matter.

They were aware that in the early 1960s Peter Higgs had published papers demonstrating that spontaneous symmetry breaking events associated with coupling to a scalar field could create new kinds of force-carrying particles, some of them massive. This led them to speculate that if the virtual particles that carry the electromagnetic and weak forces (known collectively as the intermediate vector boson W and Z particles) were related by such a broken symmetry, it might be possible to estimate their masses in terms of the unified, more symmetrical force from which the two forces were thought to have arisen. Working independently, each constructed a unified quantum field theory of electromagnetic and weak interactions (i.e., a quantum electroweak theory) that could make a verifiable prediction of the approximate masses of the triplet W and singlet Z particles needed to describe the weak interactions. However, at first, physicists ignored the electroweak theory, which, when used to calculate the properties of the W and Z particles and other physical quantities, predicted nonsensical infinite results.

Meanwhile, in the Netherlands, Veltman had not given up hope of renormalizing theories like the electroweak theory. Twenty years earlier, Feynman had systematized the calculation problem with his diagrams. Veltman hoped to find a way of renormalizing the theory by using "Schoonschip," which was capable of performing algebraic simplifications of the complicated expressions that all quantum field theories result in when quantitative calculations are made. When 't Hooft chose a topic for his Ph.D. thesis in 1969, he chose to apply himself to the problem Veltman was working on: renormalization of the so-called Yang–Mills non-Abelian gauge field theories, which at the time were being applied to the study of the weak interactions.

In 1971, 't Hooft succeeded in this task and published two articles describing his break-

through. However, it was the second paper, in which he renormalized massless Yang–Mills fields for theories by using Higgs spontaneous symmetry breaking mechanism to generate the masses of the fields, that attracted world attention. After using Veltman's computer program to verify 't Hooft's results, the two men were able to work out a detailed calculation of the renormalization method. Using spontaneous symmetry breaking, the renormalized non-Abelian gauge theory of electroweak interaction was now a functioning theoretical machinery capable of performing precise calculations. They presented their results at a 1971 international particle physics conference in Amsterdam.

The impact was enormous. From 1971 on, all theories for the weak interactions that were proposed were Yang–Mills theories that used spontaneous symmetry breaking, and over the next decade it became clear that the Glashow, Weinberg, and Salam's model was correct. The decisive experiments were first made by CARLO RUBBIA and his team at CERN in 1981. Two years later, in 1983, they announced the discovery of the triplet W and singlet Z particles, which was based on signals from detectors specially designed for this purpose.

Veltman and 't Hooft went on to calculate the mass of the top quark, the heavier of the two quarks in the third family of the Standard Model. Many years later, in 1995, the top quark was observed directly for the first time at the Fermi Laboratory in Batavia, Illinois. Another highly important element of the model is the existence of a massive Higgs particle responsible for the spontaneous symmetry breaking, which has not yet been observed. Physicists hope that the Large Hadron Collider (LHC), to be completed at CERN around 2005, will succeed in finding it.

As Veltman and 't Hooft both moved on to different research interests, their collaboration dissolved. Then, in 1981, after spending a sabbatical at the University of Michigan in Ann Arbor, Veltman decided to accept the university's offer to remain there. From 1981 to 1996, Veltman occupied the John D. and Catherine MacArthur Chair, working on gauge theories and their applications to elementary particle physics. In particular he studied radiative corrections in the Standard Model, the Higgs sector and its relationship to the vacuum structure in quantum field theories, and the implications for phenomenology and for new physics beyond the Standard Model. While in the United States, he served on experiment-planning committees at the big laboratories—SLAC, Brookhaven, and the Fermi Laboratory in Batavia, Illinois.

Upon retiring in 1996, Veltman became professor emeritus at Michigan and retired with his wife to the town of Bilthoven in the Netherlands, where they had previously lived.

Veltman's landmark achievement made it possible for physicists to use gauge theories to make specific predictions of particle properties capable of experimental verification. Ultimately, it was the tool that validated the Weinberg–Glashow–Salam electroweak theory.

See also BETHE, HANS ALBRECHT; DYSON, FREEMAN; GELL-MANN, MURRAY

Further Reading

'T Hooft, Gerard. *In Search of the Ultimate Building Blocks*. Cambridge: Cambridge University Press, 1997.

Veltman, Martinus J. G. "The Higgs Boson." *Scientific American* 259, no. 9 (November 1986): 88–94.

⊠   **Volta, Alessandro Guiseppe Antonio Anastasio**
(1745–1827)
Italian
Experimental Physicist (Classical Electromagnetism)

Alessandro Volta was the pioneer in the field of electricity who built the voltaic pile, the forerunner of the modern electric battery. In a famous debate with Luigi Galvani, he successfully argued that the source of the electric current generated when metals are brought into contact with the muscles of a frog are the metals rather than the frog. An ingenious experimenter, he also invented the electrophorus, a device for producing charges of static electricity.

He was born in Como, in Lombardy, Italy, on February 18, 1745, into a noble family. When young Alessandro failed to develop speech until the age of four, his family was sure he was intellectually impaired. By age seven, however, when his father died, he had caught up to his age group. Educated at religious schools, he showed an early aptitude for science, and when he was 14, he decided to become a physicist. The theory of electricity, which he learned about by studying the work of Benjamin Franklin, so entranced him that he composed an excellent Latin poem on the subject. At the age of 20, he began experimenting with static electricity. His renown as a scientist grew rapidly, leading to his appointment in 1774 as principle of the gymnasium in Como, where, the next year, became a professor of experimental physics.

Volta's work on static electricity culminated in 1775, with his invention of the electrophorus. His extensive knowledge of the nature and quantity of electrostatic charge generated by different materials enabled him to develop a fairly simple device for the production of charges: it consisted of one metal plate covered with ebonite and a second metal plate with an insulated handle. The ebonite-covered plate is rubbed and given a negative electric charge. If the plate with a handle is placed over it, a positive electric charge is attracted to the lower surface, a negative charge repelled to the upper. The negative charge is built up in the plate with the handle. Today's electrical condensers are based on this type of charge-accumulating machine. Volta also realized from his electrostatic experiments that the quantity of charge produced is proportional to the product of its electrostatic potential, which he measured with an electrometer of his own invention, and the capacity of the conductor.

At this point, Volta's interests veered toward the study of air and gases. By isolating and examining the properties of marsh gas found in Lake Maggiore (adjacent to Lake Como), he discovered methane. In another study, he accurately estimated the proportion of oxygen in the air by exploding air with hydrogen to remove the oxygen. Volta would return to these investigations 20 years later, making the discovery that the vapor pressure of a liquid depends only on temperature and is independent of atmospheric pressure, a principle that the British chemist and physicist John Dalton would later enunciate in his law of partial pressures.

However substantial these achievements, they were only a digression in Volta's pursuit of the mysteries of electricity. In 1778, he transferred his laboratory from Como to Padua, where he accepted a position as professor of experimental physics. He would remain in Padua until 1819, surviving the conflict between Austria and France that engulfed the region. His considerable political acumen enabled him to hang on to his position no matter who was in charge and carry on his scientific work.

Volta's path to his great discovery began in 1791, when his friend Luigi Galvani sent him his papers describing some interesting results: Galvani had produced contractions in the muscles of dead frogs by placing two different metals (brass and iron) into contact with the muscle and with each other. Galvani believed that the contractions had their source in the frogs' muscles. Volta was not so sure. He successfully repeated Galvani's experiments, using different metals and different animals, and concluded that the source of the electricity lay in the junction of the metals. In the ensuing controversy between the two Italians, the French physicist CHARLES AUGUSTIN COULOMB was a strong advocate of Volta's position, which prevailed as

evidence in its favor accumulated. This included an experiment in which Volta placed the metals on his tongue and produced an unpleasant sensation. He attributed this effect to electricity and went on to compose a list of metals in order of their electricity production, based on the strength of the sensation they made on his tongue. In this way, he derived what came to be known as the electromotive series, the arrangement of chemical elements in order of their standard electrode potentials.

From here, it would be but a brief step to his discovery of the voltaic pile. In 1796, attempting to measure the electricity produced by different metals, he tried piling disks of metals on top of one another and found that they had to be separated by a moist conductor to produce a current. His work was disrupted by political upheavals, but by 1800 he had created his prototype of the modern electric battery. In that year he wrote to the president of the Royal Society, Joseph Banks, to describe two arrangements of conductors that produced an electric current. One was a pile of silver and zinc disks separated by cardboard moistened with brine, the other was a series of glasses of salty or alkaline water in which bimetallic curved electrodes were dipped. This was the first electric battery.

Volta's breakthrough, which meant that high electric currents could now be produced, was greeted with immense excitement. Volta traveled to Paris in 1801 to demonstrate his discoveries to Napoleon, who, duly impressed, made him a count and awarded him a pension. He was the recipient of many honors, including membership in England's Royal Society, which honored him with its Copley medal and decoration by the Legion of Honor. In 1810, he was made a senator of the kingdom of Lombardy and given the title of count. Volta retired in 1819. On March 5, 1827, he died in Como.

Volta's invention of the battery signaled the beginning of a century of discoveries, which would establish the dynamics of electricity and electromagnetism and harness their power in ways that would transform civilization. The unit of electric potential, or electromotive force, is named the *volt* in his honor.

Further Reading

Pera, Marcello, and Jonathan Mandelbaum. *The Galvani–Volta Controversy on Animal Electricity.* Princeton, N.J.: Princeton University Press, 1992.

W

⊠ Weber, Wilhelm Eduard
(1804–1891)
German
*Experimentalist (Classical
Electromagnetism)*

Wilhelm Eduard Weber was an ingenious experimentalist who made vital contributions to electromagnetism. With the highly sensitive apparatus he developed he found new ways to measure electricity and magnetism and to define electric and magnetic units. He is also remembered as the first physicist to propose that electricity consists of charged particles.

He was born in Wittenberg, Saxony (now Germany), on October 24, 1804, into a distinguished family. His father was professor of theology at the University of Wittenberg; his older brother, Ernst, would become a seminal figure in developing the physiology of perception. When Wilhelm was 10, the family moved to Halle, where he would enter the university eight years later. After receiving his doctorate in 1826, he stayed on at Halle, at first as a lecturer and later, from 1828 to 1831, as an assistant professor. He then accepted a position as full professor at the University of Göttingen, where he met the mathematician and physicist JOHANN CARL FRIEDRICH GAUSS. Their collaboration marked the beginning of Weber's work on magnetism.

Gauss and Weber created absolute units of magnetism, defined in terms of length, mass, and time. Working on his own, Weber built highly sensitive magnetometers, as well as a 3-km telegraph to connect his physics laboratory to Gauss's at the astronomical observatory. In Gauss's and Weber's hands, this first working telegraph was used for scientific, rather than commercial, purposes: to connect a network of observation stations in order to correlate measurements of terrestrial magnetism made in different parts of the world.

Political events disrupted Weber's academic life. In 1837, when Queen Victoria came to power, her uncle became the new ruler of Hanover and promptly suspended the constitution. Weber was one of seven professors who formally protested the action, all of whom were fired. He stayed in Göttingen, pursuing his research, until he was offered a professorship in Leipzig. Under these circumstances, he branched out into the measurement of electricity, defining an electromagnetic unit for electric current that was applied to measurements of current made by the deflection of a galvanometer.

Eventually, in 1849, he regained his position in Göttingen; he would remain there until his retirement in the 1870s. His later work there focused on electrodynamics and the electrical structure of matter.

In 1855, Weber studied the ratio between the electrodynamic and electrostatic units of charge. Weber calculated this ratio as 3.1074×10^8 m/s; however, he did not understand the physical implications of the fact that this ratio was close to the speed of light. It would fall to JAMES CLERK MAXWELL, in creating his electromagnetic theory of light, to realize that this implied that light waves were time-dependent electromagnetic fields traveling through space at a speed equal to the ratio Weber had found.

During his later years, Weber focused on research in electrodynamics and the electrical structure of matter. After his death in Göttingen on June 23, 1891, Weber's name was given to the standard international unit of magnetic flux density, the weber, in recognition of his pioneering work in electromagnetism.

Further Reading

Koch Torres Assis, André. Fundamental Theories of Physics. Vol. 66. *Weber's Electrodynamics.* Dordrecht, Netherlands: Kluwer Academic, 1994.

⊠ Weinberg, Steven
(1933–)
American
Theoretical Physicist, Quantum Field Theorist, Particle Physicist, Astrophysicist

Steven Weinberg is a giant of contemporary physics, a theorist with broad research interests whose most notable work has been in unified field theory. In 1967, he hypothesized that quantum electrodynamics can be generalized into a form that allows the electromagnetic and weak forces to be unified into a single electroweak quantum field theory at extremely high energy levels. When his theory was confirmed by particle accelerator experiments in 1983, physics moved significantly closer to the goal of finding "the theory of everything," a single quantum field

theory to describe nature's basic forces. For this work, Weinberg shared the 1979 Nobel Prize in physics with SHELDON LEE GLASHOW and ABDUS SALAM, who independently developed similar versions of the same ideas.

Steven was born on May 3, 1933, in the Bronx, New York City, to Frederick Weinberg and Eva Weinberg, who had lost much of her family in Germany during the Holocaust. Encouraged by his father, who was a court stenographer, and by his teachers at the Bronx High School of Science (where Sheldon Glashow was his close friend) to follow his innate interest in science, he already knew at age 16 that he was heading for a career in theoretical physics. Far from one-sided, however, he grew up listening to classical music, a love he would retain all his life.

Weinberg attended Cornell University, where he met his future wife, Louise, another Cornell undergraduate, who would eventually become a lawyer. The couple was married in 1954, the year Weinberg graduated. He began his life as a researcher the following year, at what is now the Niels Bohr Institute in Copenhagen, working with David Frisch. After his return to the United States, he enrolled at Princeton University to complete his graduate studies. Working under Sam Treiman, he wrote his doctoral thesis on the application of renormalization theory to the effects of strong interactions in weak interaction processes and received his Ph.D. in 1957.

His first position was at Columbia University, where he stayed for only two years before moving to the University of California, Berkeley. In 1963, his daughter, Elizabeth, was born. For the next few years, he worked on a broad spectrum of problems, including high-energy behavior of Feynman graphs, second-class weak interaction currents, broken symmetries, scattering theory, and muon physics. Highly studious and self-disciplined, Weinberg writes that, in many cases, he chose a problem "because I was trying to teach myself some area of physics." In the early 1960s, he first became interested in astrophysics; he wrote papers on the cosmic population of neutrinos and began his book *Gravitation and Cosmology*, which he would complete in 1971. Late in 1965, he began to work on current algebra and the application to strong interactions of the idea of spontaneous symmetry breaking.

The following year, Weinberg left Berkeley on what was to be a leave of absence but turned out to be a final break. From 1966 to 1969, he was Loeb Lecturer at Harvard University and then visiting professor at the Massachusetts Institute of Technology (MIT), where, in 1969, he accepted a professorship in the physics department under the chairman Victor Weisskopf. In 1967, at MIT, he did his groundbreaking work, turning his previous studies of broken symmetries, current algebra, and renormalization theory in the direction of the unification of weak and electromagnetic interactions.

Weinberg's starting point was the innovative work of CHEN NING YANG and Robert Mills. In 1950, they had shown that the quantum electrodynamic (QED) formalism developed earlier by JULIAN SEYMOUR SCHWINGER, RICHARD PHILLIPS FEYNMAN, and SIN-ITIRO TOMONAGA could be generalized to include internal dynamic symmetries that were more general than the standard spacetime C (charge), P (parity), and T (time-reversal) symmetries. In terms of these so-called non-Abelian Yang–Mills gauge theories with internal symmetries, Weinberg's quest was to find the hidden symmetry of the apparent asymmetries that occurred in particle physics. He would later write:

> Nothing in physics seems so hopeful to me as the idea that it is possible for a theory to have a very high degree of symmetry, which is hidden from us in ordinary life. The physicist's task is to find this deeper symmetry.

He was fascinated by the fact that nature was replete with so-called broken symmetries—asymmetric relations that have spontaneously arisen from the functioning of symmetrical laws (e.g., the asymmetric crystal structure of ice that freezes out from the symmetric liquid structure of water when the temperature becomes low enough).

Weinberg was aware that in the early 1960s Peter Higgs had published papers demonstrating that spontaneous symmetry breaking events could create new kinds of force-carrying particles, some of them massive. This led Weinberg to speculate that if the virtual particles that carry the electromagnetic and weak forces (known

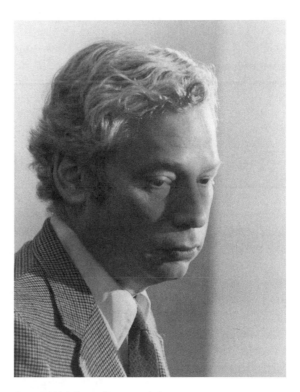

Steven Weinberg discovered that quantum electrodynamics could be generalized into a form that allowed the electromagnetic and weak forces to be unified in a single electroweak quantum field theory at extremely high energy levels. *(AIP Emilio Segrè Visual Archives)*

collectively as the intermediate vector boson W and Z particles) were related by a broken symmetry, these new theoretical ideas might make it possible to estimate their masses in terms of the unified, more symmetrical force from which the two forces were thought to have arisen.

First Weinberg tried, unsuccessfully, to apply the new theoretical ideas of symmetry breaking to the strong force, but he soon realized that the descriptions emerging from his equations—one set massless, the other massive—resembled nothing related to the strong force but fit perfectly with the particles that carry the weak and electromagnetic forces. The massless particle was the photon, carrier of electromagnetism; the massive particles were the W's and the Z's, carriers of the weak force. By accident, Weinberg had found a unified quantum field theory of electromagnetic and weak interactions (i.e., a quantum electroweak theory) that could make a verifiable prediction of the approximate masses of the required triplet W and the singlet Z particles needed to describe the weak interactions. The following year, Salam independently made the same finding.

However, over the next four years, Weinberg's unified theory attracted scant attention since, unlike quantum electrodynamics, his electroweak theory had not yet been shown to be renormalizable, that is, capable of being altered by a mathematical procedure that cancels the unwanted infinities in a quantum field theory by introducing the appropriate renormalization constants. But, in 1971, the Dutch physicist GERARD 'T HOOFT used computer algebra techniques to prove that Weinberg's electroweak theory was indeed renormalizable. After this development, the attention of the physics community shifted to testing the electroweak theory. In 1973, Weinberg accepted Schwinger's recently vacated chair as Higgins Professor of Physics at Harvard, together with an appointment as Senior Scientist at the Smithsonian Astrophysical Observatory. During the 1970s, he worked primarily with the implications of the unified theory of weak and electromagnetic interactions, development of the related theory of strong interactions known as quantum chromodynamics, and the unification of all interactions.

Although decisive experimental confirmation would not be found until 1983, Weinberg, Salam, and Glashow shared the 1979 Nobel Prize for their theoretical breakthrough. Two particle accelerator teams at the European Center for Nuclear Research (CERN) and the Fermi Laboratory in Batavia, Illinois, raced to test the predictions of the electroweak theory. The CERN team under Paul Musset found the neutral currents associated with the singlet Z particle, but these results were inconclusive, since other competing theories of the weak interactions could predict the existence of these particles.

Finally, in 1981, CARLO RUBBIA, working with Simon Van der Meer, Guido Petrucci, and Jacques Gareyte at CERN, did the decisive experiment and found the triplet W and singlet Z particles characteristic of Weinberg's electroweak theory. His team used a new type of colliding beam machine built with intersecting storage rings in which a beam of protons and antiprotons counterrotate and collide head-on. The team developed techniques for creating antiprotons, confining them in a concentrated beam and colliding them with an intense proton beam.

Hundreds of scientists were involved in these experiments in teams throughout the world. The first collisions were observed in 1981. In 1983, Rubbia and collaborating teams of scientists announced the discovery of the triplet W and singlet Z particles, which was based on signals from detectors specially designed for this purpose.

In 1982, Weinberg moved to the physics and astronomy departments of the University of Texas at Austin, where he currently holds the Josey Regental Chair of Science, and founded its theory group. His wife, Louise Weinberg, also joined the university, as a professor of law. He was extremely active in the lobbying campaign for a multibillion-dollar Superconducting Supercollider to be located near Waxahachie, Texas, which Congress killed in 1993.

Weinberg is a prolific author and fine prose stylist, who was recently in 2000 awarded the Lewis Thomas Prize, given to the researcher who best embodies "the scientist as poet." His books for general readers include the prize-winning *The First Three Minutes* (1993), which has been translated into 22 languages; *The Discovery of Subatomic Particles*; *Dreams of a Final Theory* (1993); and most recently *Facing Up: Science and Its Cultural Adversaries* (2001). He has also written many books for specialists, such as *The Quantum Theory of Fields*, volume 1, *Foundations*, and volume 2, *Modern Applications* (2000), and *Elementary Particles and the Laws of Physics*, with Richard Feynman.

An outspoken atheist, Weinberg was awarded the 1999 Emperor Has No Clothes Award from the Freedom from Religion Foundation and the 2002 Humanist of the Year Award from the American Humanist Foundation. For Weinberg, a redemption of sorts lies not in religion, but in the scientific endeavor:

> The effort to understand the universe is one of the very few things that lifts human life above the level of farce, and gives it some of the grace of tragedy.

Weinberg's discovery of the electroweak theory became the precursor to what is known today as the Standard Model of the strong and electroweak interactions, which uses a combination of three symmetry principles to unify the electromagnetic, weak, and nuclear forces into one renormalizable quantum field theoretic formalism. Weinberg is today's foremost proponent of the idea that physicists are moving toward generalizing the Standard Model into the formulation of the long-sought "final theory" that will unify all particles and fundamental forces of nature in the context of a single universal symmetry principle.

See also GELL-MANN, MURRAY; LEE, TSUNG-DAO; WU, CHIEN-SHIUNG.

Further Reading

Weinberg, Steven. *Dreams of a Final Theory: The Scientist's Search for the Ultimate Laws of Nature.* New York: Vintage Books, 1993.

———.*Facing Up: Science and Its Cultural Adversaries.* Cambridge, Mass.: Harvard University Press, 2001.

———.*The First Three Minutes: A Modern View of the Origin of the Universe.* New York: Basic Books, 1977.

⊠   ## Wheeler, John Archibald
(1911–   )
American
Quantum Theorist, Particle Physicist, Relativist, Astrophysicist

John Archibald Wheeler has worked at the cutting edge of theoretical physics for more than five decades. The scope of his research and contributions to the main fields of modern physics has been amazing. He was directly involved in the theoretical development of the atomic bomb. With his student RICHARD PHILLIPS FEYNMAN, he reformulated the theory of electricity and magnetism with insights about charged particles moving backward and forward in time. When he turned his attention to ALBERT EINSTEIN's general theory of relativity, he single-handedly reversed the opinion of most physicists that general relativity was irrelevant to all but a few minor features of physical reality. He coined the term *black hole* and contributed to the understanding of this bizarre cosmic phenomenon. In his later years Wheeler made a bold attempt to unify the theory of relativity with the measurement process in quantum mechanics.

Wheeler was born on July 9, 1911, in Jacksonville, Florida, the son of two librarians. Not surprisingly, then, it was the reading of science books when he was a boy that first aroused his interest in science. He attended Baltimore City College and went on to Johns Hopkins University, in Baltimore, Maryland, which awarded him a Ph.D. in 1933, when he was only 21. The following year he married an old acquaintance, Janette Hegner, after only three dates. He then served one-year apprenticeships with Gregory Breit at New York University and with NIELS HENRIK DAVID BOHR, the father of quantum mechanics, at Bohr's institute in Copenhagen. Then followed three years of teaching physics at the University of North Carolina. In 1938, he joined the faculty of Princeton University, where, apart from numerous leaves of absence, he would spend the greater part of his long career.

In 1939, Bohr visited the United States and asked Wheeler to collaborate with him on a problem of considerable interest. That year, LISE MEITNER and Otto Frisch, interpreting puzzling results obtained by Otto Hahn and Fritz Strassmann, had discovered nuclear fission, launching the field of fission physics. But a solid theoretical underpinning for the phenomenon was still missing. Bohr and Wheeler's highly compatible brainstorming resulted in publication of the paper "The Mechanism of Nuclear Fission" (1939), in which they explained fission in terms of a liquid-drop model. In this theory, a slow neutron entering a uranium-235 nucleus modeled in terms of a liquid drop caused it to split as a drop of liquid into two smaller drops representing the nuclei of a tellurium-137 atom and a zirconium-97 atom, while emitting two neutrons. The energy released by this type of reaction formed the basis of nuclear fission theory.

Wheeler later recalled, "Like most physicists, I was interested in nuclear fission for what it revealed about basic science, not for what it

might have to do with reactors or bombs." Nevertheless, his work with Bohr proved extremely important to the Manhattan Project, singling out uranium-235 for use in the development of an atomic bomb. After the Japanese attack on Pearl Harbor, Wheeler worked at the Metallurgical Laboratory at the University of Chicago, where he identified and proposed countermeasures to the problem of reactor poisoning, and in Richmond, Washington, where he was involved in the development of the giant nuclear reactors at nearby Hanford, designed to produce plutonium for atomic weapons. Despite these contributions, having lost his younger brother, Joe, in the final year of World War II, he always regretted that neither he nor the United States had done more to accelerate the production of the first atomic bomb.

After the war, Wheeler worked with his graduate student Richard Feynman and returned to an earlier research interest on the problem of relativistic action at a distance. The concept struck him as a simpler, more satisfying description of electrodynamics than the standard field theory of electrodynamics, which assigned "substance" to electric and magnetic fields existing in space. The resulting theory published in 1945 became known as Wheeler–Feynman electrodynamics. It was based on the amazing idea, suggested by Wheeler to Feynman in one of their many conversations, that the theory of relativity allowed the possibility that the electrodynamic interaction between charged particles might be able to propagate both backward and forward in time.

Later in the early 1950s, while on a sabbatical in Paris, Wheeler responded to a call from his good friend EDWARD TELLER and joined the effort to develop the hydrogen bomb at Los Alamos. From 1951 to 1953, he was director of Princeton's Project Matterhorn, which was instituted in order to design thermonuclear weapons.

Returning to academia after the successful conclusion of the hydrogen bomb project in the late 1950s, Wheeler delved into Einstein's general theory of relativity, which was based on the radical idea that the force of gravity was due to the warping or curvature of the geometry of spacetime caused by the presence of mass–energy. He was appalled by the prediction, contained in the theory's equations, that a dead star could collapse into a mass so dense that not even light could escape from it, leading it eventually to be squeezed out of existence. The final state of this collapsed stellar object had the bizarre property that at its center there existed a singularity where spacetime would be infinitely curved. At this point, as Wheeler would say, "smoke pours out of the computer."

In 1956, Wheeler studied articles by SUBRAMANYAN CHANDRASEKHAR, LEV DAVIDOVICH LANDAU, J. ROBERT OPPENHEIMER, and George Volkoff on the collapse of dead stars, hoping to continue their investigations. He set himself the task of comprehending all "cold dead matter," matter that had completely burned its nuclear fuel, and, with his student B. Kent Harrison, worked out the equation of state for such matter. In the process of doing so, he revived interest in general relativity, which had been neglected because it could not be tested in the lab, and made Princeton the center of research in the field.

During the same period, in his book *Geometrodynamics* (1962), he studied the interaction of gravitational and electrodynamic fields in general relativity, developing the radical concept of geons, concentrations of electromagnetic radiation energy so intense that they are held together by their own gravity.

In the following years Wheeler continued to resist the idea of a final state of stellar collapse with its inherent spacetime singularity. Nonetheless, he eventually accepted the idea and, in 1967, called it by the name for which it is now famous, a black hole. During this period, Wheeler also formulated the "no-hair" theorem, which states that black holes are "bald": that is, their only known physical properties are their mass, charge, and angular

John Archibald Wheeler single-handedly reversed the opinion of most physicists that Albert Einstein's general theory of relativity was irrelevant to all but a few minor features of physical reality. *(AIP Emilio Segrè Visual Archives)*

momentum. He derived a harsh lesson from this work, noting in his 1998 autobiography that the black hole "teaches us that space can be crumpled like a piece of paper into an infinitesimal dot, that time can be extinguished like a blown-out flame, and that the laws of physics that we regard as 'immutable' are anything but."

In analyzing the connection between quantum mechanics and general relativity Wheeler thought that this kind of breakdown of physics was not limited to dead, distant stars, because even spacetime was subject to the uncertainty principle. Ultimately, it, too, was discontinuous, dissolving into a chaos of unconnected points and wormholes, what Wheeler called quantum foam.

In 1976, he took a 10-year leave from Princeton, where he was Joseph Henry Professor of Physics, to become director of the Center for Theoretical Physics at the University of Texas, Austin. There he returned to the study of quantum theory, which he regarded as a greater challenge than relativity for the 20th century. One idea he explored with his Texas colleagues was the notion that the universe is a giant computer and that quantum theory can somehow be derived from information theory.

In 1978 Wheeler proposed a "delayed choice" double-slit experiment, successfully carried out in 1984, in which a difference in what one measures on a photon *now* makes an irretrievable difference in what one has the right to say the photon *already did* in the past. Specifically, in a delayed choice double-slit experiment, in which photons pass through two slits and create an interference pattern, the experimenter could wait until after the photon had passed the slits to determine which detector to employ and thus determine whether it had been a particle or a wave in the past. Wheeler called this process in which the physicist observer acting in the present would be participating in creating the past of the photon observer-participancy. On hearing about the success of the experiment he claimed:

> The experimental verdict is in: the weirdness of the quantum world is real, whether we like it or not. . . . The very building blocks of the universe are these acts of observer-participancy. You wouldn't have the stuff out of which to build the universe otherwise. The participatory principle takes for its foundation the absolutely central point of the quantum: No elementary phenomenon is a phenomenon until it is an observed (or registered) phenomenon.

In 1979, he became the center of a prolonged controversy, when he called for the expulsion of the recently admitted parapsychologists from the American Association for the Advancement of Science. For Wheeler, who believed that a phenomenon must be observed to be taken seriously, parapsychology had produced no hard results and was unworthy of membership in a respectable scientific society.

Wheeler was a beloved teacher, characterized by former students as a physicist who was never afraid to think about Really Big Questions and who "brought the fun back into physics." He is equally gifted in explaining complex ideas in simple language. His much-praised books include *Spacetime Physics* with Edwin Taylor (1963 and 1992); *Gravitation Theory and Gravitational Constants* (1965); *Einstein's Vision* (1968); *Gravitation*, with his former students Kip Thorne and Charles Misner (1973); *Frontiers of Time* (1979); *Quantum Theory and Measurement* (1983); *A Journey into Gravity and Spacetime* (1990); *At Home in the Universe* (1993); and *Exploring Black Holes* with Edwin Taylor (2000).

Personally modest, Wheeler is possessed of a contagious enthusiasm and charming informality. He is a scientist–philosopher, whose thoughts have ranged from the smallest microstructure to the largest structures in the cosmos. He created the phrase "the gates of time," to express the beginning of the cosmos (the big bang) and its ending (the big crunch).

Married for more than 67 years, John and Janette Wheeler have three children, eight grandchildren, and nine great-grandchildren. The establishment of a chair in his name at Princeton commemorated his 90th birthday. After heart surgery in 1986, Wheeler moved to a retirement home near Princeton. At age 91, he commuted to his office at Princeton twice a week; there he dictated to his secretary his still lively thoughts about the nature of the cosmos. In a 2002 interview he said:

> The creation question is so formidable that I can hardly hope to answer it in

the time left to me. But each Tuesday and Thursday I will put down the best response that I can, imagining that I am under torture.

Believing that "We will understand how simple the universe is when we recognize how strange it is," Wheeler continues to explore ideas that would intimidate less adventurous minds.

See also GAMOW, GEORGE; HAWKING, STEPHEN.

Further Reading

Aczel, Amir D. *Entanglement: The Greatest Mystery in Physics*. New York: Four Walls Eight Windows, 2002.

Thorne, Kip S. *Black Holes and Time Warps: Einstein's Outrageous Legacy*. New York and London: W. W. Norton, 1994.

Wheeler, John Archibald, with Kenneth Ford. *Geons, Black Holes, and Quantum Foam: A Life in Physics*. New York: W. W. Norton, 1998.

⊠ Wien, Wilhelm (Carl Werner Otto Fritz Franz)
(1864–1928)
Prussian
Theoretical and Experimental Physicist (Electrodynamics, Thermodynamics)

Wilhelm Wien is most famous for his contributions to the study of blackbody radiation. His formulation of a displacement law for radiation emitted by a perfectly efficient blackbody earned him the 1911 Nobel Prize in physics and led directly to MAX ERNEST LUDWIG PLANCK's discovery of quantum theory.

Wilhelm Wien was born a landowner's son at Gaffken, East Prussia (now Poland), on January 13, 1864, and seemed destined for the life of a gentleman farmer. He received his secondary schooling in Rasternburg and Konigsberg, followed by university studies, from 1882 to 1886,

mainly at the University of Berlin, where he was a student of HERMANN LUDWIG FERDINAND VON HELMHOLTZ; he also studied briefly in Göttingen and Heidelberg. In 1886, he was awarded his Ph.D. for an experimental thesis on the diffraction of light on sections of metals and the influence of materials on the color of refracted light. When his father became ill, he returned home to manage the family estate. But in 1890, when economic problems led to its sale, Wien returned to Berlin, where he resumed his scientific work as an assistant to Helmholtz.

In Berlin, in 1893, he extended the theories of LUDWIG BOLTZMANN, who in 1884 had deduced a law for the total energy emitted by a blackbody, that is, a surface that absorbs all radiant energy impinging on it. Since a perfect blackbody does not exist in nature, Wien devised a method to create a good approximation of one experimentally, by using an oven with a tiny pinhole. Any radiation entering the pinhole would be scattered and reflected from the inner walls of the oven so often that nearly all incoming radiation would be absorbed and the chance of some of it finding its way out of the hole again would be extremely small. Thus the radiation emerging from this hole would be very close to the equilibrium blackbody electromagnetic radiation corresponding to the oven temperature.

Using this experimental method, Wien discovered his law of displacement, which states that the frequency at which the maximal energy is radiated is proportional to the absolute temperature of the blackbody. The basis of this work was the assumption that the blackbody contained a very large number of electromagnetic oscillators having all possible frequencies and in thermal equilibrium. Wien's law proved applicable at high frequencies but had serious limitations at low ones. LORD RAYLEIGH (JOHN WILLIAM STRUTT) found another formula that satisfied low frequencies but not high ones. In 1900, in the effort to develop a treatment valid for the whole range of frequencies, Planck enunciated his quantum theory. Wien was awarded the Nobel Prize in physics for his contributions, in 1911.

In 1896, Wien was appointed professor of physics at Aix-la-Chapelle. It was at Aix, two years later, that he met his future wife, Luise Mehler, with whom he had four children. He also found a laboratory equipped for the study of electrical charges in vacuo and began to study the nature of cathode rays, that is, particles that seemed to emanate from hot filaments of wire. He confirmed an earlier discovery that negative cathode rays are composed of rapidly moving, negatively charged particles (electrons). Almost at the same time as JOSEPH JOHN (J. J.) THOMSON in Cambridge, England, but with a different method, he measured the relation of the electric charge on these particles to their mass and found, as Thomson did, that they are about 2000 times lighter than hydrogen atoms. Wien also presented valuable ideas on positive cathode rays, by investigating their deflection while studying streams of ionized gases in the presence of electrostatic and magnetic fields. He concluded that positive cathode rays are composed of positively charged particles, with a mass approximately equal to that of the hydrogen atom. With this work he laid the foundation of mass spectroscopy. J. J. Thomson refined Wien's apparatus and conducted further experiments in 1913. After work by ERNEST RUTHERFORD in 1919, Wien's particle was accepted and named the proton.

In 1899, Wien became a professor of physics at Geissen but remained there only a year before accepting a similar position at Würzburg. In 1900, he published his *Textbook of Hydrodynamics* and a theoretical paper on the possibility of an electromagnetic basis for mechanics. During his Würzburg years he continued his work on cathode rays, showing that if the pressure is not extremely weak, these rays lose and regain their electric charge as they move along by collision with atoms of residual gas. In addition, he mea-

sured the progressive decrease of the luminosity of cathode rays after they leave the cathode. From these experiments he deduced what classical physics calls the decay of luminous vibrations in atoms.

In 1920, Wien succeeded WILHELM CONRAD RÖNTGEN as professor of physics in Munich, where he remained until his death on August 30, 1928.

Wien's work on blackbody radiation, which demonstrated the inability of classical physics to explain the phenomenon of radiation, stimulated Planck to develop revolutionary new ideas. In the words of MAX THEODOR FELIX VON LAUE, "his immortal glory" was that "he led us to the very gates of quantum physics."

Further Reading
Kuhn, Thomas S. *Black-Body Theory and the Quantum Discontinuity.* Chicago: University of Chicago Press, 1987.

⊠ **Wigner, Eugene Paul**
(1902–1995)
Hungarian/American
Quantum Theorist, Nuclear Physicist, Mathematical Physicist

A giant of 20th-century physics, Eugene Paul Wigner is most famous for the work that earned him the 1963 Nobel Prize in physics: the introduction of symmetry theory to quantum mechanics and chemistry. Yet he was a scientist of amazing breadth, who also pioneered the application of quantum mechanics in the fields of chemical kinetics and the theory of solids, formulated many of the basic ideas in nuclear physics and nuclear chemistry, and did seminal work in quantum chaos.

He was born in Budapest, Hungary, on November 17, 1902, with the Hungarian name Jeno Pal Wigner, into an upper-middle-class Jewish family. He was the middle of three children of Antal Wigner, director of a leather factory, and Erzsebet Wigner, a homemaker, both of whom were nonobservant Jews. The conversion of his family to Lutheranism when he was in his late teens had a minimal effect on Wigner, who would later refer to himself as "only mildly religious." Eugene was tutored at home from the age of five to 10, when he entered an elementary school. By the time he went on to the Lutheran High School in Budapest, he had already developed a deep interest in mathematics. In high school, where he gained a solid grounding in mathematics, literature, the classics, and religion, he met John von Neumann, the mathematical genius who would become a lifelong friend. His student days were abruptly interrupted in March 1919 when his entire family— imperiled members of the managerial class—fled to Austria in the wake of the Communist takeover in Hungary. When, only a few months later, the Communists fell from power, the Wigners were able to return. Eugene continued his studies, taking two years of stimulating physics and mathematics courses with von Neumann. When he graduated in 1920, he was one of the outstanding students in his class.

Although his fervent desire was to become a physicist, acceding to the wishes of his father, who was determined that Eugene should join the family business, he studied for a degree in chemical engineering. He dutifully enrolled at the Technische Hochschule in Berlin, which awarded him a doctorate in engineering in 1925 for a thesis containing the first theory of the rates of association and dissociation of molecules. At the same time, however, he continued studying physics and mathematics, attending lectures by the University of Berlin's extraordinary physics faculty, which included ALBERT EINSTEIN, MAX ERNST LUDWIG PLANCK, MAX THEODOR FELIX VON LAUE, and WALTHER NERNST. There he enjoyed friendships with the Hungarian physicists Michael Polyani, Leo Szilard, and EDWARD TELLER. After receiving his

doctorate, he returned home and joined his father's tannery firm. He was so clearly unsuited to this life, however, that after a few months his father supported his decision to take a job in Berlin working as a research assistant to a crystallographer at the Kaiser Wilhelm Institute. His task was to determine "why the atoms occupy positions in the crystal lattices which correspond to symmetry axes," a problem that obliged him to study group theory (the study of the formation and properties of mathematical groups).

By this time both WERNER HEISENBERG's matrix version of quantum mechanics and ERWIN SCHRÖDINGER's competing wave theory had been published. Wigner had the idea of applying symmetry theory to quantum mechanics and soon realized that such an approach would open up vast new areas of mathematical physics. His first application was to the degenerate states of symmetrical, atomic, and molecular systems. In 1926, extending Heisenberg's work on two-electron atoms, Wigner published a paper on the spectrum of three-electron atoms. This initial work involving the application of group theory to quantum mechanics would have deep implications for fundamental physics. It led Wigner to the realization that quantum mechanics, with its inherent superposition principle for quantum states, permitted more far-reaching conclusions concerning invariant quantities associated with these states than classical mechanics. He was able to use the tools of group theory to derive new rules concerning the symmetry of atomic spectra that followed from the simple assumption of rotational symmetry of the associated quantum states. His famous work in applying group theory to quantum mechanics had begun.

In 1927, the great mathematician David Hilbert, who had become interested in quantum mechanics and needed to collaborate with a physicist, invited him to the University of Göttingen. Although Hilbert's serious illness caused their collaboration to flounder, the year in Göttingen proved highly productive for Wigner. In his 1927 paper "On the Conservation Laws of Quantum Mechanics," he formulated his law of the conservation of parity, which states that it should be impossible to distinguish left from right in the fundamental interactions of elementary particles. Wigner's conservation of parity law became an integral part of quantum mechanics. However, in 1956, the physicists TSUNG-DAO LEE, CHEN NING YANG, and CHIEN-SHIUNG WU showed that conservation of parity is violated in the weak interactions of subatomic particles by virtue of the fact that the neutrino has a left-handed spin whereas its antiparticle the antineutrino has a right-handed spin.

Eugene Paul Wigner introduced the concept of symmetry principles into the theory of quantum mechanics and chemistry. *(AIP Emilio Segrè Visual Archives, Physics Today Collection)*

Returning to Berlin, Wigner lectured on quantum mechanics and worked on what would become his classic text, *Group Theory and Its Application to the Quantum Mechanics of Atomic Spectra* (1931). In 1930, Wigner and von Neumann were hired by Princeton University, sharing a single position. Between 1930 and 1933, Wigner divided his time between Berlin and Princeton, but when the Nazis rose to power in 1933, he saw his Berlin position dissolve. As did others of his generation of brilliant Hungarian Jewish physicists, including Szilard, von Neumann, and Teller, he immigrated to the United States and became a naturalized citizen in 1937.

Wigner, who was molded by the standards of polite behavior of upper-class European society, had a difficult time adjusting to the United States and was initially lonely and isolated. His formality gave rise to a great body of "Wigner stories," such as the one about his habit of writing, "Your paper contains some very interesting conclusions!" on research papers in which he had found many errors. He was joined at Princeton by his younger sister, Margit, whom he introduced to the great British physicist PAUL ADRIEN MAURICE DIRAC during his visit to Princeton; they married in 1937. Wigner's own marriage to Amelia Frank, a member of the physics faculty, whom he met while teaching at the University of Wisconsin, from 1936 to 1938, ended in tragedy. When they had been married for less than a year, she died of cancer in 1937, plunging him into a deep depression. He managed to do important work at this time, formulating the theory of neutron absorption, which later proved useful in building nuclear reactors.

Needing to escape the scene of his grief, he returned to Princeton, where he became Thomas D. Jones Professor of Mathematical Physics in 1938. Three years later, he married Mary Annette Wheeler, a physics professor at Vassar College, with whom he would have two children. During the prewar years, he did major work in a number of fields. He pioneered the application of quantum mechanics to important aspects of solid state physics, determined that the nuclear force that binds neutrons and protons is necessarily short-range and independent of any electric charge, and developed the principles involved in applying group theory to investigate the energy level of atoms.

By 1938, when it was clear that war in Europe was imminent, Wigner persuaded his parents to immigrate to the United States. That year, Otto Hahn and Fritz Strassmann in Berlin announced the discovery of nuclear fission, revealing the huge amounts of energy released in the process. Wigner, Szilard, and ENRICO FERMI concluded that a fission-induced chain reaction could be achieved with sufficient materials. In 1939, Wigner and Szilard, fearing the consequences of the Nazis' obtaining large quantities of uranium, persuaded ALBERT EINSTEIN to join them in writing a letter to President Franklin D. Roosevelt that set in motion the race to develop an atomic bomb. Wigner worked at the Metallurgical Laboratory at the University of Chicago during World War II, from 1942 to 1945; there he helped Fermi build the first atomic pile. Putting to use his experience as a chemical engineer, Wigner headed a group that succeeded in designing large reactors that could produce the required quantities of plutonium. Although proud of his contribution to the release of nuclear energy, Wigner was loath to use the weapon on civilians and was one of the physicists who tried to persuade President Truman not to drop an atomic bomb on Japan.

When the war ended, Wigner began to think about ways of exploring peaceful uses of atomic energy. From 1946 to 1947, he was director of research and development at the Clinton Laboratories, the forerunner of Oak Ridge National Laboratory in Tennessee, where he trained young scientists and engineers in the principles involved in reactor development and assembled an expert team to design safe and effi-

cient nuclear reactors. As the debate over a national nuclear energy program heated up, Wigner decided he was not the person to direct a laboratory in such a complex, politicized environment and returned to his work at Princeton.

He plunged into a period of intense research, frequently collaborating with associates or graduate students and directing 40 doctoral theses. He conducted research on quantum mechanics, the theory of the rates of chemical reactions, and nuclear structure and made fundamental contributions to the quantum theory of chaos. He also wrote a number of philosophical essays during this period.

In 1963, along with MARIA GERTRUDE GOEPPERT-MAYER and J. Hans D. Jensen, he received the Nobel Prize in physics, for "his contributions to the theory of the atomic nucleus and the elementary particles, particularly through the discovery and application of fundamental symmetry principles."

His books include *Nuclear Structure* (1958) with L. Eisenbud, *The Physical Theory of Neutron Chain Reactors* (1958) with A. Weinberg, *Dispersion Relations and Their Connection with Causality* (1964), and *Symmetries and Reflections* (1967). He published more than 500 papers, which have been collected in an eight-volume edition of his work.

When he retired from Princeton in 1971, Wigner continued his intense involvement with physics, philosophy, and technology and accepted visiting professorships at various universities. He consulted with Oak Ridge National Laboratory on means of protecting civilians in the event of nuclear attack and worked with the Federal Emergency Management Agency (FEMA) on prevention of and provision of emergency aid in national disasters.

In 1977, his wife, Mary, died of cancer. Two years later, he married Eileen Hamilton, the widow of Princeton's dean of graduate studies, who was his close companion for the rest of his life. Near the end of his life, after 60 years in the United States, he wrote that he still felt more

Hungarian than American and that "much of American culture escapes me." When Communist repression began to weaken its hold on Hungary, he resumed ties with the country's cultural and scientific leaders, becoming a spokesperson for enhanced freedoms. In his eighties he experienced serious memory losses in all areas but science and technology. He died on January 1, 1995, in Princeton, New Jersey, at the age of 92.

The extent of Wigner's scientific legacy to physics is reflected in the many concepts and phenomena that bear his name: the Wigner–Eckart theorem for the addition of angular momenta, the Wigner effect in nuclear reactors, the Wigner correlation energy, as well as the Wigner crystal in solids, the Wigner force, the Breit–Wigner formula in nuclear physics, and the Wigner distribution in the quantum theory of chaos. His spiritual legacy is revealed in the philosophical writings in which he plumbed the nature and scope of the scientific endeavor:

> The miracle of the appropriateness of the language of mathematics for the formulation of the laws of physics is a wonderful gift, which we neither understand nor deserve.
>
> Physics does not even try to give us complete information about the world around us—it gives information about the correlations of those events.
>
> The promise of future science is to furnish a unifying goal to mankind rather than merely the means to an easy life, to provide some of what the human soul needs, in addition to bread alone.

Further Reading

Szanton, Andrew. *The Recollections of Eugene P. Wigner as Told to Andrew Szanton*. New York: Perseus Books Group, 1992.

Wigner, Eugene. *Symmetries and Reflections*. Woodbridge, Conn.: Ox Bow Press, 1979.

———."The Unreasonable Effectiveness of Mathematics in the Natural Sciences," in Timothy Ferris, ed., *The World Treasury of Physics, Astronomy, and Mathematics*. Boston: Little, Brown, 1991.

⊠ **Wilson, Charles Thomson Rees**
(1869–1959)
Scottish
Experimentalist, Atomic Physicist, Particle Physicist

Charles Thomas Rees Wilson is famous for his invention of the Wilson cloud chamber, the first instrument to detect the tracks of atomic particles. For this work, which initiated the age of modern experimental particle physics, he was awarded the 1927 Nobel Prize in physics.

He was born on February 14, 1869, in the parish of Glencorse, just outside Edinburgh, into a family who had farmed in the south of Scotland for many generations. His father died when he was four and the family moved to Manchester, where he received his early education at a private school. He went on to Owens College (now the University of Manchester), where he majored in biology, intending to become a physician. After receiving his B.Sc. in 1887, he won a scholarship to Cambridge University's Sidney Sussex College, where he became interested in physics and chemistry and earned a degree in natural sciences. After graduation, he taught for four years at Bradford Grammar School, then returned to Cambridge, where he was the Clerk Maxwell scholar from 1896 to 1899.

Wilson invented the cloud chamber at the outset of his career and, like many great discoveries, Wilson's was made inadvertently, while he was looking for something else. While vacationing in 1894, he climbed Ben Nevis, Scotland's highest mountain, where he was intrigued by the marvelous optical effects of coronas and "glories" (i.e., colored rings cast by surrounding shadows on mist and clouds). The experience sparked a desire to study atmospheric cloud formation, and when he returned to Cambridge, from 1895 to 1899, he carried out a series of groundbreaking experiments designed to produce "artificial clouds" in the laboratory. He succeeded in doing this by causing the adiabatic (i.e., with no heat loss) expansion of moist air. At the time, physicists thought that each water droplet must form around a dust particle, but in the supersaturated air of Wilson's cloud chamber microscopic droplets formed in the absence of dust particles. When Wilson exposed the cloud of water vapor to X rays, which had been discovered that same year by WILHELM CONRAD RÖNTGEN, the process of droplet formation intensified. From this observation, he concluded that water vapor condensed around ions, atoms that have become charged by gaining or losing electrons. Thus, he reasoned that the track of a positively charged alpha particle, for example, would become visible in a line of water droplets. Wilson then used illumination to make the track stand out clearly, enabling the cloud chamber both to detect ions in gases and to record them photographically.

Having achieved this much, Wilson abandoned the cloud chamber to pursue another passion spurred by the dramatic weather conditions of Ben Nevis. In 1895, while on the mountain, he heard distant thunder and, suddenly feeling his hair stand up, fled without waiting for the storm to break. The experience led to an intense interest in atmospheric electricity, which he pursued from 1900 to 1910, while working as a lecturer at Sidney Sussex. He devised the gold-leaf electroscope, a sensitive device for measuring electric charge, which enabled him to demonstrate that some electrical charge always occurs in air and that the conductivity of air inside the electroscope is the same in daylight as in darkness and is independent of the sign of the charge for leaf potential. Wilson was at a loss to explain his results; many years later they would be understood in terms of the existence of cosmic rays: radiation emitted everywhere in the universe.

The year 1911 was a pivotal one for Wilson: he married Jessie Frasier of Glasgow, with whom he would have two sons and two daughters, and he developed a working model for a more advanced cloud chamber. Since the track of a charged particle was detectable because water droplets condensed along the particle's path, he reasoned, on the basis of JAMES CLERK MAXWELL's laws of electrodynamics, that if he applied a magnetic field to the chamber, the track would curve and give a measure of the charge and mass of the particle. He built his new chamber in the form of a short cylinder in which saturation was achieved and controlled by the movement of a piston through a fixed distance. The condensation effects were monitored through the other end. The instrument immediately became essential to the study of radioactivity; it was used to confirm the classic alpha particle scattering and transmutation experiments first performed by ERNEST RUTHERFORD. However, the immense value of the Wilson cloud chamber and subsequent variations developed by others only became apparent in the early 1920s, when modifications by LORD PATRICK MAYNARD STUART BLACKETT and CARL DAVID ANDERSON led to the discovery of the positron in 1932 and the pi meson in 1936.

In 1925, Wilson was appointed Jacksonian Professor of Natural Philosophy at Cambridge, a post he held until 1934. Two years later, he retired and moved to the village of Carlops, near his birthplace. C. T. R., as his friends called him, continued to meet with friends and colleagues, continued his research, and wrote his last paper, a long-promised manuscript on the theory of thundercloud electricity, when he was 87. He died at Carlops on November 15, 1959, surrounded by his family.

Wilson's cloud chamber, celebrated by Rutherford as "the most original and wonderful instrument in scientific history," represented the beginning of experimental particle physics. Although it is not used today, the principle underlying it was incorporated by DONALD ARTHUR GLASER in 1952 into the bubble chamber, an extremely sensitive detector that uses supersaturated liquid helium and is a component of today's giant particle accelerators.

Further Reading

Wilson, John Graham. *The Principle of Cloud Chamber Technique*. Cambridge: Cambridge University Press, 1951.

⊠   Wu, Chien-Shiung
(1912–1997)
Chinese/American
Experimentalist, Nuclear Physicist

Chien-Shiung Wu, known to her scientific colleagues simply as "Madame Wu," was an extraordinary experimentalist who provided the evidence for CHEN NING YANG and TSUNG-DAO LEE's hypothesis that the law of conservation of parity, the mirror symmetry between right and left, previously thought to be universal in nature, is violated in the weak interaction beta decay of the atomic nucleus.

Wu was born in Liu-ho, a small town near Shanghai, China, on May 29, 1912. Her father, Wu Zong-yee, was a school principal. He was unusual in that time, believing that girls were entitled to the same education as boys. Wu was also encouraged to study by a leading Chinese language scholar, Hu Shi, who recognized her talents. Her love of physics began in high school and was developed at the prestigious National Central University in Nanjing, where she graduated in 1936. She went on to do her graduate work at the University of California in Berkeley, then a mecca for atomic physics, where she studied under J. ROBERT OPPENHEIMER, who would later head the U.S. effort to develop an atomic

bomb, and Ernest Lawrence, the inventor of the cyclotron. Wu received her Ph.D. in 1940 and made her reputation as a specialist in nuclear fission. She was particularly interested in the process of beta decay, in which a neutron in the nucleus of a radioactive atom spontaneously breaks apart, releasing a beta particle (a fast-moving electron) and a neutrino (a particle with no mass or charge). In this process, a proton is left behind, automatically converting the atom into an atom of a different element.

In 1942, she married Luke Yuan, another Berkeley physicist. In 1945, they had a son, Vincent, who would grow up to be a physicist. She taught physics for two years at Smith College and then at Princeton University. In 1944, Wu moved to Columbia University and joined the Manhattan Project, whose mission was to develop an atomic bomb. Her assignment was to find ways to produce more radioactive uranium. She developed the process of separating uranium-235 from the more common uranium-238 by gaseous diffusion, using ionization chambers and Geiger counters to monitor the process.

After the war, Wu remained at Columbia, becoming an associate professor in 1952 and continuing her study of beta decay. Her reputation as a meticulous experimentalist drew two other Chinese-born physicists, Yang and Lee, to consult her in 1956. In June of that year, Yang and Lee had submitted to *Physical Review* a paper, "The Question of Parity Conservation in Weak Interactions," raising the question of whether parity, the assumption that nature makes no distinction between left and right, is conserved in weak interactions and suggesting several experiments to decide the issue. At the time, physicists universally believed that physical reactions would be the same (i.e., have parity or equality) whether the particles involved in them had a right-handed or a left-handed spin, or quantized rotation property. This was known as the law of conservation of parity. In physics, to say that something is invariant is equivalent to saying it is conserved. If the physical process proceeds in exactly the same way when referred to an inverted coordinate system, then parity is said to be conserved. If, on the contrary, the process has definite left- or right-handedness, then parity is not conserved in that physical process.

In the early 1950s, applying the principle of conservation of parity to individual subatomic particles and their interactions had proved highly successful in accounting for the behavior of those particles. By the end of 1955, however, a puzzling contradiction between the parity principle and the other principles employed to order the subatomic zoo had emerged. In particular, questions were raised by results beginning to pour forth from the many high-energy accelerators built in the United States after World War II, indicating that one form of radioactive decay appeared to violate the conservation of parity law. Most physicists thought this result was due to some kind of experimental error, but Yang and Lee made the revolutionary suggestion that the universally accepted conservation of parity law might not hold true in weak nuclear interactions, which include radioactive decay.

It was Madame Wu who produced the decisive evidence, when she tested their hypothesis with an experiment that used radioactive cobalt, or cobalt-60. A strong electromagnetic field could make the cobalt atoms line up, just as iron filings do near a magnet, and spin along the same axis. She could then count the number of beta particles thrown from their nuclei in different directions as the atoms decayed. If the parity law held true for weak interactions, the number of particles thrown off in the direction of the nuclear spin would be the same as the number thrown in the opposite direction. If the law did not hold, the numbers would differ. However, since the cobalt atoms, under normal temperatures, would move around too much to line up under the magnet, the experiment had to be done at almost absolute zero, the temperature at

Chien-Shiung Wu provided experimental evidence that conservation of parity between right and left-handed spinning particles is violated in the weak interaction beta decay of the atomic nucleus. (AIP Emilio Segrè Visual Archives)

which all atomic motion becomes vanishingly small.

The only place Wu could achieve these conditions was at the National Bureau of Standards in Washington, D.C., which provided a cryogenics laboratory. Her method was to place the cobalt in a strong magnetic field to line up the north–south magnetic poles of its nuclei, supercool it to near absolute zero to minimize the atoms' random thermal motions, and watch where the electrons it emitted went. She found that the majority of the tens of thousands of electrons emitted by the cobalt every second were ejected primarily in one direction. The experiment, which had to be repeated several times, was arduous, but the results were electrifying: far more beta particles flew off in the direction opposite the nuclei's spin than in the direction that matched the spin. This was evidence that the law of conservation of parity did not apply to the weak interactions.

Wu announced this finding on January 16, 1957, causing a stir in the physics community. Within days, LEON M. LEDERMAN and Richard Garwin at Columbia confirmed the result, using the university's Nevis Cyclotron, a 385-million-electron-volt (Mev) device for accelerating particles to high energies. Her discovery generated a paradigm revolution, when physicists began to realize that parity nonconservation was the basic property of the weak interaction, one of the four basic forces of the universe. It is now known that parity violation exists nearly maximally in all weak interaction processes. Inexplicably, when the Nobel Prize in physics was awarded in 1957, it went to only Yang and Lee, excluding Wu.

After this she became full professor of physics at Columbia and was elected to the National Academy of Sciences in 1958. She would remain at Columbia until her retirement in 1981, "the reigning queen of nuclear physics," as a colleague put it. She wrote a book, *Beta Decay*, that remains a standard reference on low-energy emission of electrons by decaying atoms. Named Woman of the Year by the Association of University Women in 1962, she continued to make important discoveries and tests of others' theories. In 1963, she confirmed experimentally the theory of conservation of vector currents in beta decay proposed by RICHARD PHILLIPS FEYNMAN and MURRAY GELL-MANN. She observed that electromagnetic radiation from the annihilation of positrons and electrons is polarized, as PAUL ADRIEN MAURICE DIRAC had predicted in his theory of the electron. In 1963 she, Lee and L. W. Mo observed the magnetic equivalent of the weak nuclear force, thereby confirming the symmetry between the weak and electromagnetic currents; this was the cornerstone for the later unification of these two basic forces into a single one—the electroweak force.

Wu became the first woman president of the American Physical Society in 1975. Wu retired in 1981. She died in New York on February 16, 1997, at the age of 84, after suffering a stroke. An asteroid was named in her honor in 1990.

Madame Wu, the most distinguished woman physicist of her time, was one of the giants in her field. Her elegantly designed experiments led to the revolutionary realization that parity nonconservation is the basic property of the weak interaction, one of the four basic forces of the universe.

See also FERMI, ENRICO; WEINBERG, STEVEN.

Further Reading

Lindop, Laurie. *Dynamic Modern Women: Scientists and Doctors.* New York: Holt, 1997, pp. 95–105.

McGrayne, Sharon Bertsch. *Nobel Prize Women in Science: Their Lives, Struggles, and Momentous Discoveries.* New York: H. W. Wilson, 1959, pp. 491–492.

Y

Yang, Chen Ning (Yang Chen-ning)
(1922–)
Chinese/American
*Theoretician, Nuclear Physicist,
Particle Physicist*

Chen Ning Yang, together with TSUNG-DAO LEE, predicted that conservation of parity, the symmetry between right and left occurring in physical phenomena, is violated in the weak interactions of the atomic nucleus. On the basis of the experiments of CHIEN-SHIUNG WU, which provided evidence for this symmetry violation, Yang and Lee were awarded the 1957 Nobel Prize in physics. Later in his career, Yang did seminal work on gauge particle field theories, which greatly advanced the search for a unified theory of elementary particles.

Yang was born on September 22, 1922, in Hofei, Anwhei, China, the oldest of five children born to Ke Chuan Yang and Meng Hwa Lo Yang. After reading a biography of Benjamin Franklin, he adopted the name *Frank* for himself. His father was a professor of mathematics at Tsinghua University, near Beijing. Yang attended National Southwest Associated University in Kunming, China, earning a bachelor's degree in 1942 for his thesis "Group Theory and Molecular Structure." He went on to Tsinghua University, which had moved to Kunming dur-

ing the Sino-Japanese War (1937–1945); there he wrote his master's thesis, "Contributions to the Statistical Theory of Order–Disorder Transformations," in 1944.

At the end of the war, Yang went to the United States on a Tsinghua University Fellowship and entered the University of Chicago in 1946 to work on his doctorate in physics. He had hoped to work with ENRICO FERMI at Argonne National Laboratory, but as a foreigner he was barred from working at that facility. Instead, he worked under EDWARD TELLER, writing a thesis on nuclear physics, "On the Angular Distribution in Nuclear Reactions and Coincidence Measurements." After being awarded his Ph.D. by the University of Chicago in 1948, he worked briefly as an instructor there. In 1949, he took a postdoctoral position under J. ROBERT OPPENHEIMER at the Institute for Advanced Study, Princeton University. In 1950, he married Chih Li Tu, whom he met while teaching mathematics at her high school in China; they had three children. In 1955, he became a professor at Princeton University and remained there until 1966.

Throughout his career in physics, Yang concentrated on working in the field of the weak interactions, which were long thought to cause elementary particles to disintegrate. In June 1956, Yang collaborated with Lee on a paper that raised the question of whether parity, the assumption that nature makes no distinction between left and right, is conserved in weak interactions. At the time physicists universally believed that physical reactions would be the same (i.e., have parity, or equality) whether the particles involved in them had a right-handed or a left-handed spin, or quantized rotation property. If the physical process proceeds in exactly the same way when referred to an inverted coordinate system, then parity is said to be conserved. If, on the contrary, the process has definite left- or right-handedness, then parity is not conserved in that physical process. This law of conservation of parity was explicitly formulated in the early 1930s by the Hungarian-born

Chen Ning Yang, together with Tsung-Dao Lee, predicted that conservation of parity, the symmetry between right and left that occurs in physical phenomena, is violated in the weak interactions of the atomic nucleus. *(AIP Emilio Segrè Visual Archives, Physics Today Collection, SUNY)*

physicist EUGENE PAUL WIGNER and became a feature of quantum mechanics.

The strong forces that hold atoms together and the electromagnetic forces that are responsible for chemical reactions obey the law of parity conservation. Since these are the dominant forces in most physical processes, physicists assumed that parity conservation is an inviolable natural law. In the early 1950s, applying the principle of conservation of parity to individual subatomic particles and their interactions had proved highly successful in accounting for the behavior of those particles. By the end of 1955, however, a puzzling contradiction between the parity principle and the other principles employed to order the subatomic zoo had emerged. In particular, questions were raised by results emerging from the many high-energy accelerators built in the United States after World War II that indicated that one form of radioactive decay appeared to violate the conservation of parity law. One of the newly discovered mesons—the so-called K meson—seemed to exhibit decay modes into configurations with differing parity. Exploring this paradox from every conceivable perspective, Yang and Lee discovered that contrary to what had been assumed, there was no experimental evidence against parity nonconservation in the weak interactions. The experiments that had been done, it turned out, were not relevant to the question. In their landmark 1956 *Physical Review* paper, "The Question of Parity Conservation in Weak Interactions," Yang and Lee startled the physics community with their proposal that the universally accepted conservation of parity law might not hold true in weak nuclear interactions, which include radioactive decay.

Their suggestions for experiments capable of deciding the issue were immediately taken up by Madame Wu at the cryogenic laboratory of the National Bureau of Standards in Washington, D.C. Testing radioactive cobalt atoms at temperatures approaching absolute zero, Wu found the evidence Yang and Lee were looking for: the law of parity did not apply to weak interactions. Shortly after her announcement in January 1957, other experimentalists confirmed her results. In the wake of this paradigm revolution created by the three Chinese American physicists, Yang and Lee were awarded the Nobel Prize in physics that very year.

In other arenas, Yang, in collaboration with Lee and others, did important work in statistical mechanics, the study of systems with large numbers of particles, and later investigated the nature of elementary particle reactions at extremely high energies. In 1954, while Yang and Robert Mills were visiting physicists at Brookhaven National Laboratory, they developed the Yang–Mills gauge theory of elementary particles, which is considered to be the foundation for the current understanding of the ways subatomic particles interact. Gauge field theories contain a mathematical transformation symmetry of the fields that preserves the form of the field equations. Maxwell's equations for the electromagnetic field are of this type. As Yang explained:

> The earliest understanding of these forces was Newton's understanding of gravity. The next to be understood were Maxwell's equations for electric and magnetic forces. What the gauge [Yang–Mills] theory [covers] is the equations that govern the other two types of forces. And, furthermore, it turns out that once you understand those, Maxwell's and Newton's forces also fall into the same category. So now we realize that all fundamental forces are forces that obey gauge equations.

The Yang–Mills gauge theory, in conjunction with mass symmetry breaking, a violation of the gauge symmetry that occurs when the temperature of the particle field environment cools below a certain threshold, is currently considered to be the best candidate to explain the interaction of the weak and strong forces, electromagnetism and gravity. For this reason many now consider Yang's work on gauge theories more important than his work on parity violation.

In 1964, Yang became a U.S. citizen, and from 1965 on, he was Albert Einstein Professor at the State University of New York (SUNY), Stonybrook, Long Island. During the 1970s, he was a member of the Rockefeller University and the American Association for the Advancement of Science. In 1986, he became Distinguished Professor-at-Large at the Chinese University of Hong Kong.

Yang, who had long been involved in the advancement of Chinese science, with the opening of relations between the United States and China in 1971 returned to China for the first time, to visit his parents. He recalls:

> After I came back I thought that, as a person deeply involved in both China and the U.S. and who understands the cultures of both countries very well, it was my responsibility to promote understanding between the two.

His goal for the next 20 years became the promotion of Chinese science, because he thought "that would contribute to the well-being of the people in China and also to the peace and stability of the whole world." One of his efforts was the establishment of a visiting scientist program at the SUNY-Stonybrook Institute, where hundreds of Chinese researchers have made one-year stays since the early 1980s.

Yang's work has been characterized as an amazing combination of mathematical power and elegance. His body of research on gauge theories of elementary particles continues to be absolutely central to the development and search for a unified formulation of elementary particle theories.

See also WEINBERG, STEVEN.

Further Reading

Bernstein, Jeremy. *A Comprehensible World: On Modern Science and Its Origins*. New York: Random House, 1967.

Boorse, H. A., and L. Motz, eds. *The World of the Atom*. New York: Basic Books, 1966.

Yang, Chen Ning. *Selected Papers 1945–80 with Commentary*. New York: W. H. Freeman, 1983.

⊠   **Young, Thomas**
(1773–1829)
British
Experimentalist (Optics, Classical Mechanics)

Thomas Young devised one of the privotal experiments in the history of physics: the double-slit experiment, which, by demonstrating the principle of interference of light, showed, within the framework of classical physics, that light is a wave and not a particle. He is also famous for his discoveries in the physiology of vision and his invention of the theory of elasticity of materials.

Young was born on June 13, 1773, in the small village of Milverton, Somerset, England, the first of 10 children. A prodigy, he learned to read by age two and at six read the Bible through twice. He was tutored privately at first, then attended private school. A brilliant student of languages, who had a special interest in ancient Middle Eastern languages, he had taught himself to read Latin, Greek, Italian, Hebrew, Arabic, Persian, Turkish, and Ethiopian and had mastered mathematics, physics, and chemistry by age 19.

When Young was 19, in 1792, he began studying medicine at Saint Bartholomew's Hospital in London, where he was taught by some of the most eminent physicians of the day. He became interested in the physiology of the human eye and, in 1793, read a paper to London's Royal Society in which he explained that the mechanism by which the eye focuses on objects at different distances involves a change of shape in the eye's lens, which is composed of muscle fibers. The next year, the society honored him by election to its ranks. After studying in Edinburgh and Göttingen, he moved to Cambridge, where his intellectual prowess earned him the nickname "Phenomenon Young" and where he would receive an M.B. in 1803 and an M.D. in 1808.

Young's versatile genius and prodigious energy led him to pursue a dual career as medical practitioner and scientific investigator. When an uncle left him a fortune in 1800, he moved to London and opened a medical practice there. In 1801, he became professor of natural philosophy at the Royal Institution, where his public lectures apparently went over the heads of his listeners. His research was more successful. Continuing his earlier work on vision, in 1801, by experimenting with his own eyes, he showed that astigmatism, a condition from which he suffered, is a defect of the lens of the eye that prevents light rays from converging in a single focal point.

That same year, Young hypothesized that color vision is due to the presence of structures in the retina that respond to the three primary colors. He explained color blindness as the inability of one or more of these structures to respond to light. Later experiments would confirm the existence of three types of fibers, now called cones, which are sensitive to light of different wavelenghts. In the 1850s HERMANN LUDWIG FERDINAND VON HELMHOLTZ would further develop these ideas into what came to be called the Young–Helmholtz theory or the trichromatic theory. JAMES CLERK MAXWELL would later incorporate Young's insights into a full theory of vision, explaining how all colors are produced by adding and subtracting the primary colors, establishing the foundations of modern color photography and color television.

Young's landmark studies of vision led him naturally to consider a leading controversy of the day: is light a particle or a wave? Most scientists, following the lead of SIR ISAAC NEWTON, favored particles or "corpuscles," as they were called. But Young was attracted to the wave theory espoused by CHRISTIAAN HUYGENS, which was better able to explain the phenomena of reflection, refraction, and diffraction. He assumed that light waves are propagated in a similar way to sound waves and are longitudinal vibrations, but with a different medium and frequency. He also proposed that different colors consist of different frequencies.

In 1802, Young announced his discovery of the principle of interference. He had arrived at it through his knowledge that if two sound waves of equal intensity reach the ear 180 degrees out of phase, they cancel each other out and no sound is heard. If light beams consisted of waves, he reasoned, it should be possible to observe a similar interference effect. He hypothesized that the bright bands of fringes in effects such as Newton's rings result from light waves that interfere in such a way that they reinforce each other. Over the next two years, he proved this experimentally in his famous double-slit experiment, in which he beamed light through two narrow openings and observed the resulting interference patterns. Etienne Malus's discovery, in 1808, that light waves could be polarized led Young to suggest, in 1817, that light waves contain a transverse component of vibration. In 1821, AUGUSTIN JEAN FRESNEL developed the mathematics of Young's theory and proved that the vibration of light waves is entirely transverse.

Despite his intense interest in the physical properties of light, Young ranked the importance of this research as secondary, at best, in his hierarchy of values. He wrote, "The nature of light is a subject of no material importance to the concerns of life or to the practice of the arts, but it is in many other respects extremely interesting."

Young also invented the theory of elasticity based on the concept of absolute measurements by defining the modulus (the proportionality constant) as the weight that would double the length of a rod of unit cross section. Today it is referred to as Young's modulus and is formulated as stress divided by strain in the elastic region of the loading of a material.

In 1804, Young married Eliza Maxwell; seven years later, he became a physician at Saint George's Hospital in London, a post he would hold until his death. In his later years, he dedicated himself to his first passion, ancient languages, publishing papers on Egyptology, mostly concerned with hieroglyphics. Young was among the first to interpret the writings on the Rosetta Stone, which was found at the mouth of the Nile in 1799. In 1829, his health began to fail. He died in London, of a cardiac disorder, on May 10 of that year, at age 56.

Young's demonstration of interference as evidence of the wave nature of light was a crucial link in the chain of discoveries leading from the classical wave–particle controversy of his time to the quantum wave–particle duality identified in the 20th century.

Further Reading

Kipnis, Nahum. *History of the Principle of Light*. Science Networks Historical Studies, Vol. 5. New York: Springer-Verlag, 1991.

Shamos, M. *Great Experiments in Physics*. New York: Dover, 1987.

Wood, Alexander, Frank Oldham, and Charles E. Raven. *Thomas Young, Natural Philosopher, 1773–1829*. Cambridge: Cambridge University Press, 1954.

⊠ **Yukawa, Hideki**
(1907–1981)
Japanese
Theoretician, Particle Physicist

Hideki Yukawa is famous for his 1935 theoretical work on the forces binding the atomic nucleus, which predicted the existence of new elementary particles called pi mesons. The Nobel Prize in physics awarded to him in 1949 for this work was the first to be given to a Japanese scientist.

He was born Hideki Ogawa in Tokyo, Japan, on January 23, 1907, the third son of Takuji Ogawa, a geologist. The family moved to Kyoto when Hideki's father became professor of geology at the University of Kyoto. In his autobiography, Hideki describes himself as a lonely, introverted, and silent child. When he was a student, it was not unusual for him to spend entire days reading, without talking to anyone. He wrote, "The window of my little world opened out only to the garden of science, but from that window enough light streamed in."

While attending the Third Higher School, a renowned senior high school, he became a close friend of another future Nobel Prize winner, the quantum theorist SIN-ITIRO TOMONAGA. After graduation, both Hideki and Tomonaga entered the University of Kyoto, in 1923, and majored in physics. After earning their *rigakushi* (bachelor's degrees) in physics in 1929, Hideki and Tomonaga traveled to Tokyo to listen to a series of lectures by WERNER HEISENBERG and PAUL ADRIEN MAURICE DIRAC, which made a strong impression on them. Another important influence was Yoshio Nishina, one of Japan's leading physicists, who had studied quantum theory in Copenhagen with NIELS HENRIK DAVID BOHR and pioneered quantum research in Japan. In 1932, Hideki married Sumiko Yukawa, taking her family name; they had two sons, Harumi and Takaaki. By this time, he had moved to Osaka University, where he taught and studied for his doctorate, which he received in 1938.

While pursuing his graduate studies, in 1935, Yukawa published his paper "On the Interaction of Elementary Particles," which proposed a new field theory of nuclear forces and predicted the existence of the pi meson. At the

time, physicists were struggling to understand what holds the nucleus together. In 1932, SIR JAMES CHADWICK had discovered a second nuclear component, in addition to the positively charged proton: the neutron, which had no charge. Yukawa noted the dilemma this posed: because of the Coulomb forces, the protons should repel each other and break the nucleus apart. Seeking the required attractive force to explain why this does *not* happen, he postulated the existence of an "exchange force" that counteracts the mutual repulsion and holds the nucleus together. Such a force came to be known to physicists as the strong interaction.

Yukawa sought to understand the mechanism of the strong force by using the electromagnetic force as an analogy. Here, the long-range electromagnetic interaction between charged particles is seen as the result of the continuous exchange of a quantum or unit of energy carried by a virtual particle, that is, one whose energy and momentum cannot be precisely determined because of the uncertainty principle. The virtual particle in this case is the photon, which has a zero mass and, therefore, interacts over long ranges. Yukawa postulated that just as electrons and protons interact by exchanging photons, so nucleons interact by exchanging an appropriate virtual particle. Yukawa knew that the quantum theory predicted that the range of the force generated by a virtual particle is inversely proportional to its mass. Since the nuclear force has a very short range and acts only inside the atomic nucleus, Yukawa concluded that the virtual particle generating the nuclear force would have to have a nonzero mass. Hence, in constructing a model of the nuclear force, Yukawa required that the nuclear exchange force would (1) involve the transfer of a massive particle, (2) generate an interaction strong enough to overcome the repulsive Coulomb forces between the protons, and, (3) decrease in intensity rapidly enough to have a negligible

Hideki Yukawa is famous for his theoretical work on the forces binding the atomic nucleus, which predicted the existence of new elementary particles called pi mesons. *(AIP Emilio Segrè Visual Archives, W. F. Meggars Collection)*

effect on the innermost electrons. Using quantum theory, he calculated that his predicted nuclear exchange force particle would have a mass about 200 to 300 times that of an electron (but with the same charge as an electron)—about one-ninth the mass of a proton or neutron—and that the particle would be radioactive, with an extremely short half-life.

No particle with these characteristics had yet been discovered.

However, in 1936, CARL DAVID ANDERSON discovered the muon (or mu meson) in cosmic ray tracks, which possessed some, but not all, of the properties of Yukawa's predicted particle. Although the muon had the appropriate mass, it interacted with nucleons so infrequently that it could not possibly be the nuclear "glue" Yukawa had predicted. Meanwhile, in 1936, Yukawa predicted that a nucleus could absorb one of the innermost orbiting electrons and that this process would be equivalent to emitting a positron. These innermost electrons belong to the 1K electron shell and this process of electron absorption by the nucleus became known as K capture.

Finally, in 1947, Yukawa's predicted particle was found when Cecil Powell discovered in cosmic ray tracks the pion (or pi meson), a particle similar to Anderson's muon but fulfilling all the requirements of Yukawa's exchange particle. It had a mass 264 times that of the electron, its decay products were muons, and it interacted very strongly with the nucleons. Two years later, Yukawa traveled to Sweden to accept his Nobel Prize for his prescient theoretical work.

The following year, he returned to Kyoto, where, except for various visiting professorships (including one to Princeton's Institute for Advanced Study at J. ROBERT OPPENHEIMER's invitation, 1948–1949, and one to Columbia, 1949–1953), he remained for the rest of his career. He was professor of theoretical physics from 1939 to 1950, when he became professor emeritus. Then, in 1953, he was appointed director of the university's new Research Institute for Fundamental Physics. He was active in the Pugwash Conference and a strong advocate for banning nuclear weapons and developing peaceful uses of nuclear energy. He retired in 1970.

In addition to numerous scientific papers, he published many books, including *Introduction to Quantum Mechanics* (1946), *Introduction to the Theory of Elementary Particles* (1948), and *Tabito* (The Traveler), an autobiographical account. He edited an English-language journal, *Progress of Theoretical Physics*, which he founded in 1946.

Yukawa died of pneumonia and heart disease in 1981, at the age of 74, in his home in Kyoto. He is remembered both for his specific contribution to the embryonic field of particle physics and for his important role in the development and recognition of Japanese science in the post–World War II years. Yukawa's groundbreaking insight into the nature of the nuclear force laid the foundation for future studies, which ultimately led to the currently accepted quark model of the nuclear force.

See also FEYNMAN, RICHARD PHILLIPS; GELL-MANN, MURRAY; SCHWINGER, JULIUS SEYMOUR.

Further Reading

Yukawa, Hideki. *Creativity and Intuition: A Physicist Looks at East and West.* Tokyo and New York: Kodansha International, 1973.
———. *Tabito* (The traveler). Singapore: World Scientific, 1982.

Z

⊠ Zeeman, Pieter
(1865–1943)
Dutch
Experimentalist (Electromagnetism, Optics), Atomic Physicist

Pieter Zeeman is famous for discovering what came to be known as the Zeeman effect, the splitting of the spectral lines of atoms by an external magnetic field. This effect proved to be an invaluable tool for uncovering the structure of the atom. Zeeman shared the 1902 Nobel Prize in physics with his professor, HENDRIK ANTOON LORENTZ, who had predicted the effect on the basis of his theoretical work on the relationship between light and magnetism.

Zeeman was born in Zonnemaire, a small village in the Isle of Schouwen, Zeeland, the Netherlands, on May 25, 1865, to a local clergyman and his wife. While still a pupil in a local school, he produced a description and drawing of the aurora borealis, then visible above the Netherlands, that was deemed good enough for publication in *Nature*. Zeeman went on to the gymnasium (secondary school) in Delft, where his curriculum included the classical languages required for university entrance. In Delft, he formed a nurturing friendship with the physicist HEIKE KAMERLINGH ONNES, who was a decade older than Zeeman and amazed by the boy's pas-

sion for performing experiments. Zeeman's extensive reading included works by JAMES CLERK MAXWELL on electromagnetism. In 1885, he entered the University of Leiden, where he studied under Kamerlingh Onnes and Lorentz. He earned a Ph.D. in 1893 with a thesis on the Kerr effect, the appearance of birefringence in certain isotropic substances when placed in a strong electric field. After a semester in Strasbourg studying the propagation and absorption of electric waves in fluids, he returned to Leyden, where he would be a lecturer from 1895 to 1897.

The impetus for Zeeman's groundbreaking experiment was the work of his mentor Lorentz, who hypothesized that since light is generated by oscillating electric currents, the presence of a strong magnetic field would have an effect on the charged particles that make up the oscillating currents. Specifically, it would result in a splitting of spectral lines by causing the wavelengths of the lines to vary. In 1896, Zeeman set out to test this hypothesis. He placed a sodium flare between the poles of a powerful electromagnet and produced emission spectra by using a large concave diffraction grating. This experimental arrangement enabled him to detect a broadening of the spectral lines when the magnetic field was activated. He also showed that the shape of the flame was not responsible for this effect, since a similar broadening was achieved with the sodium absorption spectra. Lorentz's hypothesis was validated.

In 1897, Zeeman refined his experiment and successfully resolved the broadening of the narrow blue–green spectral line of cadmium, produced in a vacuum discharge, into three component lines. This experiment showed that atomic energy levels are affected by external magnetic fields, which split atomic levels into discrete substates of different angular momentum. In addition to confirming Lorentz's theory, Zeeman demonstrated that oscillating particles have negative charge, as well as an unexpectedly high charge-to-mass e/m ratio. The following year, JOSEPH JOHN (J. J.) THOMPSON discovered the existence of free electrons in the form of cathode rays, which proved to have the same charge and e/m ratio that Zeeman had found for the oscillating particles in his experiment. In this way it became clear that the electrons and oscillating particles were identical. This finding confirmed that the magnetic field was affecting the forces that control the electrons within the atom.

In 1897, in the midst of this research, Zeeman moved to the University of Amsterdam, where he became a lecturer and later, in 1900, a professor of physics. He would retain this position for the next 35 years. In 1895, he had married Johanna Elisabeth Lebret, with whom he would have one son and three daughters. Zeeman, whose interests extended to literature and theater, was known as an entertaining host, engaging his guests in lively intellectual discussion. His competence and personal kindness were said to have earned him the devotion of his students.

At Amsterdam, Zeeman developed his previous work in a new direction, suggesting that the accepted existence of strong magnetic fields on the surface of the Sun could be verified, since they should cause splitting of the spectral lines derived from light from the Sun's surface. In 1908, the astronomer George Hale, director of the Mount Wilson Observatory, confirmed Zeeman's prediction, which suggested an interrelationship between the directions of polarization and those of the magnetic field. The conclusion reached was that sunspots must be associated with intense magnetic fields within the Sun.

In 1923, he became director of the new Zeeman Laboratory, built especially for him. It included a concrete block weighing a quarter of a million kilograms erected free from the floor, as a suitable platform for vibration-free experiments. Eminent scientists from around the world visited him there. He died after a short illness, in Amsterdam, on October 9, 1943.

Today the Zeeman effect not only explains the mechanism of light radiation and the nature of matter and electricity, but offers an important means for revealing the internal structure of the atom. Further study of the Zeeman effect led to important theoretical advances in the quantum mechanics of atoms, and later investigators were able to show that a Zeeman effect associated with an external magnetic field interacting with the intrinsic spin of the electron also exists.

Further Reading

Born, Max. *Atomic Physics*. New York: Dover, 1989.

Entries by Field

ACOUSTICS
Doppler, Christian
Helmholtz, Hermann Ludwig
    Ferdinand von
Langevin, Paul
Mach, Ernst
Raman, Sir Chandrasekhara
    Venkata

ASTRONOMY
Ångstrom, Anders Jonas
Archimedes
Fizeau, Armand-Hippolyte-
    Louis
Foucault, Jean-Bernard-Léon
Fraunhofer, Joseph von
Galilei, Galileo
Gauss, Johann Carl Friedrich
Hooke, Robert
Huygens, Christiaan
Newton, Sir Isaac

ASTROPHYSICS AND COSMOLOGY
Alfvén, Hannes Olof Gösta
Bethe, Hans Albrecht
Chandrasekhar, Subramanyan
Gamow, George
Hawking, Stephen
Penzias, Arno Allan

Purcell, Edward Mills
Taylor, Joseph H., Jr.
Van Allen, James
Weinberg, Steven
Wheeler, John Archibald

ATOMIC PHYSICS
Blackett, Patrick Maynard
    Stuart, Lord
Chu, Steven
Franck, James
Landau, Lev Davidovich
Langevin, Paul
Lenard, Philipp von
Millikan, Robert Andrews
Oppenheimer, J. Robert
Perrin, Jean-Baptiste
Rutherford, Ernest
Sommerfeld, Arnold Johannes
    Wilhelm
Stark, Johannes
Stern, Otto
Thomson, Joseph John (J. J.)
Wilson, Charles Thomson
    Rees
Zeeman, Pieter

BIOPHYSICS
Chu, Steven
Franck, James

Glaser, Donald Arthur
Purcell, Edward Mills

CLASSICAL ELECTRODYNAMICS AND ELECTROMAGNETISM
Ampère, André-Marie
Boltzmann, Ludwig
Cavendish, Henry
Compton, Arthur Holly
Coulomb, Charles-Augustin de
Einstein, Albert
Faraday, Michael
FitzGerald, George Francis
Foucault, Jean-Bernard-Léon
Gauss, Johann Carl Friedrich
Henry, Joseph
Hertz, Heinrich Rudolf
Kelvin, Lord (William
    Thomson)
Kirchhoff, Gustav Robert
Laue, Max Theodor Felix von
Lenz, Heinrich Friedrich Emil
Lorentz, Hendrik Antoon
Maxwell, James Clerk
Ohm, Georg Simon
Ørsted, Hans Christian
Planck, Max Ernest Ludwig
Poynting, John Henry
Rayleigh, Lord (John William
    Strutt)

Röntgen, Wilhelm Conrad
Stefan, Josef
Tesla, Nikola
Thomson, Joseph John (J. J.)
Volta, Alessandro Guiseppe
    Antonio Anastasio
Weber, Wilhelm Eduard
Wien, Wilhelm (Carl Werner
    Otto Fritz Franz)
Zeeman, Pieter

CLASSICAL MECHANICS
Archimedes
Coulomb, Charles-Augustin de
Foucault, Jean-Bernard-Léon
Galilei, Galileo
Hooke, Robert
Huygens, Christiaan
Mach, Ernst
Newton, Sir Isaac
Young, Thomas

ELECTRICAL AND ELECTRONIC ENGINEERING
Alvarez, Luis W.
Bardeen, John
Kilby, Jack St. Clair
Tesla, Nikola

ELECTROSTATICS
Coulomb, Charles-Augustin de

GEOPHYSICS
Lenz, Heinrich Friedrich Emil

GRAVITATION
Cavendish, Henry
Einstein, Albert
Hooke, Robert
Newton, Sir Isaac
Poynting, John Henry

HIGH PRESSURE AND HIGH TEMPERATURE PHYSICS
Bridgman, Percy Williams

HOLOGRAPHY
Gabor, Dennis

HYDROSTATICS AND HYDRODYNAMICS
Archimedes
Bernoulli, Daniel
Helmholtz, Hermann Ludwig
    Ferdinand von
Landau, Lev Davidovich
Stokes, George Gabriel

INFORMATION THEORY
Gabor, Dennis

LASER SPECTROSCOPY
Bloembergen, Nicolaas
Schawlow, Arthur Leonard
Townes, Charles Hard

LOW TEMPERATURE PHYSICS AND SUPERCONDUCTIVITY
Kamerlingh Onnes, Heike
Kapitsa, Pyotr Leonidovich
Landau, Lev Davidovich

MATHEMATICAL PHYSICS
Ampère, André-Marie
Archimedes
Bernoulli, Daniel
Dyson, Freeman
Galilei, Galileo
Gauss, Johann Carl Friedrich
Huygens, Christiaan
Langevin, Paul
Newton, Sir Isaac
Rydberg, Johannes Robert
Stokes, George Gabriel
Wigner, Eugene Paul

NUCLEAR PHYSICS
Bethe, Hans Albrecht
Chadwick, Sir James
Cockcroft, John Douglas
Fermi, Enrico
Gamow, George
Geiger, Hans Wilhelm
Glaser, Donald Arthur
Goeppert-Mayer, Maria
    Gertrude
Landau, Lev Davidovich
Meitner, Lise
Oppenheimer, J. Robert
Purcell, Edward Mills
Rutherford, Ernest
Sakharov, Andrei
    Dmitriyevich
Segrè, Emilio Gino
Teller, Edward
Wigner, Eugene Paul
Wu, Chien-Shiung
Yang, Chen Ning

OPTICS
Doppler, Christian Johann
Fizeau, Armand-Hippolyte-
    Louis
Foucault, Jean Bernard Léon
Fraunhofer, Joseph von
Fresnel, Augustin Jean
Helmholtz, Hermann Ludwig
    Ferdinand von
Hooke, Robert
Huygens, Christiaan
Lorentz, Hendrik Antoon
Mach, Ernst
Maxwell, James Clerk
Michelson, Albert Abraham
Newton, Sir Isaac
Raman, Sir Chandrasekhara
    Venkata
Röntgen, Wilhelm Conrad
Snell, Willibrord
Stokes, George Gabriel

Townes, Charles Hard
Young, Thomas
Zeeman, Pieter

PARTICLE PHYSICS
Alvarez, Luis W.
Anderson, Carl David
Bhabha, Homi Jehangir
Blackett, Patrick Maynard
    Stuart, Lord
Chadwick, Sir James
Cherenkov, Pavel
    Alekseyevich
Compton, Arthur Holly
Davisson, Clinton Joseph
Dirac, Paul Adrien Maurice
Fermi, Enrico
Feynman, Richard Phillips
Fitch, Val Logsdon
Gell-Mann, Murray
Glaser, Donald Arthur
Glashow, Sheldon Lee
Lederman, Leon M.
Lee, Tsung-Dao
Mössbauer, Rudolf Ludwig
Ne'eman, Yuval
Pauli, Wolfgang
Ramsey, Norman F.
Reines, Frederick
Richter, Burton
Rubbia, Carlo
Salam, Abdus
Segrè, Emilio Gino
Ting, Samuel Chao Chung
Tomonaga, Sin-Itiro
Van de Graaff, Robert Jemison
Veltman, Martinus J. G.
Weinberg, Steven
Wilson, Charles Thomson
    Rees
Wheeler, John Archibald
Yang, Chen Ning
Yukawa, Hideki

PHILOSOPHY OF SCIENCE
Bohr, Niels Henrik David
Bridgman, Percy Williams
Einstein, Albert
Mach, Ernst

PHYSICAL CHEMISTRY
Cavendish, Henry
Curie, Marie
Faraday, Michael
Franck, James
Goeppert-Mayer, Marie
    Gertrude
Helmholtz, Hermann Ludwig
    Ferdinand von
Nernst, Walther
Ørsted, Hans Christian
Perrin, Jean-Baptiste
Stern, Otto

PLASMA PHYSICS
Alfvén, Hannes Olof Gösta
Landau, Lev Davidovich

QUANTUM ELECTRODYNAMICS AND QUANTUM FIELD THEORY
Dyson, Freeman
Feynman, Richard Phillips
Glashow, Sheldon Lee
't Hooft, Gerard
Kusch, Polykarp
Lamb, Eugene Willis Jr.
Landau, Lev Davidovich
Lenard, Philipp von
Ramsey, Norman F.
Salam, Abdus
Schwinger, Julian Seymour
Tomonaga, Sin-Itiro
Veltman, Martinus J. G.
Weinberg, Steven
Wheeler, John Archibald
Wigner, Eugene Paul

QUANTUM MECHANICS
Barkla, Charles Glover
Bloch, Felix
Bohr, Niels Henrik David
Born, Max
Broglie, Louis-Victor-Pierre,
    prince de
Dirac, Paul Adrien Maurice
Einstein, Albert
Franck, James
Heisenberg, Werner
Landau, Lev Davidovich
Landé, Alfred
Laue, Max Theodor Felix von
Oppenheimer, J. Robert
Pauli, Wolfgang
Planck, Max Ernest Ludwig
Schrödinger, Erwin
Sommerfeld, Arnold Johann
Teller, Edward

RADIOACTIVITY
Becquerel, Antoine-Henri
Curie, Marie
Meitner, Lise
Rutherford, Ernest

RADIO ASTRONOMY
Penzias, Arno Allan
Taylor, Joseph H., Jr.

RELATIVITY
Einstein, Albert
Hawking, Stephen
Landau, Lev Davidovich
Laue, Max Theodor Felix von
Lorentz, Hendrik Antoon
Taylor, Joseph H., Jr.
Wheeler, John Archibald

SOLID STATE AND CONDENSED MATTER PHYSICS
Alferov, Zhores Ivanovich
Anderson, Philip Warren

Bardeen, John
Barkla, Charles Glover
Bethe, Hans Albrecht
Bloch, Felix
Born, Max
Einstein, Albert
Josephson, Brian David
Kilby, Jack St. Clair
Kroemer, Herbert
Landau, Lev Davidovich
Langevin, Paul
Laue, Max Theodor Felix von
Lenard, Philipp von
Mott, Sir Nevill Francis
Purcell, Edward Mills
Rabi, Isidor Isaac
Schrieffer, John Robert
Schockley, William Bradford
Van Vleck, John Housbrook

SPECTROSCOPY
Ångstrom, Anders Jonas

Fraunhofer, Joseph von
Kirchhoff, Gustav Robert
Raman, Sir Chandrasekhara
    Venkata
Rydberg, Johannes Robert

STATISTICAL MECHANICS
Bloch, Felix
Boltzmann, Ludwig
Einstein, Albert
Rayleigh, Lord (John William
    Strutt)
Sommerfeld, Arnold Johannes
    Wilhelm
Stark, Johannes

TERRESTRIAL MAGNETISM
Gauss, Johann Carl Friedrich

THERMODYNAMICS
Boltzmann, Ludwig
Carnot, Nicolas Léonard Sadi

Clausius, Rudolf Julius
    Emmanuel
Helmholtz, Hermann Ludwig
    Ferdinand von
Joule, James Prescott
Kelvin, Lord (William
    Thomson)
Kirchhoff, Gustav Robert
Maxwell, James Clerk
Nernst, Walther
Rayleigh, Lord (John William
    Strutt)
Stark, Johannes
Stefan, Josef
Wien, Wilhelm (Carl Werner
    Warner Fritz Franz)

ENTRIES BY COUNTRY OF BIRTH

AUSTRIA
Boltzmann, Ludwig
Doppler, Christian
Mach, Ernst
Meitner, Lisa
Pauli, Wolfgang
Schrödinger, Erwin
Stefan, Josef

BAVARIA
Ohm, George Simon
Stark, Johannes

CHINA
Lee, Tsung-Dao
Ting, Samuel Chao Chung
Wu, Chien-Shiung
Yang, Chen Ning

DENMARK
Bohr, Niels Henrik David
Ørsted, Hans Christian

ENGLAND
Barkla, Charles Glover
Blackett, Patrick Stuart
    Maynard, Lord
Chadwick, Sir James
Cockcroft, John Douglas
Dirac, Paul Adrien Maurice

Dyson, Freeman
Faraday, Michael
Hawking, Stephen
Hooke, Robert
Joule, James Prescott
Mott, Sir Nevill Francis
Newton, Sir Isaac
Poynting, John Henry
Rayleigh, Lord (John William
    Strutt)
Thomson, Joseph John (J. J.)
Young, Thomas

ESTONIA
Lenz, Heinrich Friedrich Emil

FRANCE
Ampère, André-Marie
Becquerel, Antoine-Henri
Broglie, Louis-Victor-Pierre,
    prince de
Carnot, Nicolas Léonard Sadi
Cavendish, Henry
Coulomb, Charles Augustin
Fizeau, Armand-Hippolyte-
    Louis
Foucault, Jean-Bernard-Léon
Fresnel, Augustin Jean
Langevin, Paul
Perrin, Jean-Baptiste

GERMANY
Bethe, Hans Albrecht
Born, Max
Einstein, Albert
Franck, James
Fraunhofer, Joseph von
Gauss, Johann Carl Friedrich
Geiger, Hans Wilhelm
Goeppert-Mayer, Maria
    Gertrude
Heisenberg, Werner
Helmholtz, Hermann Ludwig
    Ferdinand von
Hertz, Heinrich Rudolf
Kirchhoff, Gustav Robert
Kroemer, Herbert
Kusch, Polykarp
Landé, Alfred
Laue, Max Theodor Felix von
Mössbauer, Rudolf Ludwig
Penzias, Arno Allan
Planck, Max Ernest Ludwig
Röntgen, Wilhelm Conrad
Stern, Otto
Weber, Wilhelm Eduard

HUNGARY
Gabor, Dennis
Lenard, Philipp von
Teller, Edward

Wigner, Eugene Paul

INDIA
Bhabha, Homi Jehangir
Chandrasekhar, Subramanyan
Raman, Sir Chandrasekhara
  Venkata

IRELAND
FitzGerald, George Francis
Kelvin, Lord (William
  Thomson)
Stokes, George Gabriel

ISRAEL
Ne'eman, Yuval

ITALY
Fermi, Enrico
Galilei, Galileo
Rubbia, Carlo
Segrè, Emilio Gino
Volta, Alessandro Giuseppe
  Antonio Anastasio

JAPAN
Tomonaga, Sin-Itiro
Yukawa, Hideki

NETHERLANDS
Bloembergen, Nicolaas
Bernoulli, Daniel
't Hooft, Gerard
Huygens, Christian
Kamerlingh Onnes, Heike
Lorentz, Hendrik Antoon
Snell, Willibrord
Veltman, Martinus
Zeeman, Pieter

NEW ZEALAND
Rutherford, Ernest

PAKISTAN
Salam, Abdus

POLAND
Curie, Marie
Rabi, Isidor Isaac

PRUSSIA
Clausius, Rudolf Julius
  Emmanuel
Michelson, Albert Abraham
Nernst, Walther
Sommerfeld, Arnold Johannes
  Wilhelm
Wien, Wilhelm (Carl Werner
  Otto Fritz Franz)

RUSSIA
Alferov, Zhores Ivanovich
Cherenkov, Pavel
  Alekseyevich
Gamow, George
Kapitsa, Pyotr
Landau, Lev Davidovich
Sakharov, Andrey
  Dmitriyevich

SCOTLAND
Maxwell, James Clerk
Wilson, Charles Thomson
  Rees

SERBIA
Tesla, Nikola

SICILY
Archimedes

SWEDEN
Alfvén, Hannes Olof Gösta
Ångstrom, Anders Jonas
Rydberg, Johannes Robert

SWITZERLAND
Bloch, Felix

UNITED STATES
Alvarez, Luis W.
Anderson, Carl David
Anderson, Philip Warren
Bardeen, John
Bridgman, Percy Williams
Chu, Steven
Compton, Arthur Holly
Davisson, Clinton Joseph
Feynman, Richard Phillips
Fitch, Val Logsdon
Gell-Mann, Murray
Glaser, Donald Arthur
Glashow, Sheldon
Henry, Joseph
Kilby, Jack St. Clair
Lamb, Willis Eugene Jr.
Lederman, Leon M.
Millikan, Robert Andrews
Oppenheimer, J. Robert
Purcell, Edward Mills
Ramsey, Norman F.
Reines, Frederick
Richter, Burton
Schawlow, Arthur Leonard
Schrieffer, John Robert
Schwinger, Julian
Shockley, William
Taylor, Joseph H., Jr.
Townes, Charles Hard
Van Allen, James
Van de Graaff, Robert Jemison
Van Vleck, John Housbrook
Weinberg, Steven
Wheeler, John Archibald

WALES
Josephson, Brian David

ENTRIES BY COUNTRY OF MAJOR SCIENTIFIC ACTIVITY

AUSTRIA
Boltzmann, Ludwig
Doppler, Christian
Mach, Ernst
Stefan, Josef

CZECHOSLOVAKIA
Mach, Ernst

DENMARK
Bloch, Felix
Bohr, Niels Henrik David
Gamow, George
Heisenberg, Werner
Ørsted, Hans Christian
Teller, Edward

ENGLAND
Barkla, Charles Glover
Bhabha, Homi Jehangir
Blackett, Lord Patrick
    Maynard Blackett
Bohr, Niels Henrik David
Born, Max
Cavendish, Henry
Chadwick, Sir James
Chandrasekhar, Subramanyan
Cockcroft, John Douglas
Dirac, Paul Adrien Maurice
Faraday, Michael
FitzGerald, George Francis

Gabor, Dennis
Gamow, George
Geiger, Hans Wilhelm
Hawking, Stephen
Hooke, Robert
Josephson, Brian David
Joule, James Prescott
Kelvin, Lord (William
    Thomson)
Maxwell, James Clerk
Mott, Sir Nevill Francis
Ne'eman, Yuval
Newton, Sir Isaac
Poynting, John Henry
Rayleigh, Lord (John William
    Strutt)
Rutherford, Ernest
Salam, Abdus
Stokes, George Gabriel
Thomson, Joseph John (J. J.)
Wilson, Charles Thomson
    Rees
Young, Thomas

FRANCE
Ampère, André-Marie
Becquerel, Antoine-Henri
Broglie, Louis-Victor-Pierre,
    prince de
Carnot, Nicolas Léonard Sadi
Coulomb, Charles-Augustin de

Curie, Marie
Fizeau, Armand-Hippolyte-
    Louis
Foucault, Jean-Bernard-Léon
Fresnel, Augustin-Jean
Huygens, Christiaan
Langevin, Paul
Perrin, Jean-Baptiste

GERMANY
Bethe, Hans Albrecht
Bloch, Felix
Born, Max
Clausius, Rudolf Julius
    Emmanuel
Einstein, Albert
Franck, James
Fraunhofer, Joseph von
Gabor, Dennis
Gamow, George
Gauss, Johann Carl Friedrich
Geiger, Hans Wilhelm
Goeppert-Mayer, Maria
    Gertrude
Heisenberg, Werner
Helmholtz, Hermann Ludwig
    Ferdinand von
Hertz, Heinrich Rudolf
Kirchhoff, Gustav Robert
Kroemer, Herbert
Landé, Alfred

Laue, Max Theodor Felix von
Lenard, Philipp von
Meitner, Lise
Michelson, Albert Abraham
Mössbauer, Rudolf Ludwig
Nernst, Walther
Ohm, George Simon
Pauli, Wolfgang
Planck, Max Ernest Ludwig
Röntgen, Wilhelm Conrad
Schrödinger, Erwin
Sommerfeld, Arnold Johannes
    Wilhelm
Stark, Johannes
Stern, Otto
Teller, Edward
Ting, Samuel Chao Chang
Weber, Wilhelm Eduard
Wien, Wilhelm (Carl Werner
    Otto Fritz Franz)
Wigner, Eugene Paul

INDIA
Bhabha, Homi Jehangir
Raman, Sir Chandrasekhara
    Venkata

ISRAEL
Ne'eman, Yuval

ITALY
Fermi, Enrico
Galilei, Galileo
Rubbia, Carlo
Salam, Abdus
Segrè, Emilio Gino
Volta, Alessandro Giuseppe
    Antonio Anastasio

JAPAN
Tomonaga, Sin-Itiro
Yukawa, Hideki

NETHERLANDS
't Hooft, Gerard
Huygens, Christiaan
Kamerlingh Onnes, Heike
Lorentz, Hendrik Antoon
Snell, Willibrord
Veltman, Martinus
Zeeman, Pieter

PAKISTAN
Salam, Abdus

POLAND
Curie, Marie

RUSSIA
Alferov, Zhores Ivanovich
Bernoulli, Daniel
Cherenkov, Pavel
    Alekseyevich
Gamow, George
Kapitsa, Pyotr Leonidovich
Landau, Lev Davidovich
Lenz, Heinrich Friedrich Emil
Sakharov, Andrey
    Dmitriyevich

SCOTLAND
Kelvin, Lord (William
    Thomson)
Maxwell, James Clerk
Wilson, Charles Thomson
    Rees

SICILY
Archimedes

SWEDEN
Alfvén, Hannes Olof Gösta
Ångstrom, Anders Jonas
Meitner, Lisa
Rydberg, Johannes Robert

SWITZERLAND
Bernoulli, Daniel
Clausius, Rudolf Julius
    Emmanuel
Einstein, Albert
Lederman, Leon M.
Pauli, Wolfgang
Rubbia, Carlo
Schrödinger, Erwin
Ting, Samuel Chao Chung

UNITED STATES
Alvarez, Luis W.
Anderson, Carl David
Anderson, Philip Warren
Bardeen, John
Bethe, Hans Albrecht
Bloch, Felix
Bloembergen, Nicolaas
Bridgman, Percy Williams
Chandrasekhar, Subramanyan
Chu, Steven
Compton, Arthur Holly
Davisson, Clinton Joseph
Dyson, Freeman
Fermi, Enrico
Feynman, Richard Phillips
Fitch, Val Logsdon
Franck, James
Gamow, George
Gell-Mann, Murray
Glaser, Donald Arthur
Glashow, Sheldon
Goeppert-Mayer, Maria
    Gertrude
Henry, Joseph
Kilby, Jack St. Clair
Kroemer, Herbert
Kusch, Polykarp
Lamb, Willis Eugene Jr.
Landé, Alfred
Lederman, Leon M.

Lee, Tsung-Dao
Michelson, Albert Abraham
Millikan, Robert Andrews
Mössbauer, Rudolf Ludwig
Ne'eman, Yuval
Oppenheimer, J. Robert
Penzias, Arno Allan
Purcell, Edward Mills
Rabi, Isidor Isaac
Ramsey, Norman F.
Reines, Frederick

Richter, Burton
Rubbia, Carlo
Schawlow, Arthur Leonard
Schrieffer, John Robert
Schwinger, Julian Seymour
Segrè, Emilio Gino
Shockley, William
Stern, Otto
Taylor, Joseph H., Jr.
Tesla, Nikola
Teller, Edward

Ting, Samuel Chao Chung
Townes, Charles Hard
Van Allen, James
Van de Graaff, Robert Jemison
Van Vleck, John Housbrook
Veltman, Martinus J. G.
Weinberg, Steven
Wheeler, John Archibald
Wigner, Eugene Paul
Wu, Chien-Shiung
Yang, Chen Ning

ENTRIES BY YEAR OF BIRTH

Raman, Sir Chandrasekhara
  Venkata
Rutherford, Ernest
Rydberg, Johannes
Schrödinger, Erwin
Sommerfeld, Arnold Johannes
  Wilhelm
Stark, Johannes
Stern, Otto
Tesla, Nikola
Thomson, Joseph John (J. J.)
Van Vleck, John Housbrook
Wien, Wilhelm
Wilson, Charles Thomson
  Rees
Zeeman, Pieter

1900–1909
Alfvén, Hannes Olaf Gösta
Anderson, Carl David
Bardeen, John
Bethe, Hans Albrecht
Bhabha, Homi Jehangir
Bloch, Felix
Cherenkov, Pavel
  Alekseyevich
Dirac, Paul Adrien Maurice
Fermi, Enrico
Gabor, Dennis
Gamow, George
Goeppert-Mayer, Maria
  Gertrude
Heisenberg, Werner

Landau, Lev Davidovich
Mott, Sir Nevill Francis
Oppenheimer, J. Robert
Pauli, Wolfgang
Segré, Emilio Gino
Teller, Edward
Tomonaga, Sin-Itiro
Van de Graaff, Robert Jemison
Wigner, Eugene Paul
Yukawa, Hideki

1910–1919
Alvarez, Luis W.
Chandrasekhar, Subramanyan
Feynman, Richard Phillips
Kusch, Polykarp
Lamb, Eugene Willis Jr.
Purcell, Edward Mills
Ramsey, Norman F.
Reines, Frederick
Schwinger, Julian Seymour
Shockley, William Bradford
Townes, Charles Hard
Van Allen, James
Wheeler, John Archibald
Wu, Chien-Shiung

1920–1929
Anderson, Philip Warren
Bloembergen, Nicolaas
Dyson, Freeman
Fitch, Val Logsdon
Gell-Mann, Murray

Glaser, Donald Arthur
Kilby, Jack
Kroemer, Herbert
Lederman, Leon M.
Lee, Tsung-Dao
Mössbauer, Rudolf Ludwig
Ne'eman, Yuval
Sakharov, Andrei
  Dimitriyevich
Salam, Abdus
Schawlow, Arthur Leonard
Yang, Chen Ning

1930–1939
Alferov, Zhores Ivanovich
Glashow, Sheldon Lee
Penzias, Arno Allen
Richter, Burton
Rubbia, Carlo
Schrieffer, John Robert
Ting, Samuel Chao Chung
Veltman, Martinus J. G.
Weinberg, Steven

1940–1949
Chu, Steven
Hawking, Stephen
't Hooft, Gerard
Josephson, Brian David
Taylor, Joseph H., Jr.

CHRONOLOGY

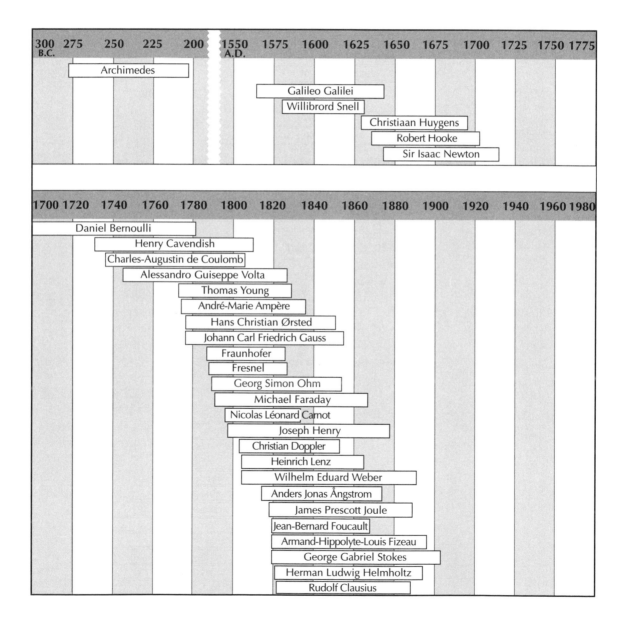

300 B.C.	275	250	225	200	1550 A.D.	1575	1600	1625	1650	1675	1700	1725	1750	1775

Archimedes

Galileo Galilei

Willibrord Snell

Christiaan Huygens

Robert Hooke

Sir Isaac Newton

1700	1720	1740	1760	1780	1800	1820	1840	1860	1880	1900	1920	1940	1960	1980

Daniel Bernoulli

Henry Cavendish

Charles-Augustin de Coulomb

Alessandro Guiseppe Volta

Thomas Young

André-Marie Ampère

Hans Christian Ørsted

Johann Carl Friedrich Gauss

Fraunhofer

Fresnel

Georg Simon Ohm

Michael Faraday

Nicolas Léonard Carnot

Joseph Henry

Christian Doppler

Heinrich Lenz

Wilhelm Eduard Weber

Anders Jonas Ångstrom

James Prescott Joule

Jean-Bernard Foucault

Armand-Hippolyte-Louis Fizeau

George Gabriel Stokes

Herman Ludwig Helmholtz

Rudolf Clausius

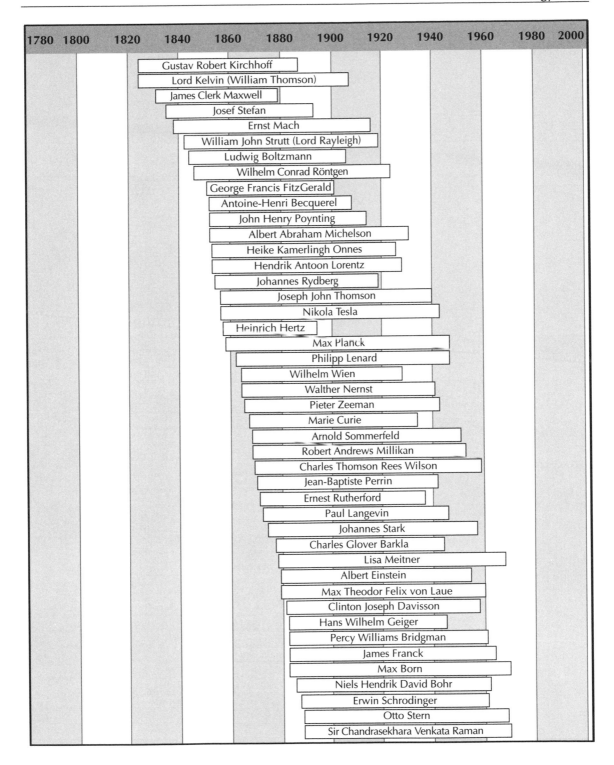

1780	1800	1820	1840	1860	1880	1900	1920	1940	1960	1980	2000

Gustav Robert Kirchhoff
Lord Kelvin (William Thomson)
James Clerk Maxwell
Josef Stefan
Ernst Mach
William John Strutt (Lord Rayleigh)
Ludwig Boltzmann
Wilhelm Conrad Röntgen
George Francis FitzGerald
Antoine-Henri Becquerel
John Henry Poynting
Albert Abraham Michelson
Heike Kamerlingh Onnes
Hendrik Antoon Lorentz
Johannes Rydberg
Joseph John Thomson
Nikola Tesla
Heinrich Hertz
Max Planck
Philipp Lenard
Wilhelm Wien
Walther Nernst
Pieter Zeeman
Marie Curie
Arnold Sommerfeld
Robert Andrews Millikan
Charles Thomson Rees Wilson
Jean-Baptiste Perrin
Ernest Rutherford
Paul Langevin
Johannes Stark
Charles Glover Barkla
Lisa Meitner
Albert Einstein
Max Theodor Felix von Laue
Clinton Joseph Davisson
Hans Wilhelm Geiger
Percy Williams Bridgman
James Franck
Max Born
Niels Hendrik David Bohr
Erwin Schrodinger
Otto Stern
Sir Chandrasekhara Venkata Raman

1780	1800	1820	1840	1860	1880	1900	1920	1940	1960	1980	2000

Alfred Landé

Sir James Chadwick

Arthur Holly Compton

Louis-Victor-Pierre de Broglie

Pyotr Leonidovich Kapitsa

John Douglas Cockcroft

Lord Blackett

Isidor Isaac Rabi

John Housbrook Van Vleck

Wolfgang Pauli

Dennis Gabor

Enrico Fermi

Robert Jemison Van de Graaff

Werner Heisenberg

Paul Adrien Maurice Dirac

Eugene Paul Wigner

Robert J. Oppenheimer

George Gamow

Pavel Alekseyevich Cherenkov

Felix Bloch

Carl David Anderson

Emilio Gino Segrè

Sir Nevill Francis Mott

Marie Goeppert-Mayer

Sin-Itiro Tomonaga

Hans Albrecht Bethe

Hideki Yukawa

Lev Davidovich Landau

John Bardeen

Hannes Olof Gösta

Edward Teller

Homi Jehangir Bhabha

William Bradford Shockley

Subramanyan Chandrasekhar

Luis W. Alvarez

Polykarp Kusch

John Archibald Wheeler

Edward Mills Purcell

Chien-Shiung Wu

Willis Eugene Lamb

James Van Allen

Norman F. Ramsey

Charles Hard Townes

Richard Phillips Feynman

1780	1800	1820	1840	1860	1880	1900	1920	1940	1960	1980	2000

Julian Seymour Schwinger

Frederick Reines

Nicolaas Bloembergen

Andrei Dmitriyevich Sakharov

Arthur Leonard Schawlow

Leon M. Lederman

Chen Ning Yang

Philip Warren Anderson

Freeman Dyson

Val Logsdon Fitch

Jack St. Clair Kilby

Yuval Ne'eman

Abdus Salam

Donald Arthur Glaser

Tsung-Dao Lee

Herbert Kroemer

Murray Gell-Mann

Rudolf Ludwig Mössbauer

Zhores Ivanovich Alferov

Burton Richter

John Robert Schrieffer

Martinus J.G. Veltman

Sheldon Lee Glashow

Arno Allan Penzias

Steven Weinberg

Carlo Rubbia

Samuel Chao Chung Ting

Brian David Josephson

Joseph H. Taylor, Jr.

Stephen Hawking

Gerard 't Hooft

Steven Chu

BIBLIOGRAPHY

Abbott, David, ed. *The Biographical Dictionary of Scientists: Physicists*. New York: Peter Bedrick Books, 1984.

Aczel, Amir D. *God's Equation: Einstein, Relativity, and the Expanding Universe*. New York: Dell, 2000.

———. *Entanglement: The Greatest Mystery in Physics*. New York: Four Walls Eight Windows, 2002.

Alfvén, Hannes. *On the Origin of the Solar System*. Westport, Conn.: Greenwood Press, 1973.

———. *Worlds–Antiworlds: Antimatter in Cosmology*. New York: W. H. Freeman, 1966.

Alvarez, Luis W. *Adventures of a Physicist*. New York: Basic Books, 1978.

American Men and Women of Science, 1995–96: A Biographical Directory of Today's Leaders in Physical, Biological, and Related Sciences. New York: R. R. Bowker, 1994.

Archimedes. *The Works of Archimedes*. New York: Dover, 2002.

Isaac, Asimov. *Asimov's Biographical Encyclopedia of Science and Technology: The Lives and Achievements of 1510 Great Scientists from Ancient Times to the Present Chronologically Arranged*. New York: Doubleday, 1982.

———. *Atom: Journey Across the Subatomic Cosmos*. New York: Dutton/Plume. 1992

Badash, Lawrence. *Kapitsa, Rutherford, and the Kremlin*. New Haven, Conn.: Yale University Press, 1985.

Baird, David, Alfred Nordman, and R. I. Hughes, eds. *Heinrich Hertz: Classical Physicist, Modern Philosopher*. Boston: Kluwer Academic, 1997.

Balibar, Francoise. *Einstein: Decoding the Universe*. New York: Harry N. Abrams, 2001.

Barkan, Diana. *Nernst: Architect of Physical Revolution*. London: Cambridge University Press, 1999.

Beckman, Olof. *Ångstrom: Father and Son*. Acta Universitatis Upsaliensis, C. Organisation Och Historia, No. 60. Philadelphia: Coronet Books, 1997.

Berlinski, David. *Newton's Gift: How Sir Isaac Newton Unlocked the System of the World*. New York: Free Press, 2000.

Bernstein, Jeremy. *A Comprehensible World: On Modern Science and Its Origins*. New York: Random House, 1967.

———. *Albert Einstein and the Frontiers of Physics*. Oxford: Oxford University Press, 1996.

———. *Three Degrees Above Zero: Bell Labs in the Information Age*. New York: Scribner & Sons, 1984.

———. *Hans Bethe, Prophet of Energy*. New York: Basic Books, 1980.

Bethe, Hans Albrecht. *The Road from Los Alamos.* Master of Modern Physics. New York: Springer-Verlag, 1991.

"Biographical Memoirs," Washington, D.C. National Academy Press, 1994, 2000. Available on-line: URL: http://books.nap.edu/books.

Boag, J. W., P. E. Rubinin, and D. Shoenberg, eds. *Kapitsa in Cambridge and Moscow: Life and Letters of a Russian Physicist.* Amsterdam: Elsevier Science, 1990.

Bodanus, David. *E = Mc²: A Biography of the World's Most Famous Equation.* New York: Walker, 2000.

Bonner, Yelena. *Alone Together.* New York: A. A. Knopf, 1986.

Boorse, H. A. and L. Motz, eds. *The World of the Atom.* New York: Basic Books, 1966.

Born, Max. *Atomic Physics.* New York: Dover, 1989.

———. *My Life.* New York: Scribner's, 1978.

Brennan, Richard P. *Heisenberg Probably Slept Here: The Lives, Times and Ideas of the Great Physicists of the 20th Century.* New York: John Wiley & Sons, 1997.

Brian, Denis. *Einstein: A Life.* New York: John Wiley & Sons, 1997.

Bridgman, Percy Williams. *The Logic of Modern Physics.* New York: Macmillan, 1961.

Brown, Andrew P. *The Neutron and the Bomb: A Biography of Sir James Chadwick.* Oxford: Oxford University Press, 1997.

Buchwald, Jed Z. *The Creation of Scientific Effects: Heinrich Hertz and Electric Waves.* Chicago: Chicago University Press, 1994.

———. *Histories of the Electron.* Dibner Institute Studies in the History of Science and Technology. Cambridge, Mass.: MIT Press, 2001.

———. *The Rise of the Wave Theory of Light: Optical Theory and Experiment in the Early Nineteenth Century.* Chicago: University of Chicago Press, 1989.

Bühler, W. K. *Gauss: A Biographical Study.* New York: Springer-Verlag, 1987.

Cahan, David, ed. *Hermann Von Helmholtz and the Foundations of Nineteenth-Century Science.* California Studies in the History of Science, No. 12. Berkeley: University of California Press, 1993.

Calder, Nigel. *Einstein's Universe.* New York: Random House, 1990.

Cantor, Geoffrey, David Gooding, and Frank A. James. *Michael Faraday.* Amherst, N.Y.: Prometheus Books, 1996.

Cardwell, D. S. L. *From Watt to Clausius: The Rise of Thermodynamics in the Early Industrial Age.* Cornell, N.Y.: Cornell University Press, 1971.

Cassidy, David. *Uncertainty: The Life and Science of Werner Heisenberg.* New York: W. H. Freeman, 1992.

Cercignani, Carlo. *Ludwig Boltzmann: The Man Who Trusted Atoms.* Oxford: Oxford University Press, 1998.

Cheney, Margaret. *Tesla: Man Out of Time.* New York: Dell, 1981.

Cohen, R. S., ed. *Boston Studies in the Philosophy of Science.* Vol. 6. *Ernst Mach: Physicist and Philosopher.* Boston: Kluwer Academic Publishers, 1975.

Compton, Arthur Holly. *Atomic Quest—a Personal Narrative.* Oxford: Oxford University Press, 1956.

Concise Dictionary of Scientific Biography. New York: Scribner's, 1981.

Cooper, Dan. *Enrico Fermi: And the Revolutions in Modern Physics.* Oxford: Oxford University Press, 1998.

Coulson, Thomas. *Joseph Henry: His Life and Work.* Princeton, N.J.: Princeton University Press, 1950.

Crease, Robert P., and Charles C. Mann. *The Second Creation: Makers of the Revolution in Twentieth Century Physics.* Revised ed. New Brunswick, N.J.: Rutgers University Press, 1995.

Curie, Eve. *Madame Curie.* New York: Doubleday, 1937.

Dahl, Per F. *Flash of the Cathode Rays: A History of J.J. Thomson's Electron.* Philadelphia: Institute of Physics, 1997.

Daintith, John, and John O. E. Clark. *The Facts On File Dictionary of Physics.* 3rd ed. New York: Facts On File, 1999.

Daintith, John, Sarah Mitchell, Elizabeth Tootill, and Derek Gjertsen, eds. *Biographical Encyclopedia of Scientists.* Philadelphia: Institute of Physics, 1994.

Darrigol, Olivier. *Electrodynamics from Ampère to Einstein.* Oxford: Oxford University Press, 2000.

Dictionary of Scientific Biography. New York: Scribner's, 1970–1980.

Drake, Ellen Tan. *Restless Genius: Robert Hooke and His Earthly Thoughts.* Oxford: Oxford University Press, 1996.

Drake, Stillman. *Galileo At Work: His Scientific Biography.* Chicago: University of Chicago Press, 1978.

Dugdale, J. S. *Entropy and Its Physical Meaning.* London: Taylor & Francis, 1996.

Dyson, Freeman. *Disturbing the Universe.* Cambridge, Mass.: Perseus Books, 2001.

Eden, Alan. *The Search for Christian Doppler.* New York: Springer-Verlag, 1992.

Einstein, Albert. *The Expanded Quotable Einstein,* Alice Calaprice, ed. Princeton, N.J.: Princeton University Press, 2000.

————. *Ideas and Opinions.* New York: Random House, 1988.

————. *Out of My Later Years.* Secaucus, N.J.: Carol, 1972.

Elliott, Clark A. *Biographical Dictionary of American Science: The Seventeenth Through the 19th Centuries.* Westport, Conn.: Greenwood Press, 1979.

Fauvel, John, ed. *Let Newton Be!* Oxford: Oxford University Press, 1988.

Fermi, Laura. *Atoms in the Family: My Life with Enrico Fermi.* Chicago: University of Chicago Press, 1994.

Ferris, Timothy. *Coming of Age in the Milky Way.* New York: William Morrow, 1988.

————, ed. *The World Treasury of Physics, Astronomy, and Mathematics.* Boston: Little, Brown, 1991.

Feynman, Richard. *The Meaning of It All: Thoughts of a Citizen Scientist.* Reading, MA: Addison Wesley, 1995.

————. *QED: The Strange Theory of Light and Matter.* Princeton, N.J.: Princeton University Press, 1985.

Feynman, Richard, and R. Leighton. *Surely You're Joking, Mr. Feynman: Adventures of a Curious Character.* New York: W. W. Norton, 1997.

Forward, Robert L. and Joel Davis. *Mirror Matter: Pioneering Antimatter Physics.* Lincoln, Nebr.: iUniverse. Com, 2001.

Franklin, Allan. *Are There Really Neutrinos? An Evidential History.* New York: Perseus Books, 2000.

Fraser, Gordon. *Antimatter: The Ultimate Mirror.* London: Cambridge University Press, 2000.

Fraunfelder, H. *The Mössbauer Effect.* New York: W. A. Benjamin, 1963.

Frisch, Otto. *What Little I Remember.* London: Cambridge University Press, 1991.

Gabor, Dennis. *Innovations: Scientific, Technological and Social.* Oxford: Oxford University Press, 1970.

Galileo. *Discoveries and Opinions of Galileo: Including the Starry Messenger (1610), Letter to the Grand Duchess Christina (1615), and Excerpts from Letters on Sunspots (1613), The Assayer (1623).* Translation by Stillman Drake. Based on Work by Galileo. New York: Doubleday, 1989.

Gamow, George. *The Great Physicists from Galileo to Einstein.* New York: Dover, 1988.

————. *My World Line: An Informal Autobiography.* New York: Viking Penguin, 1970.

————. *Thirty Years That Shook Physics: The Story of Quantum Theory.* New York: Dover, 1985.

Gell-Mann, Murray. *The Quark and the Jaguar: Adventures in the Simple and the Complex.* New York: W. H. Freeman, 1994.

Gell-Mann, Murray and Yuval Ne'eman. *The Eightfold Way*. Cambridge, Mass.: Perseus 2000.

Ghani, Abdul. *Abdus Salam*. Karachi: Ma'aref, 1982.

Glashow, Sheldon L. *The Charm of Physics*. New York: Springer-Verlag, 1991.

———. *Interactions: A Journey Through the Mind of a Particle Physicist and the Matter of This World*. New York: Warner Books, 1988.

Glasser, Otto. *Wilhelm Conrad Roentgen and the Early History of the Roentgen Rays*. Novato, Calif.: Jeremy Norman, 1992.

Gleick, Richard. *Genius: The Life and Science of Richard Feynman*. New York: Pantheon Books, 1992.

Goudaroulis, Y. and Gavrolu, K., eds. *Through Measurement to Knowledge: The Selected Papers of Heike Kamerlingh Onnes*. Dordrecht, Netherlands: Kluwer Academic, 1991.

Greene, Brian. *The Elegant Universe: Superstrings, Hidden Dimensions, and the Quest for the Ultimate Theory*. New York: W. W. Norton, 1999.

Gribben, John. *Schrödinger's Kittens and the Search for Reality: Solving the Quantum Mysteries*. Boston: Little, Brown, 1996.

———. *In Search of Schrödinger's Cat: Quantum Physics and Reality*. New York: Bantam Doubleday Dell, 1984.

Harman, Peter M. *The Natural Philosophy of James Clerk Maxwell*. London: Cambridge University Press, 1998.

Hassan, Z. and C. H. Lai, eds. *Ideals and Realities: Selected Essays of Abdus Salam*. Singapore: World Scientific, 1983.

Hawking, Steven. *A Brief History of Time: From the Big Bang to Black Holes*. New York: Bantam Books, 1998.

———. *Stephen Hawking's Universe: The Cosmos Explained*. New York: Basic Books, 1998.

———. *The Future of Spacetime*. New York: W. W. Norton, 2002.

Herken, Gregg. *Brotherhood of the Bomb: The Tangled Lives and Loyalties of Robert Oppenheimer, Ernest Lawrence and Edward Teller*. New York: Henry Holt, 2002.

———. *The Universe in a Nutshell*. New York: Bantam Books, 2002.

Heilbron, J. L. *The Dilemmas of an Upright Man: Max Planck and the Fortunes of Germany*. Cambridge, Mass.: Harvard University Press, 2000.

Helmholtz, Hermann Von and David Cahan, eds. *Science and Culture: Popular and Philosophical Essays*. Chicago: University of Chicago Press, 1995.

Hermann, A. *Wolfgang Pauli: Scientific Correspondence with Bohr, Einstein, Heisenberg*. Vol. 191. New York: Springer-Verlag, 1979.

Hewish, Antony, et al. *Pulsars as Physics Laboratories*. Oxford: Oxford University Press, 1994.

't Hooft, Gerard. *In Search of the Ultimate Building Blocks*. London: Cambridge University Press, 1997.

———. *Under the Spell of the Gauge Principle*. Singapore: World Scientific, 1994.

Hofmann, James R. *Andre-Marie Ampère: Enlightenment and Electrodynamics*. London: Cambridge University Press, 1966.

Koch Torres Assis, André. Fundamental Theories of Physics. Vol. 66. *Weber's Electrodynamics*. Dordrecht, Netherlands: Kluwer Academic, 1994.

Krauss, Lawrence. *Atom*. Boston: Little, Brown, 2002.

Kropp, W., J. Schultz, M. Moe, et al., eds. *Neutrinos and Other Matters: Selected Works of Frederick Reines*. Singapore: World Scientific, 1996.

Kuhn, Thomas S. *Black-Body Theory and the Quantum Discontinuity*. Chicago: University of Chicago Press, 1987.

Kursunoglu, Behram N. and Eugene Wigner, eds. *Paul Adrien Maurice Dirac: Reminiscences About a Great Physicist*. London: Cambridge University Press, 1987.

Lafferty, Peter. *Archimedes*. New York: Bookwright Press, 1991.

Laurikainen, Kalervo Vihtori. *Beyond the Atom: Philosophical Thoughts of Wolfgang Pauli*. New York: Springer-Verlag, 1988.

Lederman, Leon M. and David N. Schramm. *From Quarks to the Cosmos: Tools of Discovery*. New York: W. H. Freeman, 1995.

Lederman, Leon M. and Dick Teresi. *The God Particle: If the Universe Is the Answer, What Is the Question?* New York: Houghton Mifflin, 1993.

Holloway, David. *Stalin and the Bomb: The Soviet Union and Atomic Energy, 1939–1956*. New Haven, Conn.: Yale University Press, 1994.

Howard, Donald R. and John Stachel, eds. *Einstein: The Formative Years 1879–1909*. Boston: Birkhauser, 1986.

Hunt, Bruce J. *The Maxwellians*. Ithaca, N.Y.: Cornell University Press, 1995.

Hunt, Inez and Wanetta W. Draper. *Lightning in His Hand: The Life Story of Nikola Tesla*. Thousand Oaks, Calif.: Sage Books, 1964.

"Index of Inventors." Available on-line. URL: http://www.invent.org/book/book-text/indexbyname.html.

Ipsen, D. C. *Archimedes: Greatest Scientist of the Ancient World*. Hillside, N.J.: Enslow, 1988.

Jaffe, Bernard. *Michelson and the Speed of Light*. Garden City, N.Y.: Doubleday, 1960.

Jelved, Karen, ed. *Selected Scientific Works of Hans Christian Ørsted*. Princeton, N.J.: Princeton University Press, 1998.

Johnson, George. *Strange Beauty: Murray Gell-Mann and the Revolution in Twentieth-Century Physics*. New York: Alfred A. Knopf, 1999.

Jungnickel, Christa and Russell K. McCormach, eds. *Cavendish: The Experimental Life*. Bucknell, Pa.: Bucknell University Press, 1999.

Kane, Gordon. *The Particle Garden: Our Universe as Understood by Particle Physicists*. Cambridge, Mass.: Perseus Books, 1995.

Kapitsa, S. P. and S. Drell. *Sakharov Remembered*. New York: American Institute of Physics, 1991.

Kargon, Robert H. *The Rise of Robert Millikan: Portrait of a Life in American Science*. Ithaca, N.Y.: Cornell University Press, 1982.

Keithley, Joseph F. *The Story of Electrical and Magnetic Measurements: From Early Days to the Beginnings of the 20th Century*. New York: Wiley-IEEE Press, 1998.

Kipnis, Nahum. *History of the Principle of Light*. Science Networks Historical Studies, Vol. 5. New York: Springer-Verlag, 1991.

Khalatnikov, I. M., ed. *Landau, the Physicist and the Man: Recollections of L. D. Landau*. Oxford: Oxford University Press, 1989.

Lindley, David. *Boltzmann's Atom: The Great Debate That Launched a Revolution in Physics*. New York: Free Press, 2000.

Lindop, Laurie. *Dynamic Modern Women: Scientists and Doctors*. New York: Holt, 1997.

Lindsay, R. B. *Lord Rayleigh, the Man and His Works*. London: Oxford University Press, 1970.

Livanova, Anna Mikhailovna. *Landau: A Great Physicist and Teacher*. Oxford: Oxford University Press, 1980.

Livingston, Dorothy Michelson. *Master of Light: A Biography of Albert A. Michelson*. New York: Scribner, 1973.

Ludwig, Charles. *Michael Faraday: Father of Electronics*. Scottdale, Pa.: Herald Press, 1978.

Mailis, Norberto. *The Quantum Theory of Magnetism*. Singapore: World Scientific, 2001.

Marshall, Ian N., Danah Zohar, and F. David Peat. *Who's Afraid of Schrödinger's Cat: An A-to-Z Guide to All the New Science Ideas You Need to Keep Up With the New Thinking*. New York: William Morrow, 1998.

Matson, James. *The Pioneers of NMR and Magnetic Resonance in Medicine: The Story of MRI*. Jericho, NY: Dean Book, 1996.

Maxwell, James Clerk, ed. *The Electrical Researches of the Honourable Henry Cavendish.* London: Frank Cass, 1967.

McMurray, Emily J., ed. *Notable Twentieth-Century Scientists.* New York: Gale Research, 1995.

Mehra, J. and K. Milton. *Climbing the Mountain: The Scientific Biography of Julian Schwinger.* Oxford: Oxford University Press, 2000.

Mendelssohn, Kurt. *The World of Walther Nernst: The Rise and Fall of German Science, 1864–1941.* Pittsburgh: University of Pittsburgh Press, 1973.

Milburn, Gerard J. *Schrödinger's Machines: The Quantum Technology Reshaping Everyday Life.* New York: W. H. Freeman, 1997.

Millar, David, Ian Millar, John Millar, and Margaret Millar, eds. *The Cambridge Dictionary of Scientists.* New York: Cambridge University Press, 1996.

Millikan, Robert Andrews and I. Bernard Cohen. *Autobiography of Robert A. Millikan.* Manchester, N.H.: Ayer, 1980.

Moore, Walter J. *A Life of Erwin Schrödinger.* London: Cambridge University Press, 1994.

Mott, Nevill. *A Life in Science.* London: Taylor & Francis, 1987.

Moyer, Albert E. *Joseph Henry: The Rise of an American Scientist.* Washington, D.C.: Smithsonian Institution Press, 1997.

"National Inventors Hall of Fame." Available on-line. URL: http://www.invent.org/inventure.html.

Ne'eman, Yuval and Yoram Kirsh. *The Particle Hunters.* Cambridge: Cambridge University Press, 1996.

Newton, Isaac. *The Principia: Mathematical Principles of Natural Philosophy.* Translated by Bernard Cohen and Anne Whitman. Berkeley: University of California Press, 1999.

The Nobel Foundation. "Nobel Laureates." Available on-line. URL: http://www.nobel.com.

Oliphant, Mark. *Rutherford: Recollections of the Cambridge Days.* London: Elsevier, 1972.

Overbye, Dennis. *Einstein in Love: A Scientific Romance:* New York: Viking Penguin, 2000.

Pais, Abraham. *Niels Bohr's Times: In Physics, Philosophy, and Polity.* Oxford: Oxford University Press, 1991.

———. *Subtle Is the Lord . . . the Science and the Life of Albert Einstein.* Oxford: Oxford University Press, 1982.

Pais, Abraham, et al., eds. *Paul Dirac, the Man and His Work.* London: Cambridge University Press, 1998.

Pasachoff, Naomi. *Marie Curie and the Science of Radioactivity.* Oxford: Oxford University Press, 1997.

Pelletier, Paul A. *Prominent Scientists: An Index to Collective Biographies.* New York: Neal Schuman, 1994.

Penzias, Arno. *Digital Harmony: Business, Technology and Life After Paperwork.* New York: Harper Business, 1996.

———. *Ideas and Information: Managing in a High-Tech World.* New York: W. W. Norton, 1989.

Pera, Marcello and Jonathan Mandelbaum. *The Galvani-Volta Controversy on Animal Electricity.* Princeton, N.J.: Princeton University Press, 1992.

Perrin, Jean. *Atoms.* Reprint edition of 1913 French original. Woodbridge, Conn.: Ox Bow Press, 1990.

American Physical Society. "Physics Today." Available on-line. URL: http://www.aip.org/publications.html/pt.

Porter, Roy, ed. *The Biographical Dictionary of Scientists.* New York: Oxford University Press, 1994.

Quinn, Susan. *Marie Curie: A Life.* New York: Simon & Schuster, 1995.

Rabi, Isidor. *Science: The Center of Culture.* New York: World, 1970

Rabi, Isidor, et al. *Oppenheimer.* New York: Scriber's & Sons, 1969.

Ramsey, Norman F. *Spectroscopy with Coherent Radiation: Selected Papers of Norman F. Ramsey.* Singapore: World Scientific, 1997.

Reid, T. R. *The Chip: How Two Americans Invented the Microchip and Launched a Revolution.* New York: Simon & Schuster, 1984.

Reston, James. *Galileo: A Life.* New York: HarperCollins, 1994.

Rhodes, Richard. *Dark Sun: The Making of the Hydrogen Bomb.* New York: Simon & Schuster, 1995.

———. *The Making of the Atom Bomb.* New York: Simon & Schuster, 1986.

Ridgen, John S. *Rabi: Scientist and Citizen.* New York: Basic Books, 1987.

Rife, Patricia. *Lisa Meitner and the Dawn of the Nuclear Age.* Cambridge, Mass.: Birkhauser, 1998.

Riordan, Michael and Lillian Hoddeson. *Crystal Fire: The Invention of the Transistor and the Birth of the Information Age.* New York: W. W. Norton, 1998.

Ruhla, Charles. *The Physics of Chance: From Blaise Pascal to Niels Bohr.* Oxford: Oxford University Press, 1992.

Sachs, Robert. *Maria Goeppert-Mayer, 1906–1972, a Biographical Memoir.* Washington, D.C., National Academy of Sciences, 1979.

"St. Andrews, History of Science." Available on-line. URL: http://www-history.mcs.st-andrews.ac.uk/history/.

Sakharov, Andrey Dmitriyevich. *Memoirs.* New York: A. A. Knopf, 1990.

———. *Progress, Coexistence, and Intellectual Freedom.* New York: Norton, 1968.

———. *Sakharov Speaks.* New York: A. A. Knopf, 1974.

Salam, Abdus. *Science and the Third World.* Edinburgh: Edinburgh University Press, 1991.

Schlesinger, Bernard S. and June H. Schlesinger. *Who's Who of Nobel Prize Winners.* Phoenix: Oryx Press, 1991.

Schrieffer, J. Robert. *Theory of Superconductivity.* New York: Perseus, 1999.

Schrödinger, Erwin. *Space-Time Structure.* Cambridge: Press Syndicate of the University of Cambridge, 1990.

———. *My View of the World.* Woodbridge, Conn.: Ox Bow Press, 1994.

Schweber, Silvan S. *In the Shadow of the Bomb.* Princeton, N.J.: Princeton University Press, 2000.

———. *QED and the Men Who Made It: Dyson, Feynman, Schwinger, and Tomonaga.* Princeton, N.J.: Princeton University Press, 1994.

"Scientific Biographies." Available on-line. URL: http://home.achilles.net/~jtalbot/bio/biographies.html.

Segrè, Claudio G. *Atoms, Bombs and Eskimo Kisses: A Memoir of Father and Son.* New York: Viking Penguin, 1995.

Segrè, Emilio. *Enrico Fermi, Physicist.* Chicago: University of Chicago Press, 1995.

———. *A Mind Always in Motion: The Autobiography of Emilio Segrè.* Berkeley: University of California Press, 1993.

Shamos, M. *Great Experiments in Physics.* New York: Dover, 1987.

Siegbahn, Manne. *Swedish Men of Science.* Stockholm: Almquist & Wiksell International, 1952.

Silk, Joseph. *The Big Bang.* New York: Henry Holt, 2000.

Silverman, Mark P. *Waves and Grains.* Princeton, N.J.: Princeton University Press, 1998.

Sime, Ruth Lewin. *Lise Meitner: A Life in Physics.* Berkeley: University of California Press, 1996.

Smith, Alice K. and Charles Weiner, ed., *Robert Oppenheimer: Letters and Recollections.* Cambridge, Mass.: Harvard University Press, 1980.

Smith, Crosbie and M. Norton Wise. *Energy and Empire: A Biographical Study of Lord Kelvin.* Cambridge: Cambridge University Press, 1989.

Smolin, Lee. *Three Roads to Quantum Gravity*. New York: Basic Books, 2001.

Sobel, Dava. *Galileo's Daughter: A Historical Memoir of Science, Faith, and Love*. New York: Walker, 1999.

Solomey, Nickolas. *The Elusive Neutrino: A Subatomic Detective Story*. New York: W. H. Freeman, 1997.

Spangenburg, Ray and Diane Moser. *Niels Bohr: Gentle Genius of Denmark*. New York: Facts On File, 1995.

Speyer, Edward. *Six Roads from Newton: Great Discoveries in Physics*. Wiley Popular Science. New York: John Wiley, 1996.

Srinivasan, G., ed. *Chandrasekhar: The Man Behind the Legend—Chandra Remembered*. Chicago: University of Chicago Press, 2000.

Stewart, C., ed. *Coulomb and the Evolution of Physics and Engineering in Eighteenth Century France*. Princeton, N.J.: Princeton University Press, 1971.

Struik, Dirk Jan. *The Land of Stevin and Huygens: A Sketch of Science and Technology in the Dutch Republic During the Golden Century*. Dordrecht, Netherlands: D. Reidel, 1982.

Susskind, Charles. *Heinrich Hertz: A Short Life*. San Francisco: San Francisco Press, 1995.

Sutton, Christine. *Spaceship Neutrino*. London: Cambridge University Press, 1992.

Szanton, Andrew. *The Recollections of Eugene P. Wigner as Told to Andrew Szanton*. New York: Perseus Books, 1992.

Taubes, Gary. *Nobel Dreams: Power, Deceit, and the Ultimate Experiment*. New York: Random House, 1987.

Teller, Edward with Judith L. Shoolery. *Memoirs: A Twentieth-Century Journey in Science and Politics*. New York: Perseus Books, 2001.

Tesla, Nikola. *My Inventions: The Autobiography of Nikola Tesla*. Edited and with an introduction by Ben Johnston. Williston, Vt.: Hart Bros., 1944.

Thompson, Silvanus P. *The Life of Lord Kelvin*. New York: Chelsea, 1977.

Thorne, Kip S. *Black Holes and Time Warps: Einstein's Outrageous Legacy*. New York: W. W. Norton, 1994.

Tomonaga, Sin-Itiro. *The Story of Spin*. Chicago: University of Chicago Press, 1998.

Townes, Charles Hard. *How the Laser Happened: Adventures of a Scientist*. Oxford: Oxford University Press, 1999.

Trower, W. Peter, ed. *Discovering Alvarez*. Chicago: University of Chicago Press, 1987.

Van Allen, James A. *Origins of Magnetospheric Physics*. Washington, D.C.: Smithsonian Institution Press, 1983.

Venkataraman, G. *Bhabha and His Magnificent Obsessions*. Hyderabad, India: Universities Press, 1994.

———. *Journey into Light: Life and Science of C. V. Raman*. Bangalore, India: Indian Academy of Sciences, 1989.

Vidali, Gianfranco. *Superconductivity: The Next Revolution?* Cambridge: Cambridge University Press, 1993.

Von Baeyer, Hans Christian. *Taming the Atom: The Emergence of the Visible Microworld*. New York: Dover, 2000.

———. *Warmth Disperses and Time Passes: A History of Heat*. New York: Modern Library, 1999.

Warshofsky, Fred. *The Chip War: the Battle for the World of Tomorrow*. New York: Scribner & Sons, 1984.

Weinberg, Steven. *Dreams of a Final Theory: The Scientist's Search for the Ultimate Laws of Nature*. New York: Vintage Books, 1993.

———. *Facing Up: Science and Its Cultural Adversaries*. Cambridge, Mass.: Harvard University Press, 2001.

———. *The First Three Minutes: A Modern View of the Origin of the Universe*. New York: Basic Books, 1977.

Weiss, Richard J., ed. *The Discovery of Anti-Matter: The Autobiography of Carl David Anderson, the Youngest Man to Win the Nobel Prize.* Series in Popular Science, Vol. 2. Singapore: World Scientific, 1999.

Westfall, Richard S. *Never At Rest: A Biography of Isaac Newton.* London: Cambridge University Press, 1983.

Wheaton, Bruce R. *The Tiger and the Shark: Empirical Roots of Wave–Particle Dualism.* London: Cambridge University Press, 1990.

Wheeler, John Archibald with Kenneth Ford. *Geons, Black Holes, and Quantum Foam: A Life in Physics.* New York: W. W. Norton, 1998.

Whitaker, Andrew. *Einstein, Bohr and the Quantum Dilemma.* London: Cambridge University Press, 1996.

Who's Who in Science and Engineering, 1994–1995. New Providence, N.J.: Marquis Who's Who, 1994.

Who's Who in Science in Europe. Essex, England: Longman, 1994.

Wigner, Eugene. *Symmetries and Reflections.* Woodbridge, Conn.: Ox Box Press, 1979.

Williams, Trevor I., ed. *A Biographical Dictionary of Scientists.* 4th ed. Glasgow: Harper-Collins, 1994.

Wilson, D. B. *Kelvin and Stokes: A Comparative Study in Victorian Physics.* Bristol: Adam Hilger, 1987.

Wilson, E. J. N. *An Introduction to Particle Accelerators.* Oxford: Oxford University Press, 2001.

Wilson, Jane, ed. *All in Our Time: The Reminiscences of 12 Nuclear Physicists.* Chicago Bulletin of the Atomic Scientists, 1975.

Wilson, John Graham. *The Principle of Cloud Chamber Technique.* London: Cambridge University Press, 1951.

Wolpert, Lewis and Alison Richards. *Passionate Minds: The Inner World of Scientists.* Oxford: Oxford University Press, 1998.

Wood, Alexander, Frank Oldham, and Charles E. Raven. *Thomas Young, Natural Philosopher, 1773–1829.* London: Cambridge University Press, 1954.

Yang, Chen Ning. *Selected Papers 1945–80 with Commentary.* New York: W. H. Freeman, 1983.

Yoder, Joella G. *Unrolling Time: Christiaan Huygens and the Mathematization of Nature.* Cambridge University Press, 1988.

Yount, Lisa. *A to Z of Women in Science and Math.* New York: Facts On File, 1998.

Yukawa, Hideki. *Tabito* (The Traveler). Singapore: World Scientific, 1982.

———. *Creativity and Intuition: A Physicist Looks at East and West.* Tokyo and New York: Kodansha International, 1973.

INDEX

Note: **Boldface** page numbers indicate main headings. *Italic* page numbers indicate illustrations.

DATE DUE		
9/16		
9/21		
FEB 07		
FEB 09		
FEB 13		

Pike High School
5401 W. 71st Street
Indianapolis, IN 46268